Lecture Notes in Computer Science   14931

Founding Editors

Gerhard Goos
Juris Hartmanis

The series Lecture Notes in Computer Science (LNCS), including its subseries Lecture Notes in Artificial Intelligence (LNAI) and Lecture Notes in Bioinformatics (LNBI), has established itself as a medium for the publication of new developments in computer science and information technology research, teaching, and education.

LNCS enjoys close cooperation with the computer science R & D community, the series counts many renowned academics among its volume editors and paper authors, and collaborates with prestigious societies. Its mission is to serve this international community by providing an invaluable service, mainly focused on the publication of conference and workshop proceedings and postproceedings. LNCS commenced publication in 1973.

Toshimitsu Masuzawa · Yoshiaki Katayama ·
Hirotsugu Kakugawa · Junya Nakamura ·
Yonghwan Kim
Editors

Stabilization, Safety, and Security of Distributed Systems

26th International Symposium, SSS 2024
Nagoya, Japan, October 20–22, 2024
Proceedings

 Springer

Editors
Toshimitsu Masuzawa ⓘ
Osaka University
Suita, Osaka, Japan

Hirotsugu Kakugawa ⓘ
Ryukoku University
Otsu, Shiga, Japan

Yonghwan Kim ⓘ
Nagoya Institute of Technology
Nagoya, Aichi, Japan

Yoshiaki Katayama ⓘ
Nagoya Institute of Technology
Nagoya, Aichi, Japan

Junya Nakamura ⓘ
Toyohashi University of Technology
Toyohashi, Aichi, Japan

ISSN 0302-9743　　　　　ISSN 1611-3349 (electronic)
Lecture Notes in Computer Science
ISBN 978-3-031-74497-6　　　ISBN 978-3-031-74498-3 (eBook)
https://doi.org/10.1007/978-3-031-74498-3

This Springer imprint is published by the registered company Springer Nature Switzerland AG
The registered company address is: Gewerbestrasse 11, 6330 Cham, Switzerland

If disposing of this product, please recycle the paper.

Preface

The papers in this volume were presented at the 26th International Symposium on Stabilization, Safety, and Security of Distributed Systems (SSS 2024), held on October 20–22, 2024, at Nagoya International Center, Aichi, Japan.

SSS is an international forum for researchers and practitioners in the design and development of distributed systems with a focus on systems that are able to provide guarantees on their structure, performance, and/or security in the face of an adverse operational environment.

SSS started as a workshop dedicated to self-stabilizing systems, and the first two editions were held in 1989 and 1995, in Austin, USA and Las Vegas, USA, respectively. From then, the workshop was held biennially until 2005 when it became an annual event. It broadened its scope and attracted researchers from other communities. In 2006, the name of the conference was changed to the International Symposium on Stabilization, Safety, and Security of Distributed Systems (SSS).

This year the Program Committee was organized into four tracks, reflecting major trends related to the conference: (i) Track A. Self-Stabilizing and/or Dynamic Systems: Theory and Practice, (ii) Track B. Distributed and Concurrent Computing: Foundations, Fault-Tolerance and Scalability, (iii) Track C. Cryptography and Security, and (iv) Track D. Moving and Computing.

We received 69 submissions. Each submission was double-blind reviewed by at least three program committee members with the help of external reviewers. Out of the 69 submitted papers, 4 were (reviewed) invited papers, and 22 papers were selected as regular papers. The symposium also included 6 brief announcements. Selected papers from the symposium will be published in a special issue of the journal *Theoretical Computer Science* (TCS).

This year, we were very fortunate to have two distinguished keynote speakers: Michel Raynal and Koutarou Suzuki. We were happy to award the best paper award to Umesh Biswas, Trisha Chakraborty, and Maxwell Young for their paper "Softening the Impact of Collisions in Contention Resolution" and the best student paper award to Haruki Kanaya and Yuichi Sudo for their paper "Complete Graph Identification in Population Protocols."

We are grateful to the Program Committee and the External Reviewers for their valuable and insightful comments. We also thank the members of the Steering Committee for their invaluable advice. Last but not least, on behalf of the Program Committee, we thank all the authors who submitted their work to SSS 2024.

October 2024   Toshimitsu Masuzawa
                                                                   Yoshiaki Katayama
                                                                  Hirotsugu Kakugawa
                                                                    Junya Nakamura
                                                                    Yonghwan Kim

Organization

General Co-chairs

Toshimitsu Masuzawa   Osaka University, Japan
Yoshiaki Katayama   Nagoya Institute of Technology, Japan
Hirotsugu Kakugawa   Ryukoku University, Japan

Organizing Chair

Yonghwan Kim   Nagoya Institute of Technology, Japan

Publicity and Proceedings Chairs

Junya Nakamura   Toyohashi University of Technology, Japan
Yonghwan Kim   Nagoya Institute of Technology, Japan

Treasurer

Hirotsugu Kakugawa   Ryukoku University, Japan

Steering Committee

Anish Arora   Ohio State University, USA
Shlomi Dolev   Ben-Gurion University of the Negev, Israel
Sandeep Kulkarni   Michigan State University, USA
Toshimitsu Masuzawa   Osaka University, Japan
Franck Petit   Sorbonne Université, France
Sébastien Tixeuil   Sorbonne Université, France (Chair)
Elad Michael Schiller   Chalmers University of Technology, Sweden

Advisory Committee

Sukumar Ghosh   University of Iowa, USA
Ted Herman   University of Iowa, USA

In Memory of

Ajoy Kumar Datta
Edsger W. Dijkstra
Mohamed Gouda

Program Committee

Track A. Self-Stabilizing and/or Dynamic Systems: Theory and Practice

Arnaud Casteigts (Co-chair)   University of Geneva, Switzerland
Sayaka Kamei (Co-chair)   Hiroshima University, Japan
Karine Altisen   Verimag, France
Giuseppe Antonio Di Luna   University of Rome—Sapienza, Italy
Luciana Arantes   Sorbonne Université—LIP6, France
Lelia Blin   IRIF, Université Paris Cité, France
Swan Dubois   Sorbonne Université and Inria, France
Anissa Lamani   Université de Strasbourg, France
Pierre Leone   University of Geneva, Switzerland
Franck Petit   LiP6 CNRS-INRIA Sorbonne Université, France
Christian Scheideler   University of Paderborn, Germany
Yuichi Sudo   Hosei University, Japan

Track B. Distributed and Concurrent Computing: Foundations, Fault-Tolerance and Scalability

Fukuhito Ooshita (Co-chair)   Fukui University of Technology, Japan
Andrea Richa (Co-chair)   Arizona State University, USA
Rida Bazzi   Arizona State University, USA
Carole Delporte-Gallet   Université Paris Cité, France
Panagiota Fatourou   University of Crete, Greece
Laurent Feuilloley   University of Lyon, France

Olga Goussevskaia	Federal University of Minas Gerais, Brazil
Naoki Kitamura	Osaka University, Japan
Moti Medina	Bar-Ilan University, Israel
Sathya Peri	Indian Institute of Technology Hyderabad, India
Michel Raynal	University of Rennes, France
Christian Scheideler	University of Paderborn, Germany
Gregory Schwartzman	JAIST, Japan
Gokarna Sharma	Kent State University, USA
Jamison Weber	Arizona State University, USA
Yingjie Xue	Hong Kong University of Science and Technology, China
Maxwell Young	Mississippi State University, USA

Track C. Cryptography and Security

Quentin Bramas (Co-chair)	University of Strasbourg, France
Pascal Felber (Co-chair)	Université de Neuchatel, Switzerland
Emmanuelle Anceaume	CNRS IRISA, France
Pierre-Louis Aublin	IIJ Research Laboratory, Japan
Christian Cachin	University of Bern, Switzerland
Pandu-Rangan Chandrasekaran	IIT Madras, India
Naohiro Hayashibara	Kyoto Sangyo University, Japan
Rüdiger Kapitza	FAU Erlangen-Nürnberg, Germany
Pascal Lafourcade	Université Clermont Auvergne, France
Miguel Matos	IST Lisbon, Portugal
Atsuko Miyaji	Osaka University, Japan
Raoul Strackx	Fortanix, Netherlands
Sara Tucci Piergiovanni	CEA LIST, France
Osman Unsal	Barcelona Supercomputing Center, Spain
Susanne Wetzel	Stevens Institute of Technology, USA

Track D. Moving and Computing

Konstantinos Georgiou (Co-chair)	Toronto Metropolitan University, Canada
Masahiro Shibata (Co-chair)	Kyushu Institute of Technology, Japan
John Augustine	IIT Madras, India
Doina Bein	California State University, USA
François Bonnet	Tokyo Institute of Technology, Japan
Fabien Dufoulon	Lancaster University, UK
Ryota Eguchi	Nara Institute of Science and Technology, Japan

Darya Melnyk	TU Berlin, Germany
Othon Michail	University of Liverpool, UK
Alfredo Navarra	University of Perugia, Italy
Aris Pagourtzis	National Technical University of Athens, Greece
Denis Pankratov	Concordia University, Canada
Partha Sarathi Mandal	IIT Guwahati, India
Ramachandran Vaidyanathan	Louisiana State University, USA
Yukiko Yamauchi	Kyushu University, Japan

External Reviewers

Almalki, Nada
Antoniadis, Karolos
Bhattacharya, Adri
Chatterjee, Soumyottam
Daymude, Joshua
Di Fonso, Alessia
Durand, Anaïs
Jana, Saswata
Karmakar, Sushanta
Katsarakis, Antonios
Konstantinidis, Orestis
Kumar, Manish
Manaswini, Piduguralla
Mostarda, Leonardo
Olivier-Anclin, Charles
Pattanayak, Debasish
Pergaminelis, Chris
Pramanick, Subhajit
Puys, Maxime
Ravish, Ankit
Spyrakou, Marianna
Thomas, Samuel
Van Strydonck, Thomas
Venkat, Rakesh
Vial Prado, Francisco José
Viglietta, Giovanni

Contents

Keynotes

On Distributed Computing: A View, Physical Versus Logical Objects, and a Look at Fully Anonymous Systems

Michel Raynal[✉]

IRISA, CNRS, Inria, Univ Rennes, 35042 Rennes, France
michel.raynal@irisa.fr

Abstract. This article presents a short (and partial) history of synchronization in systems made up of asynchronous sequential processes (automata). Among other points, it shows that synchronization (which consists in ordering operations issued by processes on shared objects) has a different flavor according to the fact that the objects are physical objects (such as a printer or a disk) or logical objects (immaterial objects represented by sequences of bits). It then follows from this physical/logical nature of computing objects that mutual exclusion is to physical objects what consensus is to logical objects. The article also addresses recent results on process synchronization in fully anonymous systems (systems in which processes cannot be distinguished one from the other, and where there is a disagreement on the addresses of the memory registers.

Keywords: Agreement · Algorithm · Anonymity · Asynchrony · Atomicity · Concurrency · Consensus · Distributed computing · Informatics · Liveness · Logical (immaterial) object · Memory anonymity · Mutual exclusion · Parallel computing · Process anonymity · Physical object · Safety · Sequential process · Sequential specification · Synchronization · Total order

The article is made up of three parts. The first part (Sects. 1, 2, and 3) describes the view of the author on what are informatics and distributed computing. The second part (Sects. 4, 5, 6, and 7) focuses on the most important distributed computing problems (namely mutex and consensus) and the fact that there are the two sides of the same coin. The third part (Sect. 8) focuses on synchronization in systems where both the processes and the memory registers are anonymous (topic recently introduced by G. Taubenfeld in [67,68]). .

T. Masuzawa et al. (Eds.): SSS 2024, LNCS 14931, pp. 3–19, 2025.
https://doi.org/10.1007/978-3-031-74498-3_1

1 Algorithms Are the Heart of Informatics

Given a computing model (set of primitive operations), a sequential *algorithm* is a sequence of operations (provided by the model), that solves a problem that has been previously defined by a set of properties (specification). The proof of an algorithm consists in showing that, given correct inputs, all the executions of the algorithm satisfy the properties.

On data and algorithms At the center of informatics reside algorithms [29]. This is a direct consequence of the fact that, if we suppress algorithms there is no more computation and so there is no more informatics. A schematic view is presented in Fig. 1. *Informatics* is a science of abstractions. And –at its center– *algorithmics* consists in building higher and higher level computing abstractions (with appropriate data representation).

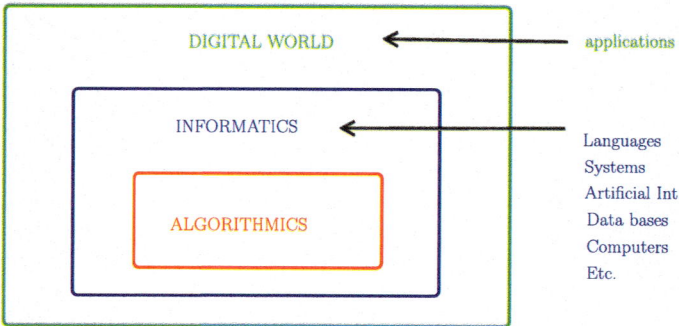

Fig. 1. From algorithms to applications

From equations to algorithms When looking at the past evolution, the main resource until World War II was the pair matter/energy. Now, the main resource is data, which, once processed, provides us with information. As matter/energy, data/information can be collected (extracted), consumed, transformed, stored, carried, etc. But there a fundamental difference between matter/energy and data/information: differently from matter/energy, data/information is abstract, does not burn, and can be copied at "zero cost".

Moreover, while the aim of science was to "put the world into equations", we know today that everything cannot be captured by a formula. So, the aim of today science is to *put the world into equations and algorithms*. Said in another way, Galileo Galilei said "the great book of nature is written in mathematical language". This is true for physics, not for other sciences such as life sciences. Today, the pair made up of *mathematics + informatics* seems to be the language of all sciences [6].

A few differences between informatics and mathematics In addition to being the domain of algorithms, informatics and mathematics present an important difference, namely mathematics do not have the touch "run" which transforms a text describing an algorithm (written in some programming language) into an execution! This has a fundamental consequence, namely informatics evolves in the field of "finite". As an example, while mathematics consider the real number π as a number with an infinite number of digits, any algorithm that uses π is restricted to consider it as made up of a finite number of digits. More generally, any algorithm uses a finite memory and a finite time.

To illustrate the previous claim let us consider the following two objects: the object denoted "1+2" and the object denoted "3" with their usual meaning, and let us ask the question: "are these two objects the same object or are they different objects?". Nearly all mathematicians answer "it is the same object", while nearly all informaticians answer "they are different objects". This is due to the simple observation that informaticians see "1+2" as an algorithm which need to be *executed* to obtain a result, namely the value 3.

2   Concurrent Computing

While a sequential process describes the behavior of a given state machine [47], *concurrent computing* is about the study of asynchronous sequential processes that execute concurrently (i.e., possibly at the same time) but not fully independently from each other.[1] Asynchronous means that each process proceeds to its own speed, which can vary with time and remains always unknown to the other processes. The code executed by the processes can be specific or not to each process.

Considering a set of sequential processes, the concept of *concurrent processes* (multi-process program) captures the fact that the individual behavior of each process must be controlled so that the global behavior of the set of processes remains consistent (which can be captured by predicates and invariants, e.g., [24, 25, 33, 34]).

Fundamental notions related to synchronization have been introduced by Edsger Dijkstra in the sixties [20–23] (including the concept of a process and the concept of mutual exclusion). Later in the seventies and the eighties Leslie Lamport investigated the concept of atomicity and introduced read/write registers weaker that atomic registers on top of which he solved the mutual exclusion problem [37]. Lamport also introduced a new formalism to capture the timing of operations issued by different processes [40–43].[2] The interested reader will also find more historical and technical developments in [3, 8, 36, 44, 50, 59, 62, 64–66].

---

[1] At the very beginning, a multiprocessor machines were simulated on a single monoprocessor enriched with peripheral devices. Then it was a real physical multiprocessor. Today it is provided by what is sometimes called an *Internet machine* covering the world.

[2] More precisely he introduced a two-arrow notation ($\rightarrow$ and $-\rightarrow$) that allows the capture of the order on the start and end of operation executions; $op \rightarrow op'$ is true iff operation op ends before operation op' starts, and $op -\rightarrow op'$ is true iff true iff operation op begins before op' ends.

3   Parallel Computing Versus Distributed Computing

Parallel Computing Parallel computing is a natural extension of sequential computing in the sense that it seeks to detect and exploit *data independence* to obtain efficient programs: once identified, independent sets of data can be processed independently from each other on a multiprocessor. It is nevertheless important to notice that, while independent data can be processed in parallel, any parallel program could be executed on a single processor with an appropriate scheduler (the corresponding sequential execution would be of course highly inefficient!).

Distributed computing The nature of distributed computing is totally different. Namely, distributed computing is characterized by the fact that there is a set of predefined (and physically distributed) computing entities (processes) that are *imposed* to programmers and these entities need to *cooperate to a common goal*. Moreover the behavior of the underlying infrastructure (also called environment) on which the distributed application is executed is not under the control of the programmers who have to consider it as an *hidden input*. Asynchrony and failures are the most frequent phenomenons produced by the environment that create a "context uncertainty" distributed computing has to cope with. In short, distributed computing is characterized by the fact that, in any distributed run, *the run itself is one of its entries* [60].

A duality To summarize, parallel computing is the exploitation of data independence to obtain efficient algorithms (programs), while the aim of distributed computing is to allow predefined computing entities to cooperate to a common goal. So the aim of distributed computing is to define basic cooperation abstractions that make easier the design of upper layer applications in the presence of asynchrony and/or failures.

4   1965: Mutual Exclusion

The very first objects that were shared by concurrent processes were *physical* objects (resources) such as discs, tapes, and shared memory. Mutual exclusion was then introduced to make their accesses by the processes consistent (what does happen if several processes simultaneously access such a physical object?). This is the very first distributed computing problem identified and solved by E. Dijkstra who proposed the "mutual exclusion" object (in short mutex) that allows a process to bracket a sequence of code (denoted *critical section*, CS in short), so that it can be executed by at most one process at a time [21]. So, a mutex object provides the processes with two operations denoted acquire() and release() that allow to bracket the critical section code as described by the following pattern: "acquire();   CS;   release()". As any computing object, mutex is specified by a set of properties that defines all its correct behaviors, namely:

- Mutual exclusion (safety). At most one process at a time executes the CS code.
- Starvation freedom (liveness). Any invocation of acquire() and release() terminates.[3]

Encapsulation and sequential execution The operations acquire() and release() can be made invisible at the application level by defining a higher level operation op() such that

operation op() **is** acquire(); CS; release() **end operation**.

It is easy to see that mutual exclusion allows processes to execute sequentially (we also say linearize [32]) predefined parts of code concerning their cooperation. So, mutual exclusion allows the processes to build a total order on the execution of the critical section codes protected by the same mutex object. From a historical point of view, mutex can be considered as the first distributed computing problem: it allows a predefined set of processes to cooperate to a common goal, namely preserve the consistency of an object in the presence of concurrency. A rigorous exposition of the mutex theory is presented in [42].

Instantiating a mutex algorithm Let us consider a n-process system. While it is possible to design mutex algorithms tailored for ad'hoc values of n (for example there are mutex algorithms specifically designed for two processes only, e.g. [56]), and consequently such algorithms do not work for more than two processes. Nearly all mutex algorithms are designed to work for any number $n \geq 2$ of processes, i.e., n is a parameter that can differ in each instance of the algorithm.[4]

On the fault-tolerance side A process crashes when it unexpectedly and definitively halts. Usual algorithms that solve mutex allow a process to crash when it is not executing acquire(), release() or the code in the critical section. Unfortunately mutual exclusion cannot be solved if a process may crash at any time. This is due to the fact that if a process crashes while executing acquire(), release() or code inside the critical section, due to asynchrony, no other process can be informed of its crash. To solve this issue, the system must be enriched with additional computational power.

An approach consists in providing processes with information on failures. This is the *failure detector* approach introduced in [14]. The integer n being the number of processes, let us consider a model that allows up to t processes to crash. When considering systems where processes communicate through read/write registers, the weakest failure detector (denoted QP for Quasi-Perfect) that allows mutex to be solved has been introduced in [19]. Weakest means that

---

[3] When he introduced mutual exclusion [21], Dijkstra considered a weaker liveness property, named deadlock-freedom: If one or several processes invoke acquire(), at least one of them will enter the critical section.

[4] As we will see in Sect. 5, due to computability issues, the situation is different for consensus where a consensus algorithm for n processes does not work for $(n + x)$ processes for $x \geq 1$. This is related to the additional computability power needed to solve consensus in crash-prone systems.

no failure detector that provides processes with less information on failures than QP allows mutex to be solved in asynchronous read/write systems. Assuming $t < n/2$ (i.e. the system is partition-free), the weakest failure detector (denoted T) that allows mutual exclusion to solved in asynchronous message-passing systems has been introduced in [18].

These two failure detectors have close but different definitions. Both are weaker that the perfect failure P and stronger than the eventually perfect failure detector $\Diamond P$ defined in [14]. Moreover, both are stronger than the weakest failure detector (eventual leader denoted Ω) that allows consensus to be solved in read/write systems when $t < n$ [48] and in message-passing when $t < n/2$ [13].

From safe read/write bits to atomic read/write registers Dijkstra's mutex algorithm and all other mutex algorithms (except [37]) assume that the underlying base registers on top of which mutex is solved are atomic read/write registers [64]. So, they build upper layer atomicity on top of lower layer read/write atomic registers.

In a very interesting way (and in some sense very surprisingly), it has been shown by L. Lamport that, while the *atomicity* of basic read/write registers is sufficient to solve mutex, it is not necessary. More precisely, mutual exclusion can be solved on top of single-writer multi-reader *safe* registers (see [37, 41]). A safe register is a register that can be written by a single process and read by any number of processes. A write defines the new value of the register. A read whose execution is not concurrent with a write returns the last value written in the register. A read concurrent with a write returns *any value* that the register can contain (so it can return a value that has never been written in the register!). In a non-trivial way, multi-writer multi-reader atomic registers can be built on top of single-writer single-reader safe bits (despite asynchrony and process failures. A survey of such constructions is presented in Section V of [59].

Another great advance in breaking atomicity notion was the notion of concurrent writing and reading introduced Lamport in [38], and later investigated Peterson in [57].

5   1980: Consensus (Fault-Tolerant Distributed Agreement)

Definition The *consensus* problem was introduced by S. Pease, R. Shostak and L. Lamport in [46, 54] in the context of synchronous distributed systems prone to Byzantine process failures (arbitrary misbehavior of a process). This problem is at the core of distributed computing agreement problems. We consider here asynchronous read/write or message-passing systems prone to crash failures. Let a process be *correct* in a run if it does not crash during that run. In such a context a consensus object is defined by a single operation denoted propose() that takes a value as input parameter and returns a value. When a process invokes propose(v) and obtains the value v' we say that it proposes v and decides v'. Consensus is defined by the following properties.

- Validity (safety). If a process decides v then a process proposed v.
- Agreement(safety). No two processes decide different values.
- Termination (liveness). If a process invokes **propose**() and does crashes, it decides.

Impossibility Unfortunately consensus is impossible to solve in the presence of asynchrony even a single process may crash, be communication by message-passing [27] or read/write registers [49]. This means that the system has to be enriched with additional *computability power* to make consensus solvable. Several enrichments are possible.

- Enrich the system with synchrony assumptions (e.g. [26]).
- Enrich the system with scheduling assumptions (e.g. [7]).
- Enrich the system with randomization (e.g. [5,51]).
- Restrict the set of input vectors that can be proposed (e.g. [53]). (An input vector has one entry per process containing the value it proposes. Of course a process knows only the value of its entry).
- Enrich the system with information on failures (failure detector approach [13, 14]).
- Enrich the system with asynchronous rounds such that, for each round r and each process p, the model provides the set of processes that p hears of at round r. The features of a specific system is then captured as a whole, just by a predicate over the collection of heard-of sets [15].

Consensus number of an object Let us consider an asynchronous crash prone system in which the processes communicate by reading and writing atomic registers (RW type). As just noticed, the previous impossibility results states that consensus cannot be solved in such a system. So, a fundamental question is: which additional computability power (defined not in terms of system behaviors but in terms additional object types) needs to be added to the system model so that the consensus can be solved. To this end, M. Herlihy introduced the notion of *consensus number* [30].

The consensus number of an object type T, denoted $CN(T)$, is the greatest number of processes for which consensus can be solved from any number of atomic read/write registers and any number of objects of type T. If there is no such greatest number, the consensus number of T is $+\infty$.

Let RW_TS be the type of RW registers accessed with Test&Set() operation, and RW_CS be the type of RW registers accessed with Compare&Swap() operation.[5] It has been shown in [30] that $CN($RW_TS$)= 2$ and $CN($RW_CS$)= +\infty$. More generally, [30] introduces an infinite hierarchy of objects, that cover all possible consensus numbers. The interested reader can look at [55] where is defined the notion of k-sliding window RW register. This object family spans

---

[5] Roughly speaking both operations return the current value of the register and write a new value in it. The difference lies in the fact that Test&Set() is an unconditional write of a predefined value, while Compare&Swap() is a conditional write of a value.

the whole consensus hierarchy: the consensus number of the k-sliding window RW register is exactly k.

Let us observe that there is no notion of consensus number defined for message-passing systems. The interested reader will find a first step of such an approach (based on message patterns) in [52].

6   Consensus: Agree on a Total Order

Observation: Physical Objects versus Logical Objects A *physical* object is an object that cannot be replicated by software (e.g., a printer), while a *logical* (or *immaterial*) object is an object the value of which can by replicated by software (data). Said differently, at the basic level the value of a logical object is a structured set of bits while a physical object is a hardware device. Let us notice that logical objects can be duplicated for free.[6]

Ordering object operations Let us consider a logical object defined by a sequential specification, e.g., a stack with its two operations push() and pop(). To cope with asynchrony and failures, the stack can be replicated on each process. So the main issue consists in ensuring that the push() and pop() operations issued by the processes are applied in the same order to all the local copies of the stack. This, is the "distributed state machine" approach [39]. A simple way to attain this goal consists for each process in:

1. announce the operation it wants to execute,
2. regularly defines a sequence on the operations it sees as announced and not yet executed,
3. and proposes this sequence as input to a consensus instance.

Combined with a sequence of consensus instances (in which all processes agree a priori), this allows all the local copies of the stack to progress the same way [13, 30].

Hence, as it allows to build a total order on operations, consensus lies at the core of fault-tolerant implementations for the objects defined by a sequential specification. (For objects not defined by a sequential specification, i.e. concurrent objects, the reader can consult [9–11,58].)

Consensus versus mutex: illustration Let us consider money transfer as an object providing its users (a user is a process associated with one and only one money account) with two operations transfer() that allows a process to transfer money from its account to another account, and balance() that allows a user to read an account. Let us observe that an account is a logical object.

It has recently been shown that money transfer among a set of processes, each having its own account, does not need consensus [4, 16, 28].[7] It is an announcement/broadcast problem that must satisfy causality requirements.

---

[6] One of very early observation on the physical versus logical nature of a file object was done in [17] where was introduced the reader/writer problem.

[7] It is pleasant to observe that the heavy Blockchain machinery was introduced to built a total order on the cryptocurrency operations issued by users, and this is

When several persons share the same account, the associated process consists of several threads, one per person co-owner of the account. The invocations of the operations transfer() issued by the threads that are co-owners of the same account must then be ordered in order to prevent double-spending from the corresponding account. This could be realized with mutex (enriched with an appropriate failure detector or random numbers if the system is crash-prone).

But, as an account is a logical object this ordering can be realized (despite process failures and asynchrony) with the help of consensus. It follows that if each account can be accessed by at most k threads, an object the consensus number of which is k is sufficient to realize money transfer (this was first noticed in [28]).

7   Mutex Versus Consensus: Both Sides of the Same Coin

When considering objects the consistency of which is defined by a sequential specification (i.e., objects whose operations must appear as being executed sequentially), it follows from the previous simple observations that, while both mutex and consensus can be used to build a total order, mutex is for physical objects (which by nature cannot be replicated), and consensus is for logical objects (structured sets of bits which can be replicated)[8]. In this sense, mutex and consensus are the two sides of the same coin [61] as summarized in Table 1.

Table 1. Total order: mutex versus consensus

Nature of the object	Possible replication	Total order obtained from	Underlying coordination	Helping needed	Weakest FD
Physical	No	Mutex	Strong	Yes	QP, T
Logical	Yes	Consensus	Weak	Yes	Ω

The "Underlying coordination" column refers to the type of synchronization needed to implement mutex or consensus, namely, mutex ensures that the concerned object can be physically accessed by at most one process at a time, while consensus does not prevent several processes from invoking and simultaneously executing object operations (after these operations have been totally ordered by a consensus instance). The column "Helping needed" refers to the fact the algorithms implementing mutex or consensus need specific helping mechanisms

---

not needed! For the interested reader, [16,28] consider money transfer as an object defined by a sequential specification, while [4] considers money transfer as an object defined by a concurrent specification.

[8] Of course, in some specific contexts, it can be interesting to use mutex for logical objects, but this is another issue not addressed in this note.

to ensure the liveness of the operations on the object that is built [1, 12, 30, 60].[9]
The last column "Weakest FD" concerns the weakest failure detectors that allows
mutex or consensus to be solved. As already indicated, for read/write systems
it is the failure detector QP for mutex [19] and Ω for consensus [48], while,
for message-passing systems such that $t < n/2$, it is the failure detector T for
mutex [18] and the eventual leader failure detector Ω for consensus [13]. It is
worth noticing that the weakest information on failures that allows mutex to
be solved includes a perpetual property [18, 19], while that the weakest information on failures needed to solve consensus needs to satisfy an eventual property
only [13]. This is strongly related to the underlying nature of the object (physical
versus logical).

Let us again insist on the fact that, in a crash-prone system where the processes communicate through read/write atomic registers (resp. message-passing
when assuming $t < n/2$), the weakest failure detectors QP (resp. T) that allows
mutex to be solved is stronger than the weakest failure detector Ω that allows
consensus to be solved. As previously noticed, this is due to the fact that the
implementation of mutex requires a stronger underlying synchronization than
the one needed to implement consensus. More precisely, this is the main difference between mutex and consensus, because of their very definitions mutex does
not allow concurrency at the implementation level, whereas consensus does.

Last but not least, let us notice that a recent paper by L. Lamport [45] presents
a deconstruction of his famous Bakery mutex algorithm [37] from which is built a
distributed state machine as defined in [39] (i.e., any object defined by a sequential specification).

8   Synchronization in Fully Anonymous Systems

Process anonymity An anonymous process system is a system in which no two
processes can be distinguished one from the other. So processes have no names,
have the same code, and the same initialization of their local variables. This
notion was introduced a long time ago (1980) in the context of message-passing
systems [2].

Memory anonymity The notion of anonymous memory has been recently
introduced by G. Taubenfeld [67, 68]. It means that "there is no a priori agreement between the processes on the names of the shared memory location" (register). Considering a shared memory defined as an array $SM[1..m]$ of memory
locations (registers) memory anonymity means that, while the same location
identifier $SM[x]$ always denotes the same memory location for a process p_i,
it does not necessarily denote the same memory location for two different processes p_i and p_j.

---

[9] As far liveness properties are concerned, wait-freedom [30] and non-blocking [32] for
consensus correspond to starvation-freedom and deadlock-freedom for mutex. Differently obstruction-freedom [31] for consensus has no corresponding liveness property
that could be associated with mutex (this is due to the fact that mutex implicitly
considers the object to with it is applied as a "physical" object.

Definition More formally, an anonymous memory $AM[1..m]$ is defined as follows.

- For each process p_i an adversary defined a permutation $f_i()$ over the set $\{1, 2, \cdots, m\}$, such that when p_i uses the address $AM[x]$, it actually accesses $AM[f_i(x)]$,
- No process knows the permutations, and
- All the registers are initialized to the same default value denoted $\bot$.

An example of anonymous memory is described in Fig. 2.

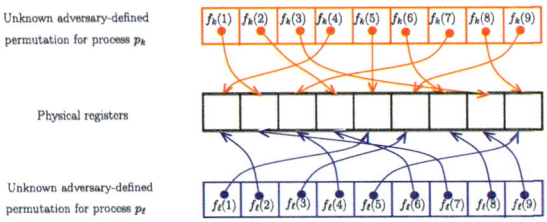

Fig. 2. An example of an anonymous memory

Full anonymity A fully anonymous shared memory system is a system in which both the processes and the memory registers are anonymous.

Mutex in failure-free fully anonymous shared memory system Let $M(n)$ be the set of all the integers that are relatively prime with all the integers smaller or equal to n:

$$M(n) = \{m \text{ such that } \forall \ell \in \{1, \cdots, n\} \ : \ \gcd(\ell, m) = 1\}.$$

Assuming $m \in M(n)$, reference [63] presents a deadlock-free mutex algorithm for systems made up of n anonymous processes communicating through m anonymous registers accessed by read, write and Compare&Swap operations. The same article shows that the condition $m \in M(n)$ is also necessary. This means that $m \in M(n)$ is the minimal asymmetry seed needed to solve mutex despite the net effect of asynchrony and full anonymity.

Elections in failure-free fully anonymous shared memory system This problem has been posed and solved in [35].

In a process anonymous system made up of n processes election consists in providing each process p_i with a local Boolean variable $elected_i$, which, initialized to false, is set to true if and only if p_i is elected.

The election problem can be decomposed into four classes according to the constraint on the number of processes that can/must be elected and the fact that all processes are assumed to participate or not to the election.

Here we present an algorithm from [35] in which all the processes participate and exactly d processes are elected where $1 \leq d < n$. This algorithm is based on the Compare&Swap operation.

Theorem 1 (Bezout, 1730-1783). *Let m and n be two positive integers and let $d = \gcd(m, n)$. There are two positive integers u and v such that $u \times m = v \times n + d$.*[10]

Consider a rectangle made up of $u \times m$ squares. On one side, this means that u squares are associated with each of the m anonymous registers. On another side, each of the n processes progresses until it has "captured" v squares (from an operational point of view, the capture of a square is a successful invocation of Compare&Swap($R[x], val, val + 1$).

Then, when $v \times n$ squares have been captured by the processes, each process competes to capture one more square. As it remains only $d = u \times m - v \times n$ squares, the processes that succeed in capturing one more square are the d leaders.

Local variables at each process p_i

- won_i (initialized to 0): number of squares captured by p_i.
- sum_i (initialized to 0): local view of the numbers of squares captured by all the processes.
- $myview_i[1..m]$: local copy (non-atomically obtained) of the anonymous memory $R[1..m]$.

ALGORITHM OF ANY PROCESS p_i IN THE FULLY ANONYMOUS **RMW** MODEL

u and v: smallest positive integers such that $u \times m = v \times n + d$
The initial value of all the shared registers is 0.

operation elect() **is**
1 **repeat**
2   **for each** $j \in \{1, ..., m\}$ **do** $myview_i[j] \leftarrow R[j]$ **end_do** // read of the anony. mem.
3    $sum_i \leftarrow myview_i[1] + \cdots + myview_i[m]$ // # successful Compare&Swap() seen
4   **if** $\exists\, x \in \{1, ..., m\}\, :\, myview_i[x] < u$ **then**
5   **if** $won_i < v$ **then**
6   **if** Compare&Swap($R[x], myview_i[x], myview_i[x] + 1$) **then** $won_i \leftarrow won_i + 1$ **fi fi**
7   **if** $sum_i \geq v \times n$ **then**   // $sum_i \geq v \times n$ implies $won_i = v$
8   **if** Compare&Swap($R[x], myview_i[x], myview_i[x] + 1$) **then return** (leader) **fi fi**
9   **fi**
10 **until** $sum_i = u \times m$ **end repeat**
11 **return** (not leader).

Fig. 3. Exact d-election for n anonymous processes and m RMW anonymous registers

Description of the algorithm Assuming d is a multiple of $\gcd(m, n)$ and all the processes participate, Algorithm 2 (described in Fig. 3) solves exact d-election for n anonymous processes and m RMW anonymous registers.

---

[10] The pair $\langle u, v \rangle$ is not unique. Euclid's $\gcd(m, n)$ algorithm can be used to compute such pairs.

When it invokes elect(), a process p_i enters a repeat loop lines 1–10. Each time it enters the loop, p_i asynchronously reads the anonymous memory non-atomically (line 2) and then counts in sum_i the number of squares that have been captured by all processes as indicated by the previous asynchronously reads (line 3).

If p_i sees a register $R[x]$ that has been captured less than u times (line 4), there are two cases.

- If $won_i < v$, p_i tries to capture one of the u squares of $R[x]$. To this end p_i uses the RMW operation: it invokes Compare&Swap($R[x], myview_i[x], myview_i[x] + 1$). If it is successful, it increases won_i, the number of squares it has captured so far (line 6).
- If $sum_i \geq v \times n$ (we have then $won_i = v$), p_i strives to capture one more square (line 8). If it is successful, it is elected as of the d leaders.
  In the other case, if $sum_i = u \times m$, all the squares have been captured, so p_i is not a leader (line 11). Otherwise, p_i re-enters the repeat loop.

Theorem. Let m, n and d be such that gcd(m, n) divides d, and assume all the processes invoke elect(). Algorithm 2 (Fig. 3) solves exact d-election in a fully anonymous system where communication is through RMW registers.

Proof. Let us first observe that, due to the atomicity of Compare&Swap(), if several processes invoke Compare&Swap($X, v, v + 1$) on the very same register X whose value is v, exactly one of of them succeeds in writing $v + 1$. It follows that each of the $u \times m$ squares is captured by only one process. Moreover, due to the predicate of line 5, each process eventually captures v squares. Once this occurs, it remains d squares, which are captured by d distinct processes at line 8 (these processes are distinct because, once a process captured such a square, it returns the value **leader** and stops executing). Moreover, a process can capture one of the d remaining squares only after each process has captured v squares at line 6. It follows that exactly d processes exit the algorithm at line 7 with a successful Compare&Swap(), and the $(n - d)$ other processes exit the algorithm at line 11.

Acknowledgments

The author would like to thank the conference co-chairs Toshimitsu Masuzawa, Yoshiaki Katayama, and Hirotsugu Kakugawa for their kind invitation as keynote speaker at SSS 2024 in Nagoya, Japan.

He wants also to thank Gadi Taubenfeld for having introduced him to the world of anonymous memory systems, and all the people cited in the references (without them this article would not exist!).

References

1. Afek, Y., Attiya, H., Dolev, D., Gafni, E., Merritt, M., Shavit, N.: Atomic snapshots of shared memory. J. ACM **40**(4), 873–890 (1993)
2. Angluin D., Local and global properties in networks of processes. In: Proceedings 12th Symposium on Theory of Computing (STOC'80), pp. 82–93. ACM Press (1980)
3. Apt, K.R., Hoare, C.A.R. (eds.): Edsger Wybe Dijkstra: his life, work, and legacy. In: Association for Computing Machinery, p 550. Morgan and Claypool Publishers (2022)
4. Auvolat, A., Frey, D., Raynal, M., Taïani, F.: Money transfer made simple: a specification, a generic algorithm, and its proof. Electron. Bull. EATCS (European Association of Theoretical Computer Science) **132**, 22–43 (2020)
5. Ben-Or, M.: Another advantage of free choice: completely asynchronous agreement protocols. In: Proceedings of 2nd ACM Symposium on Principles of Distributed Computing (PODC'83), pp. 27–30. ACM Press(1983)
6. Berry, G.: Talks at Collège de France (2012–2019)
7. Bracha, G., Toueg, S.: Asynchronous consensus and broadcast protocols. J. ACM **32**(4), 824–840 (1985)
8. Brinch Hansen, P.: (ed.) The Origin of Concurrent Programming, p. 534. Springer (2002)
9. Castañeda A., Rajsbaum S., Raynal M.: Unifying concurrent objects and distributed tasks: Interval-linearizability. J. ACM **65**(6), 42, Article 45 (2018)
10. Castañeda, A., Rajsbaum, S., Raynal, M.: A linearizability-based hierarchy for concurrent specifications. Commun. ACM **66**(1), 60–71 (2023)
11. Castañeda, A., Rajsbaum, S., Raynal, M.: Set-linearizable implementations from read/write operations: sets, fetch & increment, stacks and queues with multiplicity. Distrib. Comput. **36**(2), 89–106 (2023)
12. Censor-Hillel, K., Petrank, E., Timnat, S.: Help! In: Proceedings of 34th ACM Symposium on Principles of Distributed Computing (PODC'15), pp. 241–250. ACM Press (2015)
13. Chandra, T.D., Hadzilacos, V., Toueg, S.: The weakest failure detector for solving consensus. J. ACM **43**(4), 685–722 (1996)
14. Chandra, T., Toueg, S.: Unreliable failure detectors for reliable distributed systems. J. ACM **43**(2), 225–267 (1996)
15. Charron-Bost, Schiper A.: Computing in distributed systems with benign faults. The heard-of model. Distrib. Comput. **22**, 49–71 (2009)
16. Collins, D., Guerraoui, R., Komatovic, J., Monti, M., Xygkis, A., Pavlovic, M., Kuznetsov, P., Pignolet, Y.-A., Seredinschi, D.A., Tonlikh, A.: Online payments by merely broadcasting messages. In: Proceedings 50th IEEE/IFIP International Conference on Dependable Systems and Networks (DSN'20), pp. 26–38 (2020)
17. Courtois, P.J., Heymans, F., Parnas, D.L.: Concurrent control with readers and writers. Commun. ACM **14**(5), 667–668 (1971)
18. Delporte-Gallet, C., Fauconnier, H., Guerraoui, R., Kouznetsov, P.: Mutual exclusion in asynchronous systems with failure detectors. J. Parallel Distrib. Comput. **65**, 492–505 (2005)
19. Delporte-Gallet, C., Fauconnier, H., Raynal, M.: On the weakest information on failures to solve mutual exclusion and consensus in asynchronous crash prone read/write systems. J. Parallel Distrib. Comput. **153**, 110–118 (2021)

20. Dijkstra, E.W.: Over de sequentialiteit van procesbeschrijvingen (on the nature of sequential processes). In: EW Dijkstra Archive (EWD-35), Center for American History, University of Texas at Austin (Translation by Martien van der Burgt and Heather Lawrence) (1962)
21. Dijkstra, E.W.: Solution of a problem in concurrent programming control. Commun. ACM **8**(9), 569 (1965)
22. Dijkstra, E.W.: Cooperating sequential processes. In: Genuys, F. (ed.) Programming Languages, pp. 43–112 . Academic Press (1968)
23. Dijkstra, E.W.: Hierarchical ordering of sequential processes. Acta Informatica **1**(1), 115–138 (1971)
24. Dijkstra, E.W.: Guarded commands, non-determinacy and formal derivation of programs. Commun. ACM **8**, 453–457 (1975)
25. Dijkstra, E.W., Dahl, O.-J., Hoare, C.A.R.: Structured Programming, p. 220. Academic Press (1972)
26. Dolev, D., Dwork, C., Stockmeyer, L.: On the minimal synchronism needed for distributed consensus. J. ACM **34**(1), 77–97 (1987)
27. Fischer, M.J., Lynch, N.A., Paterson, M.S.: Impossibility of distributed consensus with one faulty process. J. ACM **32**(2), 374–382 (1985)
28. Guerraoui, R., Kuznetsov, P., Monti, M., Pavlovic, M., Seredinschi, D.A.: The consensus number of a cryptocurrency. Distrib. Comput. **35**, 1–15 (2022)
29. Harel, D., Feldman, Y.: Algorithmics: The Spirit of Computing, third edition, p. 572. Springer (2012)
30. Herlihy, M.P.: Wait-free synchronization. ACM Trans. Program. Lang. Syst. **13**(1), 124–149 (1991)
31. Herlihy, M.P., Luchangco, V., Moir, M.: Obstruction-free synchronization: double-ended queues as an example. In: Proceedings 23th International IEEE Conference on Distributed Computing Systems (ICDCS'03), pp. 522–529 . IEEE Press (2003)
32. Herlihy, M.P., Wing, J.M.: Linearizability: a correctness condition for concurrent objects. ACM Trans. Program. Lang. Syst. **12**(3), 463–492 (1990)
33. Hoare, C.A.R.: An axiomatic basis for computer programming. Commun. ACM **12**(10), 576–580 (1969)
34. Hoare, C.A.R.: Programming: sorcery or science? IEEE Softw. **1**(2), 5–16 (1984)
35. Imbs, D., Raynal, M., Taubenfeld, G.: Election in fully anonymous shared memory systems: tight space bounds and algorithms. In: Proceedings 29th International Colloquium on Structural Information and Communication Complexity (SIROCCO'22), pp. 174–190 . Springer LNCS 13298 (2022)
36. Jones, C.B., Misra, J. (eds.): Theories of Programming: The Life and Works of Tony Hoare, p. 430. Association for computing machinery, Morgan & Claypool Publishers (2021)
37. Lamport, L.: A new solution of Dijkstra's concurrent programming problem. Commun. ACM **17**(8), 453–455 (1974)
38. Lamport, L.: Concurrent reading and writing. Commun. ACM **20**(11), 806–811 (1977)
39. Lamport, L.: Time, clocks, and the ordering of events in a distributed system. Commun. ACM **21**(7), 558–565 (1978)
40. Lamport, L.: On inter-process communications, part I: basic formalism. Distrib. Comput. **1**(2), 77–85 (1986)
41. Lamport, L.: On inter-process communications, part II: algorithms. Distrib. Comput. **1**(2), 86–101 (1986)
42. Lamport, L.: The mutual exclusion problem: part I- a theory of interprocess communication. J. ACM **33**(2), 313–326 (1986)

43. Lamport, L.: The mutual exclusion problem: Part II: statement and solutions. J. ACM **33**(2), 313–326 (1986)
44. Lamport, L.: The computer science of concurrency: the early years (Turing lecture). Commun. ACM **58**(6), 71–76 (2015)
45. Lamport, L.: Deconstructing the Bakery to build a distributed state machine. Commun. ACM **65**(9), 58–66 (2022)
46. Lamport, L., Shostak, R., Pease, S.: The Byzantine generals problem. ACM Trans. Program. Lang. Syst. **4**(3), 382–401 (1982)
47. Lewis, H.R., Papadimitriou, C.H.: Elements of the Theory of Computation, p. 361. Prentice Hall Int. Editions (1998)
48. Lo, W.K., Hadzilacos, V.: Using failure detectors to solve consensus in asynchronous shared memory systems. In: Proceedings 8th International Workshop on Distributed Algorithms (WDAG'94), pp. 280–295. Springer LNCS 857 (1994)
49. Loui, M., Abu-Amara, H.: Memory requirements for agreement among unreliable asynchronous processes. Adv. Comput. Res. **4**, 163–183. JAI Press Inc. (1987)
50. Malkhi, D. (ed.): Concurrency: The Works of Leslie Lamport, p. 345. Association for Computing Machinery, Morgan & Claypool Publishers (2019)
51. Mostéfaoui, A., Moumen, H., Raynal, M.: Signature-free asynchronous binary Byzantine consensus with $t<n/3$, $O(n^2)$ messages, and $O(1)$ expected time. J. ACM **62**(4), 21, Article 31 (2015)
52. Mostéfaoui, A., Perrin, M., Raynal, M.: Send/receive patterns versus read/write patterns in crash-prone asynchronous distributed systems. In: Proceedings 37th International Symposium on Distributed Computing (DISC'23), Lipics, vol. 281, pp. 16:1–16:24 (2023)
53. Mostéfaoui, A., Rajsbaum, S., Raynal, M.: Conditions on input vectors for consensus solvability in asynchronous distributed systems. J. ACM **50**(6), 922–954 (2003)
54. Pease, M., Shostak, R., Lamport, L.: Reaching agreement in the presence of faults. J. ACM **27**, 228–234 (1980)
55. Perrin, M., Mostéfaoui, Raynal, M.: A simple object that spans the whole consensus hierarchy. Parallel Process. Lett. **28**(2), 9, 1850006 (2018)
56. Peterson, G.L.: Myths about the mutual exclusion problem. Inf. Process. Lett. **12**(3), 115–116 (1981)
57. Peterson, G.L.: Concurrent reading while writing. ACM Trans. Program. Lang. Syst. **5**, 46–55 (1983)
58. Rajsbaum, S., Raynal, M.: Mastering concurrent computing through sequential thinking. Commun. ACM **63**(1), 78–87 (2020)
59. Raynal, M.: Concurrent Programming: Algorithms, Principles and Foundations, p. 515. Springer. ISBN 978-3-642-32026-2 (2013)
60. Raynal, M.: Fault-Tolerant Message-Passing Distributed Systems: An Algorithmic Approach, p. 550. Springer (2018)
61. Raynal, M.: Mutual exclusion versus consensus: both sides of the same coin? Bull. EATCS **140**, 10 (2023)
62. Raynal, M.: Concurrent crash-prone shared memory systems. In: Synthesis Lectures on Distributed Computing Theory, p. 115. Morgan & Claypool (2023)
63. Raynal, M., Taubenfeld, G.: Mutual exclusion in fully anonymous shared memory systems. Inf. Process. Lett. **158**, 105938, 7 (2020) Corrigendum in Information Processing Letters, vol. 179:106304 (2023)
64. Raynal, M., Taubenfeld, G.: A visit to mutual exclusion in seven dates. Theoret. Comput. Sci. **919**, 47–65 (2022)

65. Taubenfeld, G.: Synchronization Algorithms and Concurrent Programming, p. 423. Pearson Education/Prentice Hall. ISBN 0-131-97259-6 (2006)
66. Taubenfeld, G.: Concurrent programming, mutual exclusion. Springer Encyclopedia of Algorithms, pp. 421–425 (2016)
67. Taubenfeld, G.: Coordination without prior agreement. In: Proceedings 36th ACM Symposium on Principles of Distributed Computing (PODC'17), pp. 325–334. ACM Press (2017)
68. Taubenfeld, G.: Anonymous shared memory. J. ACM **69**(4), 30, Article 24 (2022)

Invited Papers

Invited Paper: A Survey of the Impact of Knowledge on the Competitive Ratio in Linear Search

Evangelos Kranakis[✉]

School of Computer Science, Carleton University, Ottawa, ON, Canada
kranakis@scs.carleton.ca

Abstract. We give an outline of recent results and propose several open problems in linear search by searchers, modelled as autonomous mobile agents. The trajectories of the searchers are continuous and the search domain is the infinite real line or generalizations thereof (e.g., star graph). The design and analysis of the search algorithms proposed takes into account the impact of the knowledge the searchers have about

- the communication model employed by the searchers,
- the potentially faulty behaviour of searchers, and
- the starting distance and speed of the target being sought

in order to obtain an optimal competitive ratio. For group-search involving multiple searchers the approach emphasizes agent co-operation and distributed algorithm design principles. The overall approach considered is based on understanding the impact of the knowledge the searchers have about the system settings on the competitive ratio (which is the supremum of the ratio between the time the searcher travels and the time he would have taken if he had known the location of the target) of the linear group-search algorithms.

Keywords: Agent · Autonomous · Crash · Byzantine · Faulty · Mobility · Oblivious · Searcher · Speed · Target

1 Introduction

Search is of central importance to many areas of computer science including data structures, computational geometry, and artificial intelligence. Linear search (also known as cow-path problem) is carried out over the real line and is concerned with an autonomous agent (or searcher), starting at the origin and looking for a target (either static or mobile) placed at an unknown location also on the infinite line. The searcher moves with uniform (constant) speed and can change direction (without any loss in time). The ultimate goal is to find the target while at the same time minimizing the resulting competitive ratio.

Research supported in part by NSERC Discovery grant.

T. Masuzawa et al. (Eds.): SSS 2024, LNCS 14931, pp. 23–38, 2025.
https://doi.org/10.1007/978-3-031-74498-3_2

Linear search for stochastic and game theoretic systems was first proposed, independently, in [8,10]. A target is located at a given point on the infinite real line with a (possibly) known (to the searcher) probability distribution. A searcher, whose maximal velocity is one, starts at the origin of the real line and wants to find the target in minimal expected time. Deterministic search by a single mobile agent operating on the real line or the plane were subsequently investigated by several researchers including [5,6,29] for search on a star. Additional work on search can also be found in [1,2,39].

More recently, fault-tolerant search with multiple robots has attracted the attention of researchers in distributed computing and theoretical computer science. Group Search is an important task arising from the need to design algorithms for multi-agent systems in distributed computing. We are interested in designing algorithms and analyze tradeoffs involving time, mobility, and fault-tolerant communication for finding an unknown target. The proposed algorithms will employ cooperating, communicating autonomous mobile agents in a distributed setting and can operate over continuous (as well as discrete) domains. Designed algorithms (also called strategies) will be fault-tolerant and maintain optimal characteristics as measured by the competitive ratio despite hardware and/or software failures during their execution.

2 Outline of the Paper

Each section begins with a brief description of the main question or theme to be discussed. This is followed by a brief survey of recent research outcomes. The text is sprinkled with open problems and directions for future research. Throughout the paper the purpose is to discuss results relevant to group-search. An outline of the paper is as follows. In Sect. 3 we give details of the main characteristics of the computation model which is used throughout the paper. In Sect. 4 we focus on search with fault-tolerant characteristics. In Sect. 5 we outline results and tradeoffs for capturing a moving target. In Sect. 6 we concentrate on models concerning the energy cost of search. We summarize our discussion in the concluding Sect. 7.

3 Computation Model

We begin by discussing all aspects of the computation model and define the basic concepts of Searchers, Target, Trajectories, Competitive Ratio, Communication, and Faults.

Searchers. A group of autonomous mobile agents with communication capabilities can move with their own respective speeds, which may not exceed the maximum value of 1, within a continuous (e.g., segment, infinite line, plane, geometric graph) or discrete domain. The domain also indicates the degree of freedom the agent has for its movement and may be described either in one or two dimensions (e.g., line segment, infinite line, unit disk, plane, geometric graph). Agents are autonomous in that they can act on their own capabilities following a predefined algorithm (or strategy); after taking into account either available or recently discovered information, they may traverse a well-defined trajectory making instantaneous mobility decisions, like stop, start, wait, change direction, change speed, as well as collect and exchange information with other agents.

Target. A moving target is either a mobile or static agent, represented as a point, moving with a certain constant speed v. It can be recognized by the (non-faulty) searchers when they are co-located with it, but it is otherwise oblivious in that it cannot change its speed or direction of movement and has no communication capabilities. Its movement is either away or towards the origin. The target is initially placed on the real line at a certain distance d from the origin. When $v = 0$ the target is static. Searchers and targets are represented as points on the real line. Sometimes we refer to a static target as exit.

Trajectories. The trajectory of an agent is a continuous function from the non-negative reals to the domain on which its movements are taking place. If an agent has maximum speed u then its trajectory X must satisfy $|X(t') - X(t)| \leq u|t' - t|$, $\forall t', t \geq 0$. Agents are assumed to begin their search at the origin and so we must also have $X(0) = 0$. Evacuation refers to search by a group of collaborating searchers; each robot in the group follows a certain trajectory and termination time is measured by the last searcher to find the target. Capturing a moving target is similar to evacuation except that the target is mobile and the searchers capturing the target must be colocated. Our goal is to design trajectories leading to optimal competitive ratios. Such trajectories may be "zigzag" and appear rather awkward in that they must start with an infinite sequence of infinitesimal steps. In order to simplify proofs and avoid this "infinitesimal step" issue, without loss of generality, we will consider "first step" trajectories whereby there is a known lower bound, usually 1, on the starting distance of the target from the origin [28].

Competitive Ratio. For an algorithm A and an instance I of the problem considered (e.g., search, evacuation, delivery, etc.) $T_A(I)$ is the time it takes algorithm A to solve instance I. If $T_{opt}(I)$ is the optimal time of an offline algorithm for the same instance I, then the competitive ratio of an online algorithm A is defined by the ratio $CR_A := \sup_I T_A(I)/T_{opt}(I)$. If $\mathcal{A}$ is a class of algorithms solving an online version then its competitive ratio is defined by $CR_{\mathcal{A}} := \inf_{A \in \mathcal{A}} CR_A$.

The competitive ratio for evacuation in the linear search model for n agents at most f of which may be faulty is defined as $CR_f = \sup_x E_{f}^{x}/|x|$, where E_f^x is the worst case time required until the last reliable agent reaches a target at location x. CR_f represents the worst case ratio of the evacuation time to the lower bound $|x|$ on the time required to find the target.

Communication. To share their findings with the rest of the group the mobile agents employ a specific form of communication which reflects their transmission/reception capabilities. Communication and infrastructure faults affect the agents' behaviour. Two basic communication models are considered: Wireless (WiFi), and Face-to-Face (F2F). In the former, the agents are equipped with wireless transceivers and can communicate instantaneously either directly with each other or via a control centre (or base station) at any distance. In the latter, the agents can communicate only if they are colocated. The F2F model provides the agents with the most basic type of communication ability and may itself also be considered as the result either of malfunctioning wireless transceivers between individual agents or a massive breakdown of the control centre thus imposing communication with limited range on the agents.

Faults. For simplicity, the agents' faulty behaviour can be divided into three types. *Crash* faults render the agents inoperable (silent) and unable to function, but are otherwise unable to hinder or confuse the communication of the other agents. *Byzantine* faults may cause an infected agent to distribute unreliable and/or malicious information to the rest of the agents in a group and affect the correct execution of a task. Finally, *S/R* faults (a mixed type of communication model) in which either one of the "send" or "receive" operations of the agent in WiFi mode is disabled but otherwise the agent never loses its ability to communicate in F2F mode. An agent that loses its ability to *send* or *receive* wirelessly is called *receiver* or *sender*, respectively. Sometimes we also refer to a *mixed* fault communication model whereby all types of faults my be possible. Agents will be searching for a target that may be either static or mobile and is otherwise oblivious to the environment.

4   Fault Tolerance

The underlying search domain is the (infinite) real line. There are n searchers at most f of which are faulty. All searchers start at the origin and the exit is placed at an unknown location on the real line. The main problem of interest is the following.

Problem 1. *Study linear search (evacuation, capturing) strategies for n searchers at most f of which are faulty.*

4.1   Crash Faults

Assume that some agents may be crash faulty. If $n \geq 2f + 2$, there is a simple algorithm with competitive ratio 1. For $f < n < 2f + 2$ a new class of algorithms is developed in [26], called *proportional schedule algorithms*, whose competitive ratio is

$$1 + \left(\frac{4f + 4}{n}\right)^{\frac{2f+2}{n}} \left(\frac{4f + 4}{n} - 2\right)^{1 - \frac{2f+2}{n}}.$$

The algorithm ensures that at least $f + 1$ searchers (and therefore at least one of them is not crash-faulty) will reach the target.

In a proportional schedule algorithm, the movement of the robots is of zigzag type and defined by a "planar cone" determined by the turning points of the searchers in a space-time diagram as depicted in Fig. 1, which gives rise to a sequence $x_0, x_1, x_2, \ldots$ of *turning points* on the real line at which a robot changes the direction of its movement. For a fixed real number $\beta > 1$ one defines C_β to be the (inverted) cone delimited by a pair of lines $t = x\beta$ for $x \geq 0$, and $t = -x\beta$ for $x \leq 0$ in the space-time diagram. Starting with a point $(x_0, |x_0\beta|)$ on the boundary of the cone C_β, the turning points of the searcher are given by the formula

$$x_i = x_0 \left(\frac{\beta + 1}{\beta - 1}\right)^i (-1)^i$$

with corresponding expansion factor $(\beta + 1)/(\beta - 1)$. Using this formula one can define trajectories for all n searchers see [26][Fig. 3]. Additional details of the final algorithm, and its proof of correctness can be found in [26]. It turns out that proportional schedule algorithms as defined above have optimal competitive ratio for crash faulty searchers; as this has been shown in [36].

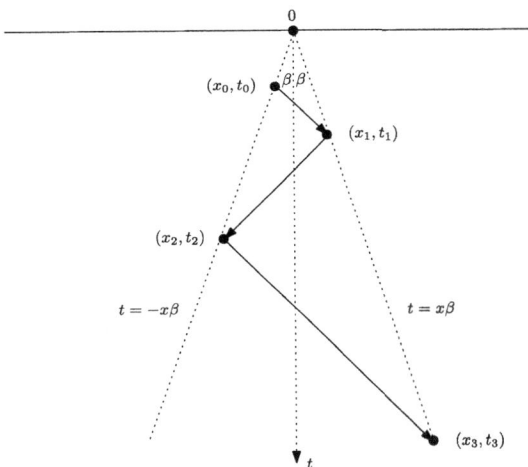

Fig. 1. The cone C_β and a general zig-zag strategy with turning points (x_i, t_i), $0 \leq i \leq 3$, on the perimeter of the cone, where x_i is distance from the origin and t_i the time that the searcher changes direction.

4.2   Byzantine Faults

The case of byzantine faulty searchers turns out to be much harder because, unlike crash faulty searchers which are passive, byzantine searchers are malicious liars and consensus decisions by the non-faulty searchers have to be based on the majority rule. Algorithms for this case were first proposed in [23] and Table 1 displays bounds obtained for $n \leq 6$. Several upper bounds were improved by [40], most notably the competitive ratio of evacuation was shown to be 8.653055 for $(n, f) = (3, 1)$.

Significant work and improvements for the Byzantine faulty case can be found in [25, 35]. In general, it turns out that the hardest instances for the Byzantine

Table 1. Upper (UB) and lower (LB) bounds on the competitive ratio of search for $n \leq 6$ of robots and $f = 1, 2$ Byzantine faults

n	f	UB	LB
3	1	9	5.23
4	1	3	3
5	1	2	2
6	1	1	1
5	2	9	4.43
6	2	$2 + \sqrt{5}$	3

case is when $n = 2f + 1$. The resulting competitive ratios of evacuation for $n = 2f + 1 \leq 9$ are displayed in Table 2. Most importantly, it was possible to show in [25, 35] an asymptotic upper bound of $4 + 2\sqrt{2}$, for larger values of $n = 2f + 1$, as $n \to \infty$. It is conjectured that $4 + 2\sqrt{2}$ is asymptotically tight but so far very little is known about its optimality. Although the previous results constituted significant progress, other tight bounds for the general case of (n, f) byzantine faults [23] remain largely open.

Table 2. Best known upper bounds (UB) on the competitive ratio of evacuation by $n = 2f + 1$ agents at most f of which are Byzantine faulty

n	f	UB
3	1	7.437011
5	2	7.253767
7	3	7.253767
9	4	7.147026

4.3 Mixed Faults

S/R (Sender/Receiver) faults are another type of fault that were considered in [24] in which a searcher may be either a sender or a receiver (for the definitions, see Sect. 3). An optimal algorithm for the real line with competitive ratio exactly $3 + 2\sqrt{2}$ was given for evacuating two searchers one of which is a sender and the other a receiver. The case of multiple searchers with such faults has never been considered. For example, consider a group of searchers with mixed faults. For example, one may consider groups of $n = s + r + c + b$ searchers with mixed faults, namely s senders, r receivers, c crash, and b byzantine faulty. No systematic algorithmic analysis has ever been done for the "non-trivial" instances of a group of three or more for either a static or moving target.

4.4   Different Speeds

A generalization of the problems discussed so far is when the agents do not necessarily have identical speeds. The authors of [7] were the first to give an optimal algorithm for the case of two agents with different speeds. They consider the linear search problem for two robots equipped with distinct maximal speeds. They scale their speeds so that the faster robot has speed 1 and the slower one $0 < v \leq 1$. In the F2F communication model they give a search strategy with optimal competitive ratio $\frac{1+3v}{1-v}$. If the communication is wireless they give a search strategy with competitive ratio $\frac{2+v+\sqrt{v^2+8v}}{2}$. Further they conclude with the surprising result that for $v \leq \sqrt{17} - 4$ the wireless communication model gives no advantage over F2F communication.

There has never been a systematic analysis of linear group search by robots with possibly unequal maximum speeds and to the best of our knowledge publication [7] is the first one to do so for two searchers. An additional publication is [15] which designs an algorithm with optimal competitive ratio in the F2F model and uses speed slowdown to facilitate communication, a well-known technique that sometimes may lead to improved competitive ratios in evacuation problems.

4.5   Capturing a Moving Target

An additional interesting twist leading to an extension of all the problems discussed in this section so far is when the target is an oblivious mobile agent with speed v. This leads to problems of capturing a moving target which in a way resemble the well-known problems of cops and robers on graphs [12].

An open problem related to [26] is when the exit is an oblivious agent moving away from the origin with speed $v < 1$ while the searchers can move with speed 1 and may suffer crash, byzantine or any other kind of mixed faults (see also [2][Section 8.5] for the non faulty case). In general, the case of capturing a moving target by faulty searchers is wide open and it would also be interesting to investigate the impact on the competitive ratio of the communication model used, namely, either of F2F, WiFi, or mixed in the sense that the searchers in the same group may suffer from different kinds of faults.

As will be seen in Sect. 5 the concept of moving target enriches the search environment and provides an additional class of interesting questions that take into account what knowledge the searchers have about the target.

4.6   Search with Priorities and Weights

The goal of priority search, introduced in [20,21] for the unit disk search domain, is for a distinguished searcher (a single one) among $n + 1$ searchers to reach and evacuate from an exit that is hidden on the search domain in as little time as possible. The remaining n searchers are there to facilitate the objective of the distinguished searcher and may not necessarily (depending on the model considered) be required to reach the hidden exit. In a way, the goal is for the distinguished searcher to evacuate first. In a generalization of priority search we may also define a *weighted evacuation* problem in which one differentiates on searchers' preferences by assigning a weight w_i to each searcher i and require to evacuate a subset of searchers of total weight $\geq W$ in minimum time. Very little has been established on this problem on an infinite line and nothing is known for a multiple, possibly faulty searchers or for capturing a mobile target.

A different approach to weighted search on the line is the recent work in [31] whereby (multiple) agents are looking for a target either in the wireless or the F2F model. For two agents the cost of the solution is the arithmetic weighted average of the times that the two agents need to reach the target. Without loss of generality one agent has weight 1 and the other has weight $w \geq 1$. The competitive ratio of the trajectory is defined as $\sup_{|x| \geq 1} \frac{T_1(x) + w \cdot T_2(x)}{(1+w)x}$, where x is the unknown location of the target, and $T_i(x)$ the evacuation time of the i-th robot. An interesting point is the knowledge of the weight as well as the identity of the searchers which affects the competitive ratio. There are many open problems depending on whether the weight w is known, as well as to whether it is known whose weight is w. The multi-robot case is also interesting.

5   Searching for a Moving Target

The additional consideration in this section is an oblivious moving target whose speed is v. When $v = 0$ the target is static which is of curse the standard case. The target may be moving either away or toward the origin of the real line and this known to the searcher(s). The first case is called *away* model and the latter *toward*. The main problem of interest is the following.

Problem 2. *Design optimal linear search and evacuation strategies for n searchers when the target is moving.*

5.1   Search and Mobility

McCabe [38] was the first to investigate the problem of searching for an oblivious mobile target in a stochastic setting whereby the target follows a Bernoulli random walk on the integers. In a deterministic (continuous) setting Alpern and Gal [2][p. 134, Eq. 8.25] were the first to show that the optimal competitive ratio of search for a target moving with speed $0 \leq v < 1$ (where 1 is the speed of the searcher) and starting at an unknown distance d away from the origin is exactly $1 + 8\frac{1+v}{(1-v)^2}$, provided that the speed v of the target is known to the searcher.

An extension of this problem was investigated by [17] in which the authors investigate knowledge, competitive ratio tradeoffs depending on whether the target is moving toward or away from the origin. In each of these two cases four subclasses arise depending on which of the four parameters d and v are known to the searcher.

In the toward model, if the searcher knows v but not d then it is shown in [17] that $1 + 8\frac{1-v}{(1+v)^2}$ is a tight bound for $v \leq \frac{1}{3}$, while it is $1 + \frac{1}{v}$ for $v \geq \frac{1}{3}$, when the target is moving toward the origin. The implication is that when $v \geq \frac{1}{3}$ the target is moving too fast and the searcher should wait.

The main optimal algorithm from [17] when d is unknown for $v < 1$ in the away model and $v \leq \frac{1}{3}$ in the toward model is the following:

Algorithm 1 Algorithm for `Away` and `Toward` Models

1: **input**: target speed v and expansion ratio a
2: $i \leftarrow 0$
3: **while** target not found **do**
4:     **if** at origin **then**
5:         $d \leftarrow (-a)^i$
6:         $i \leftarrow i + 1$
7:     **else if** at d **then**
8:         $d \leftarrow 0$
9:     move toward d

Algorithm 1 was analyzed in [17] and parameter a is the growth ratio which is chosen to minimize the competitive ratio.

When d is known but v is not then the competitive ratio is shown to be bounded from above by $1 + \frac{(\log(1-v))^2}{(1-v)^4}$, for $v \geq \frac{1}{2}$, and by 5, for $v \leq \frac{1}{2}$. The proposed algorithm from [17] for this case is different.

Algorithm 2 Online Algorithm for `NoSpeed/Away` Model

1: **input:** target initial distance d
2:         integer sequence $\{f_i : i \geq 0\}$ such that $f_i < f_{i+1}$, for $i \geq 0$ and $f_0 = 1$;
3: $t \leftarrow 0$
4: **for** $i \leftarrow 0, 1, 2, \ldots$ until target found **do**
5:     $v_i \leftarrow 1 - 2^{-f_i}$
6:     $x_i \leftarrow (-1)^i \cdot \frac{d + t v_i}{1 - v_i}$
7:     move to x_i and back to the origin
8:     $t \leftarrow t + |x_i|$

Algorithm 2 was analyzed in [17] and the upper bound above was proved. To prove a lower bound the concept of evasiveness was developed in [16]. Represented by the fraction $u = \frac{1}{1-v}$ it represents, in a way, the minimum amount of time that it takes for the robot to close a gap between itself and the target. Note that $v = 1 - \frac{1}{u}$ and by the bounds of v, we have $1 \leq u < \infty$. Algorithm 2 was later shown to be optimal in [16]. The method employed in order to prove the stronger lower bound uses the evasiviness in order to reduce the lower bound to a lemma presented in [9] about a sequence that is both positive and concave down.

Similarly when both d, v are unknown an upper bound on the competitive ratio can be shown by guessing both d, v. One must then determine the value of the growth parameter a which optimizes the resulting competitive ratio. Details of this algorithm can be found in [17] and [16].

Known results for a target moving away from the origin are displayed in Table 3. It is important to note that each row of Table 3 represents a different "amount" of knowledge that the searcher has about v, d and one can see in the second column how this affects the resulting competitive ratio. The upper bound in the second to last row of Table 3 is valid for $ud > 4$, while $CR \leq 1 + \frac{8}{d}$ for $ud \leq 4$, thus improving the previous best upper bound: $O\left(\frac{1}{d}M^8 \log_2^2 M \log_2 \log_2 M\right)$, where $M = \max(u, d)$, in [17]. It should be noted that the optimal competitive ratio when both d, v are unknown is still an open problem.

Table 3. Knowledge, Competitive Ratio tradeoffs for capturing a target moving away from the origin. The speed v of the target satisfies $0 \leq v < 1$. We use the notation $u = \frac{1}{1-v}$ for the evasiviness. The constant in the O-notation below is 56.18

Knowledge	Competitive ratio	Publication
v, d	$1 + \frac{2}{1-v}$	[17]
v	$1 + \frac{8(1+v)}{(1-v)^2}$	[2]
d	$O(u^{4-(\log_2 \log_2 u)^{-2}})$ if $u > 4$ $\Omega(u^{4-\epsilon})$ if $\epsilon > 0$	[16, 17]
$\emptyset$	$1 + \frac{1}{d}O((ud)^{4-(\log_2 \log_2 (ud))^{-2}} - 1)$ if $ud > 4$ $1 + \frac{8}{d}$ if $ud \leq 4$	[16]

Known results when the mobile target is moving toward the origin can be found in the article [17] and are displayed in Table 4. An interesting observation is the resemblance of the formulas for the competitive ratios of the zigzag algorithms in the first two rows of Tables 1 and 3, namely $1 + \frac{2}{1-v}$ and $1 + \frac{8(1+v)}{(1-v)^2}$ for the away model, and $1 + \frac{2}{1+v}$ and $1 + \frac{8(1-v)}{(1+v)^2}$ for the toward model, respectively. This is not a coincidence since the direction of movement of the target in the away model is the opposite to the one in the toward model.

The search problem involving a moving target is also interesting on more general topologies like star graphs (the traditional cow-path topology) [37]. It is even more interesting to look at the multi-searcher case and investigate tradeoffs between k, the number of searchers, and n, the number of infinite half lines of the star graph.

A different setting is when the speed of the searcher depends on physical variations on the terrain. For example, the speed depends on the direction of movement, the incline of the domain, or there is constant acceleration and/or variability depending on whether a certain segment has already been searched [27]. Additional related work on this theme can also be found in [13, 14].

Table 4. Knowledge, Competitive Ratio tradeoffs for capturing a target moving toward the origin. The speed v of the target satisfies $v \geq 0$

Knowledge	Competitive ratio
v, d	$1 + \frac{2}{1+v}$ if $v \leq 1$ $1 + \frac{1}{v}$ if $v \geq 1$
v	$1 + \frac{8(1-v)}{(1+v)^2}$ if $v \leq 1/3$ $1 + \frac{1}{v}$ if $v \geq 1/3$
d	3
$\emptyset$	$1 + \frac{1}{v}$

An alternative setting is evacuation of two searchers from an unknown exit with the assistance of an oblivious assistant (also called bike) which the searchers may share alternately (i.e., either searcher can ride the bike but only one at a time) in order to complete the evacuation task. This search problem may be considered in any communication model. The authors in [32] analyze the competitive ratio of search for two robots and a bike in the F2F model. This is extended to the Sender/Receiver model in [33]. Although there are certain resemblances, the above model is different from the work in [7] which is concerned with linear search by two distinct-speed searchers. There has never been any systematic study of bike assisted search either when the target is mobile or fault tolerance for multiple searchers.

5.2   Fault Tolerance and Target Mobility

No optimal algorithms are known for linear search of a moving target by multiple searchers which are either crash or byzantine faulty. For related work on this when the target is static we refer the reader to [26] for crash and [23] for byzantine faulty searchers. Note the assumption made above for the target to be moving away from the origin. An open problem also arises when one studies speed/distance and knowledge tradeoffs for search algorithms in the remaining three possible situations depending on knowledge about d and v as well as knowledge of whether the target is moving either away or towards the origin. For additional discussion see also Sect. 4.3.

5.3   Randomized Algorithms

The last problem is motivated by the randomized algorithm investigated in the seminal paper [34] which refers only to the case of a static target (i.e., $v = 0$). The algorithm is as follows:

Algorithm 3 RandomSearch Algorithm

1: Use $a > 1$ as expansion factor;
2: Choose $0 \leq \epsilon < 1$ random real;
3: Choose starting direction $dir \in \{left, right\}$ at random;
4: $x \leftarrow a^\epsilon$;
5: **Repeat**
6:    Explore in direction dir up to distance x;
7:    If Target not found
8:    $x \leftarrow xa$;
9:    change dir;
10: **Until** Target found;

An interesting question is to analyze this randomized algorithm for the case of a target moving with known speed $v < 1$ but unknown starting distance d. It would also be interesting to analyze, like in the paper [17], knowledge competitive ratio tradeoffs using a subset of the parameters d and v. Further, no systematic analysis has been done concerning fault tolerance in this setting.

A generalization of this problem is presented in [30], whereby the agents are p-faulty, in the sense that every attempt to change direction is an independent Bernoulli trial with known probability p, where p is the probability that a turn fails. As in the previous randomized model, one is looking for agent trajectories minimizing the worst-case expected termination time, relative to the distance of the target to the origin. Very liitle is known regarding fault-tolerance (for crash, byzantine or mixed faults) for a group of searchers, but nothing at all for capturing a moving target. For example, the research in [11] considers the case when the searcher is probabilistically faulty in that it may not realize it had found the exit (with some fixed probability so every trial is an independent Bernoulli trial). Also relevant is [3] which considers settings whereby the searcher has a "hint" about the target and studies the Pareto-efficiency of strategies. A hint may suggest either the exact position of the target on the line or the direction of optimal search with respect to the origin, or a general k-bit string encoding some information about the target.

6   Time/Energy Tradeoffs

The problem discussed here are related to measuring the energy consumption required to execute a search algorithm.

Problem 3. *Analyze Time/Energy tradeoffs for group search in the presence of faults when the target is mobile.*

This problem aims to relate time and energy consumption for evacuation of a group of agents. It was introduced in [19] for two agents, where the energy loss experienced by an agent traveling a distance x at constant speed s is given by $s^2 x$, as motivated by energy consumption models in physics. A novel search algorithm was proposed that simultaneously achieves search time $9d$ and consumes energy $8.42588d$, where d is the unknown distance of the exit from the starting position of the agents.

Another concept relating to energy consumption in search is the change of direction of a searcher (also known as "turn") which is a requirement for implementing zigzag type search algorithms. In order to execute a turn a searcher first needs to stop then turn around to change direction, and finally restart and continue moving. Such operations require energy consumption which has to be amortized in the overall cost of the competitive ratio. The turn cost of a search algorithm was first proposed and investigated by [2] and further explored in [28] and later in several other papers including in [4]. A related energy cost concept that could also be taken into account is "speedup" and/or "slowdown" in which a searcher needs either to increase or reduce its speed.

Nothing is known for more general models of energy consumption for linear search with multiple agents. The case of faulty agents has never been investigated before and the tradeoffs will undoubtedly depend on the type of fault considered. Additional open problems are related to knowledge and competitive ratio tradeoffs for capturing a moving target.

7 Conclusion

We presented a survey of recent results on linear search emphasizing the importance of the knowledge the searchers on the resulting competitive ratio. The underlying vision is to understand how competitive ratios depend on the number of agents involved, maximum number of faults and type thereof, type of agent fault and communication model, as well as knowledge about the target (starting coordinate, speed, etc.).

One approach to addressing the above problems is by building a unified, general methodology from the bottom-up after understanding the special solutions that will be developed. Although some of the problems presented may be quite challenging one may be hopeful that their solution may not only lead to improved design of search algorithms but also establish a hierarchy of competitive ratios that clarifies the impact of the parameters involved. It is also worth pointing out that similar problems occur when searching in other bounded domains like the Unit-Disk, Equilateral Triangle, etc., models. Due to space limitations these problems were not discussed in the present article, but we refer the reader to the original paper [18], and for relevant, follow-up literature to [22].

Acknowledgements. Many of the results presented carry ideas and suggestions from discussions with many researchers and collaborators. I would especially like to offer my gratitude to Jared Coleman, Jurek Czyzowicz, Konstantinos Georgiou, Dmitry Ivanov, Khaled Jawhar, Ryan Killick, Danny Krizanc, Manuel Lafond, Oscar Morales-Ponce, Lata Narayanan, Jarda Opatrny, Denis Pankratov, and Sunil Shende for numerous useful and inspiring conversations on these and related topics over many years. I apologize if I forgot to mention any important contributors and/or contributions and, of course, I take full responsibility for all errors. Last, but not least, I would like to take this opportunity to dedicate this article to the memory of Jurek Czyzowicz. I am sure Jurek would have enjoyed working on any of the problems presented in this article.

References

1. Ahlswede, R., Wegener, I.: Search Problems. Wiley-Interscience (1987)
2. Alpern, S., Gal, S.: The Theory of Search Games and Rendezvous, vol. 55. Springer (2003)
3. Angelopoulos, S.: Online search with a hint. Inf. Comput. **295**, 105091 (2023)
4. Angelopoulos, S., Arsénio, D., Dürr, C.: Infinite linear programming and online searching with turn cost. Theoret. Comput. Sci. **670**, 11–22 (2017)
5. Baeza-Yates, R., Culberson, J., Rawlins, G.: Searching in the plane. Inf. Comput. **106**(2), 234–252 (1993)
6. Baeza-Yates, R., Schott, R.: Parallel searching in the plane. Comput. Geom. **5**(3), 143–154 (1995)
7. Bampas, E., Czyzowicz, J., Gasieniec, L., Ilcinkas, D., Klasing, R., Kociumaka, T., Pajak, D.: Linear search by a pair of distinct-speed robots. Algorithmica **81**(1), 317–342 (2019)
8. Beck, A.: On the linear search problem. Israel J. Math. **2**(4), 221–228 (1964)
9. Beck, A., Newman, D.J.: Yet more on the linear search problem. Israel J. Math. **8**(4), 419–429 (1970)
10. Bellman, R.: An optimal search. SIAM Rev. **5**(3), 274–274 (1963)
11. Bonato, A., Georgiou, K., MacRury, C., Prałat, P.: Algorithms for p-faulty search on a half-line. Algorithmica **85**(8), 2485–2514 (2023)
12. Bonato, A., Nowakowski, R.: The Game of Cops and Robbers on Graphs. American Mathematical Society (2011)
13. Bose, P., De Carufel, J.-L.: A general framework for searching on a line. Theoret. Comput. Sci. **703**, 1–17 (2017)
14. Bose, P., De Carufel, J.-L., Durocher, S.: Revisiting the problem of searching on a line. In: Algorithms–ESA 2013: 21st Annual European Symposium, Sophia Antipolis, France, Sep. 2-4, 2013. Proceedings 21, pp. 205–216. Springer (2013)
15. Chrobak, M., Gasieniec, L.G.T., Martin, R.: Group search on the line. In SOFSEM, pp. 164–176. Springer (2015)
16. Coleman, J.R., Ivanov, D., Kranakis, E., Krizanc, D., Morales-Ponce, O.: Linear search for an escaping target with unknown speed. In: Rescigno, A.A., Vaccaro, U. (eds.) IWOCA 2024: 35th International Workshop on Combinatorial Algorithms, 1-3 July 2024, Ischia, Italy. LNCS, vol. 14764, pp. 396–407. Springer (2024)
17. Coleman, J.R., Kranakis, E., Krizanc, D., Morales-Ponce, O.: Line search for an oblivious moving target. In Hillel, E., Palmieri, R., Rivière, E. (edis.) 26th International Conference on Principles of Distributed Systems, OPODIS 2022, December 13-15, 2022, Brussels, Belgium. LIPIcs, vol. 253, pp. 12:1–12:19. Schloss Dagstuhl - Leibniz-Zentrum für Informatik (2022)
18. Czyzowicz, J., Gasieniec, L., Gorry, T., Kranakis, E., Martin, R., Pajak, D.: Evacuating robots via unknown exit in a disk. In DISC 2014, Austin, TX, USA, October 12-15, pp. 122–136 (2014)
19. Czyzowicz, J., Georgiou, K., Killick, R., Kranakis, E., Krizanc, D., Lafond, M., Narayanan, L., Opatrny, J., Shende, S.M.: Time-energy tradeoffs for evacuation by two robots in the wireless model. TCS **852**, 61–72 (2021)

20. Czyzowicz, J., Georgiou, K., Killick, R., Kranakis, E., Krizanc, D., Narayanan, L., Opatrny, J., Shende, S.M.: Priority evacuation from a disk: the case of $n = 1, 2, 3$. TCS **806**, 595–616 (2020)
21. Czyzowicz, J., Georgiou, K., Killick, R., Kranakis, E., Krizanc, D., Narayanan, L., Opatrny, J., Shende, S.M.: Priority evacuation from a disk: the case of $n \geq 4$. TCS **846**, 91–102 (2020)
22. Czyzowicz, J., Georgiou, K., Kranakis, E.: Group search and evacuation. In Flocchini, P., Prencipe, G., Santoro, N. (eds.) Distributed Computing by Mobile Entities: Current Research in Moving and Computing. Springer International Publishing, Cham (2019)
23. Czyzowicz, J., Georgiou, K., Kranakis, E., Krizanc, D., Narayanan, L., Opatrny, J., Shende, S.M.: Search on a line by byzantine robots. Int. J. Found. Comput. Sci. **32**(4), 369–387 (2021)
24. Czyzowicz, J., Killick, R., Kranakis, E., Krizanc, D., Narayanan, L., Opatrny, J., Pankratov, D., Shende, S.M.: Group evacuation on a line by agents with different communication abilities. In: Proceedings of 32nd ISAAC, Dec. 6-8, 2021, Fukuoka, Japan (2021)
25. Czyzowicz, J., Killick, R., Kranakis, E., Stachowiak, G.: Search and evacuation with a near majority of faulty agents. In: SIAM ACDA21 (Applied and Computational Discrete Algorithms), Pruceedings, pp. 217–227, Seattle, USA, July 19 to 21. SIAM (2021)
26. Czyzowicz, J., Kranakis, E., Krizanc, D., Narayanan, L., Opatrny, J.: Search on a line with faulty robots. Distrib. Comput. **32**(6), 493–504 (2019)
27. Czyzowicz, J., Kranakis, E., Krizanc, D., Narayanan, L., Opatrny, J., Shende, S.: Linear search with terrain-dependent speeds. In: 10th International Conference, CIAC 2017, Athens, Greece, May 24-26, 2017, Proceedings, vol. 10236, p. 430. Springer (2017)
28. Demaine, E.D., Fekete, S.P., Gal, S.: Online searching with turn cost. Theoret. Comput. Sci. **361**(2), 342–355 (2006)
29. Gal, S.: Minimax solutions for linear search problems. SIAM J. Appl. Math. **27**(1), 17–30 (1974)
30. Georgiou, K., Giachoudis, N., Kranakis, E.: Overcoming probabilistic faults in disoriented linear search. In: Rajsbaum, S., Balliu, A., Daymude, J.J., Olivetti, D. (eds.) SIROCCO 2023: June 6th to 9th, 2023, Alcala de Henares, Spain. LNCS, vol. 13892, pp. 520–535 (2023). Full version in TCS (2024)
31. Georgiou, K., Lucier, J.: Weighted group search on a line & implications to the priority evacuation problem. Theoret. Comput. Sci. **939**, 1–17 (2023)
32. Jawhar, K., Kranakis, E.: Bike assisted evacuation on a line. In: SOFSEM 2021, Bolzano-Bozen, Italy, January 25-29, 2021, Proceedings. LNCS, vol. 12607, pp. 104–118. Springer (2021)
33. Jawhar, K., Kranakis, E.: Bike assisted evacuation on a line of robots with communication faults. In: ALGOWIN 2024, Royal Holloway, University of London in Egham, United Kingdom, Sep. 5-6, 2024, Proceedings. LNCS. Springer (2024)
34. Kao, M.-Y., Reif, J.H., Tate, S.R.: Searching in an unknown environment: An optimal randomized algorithm for the cow-path problem. Inf. Comput. **131**(1), 63–79 (1996)
35. Killick, R.: Search and Rendezvous by Mobile Robots in Continuous Domains. PhD thesis, School of Computer Science, Carleton University (2021)
36. Kupavskii, A., Welzl, E.: Lower bounds for searching robots, some faulty. In: PODC 2018, pp. 447–453. ACM, Egham, UK (2018)

37. López-Ortiz, A., Schuierer, S.: The ultimate strategy to search on m rays? Theoret. Comput. Sci. **261**(2), 267–295 (2001)
38. McCabe, B.J.: Searching for a one-dimensional random walker. J. Appl. Prob. 86–93 (1974)
39. Stone, L.: Theory of Optimal Search. Academic Press, New York (1975)
40. Sun, X., Sun, Y., Zhang, J.: Better upper bounds for searching on a line with byzantine robots. In: Complexity and Approximation, pp. 151–171 (2020)

Invited Paper: Gathering Oblivious Robots in the Plane

Fabian Frei[1,2]([✉]) [iD] and Koichi Wada[3] [iD]

[1] Department of Computer Science, ETH Zurich, Zürich, Switzerland
[2] CISPA Helmholtz Center for Information Security, Saarbrücken, Germany
fabian.frei@inf.ethz.ch
[3] Hosei University, Tokyo, Japan
wada@hosei.ac.jp

Abstract. The question what kind of tasks *autonomous mobile robots* can accomplish has been investigated for multiple decades by now. The robots are *mobile* because they can move in some metric space, by default the Euclidean plane. They are *autonomous* due to the lack of a central entity controlling all movements. Instead, the well-established model of Look-Compute-Move (LCM) cycles lets each robot, independently and repeatedly, first observe the surrounding robots, then use a shared algorithm to compute a target location, and then move there. The robots are by default assumed to be *oblivious*, which means that they lose all their memory whenever they finish an LCM cycle. The arguably most fundamental task for a group of mobile robots is to *gather* in a single point. Whether autonomous mobile robots, arbitrarily scattered in the plane, are able to achieve a gathering depends on the details of the model. In this survey, we introduce the standard LCM model for oblivious robots in the Euclidean plane and discuss important modeling details and variants. We then present known results about gathering for various variants of this model using proof sketches, briefly discussing them and extending their scope in some cases, and finally highlight some remaining open questions.

Keywords: Autonomous mobile robots · Euclidean plane · Anonymous robots · Oblivious robots · Non-rigid movement · Distinct gathering · Self-stabilizing gathering

1 Introduction

Over the last few decades, the study of *autonomous mobile robots* has received much attention from various areas, including robotics, artificial intelligence, and, last but not least, distributed computing. Indeed, contributions from a wide range of researchers [1–3,5,9,10,12], helped it mature into an established research field already by 2012, the year when Flocchini et al. published the first comprehensive textbook [8] on the topic. (A newer book on approximately the same topics was published in 2019 [7].) In this paper, we aim to review the model

T. Masuzawa et al. (Eds.): SSS 2024, LNCS 14931, pp. 39–54, 2025.
https://doi.org/10.1007/978-3-031-74498-3_3

with its important variants in Sect. 2, then review both old and recent results while strengthening some of them with small proof modifications in Sect. 3, and conclude the paper by highlighting some interesting open questions.

2   Model Description

In this section, we first provide a detailed review of the model and the various variants considered in this paper.

2.1   Base Assumptions

Movement in the Euclidean plane. The name *autonomous mobile robots* already reveals that we are dealing with robots that are *mobile*, i.e., moving in some space, and *autonomous*, i.e., operating individually, without an external control. The space in which the robots move can be discrete (see the last section of the last chapter of the mentioned textbook for a short summary of the research on this [8, Chapter 7.3]). However, the default assumption, and the only one we consider in this paper, is that they move in the Euclidean plane. Moreover, the robots are assumed to be *dimensionless*, which means that they can be represented by points in the plane.

Anonymous robots. The robots are *anonymous*. This means that they do not have any identifiers that could help them distinguish each other. They *a priori* all look identical, both externally to each other and internally to themselves. This includes the fact that they all run the same algorithm. This is sometimes described by saying that the robots are *uniform* or have a *uniform algorithm.* However, note that these notations are sometimes also used the express the algorithm's independence of the number of robots.)

Oblivious robots. The robots are *oblivious*. This means that they do not have any persistent memory and plan their movements without knowing the past.

Gathering. There are different tasks that we might want the robots to solve. The goal is to equip the robots, initially scattered arbitrarily in the plane, with an algorithm that lets them move and interact in a way that accomplishes the given task. Possible tasks include *forming patterns* such as circles [8, Chapter 4], the more complicated *flocking* [8, Chapter 6], which is a coordinated movement of robots that maintains some pattern, and *scattering* [8, Chapter 5], for which the robots have to disperse as much as possible in certain prescribed ways. The opposite of scattering is *gathering* (sometimes also called *point formation*) [8, Chapter 2], which requires all robots to gather in a single location. (The location is not prescribed by the task, i.e., the robots are free to gather in an arbitrary point.) Gathering is arguably the most fundamental cooperation task we can expect a group of mobile robots to perform. In this paper, we consider only this task of gathering (in two variants) and examine what decisions about the model details determine whether the robots are able to ensure a gathering or not.

2.2   Detailed Presentation of the Model

Basics Notions and Notation. The number of robots is a fixed, at least 2, and denoted by n. This number may or may not be known to the robots. The latter case is sometimes also described by saying that the robots have a *uniform* algorithm. (As mentioned above, this renders the term *uniform* ambiguous.) For rare cases, combining specific models, tasks, and robot numbers, knowledge of n can make the robots more powerful. The common case, however, is that this choice does not affect the solvability of a task. Each robot *occupies* (i.e., is positioned at) a *location*, i.e., a point in the plane. (The terms *location, position,* and *point* are used interchangeably in the literature for the most part.) It is allowed for multiple robots to occupy the exact same location. (This is indeed necessary for the robots to achieve a gathering.) A location occupied by k robots is said to have multiplicity k.

We refer to the multiset of all the positions of all robots as a *configuration*. A *k-point configuration* is a configuration that contains exactly k distinct locations (i.e., exactly k points when ignoring multiplicities). A configuration is called *distinct* if it does not contain multiplicities (i.e., if it contains exactly n distinct locations). The *initial configuration* is the given configuration of the robots when none of them has moved yet.

The robots are by default *inactive*, which means that they neither move nor observe nor compute anything. The only thing that they do is to wait for an activation by the so-called *(activation) scheduler*, which some also have called *daemon*. There are different types of schedulers, which we discuss in detail later.

Whenever activated by the scheduler, a robot immediately starts to execute a so-called *Look-Compute-Move (LCM) cycle*.

2.3   LCM Cycle

An LCM cycle consists of the following three operations, which are always executed in the order given here:

Look. The robot sees or senses all robot positions in the current configuration. The sensing capabilities work over unlimited distance and there is no obstruction even in the case of collinear robots. The latter property is sometimes referred to as the robots being *transparent*. However, the robot positions are not observed as absolute coordinates, but only with respect to the robot's *(local) coordinate system*, which is also called its *(local) map* (or sometimes *(local) compass*). The details about these maps and what exactly the robot can observe are discussed later. All information gained by the robot about the current configuration is stored in a *snapshot* and also referred to as the robot's *observed configuration*. (We remark that there are model variants that consider various forms of limited view, which imply that some robots may sometimes miss some other robots, but we exclude these models from consideration in this paper.)

Compute. Solely based on the snapshot just obtained, the robot *deterministically* computes a *target position* that it would like to move to. This target is necessarily computed relative to the robot's map. The robot can choose any trajectory in the plane on which it would like to move from its current position to the target. (For so-called *rigid* movement, defined below, we can assume the straight geodesic trajectory directly to the target.) A priori, no computational resource restrictions are imposed on the robots. The temporarily stored snapshot and all computations are assumed to be exact.

Move. The robot starts moving from its current location towards the target along the computed trajectory. Crossing paths is allowed: The dimensionless robots do not crash when they pass or run into each other. In fact, the movement is not affected by this at all. Robots with this property may be called *crashless*. *Rigid movement* means that the target is guaranteed to be reached. *Non-rigid movement* means that the robot may be stopped on its trajectory before reaching the target. Details about non-rigid movement are discussed later. The LCM cycle ends when the target is reached or the movement of the robot is stopped. Any LCM cycle is guaranteed to finish within constant time. As soon as the LCM cycle has ended, the robot reverts to being inactive. Only inactive robots can be activated by the scheduler.

2.4 Local Coordinate Systems

We now discuss the details and model variants for how the robots observe the current configuration. As mentioned before, each robot perceives the robot locations in the Euclidean plane only relative to its own local coordinate system, which is in some parts of the literature called the robot's compass but is often just briefly referred to as its (local) map. All maps are Cartesian (i.e., based on two orthogonal axes that use the same unit length). This implies that angles and relative distances are the same in all local maps. But an adversary may arbitrarily shift, rotate, mirror along the axes, and scale any local map with respect to the global coordinate system. This translates, respectively, to an arbitrary point of origin, an arbitrary rotation around the origin, arbitrary axes orientations and thus chirality, and an arbitrary unit length (i.e., distance measurement). Moreover, all of this might change again with each new LCM cycle, even for one and the same robot.

We can use the unlimited computing power of the robots to always recompute the observed coordinates such that the observing robot is located at the point of origin. This leaves us with the three potential uncertainties about the map: The *rotation* (in which radial direction does the positive half of the first axis point?), the binary chirality (on which side of the first axis is the positive half of the second axis?), and the unit length of each local map. The map's rotation is sometimes also called its *direction* or *compass*, making the latter term ambiguous as to whether it includes the unit distance.

Each of the three properties rotation, chirality, and unit length may be *self-consistent* – which means that for each given robot property remains unchanged across all its LCM cycles – or *consistent*, which means that the property is self-consistent and the same for all robots. For each property, we require self-consistency, or require consistency, or have neither requirement. This already yields $3^3 = 27$ model variants just for these three described properties. Even more properties are sometimes considered. For example, one might say that two maps are *aligned* if the cross defined by their axes are the same. We may called two maps *axis-aligned* if the lines defined in both maps by the first axis are parallel (and thus also the lines defined by the second axes). For two axis-aligned maps we can say that an axis has the same *axis orientation* in both maps if and only if the positive half is on the same side for both maps. If both axes have the same *axis orientation* for two maps, we could say that two maps have the same *axes orientations*, but this is the same as saying that the maps have the same rotation and the same chirality.

The precise choice of the model is sometimes not made sufficiently explicit in the literature. In particular the term *orientation* is sometimes used ambiguously or even inconsistently in the literature. For example, *orientation* may sometimes refer only to which half of each axis is the positive half. But it might also used to refer to how the axes are rotated, sometimes also including the chirality, and sometimes referring to all aspects of the local map at the same time, including the unit length.

The lack of a consistency guarantee is sometimes referred to as *(static disorientation*, while the lack of a self-consistency guarantee is sometimes called *variable disorientation*. Although the word *disorientation* may hint at a potential inconsistency with respect to only the map's *orientation*, whatever this is defined to be in the respective context, it is sometimes used to refer to any type of inconsistency or lack of self-consistency. The unspecified term *disorientation* is even more ambiguous. It is sometimes used to mean static disorientation. Terms like *fully disoriented* or *completely disoriented* are also unclear without further clarifications. They are often used to refer to the lack of any consistency (or sometimes even self-consistency) guarantee for both rotation and chirality, but the unit length is sometimes still assumed to be consistent in these cases of full disorientation. In a few cases, *full disorientation* also refers to the fact that some map properties may change even during the execution of an LCM cycle. It is thus advisable to avoid these terms unless they are clearly defined.

Among all possible combinations of consistency guarantees (in particular each of rotation, chirality, and unit length being either guaranteed to be consistent or guaranteed to be self-consistent or neither), a few cases are almost trivial. This includes in particular the case of consistent rotation, as we will discuss in Sect. 3.2.

2.5   Rigid and Non-rigid Movement

It is often assumed that the robots move rigidly, which means that they always reach their computed target within an LCM cycle. The opposite is non-rigid movement, which means that robots may be stopped adversarially before reaching their target, ending their current LCM cycle prematurely. The motivation for considering non-rigid movement is that the robots may experience some temporary failure such as overheating. Even for non-rigid movement, however, the models have some restrictions on how quickly an ongoing movement can be interrupted. Without this, the adversary could trivially keep the robots from gathering by immediately stopping any movement. Thus there is, for each robot r, a fixed minimum movement distance $\delta_r > 0$, which is self-consistent, i.e., remains unchanged over time. The adversary can stop the movement of robot r only if it has already moved at least a distance of δ_r in the current LCM cycle.

In the *harshest variant* of non-rigid movement, the minimum movement distance $\delta_r > 0$ is hidden from all robots. In this case, we may equivalently assume that all robots have the same unknown minimum movement distance δ by choosing $\delta = \min_{1 \le n} \delta_r$. (Note that the adversary is never forced to interrupt a movement, even if it is allowed to do so. Lowering the unknown limit δ_r down to $\delta = \min_{1 \le n} \delta_r$ can therefore not restrict the adversary's options.) In another variant, sometimes referred to as "non-rigid($+\delta$)," robot r knows its own minimum movement distance δ_r but not that of other robots. Note that this variant automatically yields a self-consistent unit length. In yet another variant, the minimum movement distance is the same $\delta > 0$ for all robots, and known to all robots. This is sometimes referred to as "non-rigid($+\delta=$)." Note that this variant provides a consistent unit length shared by all robots. The variant where the minimum movement distance is shared by all robots but unknown to all of them coincides with the harshest variant, as shown above.

We also mentioned above that each δ_r is assumed to be self-consistent. This is justified as follows. If it could change arbitrarily with each LCM cycle, then it could for example be halved after each step, leaving us with an adversary that can trivially preventing a gathering by stopping each movement quickly enough. In contrast, if there is a constant lower bound on δ_r, then we can choose this lower bound as our δ_r.

In conclusion, we have the following *four* variants with regard to the rigidity of movement, ordered from most restrictive to least restrictive from the adversary's point of view: The first variant is that the movement is rigid. In the second variant, non-rigid($+\delta=$), movement is non-rigid but with a consistent minimum movement distance δ, shared and known by all robots. (Note that this implies a consistent unit length without loss of generality.) In the third variant, non-rigid($+\delta$), each robot has and knows its own, self-consistent minimum movement distance, but does not know that of other robots. This variant implies a self-consistent unit length without loss of generality. For the last and harshest variant, the minimum movement distance δ is the same for all robots but unknown to all robots. This variant is referred to as *non-rigid movement with unknown δ* or, by considering it the default, just as *non-rigid movement*.

2.6   Memory and Communication

The default assumption is that the robots are *oblivious*, which means that they have no memory that persists between different LCM cycles. Whenever activated, oblivious robots remember nothing about the past, except for what they might be able to deduce from the observed configuration. The polar opposite of oblivious robots are robots with unlimited permanent memory. In between, it is possible to consider robots with some constant memory capacity.

The default assumption for autonomous robots is that they have no direct means of communicating with each other. Robots without such communication capabilities are sometimes called *silent*. They can communicate only via choosing their movement targets and observing the occurring configurations.

We mention only very briefly a newer model that grants the robots some limited communication or persistent memory. The model of robots with lights assumes that each robot is equipped with exactly one light. During each LCM cycles, the robot may assign to its own light any color from a common fixed set of colors. In the *luminous model*, abbreviated by $\mathcal{LUMI}$, all lights are visible to all robots. In the *internal-light model*, denoted by $\mathcal{FSTA}$, a light is *internal*, that is, visible only to its own robot. In the *external-light model*, denoted by $\mathcal{FCOM}$, a light is *external*, that is, visible to all robots except the robot equipped with this light itself. The model names $\mathcal{FSTA}$ and $\mathcal{FCOM}$ stand *finite number of states* and *finite communication* because the color of an internal light seen only by its own robot corresponds to one possible *state* of the robot, and externally visible light clearly give the robots some options to communicate information.

In contrast with all of the models including lights, the model without any lights (or, equivalently, with lights but only one available color) is called $\mathcal{OBLOT}$, which stands for "oblivious robots."

2.7   Schedules and Schedulers

An *(activation) schedule* describes completely when each robots is activated. A fundamental restriction for all schedules is that a schedule can only activate robots that are inactive, that is, it cannot activate robots that are currently executing an LCM cycle.

Another default assumption is that a schedule is *fair*, which means that it never leaves a robot inactive for an infinitely long time. This is a reasonable restriction because a gathering is trivially be prevented by an unfair schedule that leaves two robots in different locations inactive forever. (Note, however, that robots with memory might still be able to gather if only one robot is never activated.)

The schedule that only ever activates all robots simultaneously is called *fully synchronous*. A schedule that only activates robots when all robots are currently inactive is called *semi-synchronous*. Whenever a semi-synchronous schedule activates robots, it may activate an arbitrary nonempty subset of them. For semi-synchronous schedules we call any period between the activation of a nonempty set of robots and the time that they have all finished their respective LCM cycles and have become inactive again an *(activation step* or *time step* or simply *step*.

A schedule is *centralized* if there are never two or more simultaneously active robots. Note that any centralized schedule is semi-synchronous.

A *round-robin* schedule is a special centralized schedule that activates all n robots during the first n steps (which are together called *round 0*) and then activates them all again in the same order during each of the following rounds, where round i consists of the steps $in + 1$ through $(i + 1)n$.

For any positive integer k, a schedule is k-*bounded* if, for any robot r, between any two consecutive activations of r, every other robot is activated at most k times. This standard definition of the literature is unfortunately somewhat ambiguous, but the intended and natural choice is to include into the count all activations that are synchronous with the first of the two consecutive activations of r but not those synchronous with the second one.

A schedule may be called *asynchronous* when it is not semi-synchronous. But note that each fully synchronous schedule is also semi-synchronous, and each semi-synchronous schedule is also asynchronous.

Note that under an asynchronous schedule, a robot may be activated while another one is still executing an LCM cycle. In particular, a robot may be observed while it is still moving towards its target.

Any set of schedules is called a *scheduler*. A scheduler is said to have some property if all contained schedules have this property. The fully synchronous, round-robin, centralized, semi-synchronous, k-bounded, and asynchronous scheduler is the maximal fair scheduler with the respective property and is denoted FSYNC, RROBIN, CENT, SSYNC, k-BOUNDED, and ASYNC, respectively.

2.8   The Task of Gathering

An algorithm is said to *solve* a task under a given scheduler if the algorithm solves the task for every possible combination of a schedule contained in the scheduler with an allowed initial configuration. We are interested here only in the task of *gathering*, defined below.

An algorithm is said to *achieve a gathering* for a given schedule and a given initial configuration if after a finite number of steps all robots are collocated (i.e., gathered) in a single location and then stay there forever. In which point of the plane the robots gather is irrelevant.

The task of *self-stabilizing gathering* requires that a gathering is achieved for every possible initial configuration. The task of *distinct gathering* only requires that a gathering is achieved for every possible *distinct* initial configuration.

The task of *gathering* without a further qualifier used to default to distinct gathering when the term was introduced but has more recently be used to mean self-stabilizing gathering instead. To avoid ambiguity we always use the qualifier.

3   Review of Results

We would now like to delineate the boundaries of feasibility of gathering (both self-stabilizing gathering and distinct gathering) for different schedulers and model choices. We mainly summarize known results, usually providing a short proof sketch, but also extend many results in several small ways, often just by improving the analysis and noticing that the results hold for more models than originally stated, but sometimes also by modifying the proofs. We begin by briefly repeating our model assumptions.

3.1   Common Model Assumptions

The common properties of the models for all of the results below are that they all concern *dimensionless, crashless, anonymous, autonomous, oblivious* and *silent* (and thus *lightless*) robots operating in *LCM cycles* with an unlimited visibility range, moving in the *Euclidean plane* with potentially non-rigid movement with an unknown minimum movement distance, without a priori knowledge of the number n of robots, and tasked with solving gathering. We refer to the set of these model assumptions as the *default model*.

The following might change between the different model variants considered below: the considered scheduler, the number $n \geq 2$ of robots, whether the task is *self-stabilizing* or *distinct* gathering, whether the robots have strong or weak multiplicity detection or neither, whether rigid and movement is guaranteed or not, and which consistency and self-consistency assumptions we have.

We first consider one type of consistency that, when added to the assumption of the default model, always lets the robots solve self-stabilizing (and thus also discrete) gathering, namely consistent rotation.

3.2   Consistent Rotation

If the rotation is consistent (i.e., all robots agree on the direction of the positive half of the first axis), then a gathering can be guaranteed as follows. Each robot computes upon activation the smallest square whose diagonals align with the axes and that contains all robot locations. It then targets the corner of this square that lies in the direction of the positive half of the first axis on a straight line from the center of the square. This corner will remain the same forever during the following configurations and all robots eventually gather in this point.

Theorem 1. *There is an algorithm for self-stabilizing (and thus also distinct) gathering in the default model with consistent rotation.*

We may thus from now on exclude this case from consideration, which is sometimes done only implicitly in the literature.

3.3   The Weak and the Strong Model

For the following results, we will use two classes of consistency or self-consistency assumptions very often. The first class is the empty class. We usually use this class when proving upper bounds, to make them as strong as possible. We refer to our default model without any additional consistency or self-consistency assumptions as the *weak model*. The second class consists of the following three properties: Consistent axis alignment, consistent chirality, and consistent unit length. (Note that this means that maps are always identical except for potentially the choice of origin and potentially a rotation by 180 degrees.) The default model with these three additional consistency properties and guaranteed rigid movement is called the *strong model* in this paper.

3.4   Rendezvous

We now consider the case of gathering $n = 2$ robots, which is often called *Rendezvous* in the literature. Note that self-stabilizing gathering and distinct gathering coincide for rendezvous because any configuration that is not already a gathering is distinct. Moreover, multiplicity detection does not help at all for rendezvous because a multiplicity only occurs once a gathering is achieved.

We first consider two simple algorithms. The first algorithm always targets the location of the opposite robot. It is successful if and only if every schedule contains a step in which only one robot is activated in the case of rigid movement. In the case of non-rigid movement, an infinite number of such steps is required and sufficient to guarantee success. In particular, the algorithm is successful for all fair centralized schedulers.

Theorem 2. *Rendezvous is solvable under* CENT *(and thus* RROBIN*) in the weak model.*

The second simple algorithm always targets the midpoint between the two observed locations. It is obviously successful for the fully synchronous scheduler.

Theorem 3. *Rendezvous is solvable under* FSYNC *in the weak model.*

We now consider a scheduler that has in every step the two option to activate only one or both robots. The k-bounded scheduler is, for every positive k, a scheduler of this class. Indeed, all non-centralized schedulers introduced in this paper except for FSYNC belong to this class.

Let the robots have maps that are consistent with regard to all properties, except that one map might be rotated by 180 degrees with respect to the other. Let the chosen version of the map, rotated or not, be self-consistent for each robot. Let an arbitrary gathering algorithm be given. Let the scheduler first activate only one robot at a time, alternating between them. Eventually, one robot must target the location the other one to achieve a gathering. The scheduler can now activate once both robots to prevent a gathering. The robots will swap places instead due to the respectively rotated maps, ensuring that both robots perceive the same configuration. The scheduler can continue to use this strategy to prevent a gathering forever.

Theorem 4. *Rendezvous is unsolvable under* 1-BOUNDED *for the strong model.*

We have seen that, among the schedulers introduced in this paper, gathering (both stabilizing and distinct) is possible for $n = 2$ robots exactly for FSYNC, RROBIN, and CENT. From now on we assume that $n \geq 3$.

3.5 The Fully Synchronous Scheduler

It is not hard to see that gathering is feasible under the fully synchronous scheduler (FSYNC) for any number of robots. Indeed, it suffices for every robot to compute upon activation the center of gravity of the smallest convex polygon enclosing all robot locations and then move there. This strategy works even if there are no consistency or even self-consistency guarantees for the local maps of the robots because the center of gravity is equivariant under all possible adversarial map transformations, namely shifting, rotating, mirroring, and scaling. If the movement is rigid, the gathering happens in a single step. But even for the harshest type of non-rigid movement, the robots gather after a finite number of steps. Strong or even weak multiplicity detection are not necessary either; the algorithm can just ignore all multiplicities when computing the center of gravity.

Theorem 5. *Self-stabilizing (and thus also distinct) gathering is solvable under* FSYNC *in the weak model for any* $n \geq 3$.

We can now exclude FSYNC from the following considerations. The most natural and reliable scheduler besides FSYNC is arguably RROBIN. The strategy of targeting the center of gravity fails for this scheduler already; the robots will converge to a single location but never meet there. But there is a far simpler strategy if we have strong multiplicity detection.

3.6 Strong Multiplicity Detection

Given strong multiplicity detection, an activated robot can target a point of highest multiplicity, breaking ties by choosing arbitrarily among the closest ones. This implies that a robot located in a position of highest multiplicity stays there. The only exception to this rule is if all points have the same multiplicity. In this case a robot can target any closest robot location excluding its own location. The number of points with highest multiplicities will eventually reduce by this and never increase again. This strategy does not require any consistency or self-consistency guarantees, works even for the harshest version non-rigid movement, and indeed for any fair centralized scheduler.

Theorem 6. *Self-stabilizing (and thus also distinct) gathering is solvable under* CENT *(and thus also* RROBIN*) in the weak model with strong multiplicity detection for any* $n \geq 3$.

There is a different algorithm by Dieudonné and Petit [6] that works even for fair schedulers that are not centralized, but only when n is odd. The algorithm itself is not too complicated, but the proof of correctness is rather demanding. The proof was written under the assumption of a consistent chirality. However, it can be checked that not even self-consistent chirality is required.

Theorem 7. *Self-stabilizing (and thus also distinct) gathering is solvable under* SSYNC *in the weak model with strong multiplicity detection for any odd* $n \geq 3$.

What happens in the case of ASYNC remains unclear. Modifications to the algorithm seem to be required. In particular, the smallest enclosing circle and thus its center can change with the movement of a robot, and ASYNC can thus let robots moving towards different targets at the same time. A similar problem occurs if we try to apply the ideas leading up to Theorem 9. Combining these different ideas might help. In any case, we conjecture the following to be true.

Conjecture 1. Self-stabilizing (and thus distinct) gathering is solvable under ASYNC in the weak model with strong multiplicity detection for any odd $n \geq 3$.

We have already seen that with strong multiplicity detection self-stabilizing gathering is solvable for any (and thus also any even) n under FSYNC, RROBIN, and CENT. For the complementing lower bound, we can show that there is no algorithm for self-stabilizing gathering under the k-bounded scheduler for any positive integer k. In fact, we can do this based on the infeasibility proof for the case $n = 2$. We choose as our initial configuration an arbitrary two-point configuration where each location has exactly $n/2$ robots. We let all robots in one location have the exactly same self-consistent map, and all robots in the other location the same map rotated by 180 degrees. For any given $k \geq 1$, the k-bounded scheduler has the two options of either alternately activating first all robots in one location together and then all robots in the other location together or activating all robots in both locations together. Switching only between these two activation patterns, we can force all collocated robots to behave exactly the same, as though they they were a single robot, and thus we can simulate the hardness proof for $n = 2$. The additional multiplicity detection does not help in any way because both points always keep the same multiplicity of $n/2$ until a gathering is achieved.

Theorem 8. *Self-stabilizing gathering is unsolvable under* 1-BOUNDED *in the strong model with weak multiplicity detection for all even* $n \geq 3$.

We have now determined the feasibility of self-stabilizing gathering under all discussed schedulers with strong multiplicity detection. For distinct gathering, we will see that it is solvable even if we have only weak multiplicity detection.

3.7   Weak Multiplicity Detection

Suzuki and Yamashita have proved in their seminal paper [13, Thm. 3.4] that distinct gathering is possible for $n \geq 3$ under SSYNC. They assume a consistent chirality and non-rigidity with known minimum moving distances, but it is not hard to modify their algorithm such that it works even in our weak model. It is unclear, however, how to adapt the algorithm and proof for ASYNC. Fortunately, there is another elegant solution. For each set of points in the plane there is a unique point, called the *Weber* or *Fermat* or *Torricelli* point, that minimizes the sum of the distances from itself to all other points [11]. The Weber point is invariant under straight movements of points in this towards the Weber point. Thus it suffices for all robots to target the Weber point unless there is another robot strictly between them and the Weber point, in which case they wait instead. (We assume all computations to be exact. But since the Weber point is quite hard to compute, an alternative algorithm not relying on it has also been presented [2].)

Theorem 9. *Distinct gathering is solvable under* ASYNC *in the weak model with weak multiplicity detection for any $n \geq 3$.*

The proof is easily extended from distinct to *self-stabilizing* gathering for fair *centralized* schedulers. Namely, if the initial configuration contains any points of multiplicity except at the Weber point, then only the robots located at such points move, namely in any way that guarantees that they become unaccompanied. For this, it suffices to target any point in the punctured disc around itself with a radius of less than half the observed nonzero closest pair distance. As soon as there are no points of multiplicities left outside of the Weber point, the algorithm described for distinct gathering can be applied.

Theorem 10. *Self-stabilizing gathering is solvable under* CENT *in the weak model with weak multiplicity detection.*

We note that in the case $n = 3$, the algorithm for Theorem 9 also works for self-stabilizing gathering because all initial configurations are either distinct or have a unique point of highest multiplicity.

Theorem 11. *Self-stabilizing gathering is solvable under* ASYNC *in the weak model with weak multiplicity detection for $n = 3$.*

For $n \geq 4$ robots, it is easy to adapt the impossibility proof used for Theorem 8 by noting that having at least two robots in each of the two locations of the initial configuration suffices now.

Theorem 12. *Self-stabilizing gathering is unsolvable under* 1-BOUNDED *in the strong model with weak multiplicity detection for all $n \geq 4$.*

This completes our picture gathering with weak multiplicity detection.

3.8   No Multiplicity Detection

Recall that we have discussed the case $n = 2$ already. Rendezvous is trivial under RRobin. We will show that it is impossible to solve self-stabilizing gathering under RRobin for $n \geq 3$ robots [4, Thm. 14]. First we observe that gathering is only achievable via a two-point configuration with one unaccompanied robot that sees a configuration that lets it target the location of the others. Taking such a configuration as the initial configuration, we can choose a schedule that always activates an accompanied robot and assigns to the activated robot always the same map, except for a consistent rotation by 180 degrees for one of the two locations, to let the robots always see the same configuration and thus target the opposite location. We can check that this lower bound still holds for consistent axis alignment, consistent chirality. In contrast, the proof fails as soon as we additionally assume a self-consistent axis orientation.

Theorem 13. *Self-stabilizing gathering is unsolvable under* RRobin *in the strong model without multiplicity detection for* $n \geq 3$.

Note that all our considered schedulers except for FSync contain RRobin. Consequently, self-stabilizing gathering is impossible for all of them in the strong model without multiplicity detection for all $n \geq 3$.

Having classified the schedulers for self-stabilizing gathering, it now remains to consider distinct gathering. For both 1-Bounded and for 2-bounded Cent, we can use essentially the same proof of infeasibility again. Note that we still assume the strong model. We first let the scheduler run a round-robin schedule (which is both 1-bounded and centralized) until we arrive at a two-point configuration with one unaccompanied robot that is scheduled to be activated next and would target the other robots' location to complete the gathering when activated. Now it suffices to deviate from the round-robin schedule just once. In the case of 1-Bounded we can, for example, additionally activate together with the unaccompanied robot the robot to be activated after it according to round-robin schedule enacted so far. In the case of 2-bounded Cent, we can instead activate any of the accompanied robots alone just once more. After this, we can continue with the same round-robin schedule as before. This shows that a gathering can be prevented for essentially any scheduler that is not FSync or has the option to deviate from RRobin just once at any point.

Theorem 14. *Distinct gathering is neither solvable under* 1-Bounded *nor under* Cent *in the strong model without multiplicity detection for any* $n \geq 3$.

We note that this proof, just as the one for Theorem 13, stops working if a self-consistent axis orientation is assumed.

We already know from Theorem 5 that distinct gathering is possible for FSync, leaving us with the case of distinct gathering under the round-robin scheduler. This case has remained open for decades, despite several attempts to resolve it over the entire time span. The generally accepted conjecture [4, Conj. 1] was that distinct gathering is impossible under the round-robin scheduler for any number of robots larger than 2. This conjecture was even proven under the assumption of another plausible conjecture [4, Conj. 2]. However, a brief announcement at this year's International Symposium on Distributed Computing (DISC 2024) reports that both conjectures have been disproved. Interestingly, distinct gathering is indeed impossible for $n = 3$ robots, as suspected before, but then unexpectedly returns to being feasible for 4 and more robots. Proving this result, which is rather counter-intuitive and might even appear paradoxical at first sight, required some novel and rather subtle algorithmic ideas.

The proof of infeasibility for $n = 3$ robots in the mentioned announcement relies on the fact that the robots do not have a self-consistent unit length but works even when consistent chirality is assumed. This leaves open the question whether the task becomes solvable again with a consistent unit length, or perhaps even a just self-consistent unit length, and no further consistency assumptions.

We use this opportunity to mention that the rather natural notion of self-consistency seems to have been quite neglected by the literature in general. This might be due to a belief that it is only the inconsistencies between different robots that prevent their successful collaboration. We doubt the validity of this assumption and conjecture that granting the robots self-consistency for the right map property such as the unit length may already enable them to solve an otherwise unsolvable natural task such as gathering. It might be instructive to look at the case of three robots under RROBIN, which cannot solve distinct gathering in the weak model according to Theorem 14. We believe that having a consistent unit length for the three robots is enough to make distinct gathering feasible again. If this can be proven, it still remains to see whether assuming an only *self-consistent* unit length might also be sufficient already for three robots to solve distinct gathering in the otherwise weak model.

Once all of these questions are answered and our Conjecture 1 about the feasibility of gathering any odd number of robots under ASYNC with strong multiplicity detection is proved or disproved, the most interesting remaining question about gathering in our model might be the following: What happens if the local maps are *self-consistent* with regard to *rotation*? In particular, does this assumption alone, without a self-consistent unit length and perhaps even without self-consistent chirality, already allow us to solve distinct gathering under RROBIN for $n = 3$?

References

1. Agmon, N., Peleg, D.: Fault-tolerant gathering algorithms for autonomous mobile robots. SIAM J. Comput. **36**(1), 56–82 (2006)
2. Cieliebak, M., Flocchini, P., Prencipe, G., Santoro, N.: Distributed computing by mobile robots: Gathering. SIAM J. Comput. **41**(4), 829–879 (2012)
3. Défago, X., Gradinariu, M., Messika, S., Parvédy, P.R.: Fault-tolerant and self-stabilizing mobile robots gathering. In: Dolev, S. (ed.) Proceedings of the 20th International Symposium on Distributed Computing (DISC 2006). Lecture Notes in Computer Science, vol. 4167, pp. 46–60. Springer (2006)

4. Défago, X., Potop-Butucaru, M., Parvédy, P.R.: Self-stabilizing gathering of mobile robots under crash or byzantine faults. Distributed Comput. **33**(5), 393–421 (2020)
5. Degener, B., Kempkes, B., Langner, T., auf der Heide, F.M., Pietrzyk, P., Wattenhofer, R.: A tight run-time bound for synchronous gathering of autonomous robots with limited visibility. In: 23rd ACM SPAA. pp. 139–148 (2011)
6. Dieudonné, Y., Petit, F.: Self-stabilizing gathering with strong multiplicity detection. Theoret. Comput. Sci. **428**(13), 47–57 (2012)
7. Flocchini, P., Prencipe, G., (Eds), N.S.: Distributed Computing by Mobile Entities. Springer (2019)
8. Flocchini, P., Prencipe, G., Santoro, N.: Distributed Computing by Oblivious Mobile Robots. Morgan & Claypool (2012)
9. Izumi, T., Bouzid, Z., Tixeuil, S., Wada, K.: Brief announcement: The bg-simulation for byzantine mobile robots. In: 25th DISC. pp. 330–331 (2011)
10. Kamei, S., Lamani, A., Ooshita, F., Tixeuil, S.: Asynchronous mobile robot gathering from symmetric configurations without global multiplicity detection. In: 18th SIROCCO. pp. 150–161 (2011)
11. Kupitz, Y., Martini, H.: Geometric aspects of the generalized fermat-torricelli problem. Bolyai Soc. Math. Stud. **6**, 55–129 (1997)
12. Souissi, S., Défago, X., Yamashita, M.: Using eventually consistent compasses to gather memory-less mobile robots with limited visibility. ACM Trans. Autonomous Adapt. Syst. **4**(1), 1–27 (2009)
13. Suzuki, I., Yamashita, M.: Distributed anonymous mobile robots: Formation of geometric patterns. SIAM J. Comput. **28**, 1347–1363 (1999)

Invited Paper: The Smart Contract Model

Yackolley Amoussou-Guenou[1]([✉]), Maurice Herlihy[2][iD], Maria Potop-Butucaru[3], and Sergio Rajsbaum[4][iD]

[1] Université Paris-Panthéon-Assas, CRED, Paris, France
`Yackolley.Amoussou-Guenou@u-paris2.fr`
[2] Brown University Computer Science Dept, Providence, RI 02912, USA
[3] Sorbonne Université, LIP6, Paris, France
`maria.potop-butucaru@lip6.fr`
[4] Instituto de Mathemáticas, UNAM, Mexico City, Mexico

Abstract. Many of the problems that arise in the context of blockchains and decentralized finance can be seen as variations on classical problems of distributed computing. The *smart contract model* proposed here is intended to capture both the similarities and the differences between classical and blockchain-based models of distributed computing. The focus is on *cross-chain* protocols in which a collection of parties, some honest and some perhaps not, interact through trusted smart contracts residing on multiple, independent ledgers. While cross-chain protocols are capable of general computations, they are primarily used to track ownership of assets such as cryptocurrencies or other valuable data. For this reason, the smart contract model differs in some essential ways from familiar models of distributed and concurrent computing. Because parties are potentially Byzantine, tasks to be solved are formulated using elementary game-theoretic notions, taking into account the utility to each party of each possible outcome. As in the classical model, the parties provide task inputs and agree on a desired sequence of proposed asset transfers. Unlike the classical model, the contracts provide task outputs in the form of executed asset transfers, since they alone have the power to control ownership. The model is illustrated with an *atomic cross-chain swap* task, where parties exchange assets across ledgers.

Part of this work was done while visiting LIP6, France, supported by the CNRS Fellow Ambassadeur program.

Part of this work was done while visiting IRIF and LIP6, France.

Supplementary Information The online version contains supplementary material available at https://doi.org/10.1007/978-3-031-74498-3_4.

1   Introduction

The rise of decentralized finance and distributed ledger technology presents both an opportunity and a challenge to the distributed computing community.

The *opportunity*: decentralized finance based on blockchains has become a multi-billion dollar industry, providing what seems to be an ideal application for the kinds of fault-tolerant distributed algorithms and protocols the distributed computing community has studied for decades. Specific companies and cryptocurrencies may come and go, but the ancient problem of how mutually untrusting parties can securely and efficiently exchange valuable assets will only grow in importance as the world becomes more interconnected.

The *challenge*: classical shared-memory and message-passing models do not accurately mirror the realities of blockchains and smart contracts, as discussed below. This paper proposes the *smart contract model*, which we hope will help in adapting protocols and algorithms from classical distributed computing to the world of decentralized finance. Our focus is on *cross-chain* protocols, executed by a collection of active, parties, some honest, some perhaps not, interacting through shared smart contracts deployed on different blockchains.

The smart contract model is inspired by classical distributed computing models, but it differs in some essential ways. For example, in classical models, a task is specified by its inputs and outputs. Each party chooses an input, and after communicating with the others via shared memory or messages, it chooses an output consistent with the task specification. In shared-memory models, shared objects typically keep track of a protocol's state as it executes.

By contrast, while cross-chain protocols are capable of general computations, they are primarily used to track ownership of assets such as cryptocurrencies or other valuable data. The smart contracts implement the ledgers that control asset ownership. Contracts are trusted, deterministic automata that change state in response to authenticated messages from parties.

Because parties are autonomous, and potentially Byzantine, tasks to be solved are formulated using elementary game-theoretic notions, expressed in terms of the utility to each party of each possible outcome. As in the classical model, the parties provide the inputs, typically expressed as an agreed-upon sequence of proposed asset transfers. Because some parties may be dishonest, the transfers that actually take place may differ from those previously agreed upon. In the end, unlike in the classical model, the *contracts*, not the parties, decide task outputs in the form of executed asset transfers, since the contracts alone have the power to control ownership.

The contracts are similar to shared objects in shared-memory distributed computing. They are deterministic state machines, but unlike the parties, they are trustworthy. Contract state is observable by any party, since that state is recorded on public ledgers.

We illustrate the model through an *atomic cross-chain swap* task, where the parties exchange assets across ledgers, guaranteeing that if all parties conform to the protocol, then all swaps take place, no conforming party ends up worse off, and no coalition has an incentive to deviate from the protocol.

2   Motivation and Related Work

Various papers have described formal models of various types of ledgers, we present a brief overview below. Our focus differs from these works in two ways. First, we are interested in an abstraction at the level of smart contracts, and view the ledger as a lower level implementation technique, on top of which the contracts are implemented. For us, a smart contract exists on its own, a state machine that exports operations that can be invoked by the parties, and whose state is observable by them. It is immaterial at this level of abstraction if the smart contracts are implemented by a blockchain or some other, perhaps future technology. Second, our focus is on *cross-chain transactions*, meaning that each smart contract runs on a different ledger. At our level of abstraction this means that the parties cannot communicate directly with each other. Neither the contracts can communicate with each other of course, they are passive. The parties coordinate to execute a financial transaction by asynchronously invoking operations on the contracts.

Overview of previous work. The first formal model of a permissionless blockchain system proposed by Garay *et al.* [18,19] focused on Bitcoin. The model is not radically different from classical models, e.g. in terms of partially synchronous executions and Byzantine processes, based on previous formulations of secure multiparty computation. The goal was to faithfully model the Bitcoin backbone protocol. Also, to analyze applications that can be built on top of the backbone protocol, focusing on Byzantine agreement, a classic distributed computing problem.

Later on other papers aimed at modeling at a higher level of abstraction. Anceaume *et al.* [2] were the first to make the connection between the Bitcoin ledger and the distributed shared objects theory. It introduced the notion of a *Distributed Ledger Register* (DLR) where the value of the register has a tree topology instead of a single value as in the classical theory of distributed registers. The vertices of the tree are blocks of transactions cryptographically linked. Later, Anta *et al.* [5] introduced the *Distributed Ledger Object* (DLO) formalism that defines the ledger object as an ordered sequence of records, abstracting away from registers. This work has been extended [4] to Multi-Distributed Ledger Objects (MDLO), the result of aggregating multiple Distributed Ledger Objects - DLO (a DLO is a formalization of the blockchain) and that supports append and get operations of records (e.g., transactions) from multiple clients concurrently. In order to model the behavior of distributed ledgers at runtime, Anceaume *et al.*[3] introduced the *Blockchain Abstract Data Type* abstraction, which provides a lower-level abstraction of Distributed Ledgers, suitable for both permissioned and permissionless systems. Rajsbaum and Raynal [34] provide a discussion of how ledgers are the next logical step in a long history of implementations of a sequential specification on a concurrent system. Another line of research was opened by Zappalà *et al.* [41] who propose a game theoretical framework to formally characterize the robustness of blockchains systems in terms of resilience to rational deviations and immunity to Byzantine behaviors. The framework is sufficiently general to characterize the robustness of various blockchain protocols (e.g. Bitcoin, Tendermint, Lightning Network, a side-chain protocols and a cross-chain swap protocols). Our model goes beyond other models in the literature limited to simple payments [17,21,36] or swaps [7].

Notice that there is a large research line on formal modelling of blockchain-based contracts (see e.g. [6] for a recent work and references herein), but in contrast to our work, they aim at faithfully modeling the operation of some specific system, such as Bitcoin or Ethereum often for formal verification purposes, while our goal is to come up a model that is independent of any particular system.

In the classic *fair exchange* problem, Alice and Bob wish to exchange a digital coin, in a way that either the exchange takes place, or each party keeps its token. Fair exchange has been widely studied, it includes several other important problems, such as contract signing, and has been studied also for more than two participants. It is well-known that this problem is unsolvable without a trusted third party [31], because of its relation to solving consensus. Fair exchange protocols have received renewed attention, because blockchains can be viewed as trusted third parties; for references and applications to data storage see the recent work [37]. The problem seems similar to the cross-chain swap we study, but a crucial difference is that in our case, the coins that Alice and Bob wish to exchange reside in different blockchains. Indeed, a standard technique for protocols solving fair-exchange consist of only one part of our solution: using one smart contract, Alice locks her coin with a hash $y = H(x)$, and only if Bob presents x he gets the coin. If x is not provided, Alice gets her coin back after a timeout. We instead need two smart contracts, one for each blockchain. For a solution without timeouts see [27].

3   Rationale for the Smart Contract Model

Before presenting a formal model, we present here an informal description of the "real world" of cross-chain computations.

A *distributed ledger*[1] is a publicly-readable tamper-proof distributed database used to track ownership of *assets*, which may be cryptocurrencies, financial instruments, tokens, concert tickets, or any data of value.

Parties own assets. A party can be a person, an organization, or even a bot acting on another party's behalf. Parties are untrustworthy: they may depart from any agreed-upon protocol in arbitrary ways, including irrational ways that work against their own interests. We assume only that every interaction includes at least one honest party who faithfully follows the agreed-upon protocol. While it is reasonable to assume, for example, that a super-majority of the validators securing a blockchain are honest, it is not reasonable (or prudent) to assume the same for a small number of parties willing to participate in a one-shot financial transaction or other distributed task.

---

[1] Following common usage, we use "ledgers", "chains", and "blockchains" interchangeably in informal discourse, even though ledgers need not be implemented using blockchain technology.

Parties may own assets controlled by one or more ledgers. Parties are active agents, initiating and reacting to communication, but they do not communicate directly with each other. Interactions between parties are mediated by *smart contracts*. Each smart contract resides on a single ledger, and different smart contracts may reside on different ledgers. Contracts are automata that change state in response to authenticated messages from parties. These messages are structured as calls to functions exported by contracts. The code of a smart contract must be deterministic, because it is repeatedly re-executed by mutually-suspicious parties to check the correctness of previous executions.[2]

A smart contract's code and current state are public, so a party calling a contract knows what code will be executed, and trusts the contract to execute that code correctly.[3] Since the state of a smart contract is public on its ledger, a party can read the state of the contract at any time.

In accordance with current practice, contracts are passive: a contract on one ledger cannot send messages directly to to a contract on another ledger (such network communications cannot be deterministically replayed). A contract A on one chain can learn of a state change of a contract B on another chain only if some party explicitly informs A of B's change, perhaps accompanied by evidence that the informing party is honest. For clarity and without loss of generality, we will assume that all smart contracts reside on distinct ledgers.

A *cross-chain task* (or *task* for short) specifies the problem that multiple parties want to solve. For example, Alice, Bob, and Carol might agree to do a three-way token swap, where each token is managed by its own ledger. Alice will transfer token A to Bob, Bob will transfer token B to Carol, and Carol will transfer token C to Alice.

A *cross-chain protocol* is a sequence of actions for each party to solve a cross-chain task. Honest parties who follow the protocol are said to be *compliant*, while dishonest parties who do not follow the protocol are said to be *deviating*. A party's compliance can be monitored by the other parties, but not enforced.

In our three-way swap example, the actions that Alice, Bob, and Carol execute consist of taking turns placing their assets into escrow (by calling the dedicated contracts), and when all assets are safely escrowed, each party triggers a transfer to its counterparty. If some party deviates from the protocol, then either all transfers are cancelled, or the deviating party loses its own asset. See [23] for a complete protocol description, and Sect. 5 for a two-way swap solution.

The swap task's *inputs* describe who wants to transfer what to whom. It makes sense for these inputs to be controlled by the participating parties. The task's *outputs* describe how asset ownership has changed: who transferred what to whom. It makes sense for these outputs to be controlled by the contracts, not the parties. (In real life, only the bank, not the bank's customer, has the power to determine that customer's account balance.)

---

[2] Even so-called "probabilistic smart contracts" [13] are actually deterministic because once they are written to the ledger, they return the same results when re-executed.

[3] Concurrent calls to the same contract may be executed in a non-deterministic order.

What does it mean for a swap protocol to correctly implement a swap task? By analogy with mainstream distributed computing, we might naively require a correct protocol to be *all-or-nothing*: either all transfers take place, or none does, a property also known as *atomicity*. However, because parties can depart arbitrarily from the protocol, we cannot prevent, for example, Alice from sending Bob two tokens instead of one. Bob, being rational, will pocket the extra token and continue the protocol. (Perhaps Alice and Bob are secretly laundering assets.) In the end, Alice has one token less than the task specifies, and Bob has one token more. Here is a protocol outcome that is neither "all" nor "nothing", yet this outcome is perfectly acceptable to Alice and Bob.

It follows that defining correctness in the cross-chain world requires some elementary notions from game theory. No rational party would agree to participate in a protocol that did not satisfy the following three common-sense conditions:

1. *Feasibility*: No coalition of parties can end up "better off" by deviating from the protocol [4].
2. *Liveness*: If all parties conform to the protocol, then all agreed-upon asset transfers take place.
3. *Safety*: No party that conforms to the protocol can end up "worse off", no matter how the other parties act[5].

Operationally, a cross-chain protocol execution proceeds as follows. For simplicity, we assume communication between parties and contracts occurs in rounds.

Each round has four phases: (1)*Send:* parties send messages to contracts; (2)*Contract-local:* contracts execute a local computation based on messages received; (3)*Read:* parties read the contracts' new states; (4)*Party-local:* Parties execute a local computation based on the states read, and prepare messages for the next round.

As stated, we assume contracts cannot send messages to one another, nor can they directly observe one another's states.[6] Deviating parties may communicate with one another through channels hidden from compliant parties. Failure to send a message is detected by a timeout. In Sect. 4 we will define in more detail the model including timing assumptions.

4   Smart Contract Model

This section presents the *smart contract model*. It formally defines the notion of task and of protocol, as well as the details about the model.

---

[4] This property is sometimes called a *strong Nash equilibrium*.

[5] Some protocols compensate jilted parties for fees and wasted time [39] while some do not.

[6] In practice, contracts *on the same chain* can call one another, but we are interested in problems where the contracts reside in different chains.

4.1   Cross-Chain System Model

A cross-chain system $CCS = (\mathcal{P}, \mathcal{C})$ is composed of a finite set of *parties* $\mathcal{P} = \{P_1, \ldots, P_m\}$, $m \geq 2$ and a finite set of *smart contracts* $\mathcal{C} = \{C_1, \ldots, C_n\}$, $n \geq 2$. Both parties and smart contracts can be modeled by *interface automata* [1]. An interface automaton is a tuple $IA = (V, V^{init}, \mathcal{A}^I, \mathcal{A}^O, \mathcal{A}^H, \mathcal{T})$ where:

- V is a set of states;
- $V^{init} \subseteq V$ is a set of initial states;
- $\mathcal{A}^I$, $\mathcal{A}^O$, $\mathcal{A}^H$ are mutually disjoint sets of input, output and internal actions;
- $\mathcal{T} \subseteq V \times \mathcal{A} \times V$ is the set of steps where $\mathcal{A} = \mathcal{A}^I \cup \mathcal{A}^O \cup \mathcal{A}^H$

An action $a \in \mathcal{A}$ is *enabled* at some state $v \in V$ if there is a step $(v, a, v') \in \mathcal{T}$ for some $v' \in V$. An *execution fragment* of an interface automaton is a finite sequence of alternate states and enabled actions $v_0, a_0, v_1, a_1, v_2 \ldots v_{t-1}, a_{t-1}, v_t$ such that $(v_i, a_i, v_{i+1}) \in \mathcal{T}, \forall i \in \{0, \ldots, t-1\}$.

Let $IA_P = (V_P, V_P^{init}, \mathcal{A}_P^I, \mathcal{A}_P^O, \mathcal{A}_P^H, \mathcal{T}_P)$ be the interface automaton of party $P \in \mathcal{P}$ and $IA_C = (V_C, V_C^{init}, \mathcal{A}_C^I, \mathcal{A}_C^O, \mathcal{A}_C^H, \mathcal{T}_C)$ be the interface automaton of a smart contract $C \in \mathcal{C}$. Let $Shared(\mathcal{A}_P, \mathcal{A}_C) = \mathcal{A}_P \cap \mathcal{A}_C$ be the common actions of IA_P and IA_C. IA_P and IA_C are *composable* if the following four properties hold: (i) $\mathcal{A}_P^I \cap \mathcal{A}_C^I = \emptyset$, (ii) $\mathcal{A}_P^O \cap \mathcal{A}_C^O = \emptyset$, (iii) $\mathcal{A}_P^H \cap \mathcal{A}_C = \emptyset$ and (iv) $\mathcal{A}_C^H \cap \mathcal{A}_P = \emptyset$.

The *product* of two composable interface automata IA_P and IA_C is the interface automaton $IA_{P \otimes C} = (V_P \times V_C, V_P^{init} \times V_C^{init}, \mathcal{A}_P^I \cup \mathcal{A}_C^I \setminus Shared(\mathcal{A}_P, \mathcal{A}_C), \mathcal{A}_P^O \cup \mathcal{A}_C^O \setminus Shared(\mathcal{A}_P, \mathcal{A}_C), \mathcal{A}_P^H \cup \mathcal{A}_C^H \cup Shared(\mathcal{A}_P, \mathcal{A}_C), \mathcal{T}_P \otimes \mathcal{T}_C)$ where $\mathcal{T}_P \otimes \mathcal{T}_C$ is defined as follows.
$\mathcal{T}_P \otimes \mathcal{T}_C = TR_P \cup TR_C \cup TR_{PC}$ where

$$TR_P = \{((v, u), a, (v', u)) \mid (v, a, v') \in \mathcal{T}_P \wedge a \notin Shared(P, C) \wedge u \in V_C\}$$

$$TR_C = \{((v, u), a, (v, u')) \mid (u, a, u') \in \mathcal{T}_C \wedge a \notin Shared(P, C) \wedge v \in V_P\}$$

$$TR_{PC} = \{((v, u), a, (v', u')) \mid (v, a, v') \in \mathcal{T}_P \wedge (u, a, u') \in \mathcal{T}_C \wedge a \in Shared(P, C)\}$$

The interface automaton of a cross-chain system $CCS = (\mathcal{P}, \mathcal{C})$ is the composition of the interface automaton of parties in $\mathcal{P}$ and interface automata of smart contracts in $\mathcal{C}$ following the methodology proposed in [1].

In the following we will introduce the notions of cross-chain task and cross-chain protocols.

In the sequel if V is a vector, $V[i]$ is V's i^{th} element. If D is a domain, 2^D is the powerset of D.

4.2   Cross-Chain Tasks

Given a cross-chain system CCS $= (\mathcal{P}, \mathcal{C})$ where $\mathcal{P}$ is set of m parties and $\mathcal{C}$ is a set of n smart contracts a *cross-chain task* is a tuple $(\mathcal{I}_P, \mathcal{I}_C, \mathcal{O}_C, \Delta, U)$, where:

- $\mathcal{I}_P$ is a set of m-element *input vectors*, representing each party's input to the task,

- $\mathcal{I}_C$ is a set of n-element *initial state vectors*, representing each contract's state before executing the task,
- $\mathcal{O}_C$ is a set of n-element *final state vectors*, representing each contract's state after executing the task,
- $\Delta : \mathcal{I}_P \times \mathcal{I}_C \to 2^{\mathcal{O}_C}$ is a map carrying pairs of input and initial state vectors to sets of final state vectors, and
- $U : \mathcal{I}_P \times \mathcal{I}_C \times \mathcal{O}_C \to \mathbb{R}^m$ assigns a *utility* for each party to each final state vector, with respect to each input situation.

Notice here that the map Δ is not a transition function, but gives out all the possible final states vectors when starting in a given pair of initial state and input vectors.

The utility function captures the notion of "better off" and "worse off" mentioned in the previous section. As a baseline, a contract's initial state has utility 0 for all parties. Compared to the *status quo*, a party P considers states with positive utility to be more desirable (P is "better off"), and states with negative utility to be less desirable (P is "worse off"). A final contract state is *acceptable* to P if P's utility is non-negative, and it is *preferred* if that utility is positive.

For example, in a two-way cross-chain swap, the input vector I_P may list asset p in slot P and q in slot Q. If P owns p and Q owns q in the initial state vector I_C, then in any final state vector in $\Delta(I_P, I_C)$, P owns q and Q owns p, a final state preferred by P and Q. Otherwise, the final contract states are unchanged, a final state acceptable to but not preferred by P and Q.

4.3   Cross-Chain Protocols

To execute a task, parties agree on a sequence of contract calls called a *cross-chain protocol*. As noted, compliant parties follow the agreed-upon protocol, while deviating parties do not.

Formally, given a cross-chain system $CCS = (\mathcal{P}, \mathcal{C})$ where $\mathcal{P}$ is a set of m parties and $\mathcal{C}$ is a set of n smart contracts, a cross-chain protocol is characterized by the tuple $(\mathcal{I}_P, \mathcal{I}_C, \mathcal{O}_C, \Xi)$, where

- $\mathcal{I}_P$ is a set of m-element *input vectors*,
- $\mathcal{I}_C$ is a set of n-element *initial state vectors*,
- $\mathcal{O}_C$ is a set of n-element *final state vectors*,
- $\Xi : \mathcal{I}_P \times \mathcal{I}_C \times \{0,1\}^m \to 2^{\mathcal{O}_C}$ is a map that carries an input vector, an initial state vector, and a Boolean array indicating which parties are compliant, to a set of final state vectors.

The protocol itself is the interface automaton obtained by the composition of interface automata modeling parties in $\mathcal{P}$ and interface automata modeling smart contracts in $\mathcal{C}$.

The execution of a cross-chain protocol is an execution fragment (see Sect. 4.1 for the definition) starting in a state $(I_P, I_C) \in \mathcal{I}_P \times \mathcal{I}_C$ and terminating in a state in $\Xi(I_P, I_C, B)$ where B is the Boolean array indicating which parties were compliant during the protocol execution.

A cross-chain protocol for a cross-chain task is *correct* if each execution of the protocol satisfies the following properties:

- *Feasibility:* No coalition of parties can end up better off by deviating from the protocol. Formally, there is no $B \neq 1^m$ such that for some $O_C \in \Xi(I_P, I_C, B)$, there exists a party P, $\neg B[P]$ (P deviates from the protocol) that is better off, $U(I_P, I_C, O_C)[P] > 0$.
- *Liveness:* If all parties are compliant, that is $B = 1^m$, then every final state is allowed by the task specification: $\Xi(I_p, I_C, B) \subseteq \Delta(I_p, I_C)$.
- *Safety:* No compliant party ends up worse off: if $B[P]$, then for every $O_C \in \Xi(I_P, I_C, B)$, $U(I_P, I_C, O_C)[P] \geq 0$.

As noted, we assume there is always at least one compliant party in every protocol execution because executions without compliant parties are not interesting, but it would be reckless to assume that more than one party is compliant.

4.4   Timing Models

A cross-chain system $CCS = (\mathcal{P}, \mathcal{C})$ uses message passing to communicate. We assume each party in $\mathcal{P}$ can send messages to smart contracts in $\mathcal{C}$ via authenticated channels. We say that communication channels are *asynchronous* if there is no upper bound on the message delivery delay, and *synchronous* if messages sent are delivered with delay at most δ. Finally, communication channels are *weakly synchronous* if there exist an unknown *a priori* time τ after which the communication channels behave as synchronous. Communications are authenticated (a message's sender cannot be forged) and reliable (the network does not create, lose, or duplicate messages).

We focus on a synchronous round-based model, but other models can be considered.

Synchronous round-based communication. Time is measured by a global clock accessible to parties and contracts. Computation evolves in sequential *synchronous rounds* $\{r_0, r_1, \ldots, r_i, \ldots\}$ and each round is composed of the four *phases* described earlier in Sect. 3: the send, contract-local, read, and party-local phases.

Synchronous round-free communication. Messages sent by compliant parties at time t are delivered by time $t + \delta$.

The protocol execution at a specific party goes through the same four-phase rounds: send, contract-local, read, party-local phases. Parties execute these rounds without synchronizing: one party can be in a read phase while another party is in the send or local computation phase.

Similar to the classical distributed systems the model can be further extended to *Partially synchronous* or *Asynchronous* communication. Communication channels are *weakly synchronous* respectively asynchronous and parties execute each round's send-read-local phases in an asynchronous manner.

5   Cross-Chain 2-Swap Case Study

In this section we present an atomic swap task, and a protocol solving it, inspired by Nolan's atomic swap protocol [29]. We focus on the synchronous setting. A similar protocol for a partially synchronous setting can be found in [26].

Alice has 1 bitcoin and Bob has 1 ether that they want to exchange. To accomplish this, two contracts are used: sc_A on the Bitcoin ledger, and sc_B on the Ethereum ledger. The protocol proceeds in rounds:

(Round 1)   Alice creates a secret s and the corresponding hash $h = H(s)$. She then sends h and temporarily transfers control of her coin to sc_A. The contract holds Alice's coin *in escrow* for two rounds, meaning that Alice cannot reclaim her coin for that duration. If sc_A receives the secret s from Bob within two rounds, then sc_A irrevocably transfers Alice's coin to Bob, and otherwise it refunds that coin to Alice.

(Round 2)   Bob reads the state of contract sc_A to obtain h. He then sends h and temporarily transfers control of his coin to sc_B. The contract holds Bob's coin in escrow for one round, meaning that Bob cannot reclaim it for that duration. If sc_B receives the secret s from Alice within one round, then sc_B irrevocably transfers Bob's coin to Alice, and otherwise it refunds that coin to Bob.

(Round 3)   Alice reads the state of sc_B to confirm that Bob has escrowed his coin. If so, Alice sends the secret s to sc_B, acquiring ownership of Bob's coin, making s part of sc_B's (public) state. If, at the end of Round 3, sc_B does not receive s from Alice, then sc_B refunds Bob's coin to Bob.

(Round 4)   Bob reads the state of sc_B attempting to learn the secret s. If Alice has claimed Bob's coin, then she has revealed s, so Bob forwards s to sc_A, acquiring ownership of Alice's coin, and completing the swap. If, at the end of Round 4, sc_A does not receive s from Bob, then sc_A refunds Alice's coin to Alice.

Here is a formal specification of the atomic swap task $(\mathcal{I}_P, \mathcal{I}_C, \mathcal{O}_C, \Delta, U)$, where:

- $\mathcal{I}_P$ is a single input vector, representing that each party wants to exchange a single coin.
- $\mathcal{I}_C$ is a single initial state vector $(\perp, \perp)$, representing that the protocol did not start yet.
- $\mathcal{O}_C$ is four possible binary vectors. A 0 entry means the corresponding party did not get the coin, and 1 means that it did.
- $\Delta : \mathcal{I}_P \times \mathcal{I}_C \to 2^{\mathcal{O}_C}$ maps the initial state to the set $\{(0,0),(1,1)\}$, meaning that in this example, we consider the two outputs to be correct. (Either the exchange happened or no coin got transferred.)
- $U : \mathcal{I}_P \times \mathcal{I}_C \times \mathcal{O}_C \to \mathbb{R}^m$. Since there is only a single initial state (of the parties and the contracts), we define U only on $\mathcal{O}_C$. Let $(b_1, b_2) \in \mathcal{O}_C$:

- $U((0,0)) = (0,0)$, means that if the exchange did not take place, the utility for both parties is 0.
- $U((1,1)) = (1,1)$, means that if the exchange took place, the utility for both parties is 1.
- Otherwise, $b_1 \neq b_2$. If $b_1 < b_2$, Alice gets Bob's coin, but kept her own coin, then $U((b_1, b_2)) = (2, -1)$. Symmetrically, if $b_1 > b_2$, $U((b_1, b_2)) = (-1, 2)$.

We now specify the protocol described at the beginning of the section by the tuple $(\mathcal{I}_P, \mathcal{I}_C, \mathcal{O}_C, \Xi)$ as follows:

- $\mathcal{I}_P, \mathcal{I}_C$ and $\mathcal{O}_C$ are the sets defined in the atomic swap task above.
- $\Xi : \mathcal{I}_P \times \mathcal{I}_C \times \{0,1\}^2 \rightarrow 2^{\mathcal{O}_C}$. Since there is only a single initial state (of the parties and the contracts), we define Ξ only on the vector of compliant parties. Let $(b_1, b_2) \in \{0,1\}^2$:
  - If both parties are compliant, i.e., $b_1 = b_2 = 1$, the exchange will take place $\Xi((1,1)) = (1,1)$.
  - Otherwise, at least one of $b_1, b_2 = 0$.
    - $\Xi((0,0)) = (0,0)$ and $\Xi((0,1)) = (0,0)$. If Alice deviates from the protocol, the exchange does not happen, and both parties end up with their initial coins.
    - $\Xi((1,0)) \in \{(0,0), (0,1)\}$. If Alice is compliant, but Bob is not, there are two cases. Either Bob deviates at round 2, or at round 4.
      If Bob deviates at round 2, the exchange cannot happen, and the parties end up with their own coins. Assume instead that Bob was compliant at round 2, but not at round 4. Since Alice is compliant, at the end of round 3, she gets Bob's coin. Hence, by deviating (and sending something other than s, or nothing at all), Bob will not get Alice's coin. Therefore, Alice will get her coin back and end up with both coins, while Bob will end up with none.

We now proceed by proving the correctness of our cross-chain 2-swap protocol.

Lemma 1. (Liveness) *If $B = (1,1)$, then $\Xi(B) \subseteq \Delta(I_p, I_C)$.*

Proof. Recall that $\Xi((1,1)) = (1,1)$ and $\Delta(I_P, I_C) = \{(0,0), (1,1)\}$. Therefore, we have $\Xi((1,1)) \subseteq \Delta(I_P, I_C)$. □

Lemma 2. (Safety) *No compliant party ends up worse off.*

Proof. We proof this for each party.

- First, let us assume that Alice is compliant.
  If Bob is also compliant, then the output state vector is $\Xi((1,1)) = (1,1)$, and $U(1,1) = (1,1)$. Otherwise, if Bob deviates, then the output state vector $\Xi(1,0)$ belongs to $\{(0,0), (0,1)\}$. We know that $U(0,0) = (0,0)$ and $U(0,1) = (2,-1)$. In both cases, the first component, corresponding to Alice's utility, is non-negative. Therefore, if Alice is compliant, she does not end up worse off.

– Assume that Bob is compliant. If Alice is also compliant, $\Xi((1,1)) = (1,1)$ and $U((1,1)) = (1,1)$. Assume that Alice is non-compliant. In this case, $\Xi((0,1)) = (0,0)$ and $U((0,0)) = (0,0)$.

In both cases, the second component, corresponding to Bob's utility, is greater or equal than 0, therefore, if Bob is compliant, he does not end up worse off.   □

Lemma 3. (Feasibility) *No non-empty coalition of parties can end up better off by deviating from the protocol.*

Proof. Since we consider a task with two parties, and since by assumption at least one party is compliant, the coalitions can consist only of singletons.

If Alice deviates, there is a unique possible output, $U((0,1)) = U((0,0)) = (0,0)$. In this case, her utility is 0, which is non positive, therefore, if Alice deviates, she does not end up better off.

If Bob deviates, $\forall b \in \{0,1\}, \Xi(b,0) \in \{(0,0),(0,1)\}$, and $U((0,0)) = (0,0)$, $U((0,1)) = (2,-1)$. In that case Bob's utility is either 0 or -1, which is non positive. Therefore, if Bob deviates from the protocol, he does not end up better off.   □

6   Conclusions and Open Research Directions

This paper has proposed the *smart contract* model for cross-chain protocols in which parties, some honest and some perhaps not, interact through trusted smart contracts residing on multiple, independent ledgers. As a case study, it presented a simple two-way cross-chain swap protocol, solving a fundamental task in this setting. Many other tasks have been considered in real systems that would be interesting to formalize in the smart contract model, such as auctions, loans and options.

The smart contract model differs from classical models of distribute computing, refactoring the roles of active participants (processes, parties) and of passive communication (objects, smart contracts). In the smart contract model, active participants can display Byzantine behavior, including irrational behavior, but contracts are honest. The model requires a more nuanced, game theoretic notion of correctness than classical models.

The model assumes smart contracts cannot communicate with one another, because of the lack of a practical way to ensure that replaying such cross-contract communication would always and everywhere produce exactly the same results. Nevertheless, there are emerging real-world mechanisms that do support limited forms of cross-chain communication. An *Oracle* [12] is a mechanism that allows a contract to read data from an external source (for example, the current dollar/euro exchange rate). Token bridges [35,38,42] provide a way to effectively transfer assets from one ledger to another by "freezing" an asset at the source ledger and creating a matching "wrapped" asset at the target ledger. So-called *layer two* solutions [9,25,33] allow computations to be moved from one ledger to another. We leave it to future work to expand the model presented here to encompass these increasingly important technologies.

The model's computational power can vary depending on access to cryptographic primitives. For example, cryptographic hashes enable atomic cross-chain swaps [8,10,16,23,29,30,40,43] and payment networks [15,20,22,28,32]. Public key infrastructure supports atomic broadcast [23,24]. To fully capture both the distributed and cryptographic aspects of these applications it might be necessary to extend the model in a way similar to [11]. Moreover, it might be intriguing to extend the model to encompass probabilistic behavior, perhaps using extensions to I/O automata similar to [14].

References

1. de Alfaro, L., Henzinger, T.A.: Interface automata. In: Tjoa, A.M., Gruhn, V. (eds.) Proceedings of the 8th European Software Engineering Conference held jointly with 9th ACM SIGSOFT International Symposium on Foundations of Software Engineering 2001, Vienna, Austria, September 10-14, 2001. pp. 109–120. ACM (2001)
2. Anceaume, E., Ludinard, R., Potop-Butucaru, M., Tronel, F.: Bitcoin a distributed shared register. In: Spirakis, P.G., Tsigas, P. (eds.) Stabilization, Safety, and Security of Distributed Systems - 19th International Symposium, SSS 2017, Boston, MA, USA, November 5-8, 2017, Proceedings. Lecture Notes in Computer Science, vol. 10616, pp. 456–468. Springer (2017)
3. Anceaume, E., Pozzo, A.D., Ludinard, R., Potop-Butucaru, M., Tucci Piergiovanni, S.: Blockchain abstract data type. In: Scheideler, C., Berenbrink, P. (eds.) The 31st ACM on Symposium on Parallelism in Algorithms and Architectures, SPAA 2019, Phoenix, AZ, USA, June 22-24, 2019. pp. 349–358. ACM (2019)
4. Anta, A.F., Georgiou, C., Nicolaou, N.: Atomic appends: Selling cars and coordinating armies with multiple distributed ledgers. In: Danos, V., Herlihy, M., Potop-Butucaru, M., Prat, J., Piergiovanni, S.T. (eds.) International Conference on Blockchain Economics, Security and Protocols, Tokenomics 2019, May 6-7, 2019, Paris, France. OASIcs, vol. 71, pp. 5:1–5:16. Schloss Dagstuhl - Leibniz-Zentrum für Informatik (2019)
5. Anta, A.F., Konwar, K.M., Georgiou, C., Nicolaou, N.C.: Formalizing and implementing distributed ledger objects. SIGACT News **49**(2), 58–76 (2018)
6. Bartoletti, M., Bracciali, A., Lepore, C., Scalas, A., Zunino, R.: A formal model of algorand smart contracts. In: Borisov, N., Diaz, C. (eds.) Financial Cryptography and Data Security, pp. 93–114. Springer, Berlin Heidelberg, Berlin, Heidelberg (2021)
7. Belotti, M., Moretti, S., Potop-Butucaru, M., Secci, S.: Game theoretical analysis of cross-chain swaps. In: 40th IEEE International Conference on Distributed Computing Systems, ICDCS 2020, Singapore, November 29 - December 1, 2020. pp. 485–495. IEEE (2020)
8. bitcoinwiki: Atomic cross-chain trading, https://en.bitcoin.it/wiki/Atomic_cross-chain_trading

9. Blog, E.: Optimistic rollups (Nov 2022), https://ethereum.org/en/developers/docs/scaling/optimistic-rollups/, as of 28 December 2022

10. Bowe, S., Hopwood, D.: Hashed time-locked contract transactions, https://github.com/bitcoin/bips/blob/master/bip-0199.mediawiki

11. Canetti, R., Cheung, L., Kaynar, D.K., Lynch, N.A., Pereira, O.: Compositional security for task-pioas. In: 20th IEEE Computer Security Foundations Symposium, CSF 2007, 6-8 July 2007, Venice, Italy. pp. 125–139. IEEE Computer Society (2007)

12. Chainlink, Inc.: Chainlink 2.0 and the future of Decentralized Oracle Networks — Chainlink, https://chain.link/whitepaper

13. Chatterjee, K., Goharshady, A.K., Pourdamghani, A.: Probabilistic smart contracts: Secure randomness on the blockchain. CoRR **abs/1902.07986** (2019), http://arxiv.org/abs/1902.07986

14. Civit, P., Potop-Butucaru, M.: Dynamic probabilistic input output automata. In: Scheideler, C. (ed.) 36th International Symposium on Distributed Computing, DISC 2022, October 25-27, 2022, Augusta, Georgia, USA. LIPIcs, vol. 246, pp. 15:1–15:18. Schloss Dagstuhl - Leibniz-Zentrum für Informatik (2022)

15. Decker, C., Wattenhofer, R.: A fast and scalable payment network with bitcoin duplex micropayment channels. In: Pelc, A., Schwarzmann, A.A. (eds.) Stabilization, safety, and security of distributed systems, pp. 3–18. Springer International Publishing, Cham (2015)

16. DeCred: Decred cross-chain atomic swapping, https://github.com/decred/atomicswap

17. Frey, D., Guillou, L., Raynal, M., Taïani, F.: Consensus-Free Ledgers When Operations of Distinct Processes are Commutative. In: Malyshkin, V. (ed.) Parallel Computing Technologies, vol. 12942, pp. 359–370. Springer International Publishing, Cham (2021). https://doi.org/10.1007/978-3-030-86359-3_27, series Title: Lecture Notes in Computer Science

18. Garay, J., Kiayias, A., Leonardos, N.: The Bitcoin Backbone Protocol: Analysis and Applications. In: Oswald, E., Fischlin, M. (eds.) Advances in Cryptology - EUROCRYPT 2015, vol. 9057, pp. 281–310. Springer Berlin Heidelberg, Berlin, Heidelberg (2015). https://doi.org/10.1007/978-3-662-46803-6_10, series Title: Lecture Notes in Computer Science

19. Garay, J., Kiayias, A., Leonardos, N.: The bitcoin backbone protocol: Analysis and applications. J. ACM **71**(4) (2024). https://doi.org/10.1145/3653445

20. Green, M., Miers, I.: Bolt: Anonymous payment channels for decentralized currencies. In: Proceedings of the 2017 ACM SIGSAC Conference on Computer and Communications Security (Oct 2017). https://doi.org/10.1145/3133956.3134093

21. Guerraoui, R., Kuznetsov, P., Monti, M., Pavlovic, M., Seredinschi, D.A.: The Consensus Number of a Cryptocurrency (Extended Version). ArXiv (2019). https://doi.org/10.48550/ARXIV.1906.05574, publisher: arXiv Version Number: 1

22. Heilman, E., Lipmann, S., Goldberg, S.: The arwen trading protocols (Jan 2019), https://www.arwen.io/whitepaper.pdf

23. Herlihy, M.: Atomic cross-chain swaps. In: Proceedings of the 2018 ACM symposium on principles of distributed computing. pp. 245–254. PODC '18, ACM, New York, NY, USA (2018). https://doi.org/10.1145/3212734.3212736, number of pages: 10 Place: Egham, United Kingdom tex.acmid: 3212736

24. Herlihy, M., Liskov, B., Shrira, L.: Cross-chain Deals and Adversarial Commerce. Proceedings of the VLDB Endowment **13**(2), 100–113 (Oct 2019). https://doi.org/10.14778/3364324.3364326, arXiv: 1905.09743

25. Kalodner, H., Goldfeder, S., Chen, X., Weinberg, S.M., Felten, E.W.: Arbitrum: scalable, private smart contracts. In: Proceedings of the 27th USENIX Conference on Security Symposium. pp. 1353–1370. SEC'18, USENIX Association, USA (Aug 2018)
26. Lys, L., Micoulet, A., Potop-Butucaru, M.: R-SWAP: relay based atomic cross-chain swap protocol. In: D'Angelo, G., Michail, O. (eds.) Algorithmic Aspects of Cloud Computing - 6th International Symposium, ALGOCLOUD 2021, Lisbon, Portugal, September 6-7, 2021, Revised Selected Papers. Lecture Notes in Computer Science, vol. 13084, pp. 18–37. Springer (2021)
27. Manevich, Y., Akavia, A.: Cross chain atomic swaps in the absence of time via attribute verifiable timed commitments. In: 2022 IEEE 7th European Symposium on Security and Privacy (EuroS&P). pp. 606–625. IEEE Computer Society, Los Alamitos, CA, USA (jun 2022). https://doi.org/10.1109/EuroSP53844.2022.00044
28. Network, R.: What is the raiden network?, https://raiden.network/101.html
29. Nolan, T.: Atomic swaps using cut and choose (Feb 2016), https://bitcointalk.org/index.php?topic=1364951
30. Organization, T.K.: The BarterDEX whitepaper: A decentralized, open-source cryptocurrency exchange, powered by atomic-swap technology, https://supernet.org/en/technology/whitepapers/BarterDEX-Whitepaper-v0.4.pdf
31. Pagnia, H., Darmstadt, F.C.G.: On the impossibility of fair exchange without a trusted third party (1999), https://api.semanticscholar.org/CorpusID:11671049
32. Poon, J., Dryja, T.: The bitcoin lightning network: Scalable off-chain instant payments (Jan 2016), https://lightning.network/lightning-network-paper.pdf
33. Poon, J., Buterin, V.: Plasma: Scalable autonomous smart contracts (2017), https://www.plasma.io/plasma.pdf
34. Rajsbaum, S., Raynal, M.: Mastering concurrent computing through sequential thinking. Commun. ACM **63**(1), 78?87 (dec 2019). https://doi.org/10.1145/3363823
35. Research, A.: Axelar network: Connecting applications with blockchain ecosystems (2022), https://arxiv.org/pdf/2011.12783.pdf, as of 8 May 2023
36. Sliwinski, J., Wattenhofer, R.: ABC: Proof-of-Stake without Consensus (Jul 2020), http://arxiv.org/abs/1909.10926, arXiv:1909.10926 [cs]
37. Tas, E.N., Seres, I.A., Zhang, Y., Melczer, M., Kelkar, M., Bonneau, J., Nikolaenko, V.: Atomic and fair data exchange via blockchain. Cryptology ePrint Archive, Paper 2024/418 (2024), https://eprint.iacr.org/2024/418
38. TechTarget: Cryptocurrency platform Wormhole loses $320M after attack — TechTarget, https://www.techtarget.com/searchsecurity/news/252512957/Cryptocurrency-platform-Wormhole-loses-320M-after-attack
39. Yingjie Xue, Maurice Herlihy: Hedging Against Sore Loser Attacks in Cross-Chain Transactions. In: ACM Symposium on Principles of Distributed Computing (2021)
40. Zakhary, V., Agrawal, D., El Abbadi, A.: Atomic commitment across blockchains. CoRR **abs/1905.02847** (2019), http://arxiv.org/abs/1905.02847, arXiv: 1905.02847 tex.bibsource: dblp computer science bibliography, tex.biburl: https://dblp.org/rec/bib/journals/corr/abs-1905-02847 tex.timestamp: Mon, 27 May 2019 13:15:00 +0200
41. Zappalà, P., Belotti, M., Potop-Butucaru, M., Secci, S.: Game theoretical framework for analyzing blockchains robustness. In: Gilbert, S. (ed.) 35th International Symposium on Distributed Computing, DISC 2021, October 4-8, 2021, Freiburg, Germany (Virtual Conference). LIPIcs, vol. 209, pp. 42:1–42:18. Schloss Dagstuhl - Leibniz-Zentrum für Informatik (2021)

42. Zarick, R., Pellegrino, B., Banister, C.: Layerzero: Trustless omnichain interoperability protocol (2022), https://layerzero.network/pdf/LayerZero_Whitepaper_Release.pdf, as of 8 May 2023
43. Zyskind, G., Kisagun, C., FromKnecht, C.: Enigma Catalyst: a machine-based investing platform and infrastructure for crypto-assets, https://www.enigma.co/enigma_catalyst.pdf

Invited Paper: Using Signed Formulas for Online Certification

Julius Wenzel[1]([✉]), Andreas Berg[2], and Christof Fetzer[1]

[1] Technische Universität Dresden, Dresden, Germany
{julius.wenzel,christof.fetzer}@tu-dresden.de
[2] Gematik GmbH, Berlin, Germany
andreas.berg@gematik.de

Abstract. Certifying software-based systems is a time-consuming and expensive task that requires much manual human effort. We introduce *Online Certification*, a partly automated version of the certification process, where participants provide the necessary information dynamically. All information is cryptographically signed to ensure integrity and authorization, and a system of certificates allows for fine-grained delegation of competencies. The requirements for certification, as well as the information needed to fulfill them, are represented in a subset of first-order logic. Consequently, validation is performed using automated logic reasoning. Compared to existing approaches, Online Certification enhances flexibility and agility. In cases where automatic generation of certification data is not possible, human certification processes can be integrated.

1 Introduction

As digital society progresses, organizational processes and interactions are being moved into the digital sphere, increasing our reliance on digital services. By running services in the cloud, service providers can increase availability for their clients but become dependent on the reliability of their cloud provider.

To provide the required reliability, the industry has developed security standards for operators of digital services. In general, such standards define processes that the operator of a digital service must implement and document to be audited by a qualified certification agency. If the operator meets the standard's requirements, the certification agency will issue a certificate, which the operator can use to gain the trust of its customers. For critical or sensitive applications, procedural requirements can be extensive, both in scope and in depth. In *critical sectors*, such as healthcare, power infrastructure, or water treatment, where failures pose significant threats to national security or the overall functioning of society, exceptionally high standards are set.

This work has received support by the Deutsche Forschungsgemeinschaft (DFG, German Research Foundation)—project number 389792660—TRR 248—CPEC, see https://perspicuous-computing.science.

T. Masuzawa et al. (Eds.): SSS 2024, LNCS 14931, pp. 71–86, 2025.
https://doi.org/10.1007/978-3-031-74498-3_5

Companies often fear being classified as part of a critical sector due to the complexity and cost associated with implementation and certification according to these sectors' standards. Providers struggle to keep up with the short-lived cycles in the software industry and lose flexibility because changes to the systems entail obligations to re-certify the changed parts. This is also true when a vulnerability is discovered and patched. Moreover, the certified state of the system does not necessarily the system's state at the time of usage.

We aim to address those weaknesses by introducing *Online Certification*, a dynamic, primarily automated, zero-trust approach to ensuring providers' reliability. Instead of relying on a static certificate, we want providers to present *evidence* in real time that their services fulfill requirements. Evidence is generated by the system itself, which allows us to automate parts of the certification process, to reduce the provider's certification cost, and to make certification more agile.

To automate certification, we define it on the software level. Therefore, we call this approach *certification as code*. It requires three key elements:

1. A stable interface for information exchange is needed because we want to use online certification on a distributed system. Information used for certification will come from different participants.
2. A signature scheme to distinguish genuine evidence from evidence produced by an adversary.
3. A verification engine that checks whether the obtained information can be trusted and whether it is sufficient to fulfill the given requirements.

For the signature scheme, we use a well-known concept of information security: the *Trust Domain*. It comprises the set of entities a participant is willing to trust unconditionally. Trust can be established in multiple ways, but once established, the Trust Domain may extend the set of trusted actors by delegation. Using delimited namespaces during delegation, we can isolate responsibilities and use fine-grained authorization.

The rest of the paper is organized as follows: In Sect. 2, we will present a more detailed example and demonstrate some advantages of our approach. In Sect. 3, we will explain how we use first-order logic as our interface and cryptography to ensure integrity and authorization. Section 4 will focus on a possible implementation of a verification engine. In Sect. 5, we will combine the work of the previous sections to show how an implementation of our approach would look like in the case of the example presented in Sect. 2. Section 6 will present some related approaches, and Sect. 7 will draw a conclusion.

2   Example and Objectives

In this section, we will illustrate the need for a more flexible certification approach using an example placed in the healthcare sector. We will reference it throughout the remainder of this paper. Based on the example, we will define the objectives for *Certification as code* and present our threat model.

2.1 Example and Current Certification Mechanism

We imagine a software development team that offers a smartphone app for the healthcare sector, which exchanges data with a service running in the cloud. Since the application processes highly sensitive information, it and its infrastructure need to adhere to standards set by a regulatory body or a standardization organization.

An example of such a standard is the ISO 2700x-family. ISO 27001 [9] defines how an organization should establish an Information Security Management System (ISMS). ISO 27002 [10] defines specific criteria an ISMS should address. In the healthcare sector, ISO 27799 [11] further specifies healthcare-specific criteria. None of the standards contains specific technical measures or technical specifications. The software development team's job is to transform those requirements into concrete technical rules and criteria and form a concept addressing possible security risks.

Accredited organizations, however, can review the ISMS and the resulting concept and certify an organization. The same goes for other standards and regulations that exist in parallel. If an organization is part of an important infrastructure, laws like the European NIS-2 directive [7] usually also require an auditing process that results in certification.

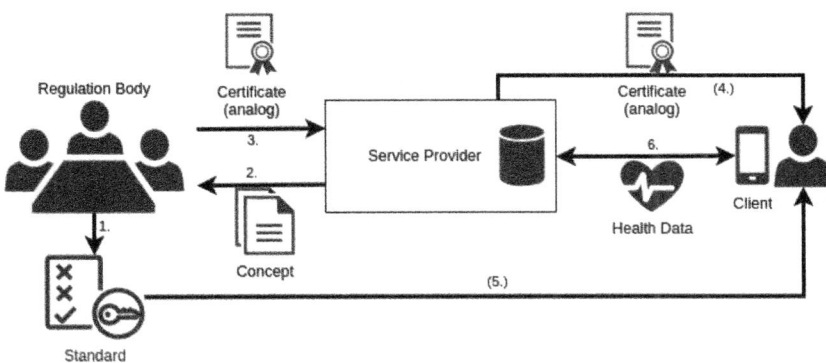

Fig. 1. Current certification and usage workflow

This leaves us with the workflow depicted in Fig. 1. The regulatory body (standards organization, legislative body, or else) formulates a standard (1.), and the software provider implements it with a concept (2.). The regulatory body or an accredited organization issues an analog certificate declaring conformity to the standard (3.). Before using the service (6.), a client can check the certificate (4.) and the standard (5.) to establish trust before exchanging data with the service provider.

2.2   Objectives

Our aim is to provide the client app with the means to verify the service's actual certification status before it submits sensitive data. This will prevent the service from, e.g.,

- using outdated software, either from us or in the form of an outdated dependency,
- running on a vulnerable operating system or
- running on server hardware with unmitigated side-channel vulnerabilities, which may allow other cloud users to eavesdrop on sensitive data.

To automate certification, we aim to describe the requirements taken from the standards as well as the provider's concept and the actual state of the system in a formal language a computer can process. Of course, not everything can be broken down into such a language, and parts of the certification (like source code auditing) will remain manual tasks. For the parts that can be automated, we need a trustworthy way to issue information about the acual state of the system. We chose to anchor our trust in specialized hardware. For the moment, we focus on *confidential computing* (CC), but we might widen our scope in a later step.

Assertions, either issued by humans or specialized hardware, need to be authenticated and integrity-protected throughout the process. We propose a signature scheme based on the delegation of trust from the Trust Domain. Finally, we aim to make our process explicable by saving all steps that lead to a decision about the conformity of the software.

2.3   Workflow

Our certification workflow is shown in Fig. 2. Trust in the system is rooted in the regulatory body, which defines the Trust Domain and issues the abstract standard. It also delegates the power to assert information about a software system to trustworthy parties, as indicated by the orange arrows in the Figure. In the digital sphere, delegation takes the form of issuing a digital certificate. Each delegation is also part of a legal contract that binds its parties in the non-digital world. Some participants might be allowed to delegate their permissions even further, such as hardware manufacturers. They certify their hardware, which in turn can issue assertions about the software at runtime. Certified natural or legal persons can still issue statements (for example, about code security) manually.

The software provider then formulates a machine-readable concept describing how to address the standards set forth. A certified verification body (such as the TÜV) asserts conformity to the standard's principles.

The standard, the concept, and all assertions, static and dynamic, are input to a verification engine that stems its trust from the regulatory body, too. The verification engine checks conformity and issues a short-lived (digital) certificate that can be sent to the client. Upon demand, it also provides the client with a trace showing all information and delegation certificates that lead to its decision. This trace can be displayed using techniques as described in [1] to convince even the most skeptical user that everything was cheked according to specifications.

2.4 Threat Model

As we want to anchor trust in our hardware, we need to assume that it works correctly; if not, our whole approach would be pointless. We therefore discard side-channel attacks and an attacker who can manipulate hardware at its production. We also suppose that our attacker is unable to guess or generate valid secret keys and/or signatures. Even if current digital signatures are usually based on hashes and hashes might collide, we ignore the possibility of a valid signature for the wrong data.

Our focus lies on the security of a distributed system running at least partially in the cloud. We, therefore, assume that our attacker can manipulate, drop, delay, and inject network traffic at will. We further assume that our attacker might collude with the cloud operator and, therefore, can turn off machines, move and modify cloud storage, and read memory that is not protected by confidential computing. We, however, assume that our attacker's objective is primarily to threaten integrity and confidentiality, not availability. Our attacker tries to be as *stealthy* as possible and, therefore, tends to avoid long outages or perma-

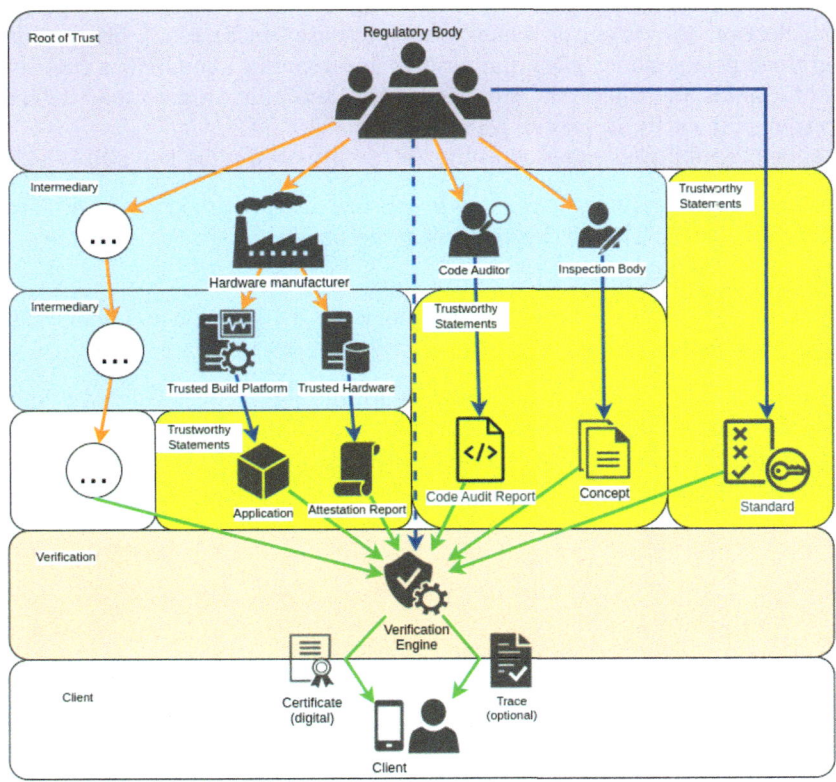

Fig. 2. Delegation and issuance workflow

nent network interruptions. Ensuring availability in a cloud-based environment is possible in principle, but it is considered to be outside the scope of this paper.

3   Concept

The following section presents the formal concept on which our approach is based. We combine elements of first-order logic, as described in Subsect. 3.2, with cryptographic signatures, as described in Subsect. 3.1. In order to verify the signatures, we create a system of public keys and certificates, also based on first-order logic and described in Subsect. 3.3

3.1   Cryptography

Before processing, the verification engine needs to verify the integrity and authenticity of the information received from the different parties. Therefore, we use public-key cryptography to sign such information, assuring two security properties. First, signing enforces authenticity and supports authorization: Only signatures from entities possessing a valid *certificate* will be accepted during verification. A system of *namespaces*, presented in Sect. 3.3, offers to grant fine-grained permissions. Second, it assures non-repudiation: Given a trace of the signed formulas and the certificates used, a user or an investigator can determine the responsible entity in case of a malfunction.

As for all public-key cryptography, we use private-public key pairs.

Definition 1. A *private-public keypair* consists of a *private key* K^s and a *public key* K^p. The latter can be derived using the one-way *key derivation primitive* genpub: $K^p = \text{genpub}(K^s)$.

We consider genpub to be a first-order logic (FOL) function and define signing and verifying as follows:

Definition 2. A *signing primitve* sign is a function that calculates a value σ known as *signature* from a *message* m and a private key: $\text{sign}(m, K^s) = \sigma$. A *verifying primitive* VerifySig is a predicate that evaluates to true according to the following formula:

$$\text{VerifySig}(m, \sigma, K^p) \Leftrightarrow \exists K^s (K^p = \text{genpub}(K^s) \wedge \sigma = \text{sign}(m, K^s))$$

3.2   Elements of Logic

In order to process the signed information, we need a common interface. If we think about regulation standards, they often come as a list of criteria an app provider has to meet for his app (identified by A, usually a URI).

$$\{ \text{CertifiedSoftware}(A), \text{UserDataProtection}(A), \dots \}$$

When writing a concept to implement the standards, the service provider usually defines a set of measures, often a checklist. Those measures work a lot like implications. The following formula

$$\forall a(\texttt{SoftwareAudited}(h_1) \land \texttt{SoftwareRunning}(a, h_1)$$
$$\Rightarrow \texttt{CertifiedSoftware}(a))$$

can be intepreded as: "If the software (identified by a hash value h_1) has been audited and is running as app a, then the provider indeed uses certified software for his app a". The provider then defines more detailed formulae describing how to fulfill `SoftwareAudited` or `SoftwareRunning`.

At the end of the day, the provider needs to assert some simple facts that do not rely on other conditions. Suppose they use a Trusted Build Tool (TBT) that guarantees the authenticity of hash values it calculates during build. In that case, it can assert that a code repository (also identified by a hash H2 has been compiled into a binary identified by H1:

$$\texttt{Compiled}(\mathsf{H2}, \mathsf{H1})$$

In the same manner, a human auditor could state that an audit for a library L, identified by another hash H3 was successful:

$$\texttt{Audited}(\mathsf{H3}, \mathsf{L})$$

We can see that we can model our system using *Horn Clauses*: *Goal Clauses* for requirements, *Definite Horn Clauses* for the rules describing how to meet them, and *Facts* to fulfill the rules. Facts and Goal Clauses use constants as predicate arguments, whereas rules use (universally quantified) variables. We define the three kinds of formulas used by our framework as follows:

Definition 3. A *fact* is a single predicate, consisting of a predicate indicator P and 0 to n arguments: $\mathsf{P}(c_1, c2, \ldots, c_n)$.

Definition 4. A *Definite Horn Clause* or *rule* is an implication.

$$\mathsf{Q}(v_1, \ldots) \land \mathsf{R}(v_2, \ldots) \land \cdots \Rightarrow \mathsf{P}(\ldots)$$

where the right-hand side is called the *head* and the left-hand side is called the *body*. The head always consists of a single predicate; the body is a conjunction of 1 to n predicates. Variables in rules are always universally qualified.

Definition 5. A *Goal Clause* or *query* consists of a negated predicate: $\neg\mathsf{P}(\ldots)$.

Cryptographic operations usually take only two types of inputs: key data or hashes with fixed length or *bitstrings* with variable length. To avoid an attacker abusing ambiguous representations, we assume that formulas always have an unequivocal textual representation that can be used as bitstring input.

We are grouping the predicates used inside our Horn clauses into *namespaces*.

Definition 6. A *namespace* is a set of predicate indicators.

Namespaces are used in certificates, as described in the next paragraphs. As an example, formulae including top-level criteria are usually issued by the same entity and therefore regrouped in one namespace:

$$\{ \text{ CertifiedSoftware}, \text{UserDataProtection } \}$$

3.3   Wrapped Formulas and Certificates

We attach a cryptographic signature and a public key to each formula to protect its integrity and provide authentication. We call the so-created triple a *Wrapped Formula* denoted by W.

Definition 7. Let ϕ be a Horn Clause and let K^s be a private key. The *Wrapped Formula* $W(\phi, K^p)$ is the tuple $\langle \phi, K^p, \sigma \rangle$ where $K^p = \text{genpub}(K^s)$ and $\sigma = \text{sign}(\phi, K^s)$.

If, for example, our auditor signs Audited with the key K^s_A, the wrapped formula is:

$$W(\text{Audited}(H3), K^p_A)$$

The signatures are authorized using *certificates*. We model certificates as 4-ary FOL predicates called Cert.

Definition 8. A *certificate* is the predicate: $\text{Cert}(K^p{}_s, Ns, T_s, T_e)$, where $K^p{}_s$ is the public key of the authorized signer, Ns is a namespace, T_s is the starting time and T_e the end time of the period in which the certificate is considered valid.

Most digital certificates like X.509 include the public key of the certificate's issuer $(K^p{}_i)$. However, since the key is already in the wrapper, we omit it inside our Cert predicate:

$$W(\text{Cert}(K^p{}_s, Ns, T_s, T_e), K^p{}_i,)$$

The unwrapping of a fact or query at a point T in time follows this formula:

$$W(\text{P}(\dots), K^p) \wedge \text{Cert}(K^p, Ns, T_s, T_e) \wedge T \in [T_s, T_e] \wedge \text{P} \in Ns \Rightarrow \text{P}(\dots)$$

Queries are unwrapped precisely the same way facts are. Just replace P by ¬P in the formula above.

3.4   Wrapping Rules

Wrapping Horn clauses is similar to wrapping facts; the difference lies in namespace checking. When unwrapping a Horn clause, we check if the clause's head is part of the namespace.

If ϕ is a definite Horn clause and $W(\phi, K^p)$ its wrapped form, we can unwrap it using the following formula:

$$W(\phi, K^p,) \wedge \text{Cert}(K^p, Ns, T_s, T_e) \wedge T \in [T_s, T_e] \wedge Head(\phi) \in Ns \Rightarrow \phi$$

3.5   Delegating Certificates

Since `Cert` is seen as a predicate, it can also be wrapped. According to the unwrapping rules, the predicate indicator `Cert` must be part of the delegator's namespace to wrap a predicate. Including this indicator in a namespace allows delegation, whereas not including `Cert` forbids it. However, care is necessary regarding the delegated namespace and validity period. To avoid an extension of granted signing powers, the namespace of the delegated certificate has to be a subset of the namespace of the delegator. In the same manner, the validity period of the delegated certificate has to be within the validity period of the delegator. Therefore, if we have already unwrapped the fact

$$\mathtt{Cert}(K^p{}_1, Ns_1, T_{s1}, T_{e1})$$

And want to unwrap

$$W(\mathtt{Cert}(K^p{}_2, Ns_2, T_{s2}, T_{e2}), K^p{}_1)$$

at time T, we can do so using the following formula:

$$W(\mathtt{Cert}(K^p{}_2, Ns_2, T_{s2}, T_{e2}), K^p{}_1)$$
$$\wedge \mathtt{Cert}(K^p{}_1, Ns_1, T_{s1}, T_{e1}) \wedge T \in [T_{s1}, T_{e1}]$$
$$\wedge \mathtt{Cert} \in Ns_1$$
$$\wedge Ns_2 \subseteq Ns_1$$
$$\wedge [T_{s1}, T_{e1}] \subseteq [T_{s1}, T_{e1}]$$
$$\Rightarrow \mathtt{Cert}(K^p{}_2, Ns_2, T_{s2}, T_{e2})$$

4   Architecture

Having described the formal foundations of our certification system, we will now show how our verification engine will implement them. Section 4.1 describes how existing files can be reused as input for a verification engine. Section 4.2 describes how the input of those files is unwraped and processed and Subsect. 4.3 explains how the result can be brought to the service and the client.

4.1   Extracting Facts from Files

We aim to reuse existing configuration files or files generated automatically by the software creation and deployment process. In the next paragraphs, we will use our running example to illustrate one possibility of reuse.

We assume a Trusted Build Tool (TBT) that compiles software and is able to output signed assertions about the process using an authorized certificate. We run it inside confidential computing (CC) to protect its integrity and the authenticity of its results.

In practice, the trustworthiness of a compiled application not only depends on the service provider's own source code, but also on that code's dependencies. Representing these dependencies as rules manually from scratch could be a time-consuming task. Luckily, the information regarding dependencies is usually available in a *Software Bill of Materials* (SBOM), a file that often already includes some signature, which we can import into our verification system. Since its format follows one of a little number of standards (SPDX and CycloneDX), we already implemented a parser that can parse rules from these.

For example, if an SBOM declares that our example app consists of the modules Core and Health and has the dependencies AppSDK and OpenCrypto, the result of the parser might be the following wrapped rule:

$$W(\texttt{AuditedAndCompiled}(\texttt{Core}, h_{10}) \land \texttt{AuditedAndCompiledHealth}, h_{11})$$
$$\land\texttt{NoVulnerabilites}(\texttt{AppSDK}, h_{20}) \land \texttt{NoVulnerabilites}(\texttt{OpenCrypto}, h_{21})$$
$$\land\texttt{Compiled}(h_{10}, h_{11}, h_{20}, h_{21}, h_1)$$
$$\Rightarrow \texttt{SoftwareAudited}(h_1), K^p{}_G)$$

$\texttt{AuditedAndCompiled}$ will be derived from facts signed by the auditors (manually) and the Trusted Build Tool (automatically) using the following rule:

$$W(\texttt{Audited}(h_1, o) \land \texttt{Compiled}(h_1, h_2)$$
$$\Rightarrow \texttt{AuditedAndCompiled}(o, h_2), K^p{}_G)$$

$\texttt{NoVulnerabilites}$ can be obtained from a (trusted) vulnerability database automatically.

4.2 Unwrapping Formulas

Signed configuration files, digital certificates, signed output of the TBT, source code audit reports, and all the other wrapped formulas for an application together form the evidence for input to our verification engine. The evidence contains the public root key from the regulatory body, but most formulas will be signed by different keys, whose certificates need to be unwrapped before the formulas can be verified. Unwrapping a fact or rule may thus extend the set of trusted certificates and, therefore, the predicates we can subsequently unwrap.

We use the following algorithm to unwrap the evidence. Since the evidence can only shrink during its execution, we are sure to reach a fix-point. E represents the evidence, U the set of unwrapped formulas and T the time for which we want to perform the check. T does not need to be the current time; we might perform verification for a moment in the past or the future.

Listing 1. The unwrapping algorithm

```
Function unwrap
   Input: E,U,T
   Output: U
Begin Function
```

```
Repeat
  E_o:=E
  For Each W(φ, K^p) In E Do Begin
    If Cert(K^p, Ns, S, E) ∈ U ∧ VerifySig(φ, K^p, σ)
      ∧φ ∈ Ns ∧ T ∈ [S, E]
      ∧((φ = Cert(K^p', Ns', S', E') ⇒ (Ns' ⊆ Ns ∧ [S', E'] ⊆ [S, E]))
    Then Begin
      E:= E − {W(φ, K^p)}
      U:=U ∪ {φ}
    End If
  End For Each
  Until E_o = E ∨ E = ∅
End Function
```

Initially, U only contains the public root key(s) representing the initial Trust Domain. Upon termination, U contains all successfully unwrapped formulas that can be used for verification. U can be imported into the logic database of a suitable reasoning engine.

The queries in U are then called upon the database to check their truthfulness: The engine tries to find the appropriate facts and rules and returns either success or failure. We implemented a simple version of this algorithm using Prolog as our logic engine and Prolog's Foreign Language Interface to interact with the database and to execute the queries.

4.3 Formula Delivery

Before a client uses the application and/or shares data with the service, we want to query the verification engine to see if the service conforms to its security standards. We assume communication between the client and the service provider to be encrypted using TLS, and we integrate verification into TLS connection creation: The verification engine has an identity in the form of an X.509 certificate from the Trust Domain authenticating it against other participants. It can also act as a Certificate Authority using a suitable signer certificate.

Upon initial start, the service connects to the verification engine and requests verification by presenting, among other things, its hardware-signed attestation report as evidence. The verification engine verifies the evidence against the standards as described above. If the requirements are met, it persists a cryptographically protected trace of its reasoning. The verification engine then issues a (short-lived) X.509 certificate for the service's domain identity and includes a reference to the trace record as an extended attribute in the certificate. We obtain a chain the client can verify against the Trust Domain's root certificate. Using the reference from the certificate, the client can choose to verify the results on their own. Before expiration of the service's X.509 certificate, (updated) evidence must be presented to renew it.

5 Application

After presenting the formal foundation and the implementation mechanism we demonstrate how the different parts come together in our running example. We focus on a single requirement—the usage of certified software—but this can easily be extended to other requirements.

5.1 Initial Setup and Requirements

The regulatory body, usually an association of different stakeholders, uses a Hardware Security Module (HSM) to create a single secret key $K_s{}^G$ in the HSM as root of trust during a so-called ceremony. This ceremony, as well as any subsequent operation involving the secret root key, requires direct interaction between each stakeholder and the HSM, thus ensuring that no single stakeholder can act unchecked. The public key $K^p{}_G = \mathrm{genpub}\, K^s{}_G$ is then included in a self-signed certificate with an unrestricted namespace (the numbers represent fictional timestamps):

$$\mathtt{Cert}(K^p{}_G, \{\dots\}, 300, 400)$$

The regulatory body then certifies the attestation service of its server's hardware manufacturer H to build on for recognizing the signatures issued by the hardware.

$$W(\mathtt{Cert}(K^p{}_H, \{\mathtt{SoftwareRunning}, \mathtt{Cert}\}, 300, 400), K^p{}_G)$$

The regulatory body also defines the need for software certification:

$$W(\neg\mathtt{CertifiedSoftware}(\mathsf{App}), K^p{}_G)$$

To fulfill this requirement and as part of the security concept, the software provider must satisfy the following rule:

$$W(\mathtt{SoftwareAudited}(h_1) \wedge \mathtt{SoftwareRunning}(a, h_1)$$
$$\Rightarrow \mathtt{CertifiedSoftware}(a), K^p{}_G)$$

5.2 Auditing and Compiling

In our scenario, the regulatory body designates trustworthy auditors by certifying them, based on some legal contracts fixing responsibilities in case something goes wrong. For auditing source code, the regulatory body designates a trustworthy auditor A like this:

$$W(\mathtt{Cert}(K^p{}_A, \{\mathtt{Audited}\}, 300, 400), K^p{}_G)$$

The regulatory body also certifies a TBT, identified by its public key $K^p{}_{TBT}$ and the hash of its binary HT. For bootstrapping reasons, this hash cannot be calculated inside CC, but we assume that the stakeholders have ways to convince themselves of its authenticity: The regulatory body also certifies a TBT, identified by its public key $K^p{}_{TBT}$ and the hash of its binary HT. For bootstrapping reasons, this hash cannot be calculated inside CC, but we assume that the stakeholders have convinced themselves of its authenticity:

$$W(\texttt{TrustedBuildTool}(\textsf{TBT},\textsf{HT},K^p{}_{TBT}),K^p{}_G)$$

To protect the build process, the TBT is only certified under the condition that it runs in confidential computing:

$$W(\texttt{TrustedBuildTool}(t,h_t,K^p{}_{TBT}) \wedge \texttt{SoftwareRunning}(t,h_t)$$
$$\Rightarrow \texttt{Cert}(K^p{}_{TBT},\{\texttt{Compiled}\},300,350),K^p{}_G)$$

For vulnerability scanning, the authority also designates a web service as source for trustworthy vulnerability data by transforming its certificate into a Cert predicate. The certificate has a shorter lifetime to limit the consequences of a service becoming malicious or malfunctioning:

$$W(\texttt{Cert}(K^p{}_V,\{\texttt{NoVulnerabilites}\},320,330),K^p{}_G)$$

5.3 Hardware

The hardware manufacturer provides a web service which certifies the processors (P) it produces:

$$W(\texttt{Cert}(K^p{}_P,\{\texttt{SoftwareRunning}\},300,400),K^p{}_H)$$

A processor can then, using its built-in secret key, issue the following signed fact:

$$W(\texttt{SoftwareRunning}(\textsf{A},\textsf{H1},K^p{}_P)$$
$$W(\texttt{SoftwareRunning}(\textsf{TBT},\textsf{HT}),K^p{}_P)$$

This mechanism is used to ensure that the TBT or any other service is running in confidential computing.

6 Related Work

Cryptographic signatures and automated reasoning for subsets of first-order logic have been well-established technologies for a long time. However, to our knowledge, those technologies have not previously been combined to verify a system's state against its requirements. This section, therefore, focuses on existing applications in the context of system security, either of first-order logic or cryptographic signatures.

6.1   Using Logic for Automated Verification

The potential of logic to verify the properties of a system without human intervention has been known for a long time [6]. For the verification of protocols, tools like ProVerif [4] work with Horn Clauses and try to falsify security protocols. To verify the properties of programs, *model checking* [8] has successfully found its way into the industry.

Our approach shows similarities to *runtime verification* [5], another use-case of logic in automated verification. However, to our knowledge, there is no runtime verification system that addresses the authenticity and integrity of its input data. This becomes a problem in distributed systems and cloud systems, where an adversary can try to intercept traffic or attack the hardware.

6.2   Other Frameworks for Software and Platform Attestation

Two existing standards try to address *runtime enforcement*: They want to provide a way to verify that the correct software runs in distributed systems. SPIFFE [2] and its canonical implementation SPIRE ties an identifier to a running piece of software (a so-called *workload*). This link can be verified automatically, but SPIRE heavily relies on external actors, such as the cloud provider and the kernel manufacturer, which are automatically part of the Trust Domain. Our approach allows us to be much more cautious: Using certificates inside rules, we can reduce our blind trust to just the hardware for automatic certification. Including CC into SPIRE has been proposed [12], but the process is still ongoing. Even when implemented, SPIRE does not know how to enrich its process with information established and signed by human actors like we do.

Another framework is the in-toto framework [3]. Its purpose is the integrity of the build process and the software supply chain. When using the framework, each building step adds metadata. This metadata can be chained together to make the build process transparent. In-toto, however, lacks a way to enforce the correctness of the metadata and chaining. Therefore, the build process needs to be run on a trusted platform, making it inconvenient for cloud-driven processes.

7   Conclusion and Future Work

This paper explored a new way of verifying a system's current security state based on first-order logic formulas characterizing the desired and current system states. We authorize and integrity-protect the formulas by signing their textual representations cryptographically. Starting with a Trust Domain, we use certificates to delegate the competence to sign formulas to human entities as well as to programs.

By comparing the system's desired state (our requirements) with its current state, we can provide assurance to users that it is safe to use the system. Our comparison uses logical reasoning, which gives us a great degree of flexibility for refinement and extension and lets us implement certification procedures currently only done by human auditors. Under our approach, software defines certification, therefore we call it *Certification as Code*. In cases where automatic validation is impossible, human certification can be integrated.

Using namespaces and validity periods to restrict the power of certificated, we showed that it is possible to define fine-grained competencies and enforce regular re-evaluation, which can be fully automated. Reusing existing configurations, confidential computing setups, attestation frameworks, hardware security modules, and more is possible.

We have implemented a prototype of a validation engine and experimented with config file parsers and with using confidential computing technology. We are currently working on implementing signed formula generation into confidential build processes. In the future, we plan to fully implement application specific workflows for modular confidential computing scenarios with mutual certification between components. Finally, we want to add support for more complex semantics than pure Horn clauses to include rules that cannot be expressed in the current semantic, thus allowing for more concise specifications.

References

1. Alrabbaa, C., Baader, F., Borgwardt, S., Dachselt, R., Koopmann, P., Méndez, J.: Evonne: interactive proof visualization for description logics (system description). In: Blanchette, J., Kovács, L., Pattinson, D. (eds.) Automated Reasoning—11th International Joint Conference, IJCAR 2022, Haifa, Israel, August 8-10, 2022, Proceedings, volume 13385 of Lecture Notes in Computer Science, pp. 271–280. Springer (2022). https://doi.org/10.1007/978-3-031-10769-6_16
2. The SPIFFE authors. Spiffe overview, 2024. URL https://spiffe.io/docs/latest/spiffe-about/overview/
3. in-toto authors. What is in-toto?, 2023. URL https://in-toto.io/in-toto/
4. Blanchet, B: An efficient cryptographic protocol verifier based on prolog rules. In: 14th IEEE Computer Security Foundations Workshop (CSFW-14), pp. 82–96
5. Cassar, I., Francalanza, A., Aceto, L., Ingólfsdóttir, A.: A survey of runtime monitoring instrumentation techniques. Electron Proc Theoret. Comput. Sci. **254**, 15–28 (2017). ISSN 2075-2180. https://doi.org/10.4204/eptcs.254.2
6. Dolev, D., Yao, A.: On the security of public key protocols. IEEE Trans. Inf. Theory **29**(2), 198–208 (1983)
7. European Parliament. Directive (eu) 2022/2555 of the european parliament and of the council of 14 december 2022 on measures for a high common level of cybersecurity across the union, amending regulation (eu) no 910/2014 and directive (eu) 2018/1972, and repealing directive (eu) 2016/1148 (nis 2 directive), 2022. https://eur-lex.europa.eu/eli/dir/2022/2555/oj?uri=CELEX:32022L2555
8. Holzmann, G.J.: The model checker spin. IEEE Trans. Software Eng. **23**(5), 279–295 (1997)
9. ISO 27001:2022. Information security, cybersecurity and privacy protection - Information security management systems - Requirements. Standard, International Organization for Standardization, Geneva, CH, January 2022
10. ISO 27002:2022. Information security, cybersecurity and privacy protection - Information security controls. Standard, International Organization for Standardization, Geneva, CH, January 2022

11. ISO 27799:2016. Health informatics - Information security management in health using ISO/IEC 27002. Standard, International Organization for Standardization, Geneva, CH, December 2016
12. Sthefano, M., da SILVA, L., et al.: Integrating spiffe and scone to enable universal identity support for confidential workloads (2021)

Papers

Optimal Asynchronous Perpetual Grid Exploration

Quentin Bramas[1]([✉]), Stéphane Devismes[2], Anaïs Durand[3],
Pascal Lafourcade[3], and Anissa Lamani[1]

[1] University of Strasbourg, ICUBE, CNRS, Strasbourg, France
{bramas,alamani}@unistra.fr
[2] Université de Picardie Jules Verne, MIS UR 4290, Amiens, France
stephane.devismes@u-picardie.fr
[3] Université Clermont Auvergne, Clermont Auvergne INP, CNRS, LIMOS, Clermont
Ferrand, France
{anais.durand,pascal.lafourcade}@uca.fr

Abstract. We address the perpetual grid exploration by a swarm of
autonomous, asynchronous, myopic, and luminous robots. We first show
that it is impossible for the robots to explore the grid regardless of their
number and the number of colors they can take if their visibility range
is one. We also show that PGE is impossible with three oblivious robots
that have a visibility range of two hops. We then present three optimal
algorithms solving the problem. The first algorithm uses four oblivious
robots with a visibility range of two, but assumes they agree on a common
chirality. For the two other algorithms, no common chirality is assumed.
The former uses three robots that have a visibility range of two and a
two-color light. The latter uses three oblivious robots under visibility
range three.

Keywords: Asynchronous myopic robots · Grids · Perpetual
exploration

1 Introduction

Swarm robotics has drawn a lot of attention in the past decade. Inspired by
natural systems, a lot of investigations focus on how to reproduce with artificial systems autonomous behaviors observed in nature. Given a collection of
autonomous mobile entities called robots, the main goal is to determine the
minimum hypotheses allowing the robots to solve a given task.

In this paper, we consider mobile robots that are autonomous (i.e., there
is no central authority to coordinate their move) and luminous (i.e., they are
endowed with lights that can take a finite number of different colors). Moreover,
they are deaf-mute (i.e., they have no direct means of communication) but they

This study has been partially supported by the French ANR project SkyData (ANR-
22-CE25-0008).

T. Masuzawa et al. (Eds.): SSS 2024, LNCS 14931, pp. 89–105, 2025.
https://doi.org/10.1007/978-3-031-74498-3_6

are endowed by visibility sensors allowing them to perceive their environment within a given distance called *visibility range*. Our robots are said to be myopic since can only sense at constant (typically small) distance. Robots operate in the well-known Look-Compute-Move (LCM) model. That is, they operate in cycles which comprise three phases: Look, Compute, and Move. During the first phase (Look), robots take a snapshot of their environment using their visibility sensors. In the second phase (Compute) and based on the previous snapshot, they decide a destination in their surrounding and update their color. Finally, in the last phase (Move), they move to the computed destination, if the destination is different from their current location. We consider the (fully) asynchronous model (ASYNC), meaning that the time between each Look, Compute, and Move phases is finite yet arbitrary long.

We study the case in which the robots have to solve the perpetual exploration problem. In this problem, robots evolve in a discrete universe and have to ensure that each location (node) is visited by at least one robot infinitely often. We consider here the universe is a finite grid, hence the problem will be referred to as the perpetual grid exploration (PGE). We focus on optimal *exclusive* solutions with respect to both the visibility range and the number of robots. Exclusiveness adds a constraint on robots behavior: they can neither occupy the same node simultaneously nor traverse the same edge at the same time.

Related Work. The exploration problem is one of the benchmark tasks when it comes to robots evolving on graphs. Various topologies have been studied: lines [13], rings [1,7,11,14,17], tori [10], grids [2,4,5,9], cuboids [3], and trees [12]. Two variants of the problem have been investigated: (i) the perpetual exploration problem [1–3,6,18], considered in this paper, which requires the robots to visit each node of the graph infinitely often and (ii) the terminating exploration problem [7,9–14,16] which requires the robots to visit each node of the graph at least once and then stop moving.

Most of the investigations consider robots with unlimited visibility range allowing them to observe every node of the system [1,2,9–14]. Robots are in this case oblivious (i.e., they cannot remember past actions) and have to solve the terminating exploration problem. Myopic robots have also been considered in both variants of the problem [4,6–8,17]. When it comes to the perpetual exploration problem, an additional assumption has an impact on the feasibility of the task and the optimality of the proposed solutions. This assumption endows the robots with a common chirality. In fact, chirality is usually assumed when robots evolve in the continuous 2D Euclidean plan but some investigations have also recently considered it in the discrete universe [5,6].

On finite and infinite grids, the exploration problem by myopic robots has been investigated almost exclusively assuming the robots are synchronous i.e., they all wake up at the same time look at their surroundings and perform their actions simultaneously. On finite grids, it has been shown that two (resp. three) synchronous robots with three colors (resp. one color) are sufficient to solve the problem when robots have visibility one and share a common chirality [6]. The case in which robots have no common chirality was investigated in [18]. It was proven that the problem is not solvable with only two robots having a finite number of colors and a finite visibility range. An optimal solution is also presented using only three robots having visibility range one, using only three colors. The case in which robots are oblivious and visibility range 2 was solved using five robots. In the case of infinite grids, assuming robots with visibility range one and few colors ($O(1)$), five (resp. six) synchronous robots are necessary and sufficient to solve the problem with (resp. without) the common chirality assumption [4,5]. Finally, in the case of cuboids, it has been shown in [3] that three synchronous robots with a common chirality endowed with five colors are necessary and sufficient to solve the perpetual exploration problem. Asynchronous myopic robots have been considered in the context of the terminating exploration problem on finite grids [16] assuming robots can start from distinct positions but allowing them to occupy the same location simultaneously during the execution (i.e., they are not exclusive). Now, in such a setting, storing at a given instant several robots at the same location can be used to remember some past actions and so may help solve the problem using fewer colors, a smaller number of robots, and/or a smaller visibility range. In the same setting as ours (perpetual exploration with exclusive, myopic robots), the case of asynchronous robots has only been considered in [6] by showing that a modified version of a synchronous algorithm also works in the asynchronous setting. This algorithm is mentioned in our summary table for comparison.

Asynchronous robots with colors and limited visibility range have been considered to solve the maximum independent set problem on grids [15]. This model is slightly different from ours as robots can cross the same link and occupy the same location at the same time. They also enter the grid initially one by one from the same location.

Contribution. We first present two impossibility results: the first one shows that perpetually exploring any finite grid is not solvable with asynchronous robots having a visibility range of one regardless of the number of robots and the number of colors for their lights. Next, we show that the problem is also unsolvable with three asynchronous robots without colored lights (i.e., with oblivious robots) and a visibility range of two.

We then present three optimal algorithms solving the problem. The first algorithm assumes a common chirality and requires four oblivious robots having visibility range two. The second algorithm uses three robots without common chirality but endowed with a light of two colors and having visibility range two. The third algorithm uses three oblivious robots without common chirality and having visibility range three. Table 1 below summarizes our contribution.

2   Model

We consider a set of $n > 0$ robots located on a *finite grid* made of $\mathcal{L} \geq n$ lines and $\mathcal{C} \geq n$ columns, i.e., robots evolve in an undirected graph $G(V, E)$ where $V = \{(i,j) \ : \ i \in [0, \mathcal{C} - 1], j \in [0, \mathcal{L} - 1]\}$ and $E = \{\{(i,j), (k,l)\} \ : \ (i,j) \in V \wedge (k,l) \in V \wedge |i - k| + |j - l| = 1\}$. The size of the grid is then $\mathcal{L} \times \mathcal{C}$. Grid coordinates are used for the analysis only, i.e., robots cannot access them.

Table 1. Summary of our results

Chirality	Visibility	# Robots	# Colors	Algorithm
Yes	1	Finite	Finite	Impossible (Theorem 2)
Yes	2	3	1	Impossible (Theorem 3)
Yes	2	3	2	[6]
Yes	2	4	1	A_1
No	2	3	2	A_2
No	3	3	1	A_3

Robots can move from a node to one of its neighbors in the grid but, in our algorithms, we prevent any two robots from being located at the same node simultaneously. In other words, any algorithm allowing such a behavior is considered as not well-defined. A node is *occupied* when a robot is located at this node, otherwise it is *empty*. Robots are *luminous*, i.e., they have lights of different colors that can be seen by robots in their surrounding. We denote by Cl the set of all possible colors. The *state* of a node is then the color of the light of the robot located at this node if it is occupied, $\perp$ otherwise.

The robots operates by executing infinitely many *asynchronous Look-Compute-Move* cycles, i.e., each robot performs its own cycles in sequence, however the time between each Look, Compute, and Move phases is finite yet unbounded and decided by an adversary. The only constraint is that both Move and Look are instantaneous (each compute phase may be arbitrarily long, yet finite). In the *Look* phase, a robot r gets a snapshot of the subgraph induced by the nodes within distance $\Phi \in \mathbb{N}^*$ from its position. Φ is called the *visibility range* of the robots. In the snapshot, nodes are labeled with their state. The snapshot is not oriented in any way as the robots do not agree on a common North. However, it is implicitly ego-centered since the robot that performs a Look phase is located at the center of the subgraph in the obtained snapshot. Notice that, since moves are instantaneous, robots are always located at nodes in a snapshot. Then, during the *Compute* phase, the robot selects a destination (either Up, Left, Down, Right, or Idle) and can change its color based only on the snapshot it received. Finally, it *moves* towards its computed destination. We say that a move is *pending* when a robot has decided to move (during its Compute phase) but has not moved yet. Note that light colors are both the only permanent memory of a robot and an indirect communication mean. The particular case where $|Cl| = 1$ corresponds to the *oblivious* assumption.

In all our algorithms, we also prevent any two robots from traversing the same edge simultaneously. Since we already forbid them to occupy the same position simultaneously, this means that we should additionally prevent robots from swapping their position. Algorithms verifying this property are said to be *exclusive*. However, to be as general as possible, we do not make this additional assumption in our impossibility results.

Configurations A *configuration* C in a grid $G(V, E)$ is a set of pairs (p, c), where $p \in V$ is an occupied node and $c \in Cl$ is the color of the robot located at p. A node p is empty if and only if $\forall c$, $(p, c) \notin C$. We sometimes just write the set of occupied nodes when the colors are clear from the context.

Views We denote by G_r the *globally oriented view* centered at the robot r, i.e., the subset of the configuration containing the states of the nodes at distance at most Φ from r, translated so that the coordinates of r is $(0, 0)$. We use this globally oriented view in our analysis to describe the movements of the robots (see, for example, Fig. 1): when we say "the robot moves Up", it is according to the globally oriented view. However, since robots do not agree on a common North, they have no access to the globally oriented view. When a robot looks at its surroundings, it instead obtains a snapshot. To model this, we assume that the *local view* acquired by a robot r in the Look phase is the result of an arbitrary *indistinguishable transformation* on G_r. The set $\mathcal{IT}$ of indistinguishable transformations depends on the assumption we make on the chirality. $\mathcal{IT}$ always contains the rotations of angle 0 (to have the identity), $\pi/2$, π and $3\pi/2$, centered at r. When we do not assume robots agree on a common chirality, we add the mirroring (robots cannot distinguish between clockwise and counterclockwise orientations) and any combination of aforementioned rotations and mirroring. Finally, we assume *self-inconsistent* robots, meaning that different transformations may be applied at different rounds.

It is important to note that when a robot r computes a destination d, it is relative to its local view $f(G_r)$, which is the globally oriented view transformed by some $f \in \mathcal{IT}$. So, the actual movement of the robot in the *globally oriented view* is $f^{-1}(d)$. For example, if $d = Up$ but the robot sees the grid upside-down (f is the π-rotation), then the robot moves $Down = f^{-1}(Up)$. In a configuration C, $V_C(i, j)$ denotes the globally oriented view of a robot located at (i, j).

Algorithm An algorithm A is a tuple $(Cl, Init, T)$ where Cl is the set of possible colors, $Init$ is a mapping from any considered grid to a non-empty set of initial configurations in that grid, and T is the transition function $Views \rightarrow \{Idle, Up, Left, Down, Right\} \times Cl$, where $Views$ is the set of possible local views.

Scheduler and Execution During an execution, the robots perform their Look-Compute-Move cycles independently and asynchronously. However, without the loss of generality, we assume that each phase of a cycle is atomic. Indeed, recall that Look and Move are already assumed to be instantaneous, moreover, we can assume each compute phase is atomic since it is only based on the previous snapshot. An adversarial scheduler selects when a robot performs the phases of its cycle. Again, without loss of generality, we assume that at most one robot performs a phase of its cycle at each time instant. Indeed, if two robots perform a phase of their cycle at the same time, we can assume that one of them performs its phase strictly before the other as the resulting configuration is the same if the order is carefully chosen.

Hence, we can consider that the time is discretized in time instants, w.l.o.g. with non-negative integers $0, 1, 2, \ldots$, that represents the instants at which one robot performs one phase of its cycle. It is easy to see that the exact values of the times are not important in our context, only the order of the events is. Hence, we define a *schedule* $S = (S_i)_{i \geq 0}$ as a sequence actions describing the order in which the robots perform their phases: $S_i = (r_i, Look|Compute|Move)$ means that at time i, the robot r_i performs the phase *Look*, *Compute*, or *Move*. A schedule is *well-defined* if, by considering the sequence of actions associated with a single robot, the order of the actions is periodic and alternates between Look, Compute, and Move starting from a Look. A schedule is *fair* if every robot is selected infinitely often. In the remainder of the paper, we assume that all the schedules are fair and well-defined.

An execution of A in the grid G is then an infinite sequence of configurations $(C_i)_{i \in \mathbb{N}}$ determined by its initial configuration C_0 and a schedule $S = (S_i)_{i \geq 0}$. Precisely, $C_0 \in Init(G)$ and $\forall i \in \mathbb{N}$, C_{i+1} is obtained by applying $S_i = (r_i, a_i)$ on C_i as follows:

- If $a_i = $ Look, then $C_{i+1} = C_i$.

Otherwise, let j the maximum integer satisfying $j < i$ and $S_j = (r_i, Look)$; let $f_j \in \mathcal{IT}$ be the transformation applied on the view of r_i in C_j; and let $p = (x, y)$ and c respectively be the node occupied by and the color of r_i in C_i.

- If $a_i = $ Compute, then $C_{i+1} = C_i \setminus \{(p, c)\} \cup \{(p, c')\}$, where $T(f_j(V_{C_j}(p))) = (\_, c')$, i.e., c' is the new color of r_i (maybe $c = c'$) computed by r_i from the view it obtained in C_j.
- If $a_i = $ Move, then $C_{i+1} = C_i \setminus \{(p, c)\} \cup \{(p', c)\}$, where
  - $p' = p$ if $f_j^{-1}(T(f_j(V_{C_j}(p)))) = (Idle, \_)$;
  - $p' = (x + 1, y)$ if $f_j^{-1}(T(f_j(V_{C_j}(p)))) = (Right, \_)$;
  - $p' = (x - 1, y)$ if $f_j^{-1}(T(f_j(V_{C_j}(p)))) = (Left, \_)$;
  - $p' = (x, y + 1)$ if $f_j^{-1}(T(f_j(V_{C_j}(p)))) = (Up, \_)$; and
  - $p' = (x, y - 1)$ if $f_j^{-1}(T(f_j(V_{C_j}(p)))) = (Down, \_)$.

Given $C_0 \in Init(G)$ and a scheduler S, we obtain an execution $e = (C_i)_{i \in \mathbb{N}}$ of A. For $0 \leq i \leq j$, we write $C_i \overset{e}{\hookrightarrow} C_j$ the fact that C_j is reached from C_i in e.

Let $r_0, \ldots, r_{n-1}$ be n robots and $Sync^{3n} = (Sync_i)_{i \in [0, 3n-1]}$ such that for all $i \in [0, n-1]$, $Sync_i = (r_i, Look)$, $Sync_{i+n} = (r_i, Compute)$, and $Sync_{i+2n} = (r_i, Move)$. The synchronous scheduler is then $Sync = (Sync^{3n})^\omega$. We denote by $C \xrightarrow{Sync^{3n}} C'$ the fact that C' is reached from C under the synchronous scheduler.

Perpetual Finite Grid Exploration An execution $(C_i)_{i \in \mathbb{N}}$ of A $= (Cl, Init, T)$ in a grid $G = (V, E)$ *achieves* the *Perpetual (Finite) Grid Exploration* (PGE) if for every node $u \in V$ and for every time t, there exists a time $t' \geq t$ such that u is occupied in $C_{t'}$. An algorithm A that uses n robots *solves* (resp., *synchronously solves*) the *Perpetual Finite Grid Exploration* (PGE) problem if for every finite grid $G = (V, E)$ with at least n lines and n columns and every initial configuration $C_0 \in Init(G)$, we have every execution (resp. every execution under the synchronous scheduler) of A in G starting from C_0 that achieves the PGE.

Well-defined Algorithms Recall that robots are assumed to be self-inconsistent. In this context, we say that an algorithm $(Cl, Init, T)$ is *well-defined* if the global destination computed by a robot does not depend on the applied indistinguishable transformation f, i.e., for every globally oriented view V, and every transformation $f \in \mathcal{IT}$, we have $T(V) = f^{-1}(T(f(V)))$. All our algorithms are well-defined. However, to be as general as possible, we do not make this assumption in our impossibility results.

An Algorithm as a Set of Rules We write an algorithm as a set of rules, where *a rule* is a triplet $(V, d, c) \in Views \times \{Idle, Up, Left, Down, Right\} \times Cl$. We say that an algorithm $(Cl, Init, T)$ includes the rule (V, d, c), if $T(V) = (d, c)$. By extension, the same rule applies to indistinguishable views, i.e., $\forall f \in \mathcal{IT}, T(f(V)) = (f(d), c)$. Consequently, we forbid an algorithm to contain two rules (V, d, c) and (V', d', c') such that $V' = f(V)$ for some $f \in \mathcal{IT}$. Hence, an algorithm corresponds to a set of rules if each destination is the result of applying one of its rules.

As an illustrative example, consider the rule R_1 given in Fig. 1. This rule is defined for robots having a visibility range of two. This rule means that, when a blue robot B sees three robots with color R, one on top at distance 2, one on the left, and one in diagonal, then the blue robot is dictated to move Up. By extension, the same rule applies if the view is rotated by π, but in that case, the destination would be Down.

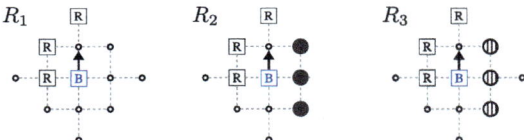

Fig. 1. Examples of rules.

In the same figure, R_2 is a rule where the three black nodes represent a part of the outer boundary of the grid, that we call *a wall* in the remaining of the paper. In our algorithms, we often define similar rule that apply regardless of the presence of a wall in some part of the view. Thus, to avoid defining several time rules with very similar views, we propose a notation to express several rules in just one picture. For example, R_3 in Fig. 1 has three nodes hatched with vertical lines, which means that the rule applies regardless of the presence of a wall located at those nodes. In practice, every rule that contains such vertical (resp. horizontal) hatched lines, represents a set of rules obtained by replacing each of those lines either by walls or by empty nodes. For example, R_3 in Fig. 1 is a concise representation of R_1 and R_2.

Algorithms having locally-defined initial configurations In a given grid, the set of possible initial configurations of an algorithm can be reduced to a singleton. In such a case, the scalability and flexibility of the algorithm is weak. To be more general, two of the algorithms we propose have *locally-defined* sets of initial configurations. Configurations in a locally-defined set of initial configurations are defined by colors and relative positions of the robots only. Hence, for a given grid, every two possible initial configurations are equal up to possible transformations applied on all robots positions. The set of possible transformations includes any combinations of translations and rotations of angle $\frac{\pi}{2}$ applied to all robot positions. Moreover, when no common chirality is assumed, the previous combinations can also be augmented with mirroring. The set of all possible initial configurations is then closed by the set of possible transformations.

Synchronous Sequential Algorithms. Let A be well-defined algorithm. Informally, A is sequential if robots move or change their color one at a time. More formally, we start by defining the notion of *seq-enabled* robots. In a configuration C, a robot r at a position p is said to be *move-enabled*, resp. *col-enabled* if we have $T(V_C(p)) = (move, \_)$, resp. $T(V_C(p)) = (\_, c)$, where $move \neq Idle$, resp. c is not the color of r in C. A robot that is move-enabled or col-enabled is said to be *enabled*. Finally, it is said to be *seq-enabled* if it is either move-enabled or col-enabled, but not both. An synchronous execution $(C_i)_{i \in \mathbb{N}}$ under is said to be *sequential* if $\forall j \in \mathbb{N}$, no robot except one is enabled in C_{3nj}, and this latter being actually seq-enabled in C_{3nj}. An algorithm is said to be *synchronous-sequential* if it is well-defined and all its synchronous executions are sequential.

We show the correctness of our algorithm by induction on the size of the grid in which they are deployed. We use Theorem 1 below to establish the base cases of these inductions using a simulator.

Theorem 1. *Let A be a synchronous-sequential algorithm using n robots and G be a grid. Every synchronous execution of A in G achieves the PGE if and only if every (asynchronous) execution of A in G achieves the PGE.*

Let $(C_i)_{i \in \mathbb{N}}$ a synchronous execution of A on some grid G. Our simulator actually computes the sequence of configurations $(C_{3ni})_{i \in \mathbb{N}}$. To apply Theorem 1, we then need to check if:

- A *is well-defined.* To that goal, we test every indistinguishable transformation (a finite number of transformations) on each of its rules (the number of rules of A is also finite).
- A *is synchronous-sequential and its synchronous execution achieves the PGE.* Since A is well-defined, all its synchronous executions are fully determined by their initial configuration. So, we should make one simulation per possible initial configuration (the set of all possible initial configurations can be computed by applying every authorized transformation on the local initial pattern). Each simulation stops as soon as a configuration appears again. This necessarily happens since the number of configurations is finite. Then, we have to check all nodes have been visited in the execution prefix (to show the correctness) and in all encountered configurations exactly one robot is enabled and this robot is actually seq-enabled (to show that A is synchronous-sequential).

3 Impossibility Results

In this section, we first establish in Theorem 2 a general impossibility result for the PGE problem with robots having visibility one. To do so, we show that a team of robots cannot cross a fence (defined below). We then give a specific impossibility result for the PGE problem with three oblivious robots having visibility range two. To show Theorem 2, we prove a variant of the test of the fence (introduced in [4]) for the case of finite grids. In our setting, a fence is the boundary of width two of a square.

More formally, Let $Square((x_1, y_1), (x_2, y_2))$ be the nodes in the square delimited by the given points, with $x_1 < x_2$ and $y_1 < y_2$: $Square((x_1, y_1), (x_2, y_2)) = \{(x, y) \in \mathbb{Z}^2 \mid x_1 \le x \le x_2 \text{ and } y_1 \le y \le y_2\}$. Then, a fence $F_{(x_1, y_1), (x_2, y_2)}$, with $x_1 + 4 < x_2$ and $y_1 + 4 < y_2$ is defined as: $F_{(x_1, y_1), (x_2, y_2)} = Square((x_1, y_1), (x_2, y_2)) \backslash Square((x_1+2, y_1+2), (x_2-2, y_2-2))$. We say a robot is outside (resp. inside) the fence if it is outside the outer square (resp. inside the inner square). We say that a set of robots has crossed a fence when they are all inside the fence at a given time. Notice that this does not mean that the robots always stay inside of the fence afterward. Formally, we say that a set of robots S *has crossed the fence* $F_{(x_1, y_1), (x_2, y_2)}$ *at Round* t if there exists $t' \le t$ such that every robot $r \in S$ is at some coordinates $(x, y) \in Square((x_1+2, y_1+2), (x_2-2, y_2-2))$ at Round t'.

We say a set of robots S *single-handed crosses the fence* F *between* t *and* t' if for every robot $r \in S$, (1) r is located outside F at Round t; (2) r is located at inside F at Round t'; and (3) only robots of S are within distance one of r between Round t and Round t'. We say that a set of robots S *has single-handed crossed* the fence F at Round t if $\exists t' < t'' \le t$ such that S single-handed crosses the fence F between t' and t''.

To be more general, we now consider any algorithm, i.e., well-defined or not. We first prove that if robots explore any finite grid, then there is a fence that is *single-handed crossed* by a subset of robots; see Lemma 1. This latter result will be used to show that, if robots have visibility one the PGE problem is impossible, whatever the number of robots is, indeed we show that the test of the fence fails; see Theorem 2.

Lemma 1. (The test of the fence in finite grids) *Let* A *be any algorithm solving the PGE using* n *robots. There is an execution of* A*, a grid* G*, a fence* F *of* G*, and a subset of robots* S *such that* S *single-handed crosses* F *within a finite number of rounds.*

Theorem 2. *There exists no algorithm solving the PGE problem with robots having visibility range one, even assuming a common chirality.*

The proof of the next theorem follows the same principles as Theorem 3.5 in [6]. We had just to adapt these principles for our specific setting. In particular, we had to show that, whatever the algorithm, the adversary can impose a single way to translate the positions of three oblivious robots by one, when they do not see any wall.

Theorem 3. *There exists no algorithm solving the PGE problem with three oblivious robots having visibility range two, even assuming a common chirality.*

4 Algorithms

4.1 Algorithm with 4 Oblivious Robots and Common Chirality

We first present an algorithm, denoted by A_1, that uses four oblivious robots, assuming a common chirality and visibility range two[1]. The algorithm we present here is for when the number of lines and columns is at least 5. The particular cases where the number of lines or columns is 4 are handled with special rules shown in the interactive simulation.

The exploration of the grid is performed alternatively rows by rows and columns by columns in a given direction. Precisely, three robots explore a line (a row or a column), while leaving the fourth robot, called *sentinel*, adjacent to a wall. When the three robots have explored the line, they perform a U-turn and move back toward the sentinel. When they meet the sentinel, the four robots perform a sequence of moves in a way that allows them to explore the next line. The lines are explored one by one in a given direction called the *exploring direction* which is initially $\downarrow$ in the illustration. After all the lines are explored, the four robots perform a sequence of moves to change the exploring direction, from $\downarrow$ to $\leftarrow$. In more detail, the four robots initially form an L-shaped pattern with a **unique** robot being adjacent to a wall (see Fig. 2) and the robot alone in its row must not see any wall. Assume without loss of generality that the sentinel is placed on Row ℓ_i and let ℓ_{i+1} and ℓ_{i+2} be the next two rows from ℓ_i in the exploring direction. The exploration of the rows ℓ_i and ℓ_{i+1} is initiated by the robot adjacent to the sentinel by moving to its adjacent node in the exploration direction $\downarrow$ by executing Rule a) of Fig. 4. While the sentinel remains idle on its node, the three other robots, being in exploration formation (Fig. 3), move in a straight line to explore Rows ℓ_i and ℓ_{i+1}. This is done by moving in a sequential manner starting from the robot that is located on ℓ_i by executing the rules defined in Fig. 4, (b – f).

Once the three explorer robots reach the opposite wall, they not only move to the next adjacent row with respect to direction $\downarrow$ but also perform a U-turn so that they move back on Rows ℓ_{i+1} and ℓ_{i+2} towards the sentinel. The U-turn is performed by executing the rules of Fig. 5. The sequence of configurations during this process is illustrated in Fig. 6. When moving towards the sentinel, a robot eventually becomes adjacent to a wall and it can see the sentinel at distance two. Robots then move to re-create the L-shaped pattern but this time on row ℓ_{i+1}. This is done by executing the rules of Fig. 8. The sequence of configurations during this process is illustrated in Fig. 7.

Let $\ell_1, \ell_2, \ldots, \ell_{\mathcal{L}}$ be the sequence of rows in direction $\downarrow$. By repeating the same process, the sentinel will eventually be located at $\ell_{\mathcal{L}-3}$ and the robots explore

---

[1] An interactive simulation of A_1 is given at https://robots.app.bramas.fr/?SSS2024/0.

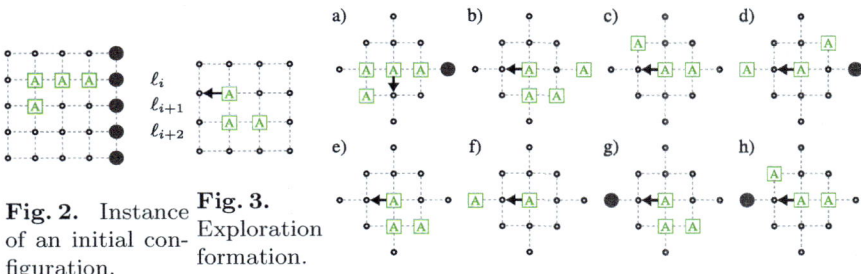

Fig. 2. Instance of an initial configuration.

Fig. 3. Exploration formation.

Fig. 4. Rules to move in a straight line.

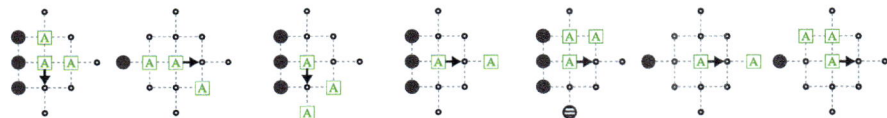

Fig. 5. Rules to perform the U-turn and initiate the exploration towards the sentinel.

Rows $\ell_{\mathcal{L}-3}, \ell_{\mathcal{L}-2}$ and $\ell_{\mathcal{L}-1}$. When robots move back towards the sentinel and reach it, they move to initiate the exploration of the grid columns by columns in direction $\leftarrow$, labeled $m_1, m_2, \ldots, m_{\mathcal{C}}$. Note that the robots cannot differentiate between rows and columns, thus, they apply the same process to the columns. That is, robots will explore the columns as they did with the rows. To switch from rows to columns exploration, the robots perform a sequence of moves first by using the rules of Fig. 8 as they are changing rows except that this time, Rule (f) will not be enabled as the robot observes another wall below, at distance two. Instead, the robot executes Rule (a) of Fig. 9 and moves towards the wall on its current row to notify the other robots that they are switching to columns exploration. The robots then create the desired L-shape on m_2 by executing the rules of Fig. 9. The sequence of configurations during this process is presented in Fig. 10. Robots explore the columns as they did for the rows and then switch again to the rows exploration in direction $\uparrow$ by placing again the sentinel adjacent to a wall on rows and so on. That is, the sentinel moves along the wall of the grid in the clockwise direction.

Theorem 4. *Algorithm* A_1 *solves the PGE problem using four oblivious asynchronous robots with visibility range two and a common chirality.*

Proof Outline. We first observe that A_1 is well-defined. Next, in the remaining of the proof we only consider a synchronous scheduler, first to prove that A_1

Fig. 6. Sequence of configurations during a change of row and a U-turn.

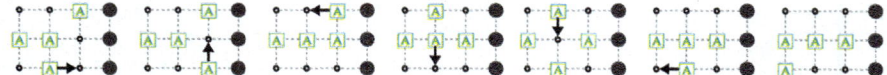

Fig. 7. Sequence of configurations during a change of row for the sentinel.

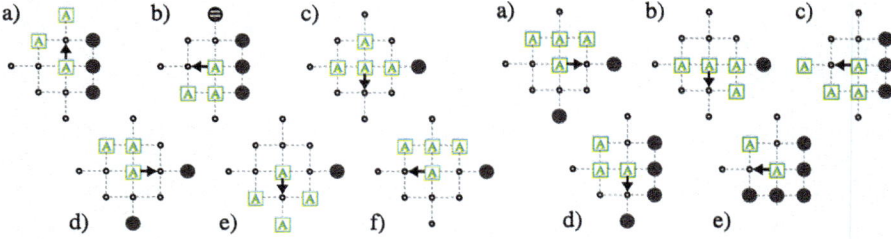

Fig. 8. Rules to move the sentinel to the next row to be explored.

Fig. 9. Rules to switch to the columns exploration.

is synchronous-sequential and then to show that A_1 solves the problem under a synchronous scheduler. These two claims are shown by induction on $n \times m$, where n is the number of lines and m is the number of columns of the grid ($n.b.$, the base case is established using our simulator). Hence, by Theorem 1, this covers the case of an asynchronous scheduler as well.   □

4.2   Algorithm with 3 2-Color Robots Without Common Chirality

We now present an algorithm, denoted by A_2, that uses three robots, each having lights of two colors, and assumes visibility range two. Note that, this time, we do not make any assumption on chirality.[2]. The exploration is similar to Algorithm

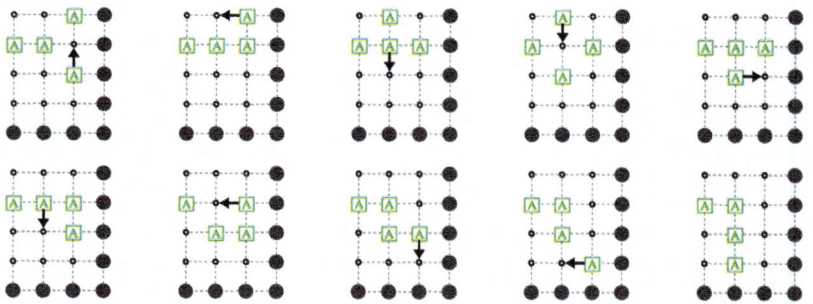

Fig. 10. Sequence of configurations during rows/columns exploration switch.

---

[2] An interactive simulation of A_2 is given at https://robots.app.bramas.fr/?SSS2024/1.

A_1, except that here two robots are used to explore a line, and these robots move back on the same line. The two exploring robots are called the *leader*, with color A, and the *follower*, with color B. They explore the grid row by row, going from one wall to the other and coming back on the same row. The third robot (of color A) plays the role of a *sentinel*. It remains at one side of the row being explored and points to the next row. The rows are explored in one direction, *e.g.*, initially from top to bottom, then the robots do a quarter turn at a corner and continue their exploration column by column, *e.g.*, from left to right, and so on.

When exploring a line, the leader and the follower move in a straight line in two steps, using the rules of Fig. 11: the leader (robot A) first moves away from its follower, then robot B follows. When they reach the opposite wall, they switch their role and colors, using the rules of Fig. 12. The leader changes its color to B, then the follower gets Color A. When their roles are switched, they start moving along the row in the opposite direction.

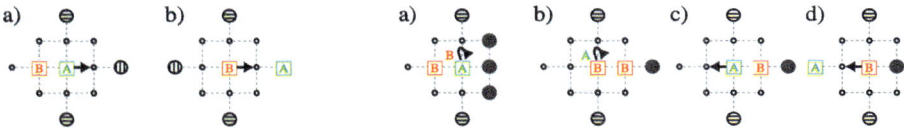

Fig. 11. Rules to move straight.   **Fig. 12.** Rules to turn around on the same row.

When they reach back the first wall, the robots, along with the sentinel, move to the next row in the direction pointed by the sentinel, using the rules of Fig. 13. The sequence of configurations during this process is illustrated in Fig. 14. After changing rows, the new leader and follower start to explore the new row as previously.

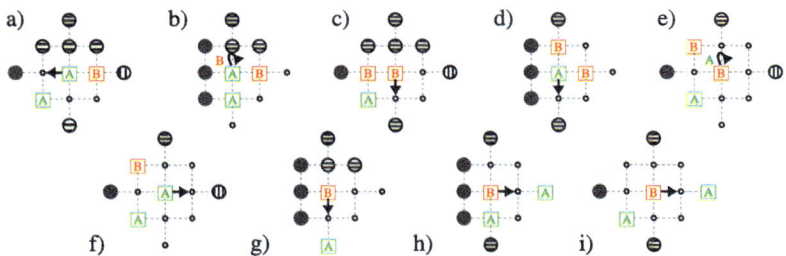

Fig. 13. Rules to perform a turn and a change of row.

When the robots have explored the penultimate row, they cannot perform the same steps as before to move to the next row since the sentinel cannot move forward. Thus, the robots perform a quarter turn and change the direction of

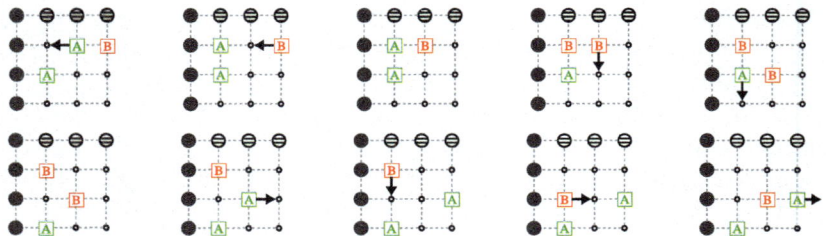

Fig. 14. Sequence of configurations during a change of row.

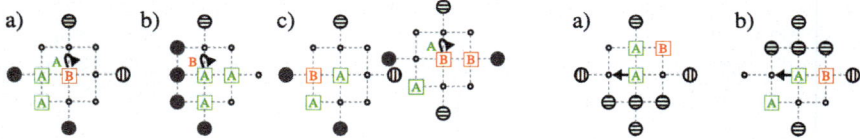

Fig. 15. Rules to perform a quarter turn and change the direction of exploration on the last row.

Fig. 16. Additional rule for 3×3 grids.

Fig. 17. Rules to reach a wall from the initial configuration.

exploration: they will now explore the grid column by column, *e.g.*, from left to right if they were going from top to bottom previously. The rules for the quarter turn are shown in Fig. 15 and the sequence of configurations reached during this process is given in Fig. 18. After the turn, the new leader and follower start to explore the second column (ordered from right to left) as previously. Notice that for grids with one side of size 3 (*e.g.*, 3×3 or 3×4), the algorithm requires an additional rule; see Fig. 16. Indeed, when the robots do the U-turn at the end of the row, the former follower sees the sentinel. This does not happen in larger grids. Nonetheless, the exploration follows the same principle.

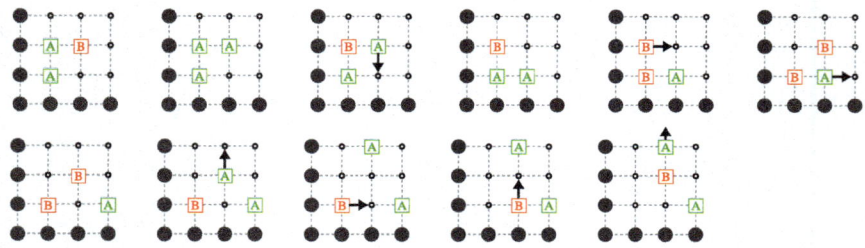

Fig. 18. Sequence of configurations during a quarter turn when robots are on the last row and change the direction of exploration.

Initial configurations. Initially, the three robots form an "L"-shape: one robot (the future sentinel) has color A and is adjacent to another robot with color

A (the future leader), which is adjacent to the third robot with color B (the future follower); as shown in Fig. 19. The three robots can be anywhere in the grid as long as they respect these relative positions. Hence, the set of initial configurations is locally-defined.

The robots will move in the direction of the wall pointed by the two robots of color A, using the rules in Fig. 17. The three following steps are repeated until reaching the wall: the sentinel moves on its row, away from the two other robots. Then, the leader moves in the same direction and the follower takes its place. When they reach the wall, they are in the same position as when they move to the next row in the exploration part. Thus, the exploration begins.

Theorem 5. *Algorithm* A_2 *solves the PGE problem using three asynchronous robots equipped with two colors and visibility range two.*

4.3   Algorithm with 3 Oblivious Robots Under Visibility Range 3 Without Common Chirality

Finally, we present an algorithm, denoted by A_3, that uses three oblivious robots and assumes a visibility range three. Again, we make no assumption on the chirality.[3] Contrary to the previous algorithms, the robots only explore the grid row by row, without switching to the column exploration. Moreover, no "sentinel" remains adjacent to a wall. The three robots explore the grid in a "snake" shape: initially the *leader* is alone on its row and column; the two *followers* are on the adjacent row, one of them being on a diagonal to the leader; as shown in Fig. 21. The three robots can be initially anywhere in the grid as long as they respect these relative positions. Hence the set of initial configurations is locally-defined. The rows are explored in one direction, *e.g.*, from top to bottom, then the robots switch directions, *e.g.*, from bottom to top.

When exploring a row, the leader and its followers move straight from one wall to the other using the rules of Fig. 20. The leader moves away from its followers. Then, the followers follow, the closest one first, the other afterward. Upon reaching the opposite wall, they perform a turn to start exploring the next row, on the opposite side from the leader, using the rules of Fig. 20. The first follower moves to the row on the opposite side from its leader. The leader follows and then, the second follower (that did not move) becomes the new leader and conversely. They start exploring the row in the opposite direction (*e.g.*, from left to right if they were exploring from right to left previously).

When the robots end the exploration of the last row and reach a corner they perform a turn in order to start exploring the rows in the opposite direction, using the rules of Fig. 23. After three moves, the second follower becomes the leader, the leader becomes the first follower and the first follower becomes the second follower. Then, the robots start the exploration of the grid in the opposite direction (*e.g.*, from bottom to top if they were exploring the rows from top to

---

[3] An interactive simulation of A_3 is given at https://robots.app.bramas.fr/?SSS2024/2.

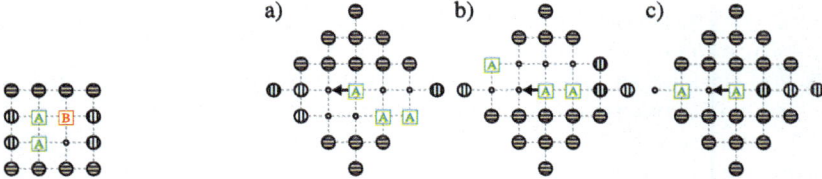

Fig. 19. Locally-defined initial configurations of A₂.

Fig. 20. Rules to move in a straight line.

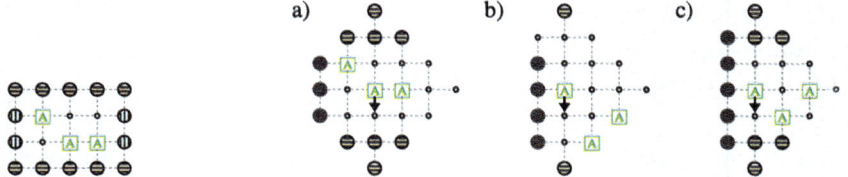

Fig. 21. Locally-defined initial configuration of A₃.

Fig. 22. Rules to move to the next row.

bottom previously). Notice that the robots have to do some special moves when turning after exploring the first row along the wall; see rules of Fig. 24.

Theorem 6. *Algorithm A₃ solves the PGE problem using three asynchronous oblivious robots with visibility range three.*

5 Conclusion

We have presented three grid exploration algorithms for swarms of luminous robots. Depending on how many they are, how many colors they have, their visibility range, and whether they share a common chirality, one can choose the most appropriate algorithm. We have proven that our algorithms are optimal as the problem becomes not solvable by weakening any of the assumptions. However, some improvements can still be made in future work. For instance, our first algorithm is not locally-defined, precisely if the robots are deployed in

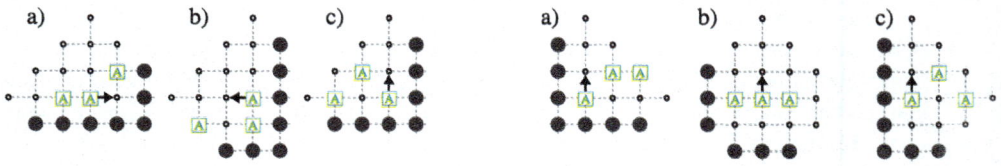

Fig. 23. Rules to turn in a corner.

Fig. 24. Rules to turn after exploring the first row.

the center of the grid, the algorithm does not work. Then, our second algorithm assumes that a robot can see another robot even if it is behind a third one (i.e., we do not assume opacity). This might not be the case in practice and designing an optimal algorithm that works in this case is a challenging open problem.

References

1. Blin, L., Milani, A., Potop-Butucaru, M., Tixeuil, S.: Exclusive perpetual ring exploration without chirality. In: DISC 2010
2. Bonnet, F., Milani, A., Potop-Butucaru, M., Tixeuil, S.: Asynchronous exclusive perpetual grid exploration without sense of direction. In: OPODIS 2011
3. Bramas, Q., Devismes, S., Durand, A., Lafourcade, P., Lamani, A.: Beedroids: how luminous autonomous swam of UAVs can save the world? In: FUN 2022
4. Bramas, Q., Devismes, S., Lafourcade, P.: Finding water on poleless using melomaniac myopic chameleon robots. In: FUN 2020
5. Bramas, Q., Devismes, S., Lafourcade, P.: Infinite grid exploration by disoriented robots. In: NETYS 2020
6. Bramas, Q., Lafourcade, P., Devismes, S.: Optimal exclusive perpetual grid exploration by luminous myopic opaque robots with common chirality. Theor. Comput, Sci (2023)
7. Datta, A.K., Lamani, A., Larmore, L.L., Petit, F.: Enabling ring exploration with myopic oblivious robots. In: IPDPS (2015)
8. Datta, A.K., Lamani, A., Larmore, L.L., Petit, F.: Ring exploration by oblivious agents with local vision. In: ICDCS (2013)
9. Devismes, S., Lamani, A., Petit, F., Raymond, P., Tixeuil, S.: Terminating exploration of A grid by an optimal number of asynchronous oblivious robots. Comput. J. (2021)
10. Devismes, S., Lamani, A., Petit, F., Tixeuil, S.: Optimal torus exploration by oblivious robots. Computing (2019)
11. Devismes, S., Petit, F., Tixeuil, S.: Optimal probabilistic ring exploration by semisynchronous oblivious robots. Theor. Comput, Sci (2013)
12. Flocchini, P., Ilcinkas, D., Pelc, A., Santoro, N.: Remembering without memory: tree exploration by asynchronous oblivious robots. Theor. Comput. Sci. **411** (2010)
13. Flocchini, P., Ilcinkas, D., Pelc, A., Santoro, N.: How many oblivious robots can explore a line. Inf. Process. Lett. **111**(20), 1027–1031 (2011)
14. Flocchini, P., Ilcinkas, D., Pelc, A., Santoro, N.: Computing without communicating: ring exploration by asynchronous oblivious robots. Algorithmica (2013)
15. Kamei, S., Tixeuil, S.: An asynchronous maximum independent set algorithm by myopic luminous robots on grids. Comput. J. **67**(1) (2024)
16. Nagahama, S., Ooshita, F., Inoue, M.: Terminating grid exploration with myopic luminous robots. Int. J. Netw. Comput. **12**(1) (2022)
17. Ooshita, F., Tixeuil, S.: Ring exploration with myopic luminous robots. In: SSS (2018)
18. Rauch, A., Bramas, Q., Devismes, S., Lafourcade, P., Lamani, A.: Optimal exclusive perpetual grid exploration by luminous myopic robots without common chirality. In: NETYS

Gathering of Robots in Butterfly Networks

Serafino Cicerone[1]([✉]), Alessia Di Fonso[1], Gabriele Di Stefano[1],
and Alfredo Navarra[2]

[1] Dipartimento di Ingegneria e Scienze dell'Informazione e Matematica, Università
degli Studi dell'Aquila, 67100 L'Aquila, Italy
{serafino.cicerone,gabriele.distefano}@univaq.it,
alessia.difonso@graduate.univaq.it
[2] Dipartimento di Matematica e Informatica, Università degli Studi di Perugia,
06123 Perugia, Italy
alfredo.navarra@unipg.it

Abstract. Robots with very weak capabilities placed on the vertices
of a graph are required to move toward a common vertex from where
they do not move anymore. The task is known as the *Gathering* problem and it has been extensively studied in the last decade with respect
to both general graphs and specific topologies. Most of the challenges
faced are due to possible isometries observable from the placement of
the robots with respect to the underlying topology. Rings, Grids, Complete graphs are just a few examples of very regular topologies where the
placement of the robots and suitable movements are crucial for succeeding in Gathering. Here we are interested in understanding what can be
done in Butterfly graphs where really many isometries are present and
most importantly unavoidable by any movement. We propose a Gathering algorithm for the so-called *leader* configurations, i.e., those where
the initial placement of the robots admits the detection (and election)
of one robot as the leader. We introduce a non-trivial technique to elect
the leader which is of its own interest. We also prove that the proposed
Gathering algorithm is asymptotically optimal in terms of synchronous
rounds required.

Keywords: Mobile robots · Synchrony · Gathering · Butterfly · Time
complexity

The work has been supported in part by the Italian National Group for Scientific
Computation (GNCS-INdAM) and by the Italian Ministry of Economic development (MISE) under the project "SICURA—Casa intelligente delle tecnologie per la
sicurezza", CUP C19C200005200004.

T. Masuzawa et al. (Eds.): SSS 2024, LNCS 14931, pp. 106–120, 2025.
https://doi.org/10.1007/978-3-031-74498-3_7

1 Introduction

A swarm of very weak—in terms of capabilities—robots, placed on the vertices of an anonymous graph, are required to move toward a common vertex (not known in advance) from where they do not move anymore. Initially, each vertex is occupied by at most one robot. The task is known in the literature as the *Gathering* and it has been extensively studied in the last decade with respect to many different assumptions on the robots' capabilities. The common theoretical approach is to detect the minimal capabilities required to solve the task.

In this paper, we consider the $\mathcal{OBLOT}$ model [19,20], where synchronized robots are assumed to be: *Anonymous*: no unique identifiers; *Autonomous*: no centralized control; *Dimensionless*: no occupancy constraints, no volume; *Oblivious*: no memory of past events; *Homogeneous*: they all execute the same *deterministic* algorithm; *Silent*: no means of direct communication; *Disoriented*: no common coordinate system, nor left-right orientation.

Each robot continuously executes a Look-Compute-Move (LCM) cycle by performing the following three operations in sequence:

- Look: the robot observes the environment. The result of this phase is a snapshot of the positions of all robots with respect to the local perception (and not to a global reference system);
- Compute: the robot executes the designed algorithm to determine its next move, using only the data sensed in the Look phase as input. For robots moving in graphs, the result of this phase either is the vertex where the robot currently resides or it is a vertex among those in the direct neighborhood (i.e., at most one edge per cycle can be traversed);
- Move: the robot moves toward the computed target. If the target is the current position, then the robot performs what is called a *nil* movement.

The time can be logically divided into global rounds. In each round, all the robots perform a LCM cycle. If $C(t)$ denotes the configuration observed by the robots at time t during their Look phase, then an *execution* of an algorithm $\mathcal{A}$ from an initial configuration C is a sequence of configurations $\mathbb{E} : C(0), C(1), \ldots,$ where $C(0) = C$ and $C(t+1)$ is obtained from $C(t)$ by moving each robot according to the result of the Compute phase as implemented by $\mathcal{A}$. On graphs, moves are always considered as *instantaneous*. This results in always perceiving robots on vertices and never on edges during Look phases. The *time complexity* of an algorithm $\mathcal{A}$ is the maximum amount of rounds required by $\mathcal{A}$ for processing any initial configuration.

It is very common (as dictated by impossibility results) that in combination with the LCM-model, robots are endowed with the so-called *multiplicity detection* capability (see e.g. [9,24]). Basically, when more than one robot resides on the same vertex x, then x is said to be occupied by a *multiplicity*. During the Look phase, robots can perceive *multiplicities*. The multiplicity detection capability might be *local* or *global*, depending on whether the multiplicity is detected only by robots composing the multiplicity itself or by any robot performing the Look phase, respectively. Moreover, the multiplicity detection can be *weak* or *strong*, depending on whether a robot can detect only the presence of a multiplicity or if it perceives the exact number of robots composing the multiplicity, respectively.

Related work The Gathering problem has been fully characterized for robots moving on the Euclidean plane [9] endowed with global weak multiplicity detection. The problem is a special case of the more general *pattern formation* problem [2,3,6,8,26].

In contrast, on graphs, it is difficult to envision general results as different challenges arise for different topologies. Apart from some impossibility results or basic conditions that guarantee the resolution of the Gathering task provided in [5,18], most of the literature focuses on specific topologies, like: Trees [10,11, 18]; Rings [12–16,18,23]; Regular Bipartite graphs [21]; Finite Grids [10]; Infinite Grids [17]; Tori [22]; Oriented Hypercubes [1]; Complete graphs [5]; Complete Bipartite graphs [5]. See [4] for a recent survey.

Most of these topologies are very symmetric, that is, vertices can be partitioned into a few classes of equivalence. The choice has been dictated by the requirement to make the design of a resolution algorithm harder as robots cannot exploit many topological properties.

Our results In this paper, we focus on synchronous robots endowed with global weak multiplicity detection, moving on Butterfly graphs.

Moving along Butterflies turns out to be really difficult as such graphs admit a lot of isometries. Moreover, even when some robots are distinguishable because of their positioning with respect to other robots, still some vertices remain *equivalent*, i.e., indistinguishable. From [18], we know that on *partitive* configurations, Gathering is unsolvable. Intuitively, a configuration is said to be partitive if its vertices can be partitioned into subsets of the same cardinality greater than one, where each subset contains equivalent vertices also with respect to the occupancy of the robots. In this paper, we propose a Gathering algorithm for the so-called *leader* configurations, i.e., those where the initial placement of the robots admits detecting (and electing) one robot as the leader. We introduce a non-trivial technique to elect the leader which is of its own interest. We also prove that the proposed Gathering algorithm is asymptotically optimal in terms of required rounds. Proofs are omitted to due space constraints.

2 Notation and Basic Concepts

A d-dimensional Butterfly $BF(d)$ (e.g., see [25]) is an undirected graph with vertices $[\ell, c]$, where $\ell \in \{0, 1, \ldots, d\}$ is the *layer* and $c \in \{0, 1\}^d$ is the *column*. The vertices $[\ell, c]$ and $[\ell', c']$ are adjacent if $|\ell - \ell'| = 1$, and either $c = c'$ or c and c' differ precisely in the ℓ-th bit. $BF(d)$ has $d + 1$ layers with 2^d vertices at each layer, which gives $(d+1) \cdot 2^d$ vertices in total. The vertices at layer 0 and d have degree 2 whereas the rest of the vertices have degree 4. We call layer 0 and d *boundary layers*. $BF(d)$ has two standard graphical representations, namely normal and diamond, see Fig. 1. Note that, $BF(d)$ can be also recursively built by using two copies of $BF(d-1)$ along with 2^d additional vertices, those forming layer 0 (or layer d). This observation leads to the following useful property (cf. diamond representation in Fig. 1):

Property 1. Let V_1 and V_2 be the sets of vertices of the two copies of $BF(d-1)$ that, along with vertices at layer 0 (level d, resp.), form $BF(d)$. If $u \in V_1$ and $v \in V_2$, then each u, v-shortest path contains a vertex at layer 0 (layer d, resp.).

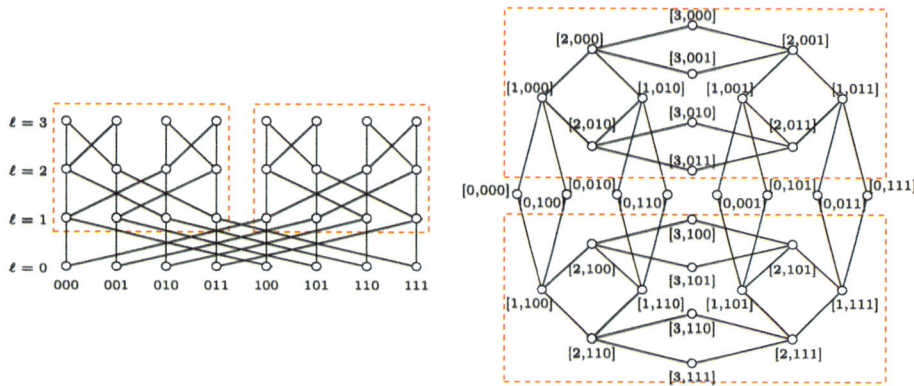

Fig. 1. Normal representation and diamond representation of $BF(3)$. The dashed rect-angles contains two copies of $BF(2)$ that, along with 2^3 additional vertices for layer 0, form $BF(3)$.

Configurations We consider configurations of robots moving in a d-dimensional Butterfly. As robots are assumed to be endowed with global weak multiplic-ity detection, a configuration is modeled as $C = (BF(d), \lambda)$, where $\lambda : V \rightarrow \{0, 1, M\}$ is a function that specifies if each vertex is *unoccupied*, *occupied* by one robot or by more than one robot (that is a multiplicity), respectively.

When the Gathering problem is concerned, two additional concepts are use-ful: C is *initial* if $\lambda(v) \in \{0, 1\}$ for each $v \in V$ (each robot lies on a different ver-tex) and $\sum_{v \in V} \lambda(v) \geq 2$ (two or more robots are required otherwise the problem is trivially solved); it is *final* if $\exists u \in V : \lambda(u) = M$ and $\lambda(v) = 0$, $\forall v \in V \setminus \{u\}$ (all the robots are on a single vertex). A *Gathering algorithm* for this problem is a deterministic distributed algorithm that, starting from any initial configuration and regardless of the adversary,[1] brings all the robots to a final configuration in a finite number of LCM-cycles, i.e., in a finite number of *rounds*. Formally, an algorithm $\mathcal{A}$ solves the Gathering problem for an initial configuration C if, for any execution $\mathbb{E} : C = C(0), C(1), C(2), \ldots$ of $\mathcal{A}$, there exists a time instant $t > 0$ such that $C(t)$ is final and no robots move after t, i.e., $C(t') = C(t)$ holds for all

---

[1] As usual in $\mathcal{OBLOT}$, the role of the adversary basically determines the worst case scenario. In our context, this will be better clarified in the subsequent sections.

$t' \geq t$. Given an initial configuration C, if there exists a Gathering algorithm for C we say that C is *gatherable*, otherwise, we say that C is *ungatherable*.

Note that any Gathering algorithm cannot exploit the labeling shown in Fig. 1 but can only use topological properties and function λ.

Automorphisms and ungatherable configurations Two undirected graphs $G = (V, E)$ and $G' = (V', E')$ are *isomorphic* if there is a bijection φ from V to V' such that $\{u, v\} \in E$ if and only if $\{\varphi(u), \varphi(v)\} \in E'$. An *automorphism* on a graph G is an isomorphism from G to itself. The set of automorphisms of a given graph, under the composition operation, forms the automorphism group of the graph denoted by $\text{Aut}(G)$. Two vertices u and v of G are *equivalent vertices* if there exists an automorphism $\varphi \in \text{Aut}(G)$ such that $\varphi(u) = v$. The equivalence classes of the vertices of G under the action of the automorphisms are called vertex *orbits*. Concerning Butterfly graphs, it is easy to observe that each orbit O_i of $BF(d)$ is composed of *complementary* layers i and $d-i$, with $0 \leq i \leq \lfloor d/2 \rfloor$. By *intermediate orbit* we mean $O_{d/2}$ when d is even. Notice that the intermediate orbit is formed by one layer only while the other orbits are formed by two layers each; moreover, O_0 corresponds to the union of the boundary layers.

The concept of isomorphism can be easily extended to configurations. In our context, $(BF(d), \lambda)$ and $(BF'(d), \lambda')$ are isomorphic if there exists $\varphi \in \text{Aut}(BF(d))$ such that $\lambda(v) = \lambda'(\varphi(v))$ for each vertex v. As for graphs, $\text{Aut}(C)$ is the automorphism group of $C = (BF(d), \lambda)$. In C, two robots r and r' are *equivalent robots* if they reside on vertices that are equivalent according to some automorphism of $\text{Aut}(C)$.

Observation 1. *Let $C = (BF(d), \lambda)$ be an initial configuration. A necessary condition for two robots r and r' to be equivalent in C is that these robots are located at the same orbit O_i, for some $0 \leq i \leq \lfloor d/2 \rfloor$.*

From an algorithmic point of view, it is important to remark that when there are equivalent vertices in a configuration, a robot r located at a vertex u cannot distinguish its position from that of a robot r' located at vertex $v = \varphi(u)$. As a consequence, no algorithm can distinguish between two equivalent robots, and then it cannot avoid that the two robots perform the same move simultaneously. Moreover, if equivalent robots are moved to any orbit of C containing more than one vertex, then no algorithm can specify the destination vertices exactly (in this case, being the destination vertices pairwise equivalent, we assume that the actual destinations are chosen by an adversary).

A configuration is said to be *partitive* if it admits an automorphism such that the vertices can be partitioned into subsets of equal size greater than one, with equivalent vertices within each subset. From [18], we know that any partitive configuration is ungatherable. To classify all the initial configurations, we introduce the following novel concept.

Definition 1. *Let $C = (BF(d), \lambda)$ be any configuration. A robot r located on a vertex v of $BF(d)$ is called potential-leader (p-leader for short) if $\lambda(v) = 1$ and $\varphi(v) = v$ for each $\varphi \in \text{Aut}(C)$.*

According to Observation 1, if O_i contains just one robot r, then r is a p-leader. We call *leader configuration* any initial configuration C in which it is possible to elect a leader. In what follows, we show that if C is initial and contains p-leaders then it is a leader configuration.

3 Leader Election

In this section, we provide an algorithm that allows robots to elect a leader in any initial configuration admitting p-leaders. The algorithm is based on an embedding of the Butterfly on the Euclidean plane. The embedding is unique up to isomorphisms.

Embedding of a Butterfly The embedding of $BF(d)$ in the Euclidean plane starts from a *root* vertex s and then proceeds level by level, where all the vertices at distance i from s in $BF(d)$ are embedded at the i-th level. In this way, we introduce a kind of *rooted representation* of $BF(d)$. We first describe Hang (cf. Algorithm 1), the algorithm that embeds $BF(d)$ by using a vertex of degree two (i.e., a vertex $s \in O_0$) as root. Then, we show how this algorithm can be adapted to embed the graph using a vertex of degree four (i.e., a vertex $s \notin O_0$) as root. The algorithm uses the following notation: for each vertex v in the Butterfly, v_ℓ and v_c represent the layer and the column, respectively, of v (hence $v = [v_\ell, v_c]$). Given a non-negative integer i, by $b_d(i)$ we denote the binary representation of i with d binary digits.

Algorithm Hang takes the dimension d of the Butterfly as input. The coordinates of the vertices of the graph are defined so that a Butterfly of dimension d is built by making two copies of a Butterfly of dimension $d-1$, one having positive x coordinates and the other with negative x coordinates. A space of width 2^d is left between the two copies of the Butterflies to place the additional vertices of layer d. Let V be the set of vertices of the graph and let X and Y be the sets of x- and y-coordinates of the vertices in V. The algorithm is recursive. The base case (see lines 2–4) defines a Butterfly of dimension 0, with a single vertex having layer 0 and an empty string as column. Both of its Cartesian coordinates are set to 0. At line 6, the function is invoked recursively over the dimension $d-1$. The call returns the set V of the vertices of the graph $BF(d-1)$ along with their coordinates X and Y. The algorithm builds two copies V_1 and V_2 of the vertices of the Butterfly of dimension $d-1$. Let v_c be the column associated to the vertices in V, the algorithm updates the column of the vertices in V_1, V_2 by concatenating to "0", "1" to v_c respectively, see lines 9 and 12. Then, the x coordinates of the vertices in V_1 (V_2, resp.) are shifted to the right (left, resp.) by 2^{d+1} with respect to the x coordinates of the vertices in V. Let V_3 be the set of 2^d vertices of layer d of the Butterfly $BF(d)$. These vertices are placed between the two Butterflies of dimension $d-1$ and evenly distributed horizontally over an interval of 2^d, half with positive, half with negative x coordinates, see line 17.

This guarantees that, at the end of the embedding algorithm, all the vertices have different x-coordinates. As for the y coordinates, vertex $[d, 0^d]$ is placed at height d, see line 18, while the other $2^d - 1$ vertices are placed at a height that

---

Algorithm 1: Hang(d)

---

1 **if** $d = 0$ **then**
2 $\quad$ $V = \{[0, \text{' '}]\};$
3 $\quad$ $X([0, \text{' '}]) = 0;$
4 $\quad$ $Y([0, \text{' '}]) = 0;$
5 $\quad$ **return** V, X, Y ;

6 $V, X, Y = $ Hang($d - 1$);
7 $V_1, V_2, V_3 = \emptyset$
8 **foreach** $v = [v_\ell, v_c] \in V$ **do**
9 $\quad$ insert $v_1 = [v_\ell, v_c +' 0']$ into $V_1;$
10 $\quad$ $X'(v_1) = X(v) + 2^{d+1};$
11 $\quad$ $Y'(v_1) = Y(v);$
12 $\quad$ insert $v_2 = [v_\ell, v_c +' 1']$ into $V_2;$
13 $\quad$ $X'(v_2) = X(v) - 2^{d+1};$
14 $\quad$ $Y'(v_2) = Y(v);$

15 **foreach** $i \in \{0, 2^d - 1\}$ **do**
16 $\quad$ insert $v = [d, b_d(i)]$ into V_3 ;
17 $\quad$ $X'(v) = -(2^{d-1} - 1/2) + i$;

18 $Y'([d, b_d(0)]) = d;$
19 **foreach** $j \in \{0, d - 1\}$ **do**
20 $\quad$ **foreach** $i \in \{2^j, 2^{j+1} - 1\}$ **do**
21 $\quad\quad$ $Y'([d, b_d(i)]) = d - 2(j + 1);$

22 **return** $V_1 \cup V_2 \cup V_3, X', Y'$;

---

corresponds to their distance from $[d, 0^d]$, see line 21. At the end of the process, root s will result in a vertex of degree two assigned with label $[d, 0^d]$. Figure 2.left shows a $BF(3)$ graph rooted at vertex $s = [0, 000]$, s has y-coordinate equal to three. Vertices of layer three are arranged between two Butterflies of dimension two at a height that corresponds to their distance from s. Vertices of layer zero are arranged at level $y = 0$.

We recall that the described embedding uses a degree-two vertex s as root and arbitrarily assigns label $[d, 0^d]$ to it (note that all the degree-two vertices form orbit O_0 and hence they are pairwise equivalent). Assume now that an embedding with a degree-four vertex s' as root is needed, and that such a vertex belongs to O_i, with $0 < i \leq \lfloor d/2 \rfloor$. In this case, the embedding algorithm first selects s' as the vertex $[d - i, 0^d]$ and then changes the embedding realized by Hang(d) by updating the y-coordinate of each embedded vertex v as follows: $Y(v) = Y(s') - d(s', v)$. This yields the requested embedding of $BF(d)$ with s' as root. Figure 2.right shows a $BF(3)$ rooted at the vertex $s' = [2000]$.

Leader election algorithm Let $C = (BF(d), \lambda)$ be any initial configuration with p-leaders. We describe here minString (cf. Algorithm 2), the algorithm designed to elect a leader among all the possible p-leaders of C. Let r be a p-leader of C and let v be the vertex where r is located. Basically, minString

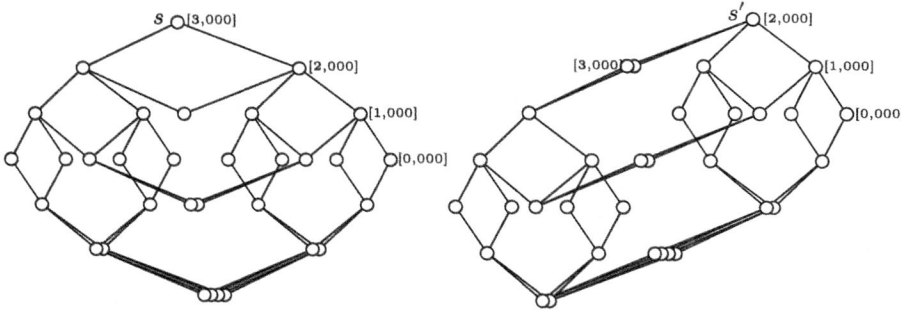

Fig. 2. *(left)* Embedding of a $BF(3)$ rooted at vertex $s = [3000]$. *(right)* Embedding of a $BF(3)$ rooted at vertex $s' = [2000]$.

Algorithm 2: minString(I, v, l)

---

1 Compute $N_l(v)$, the set of vertices at distance l from v produced by Hang(d);

2 **if** $N_l(v) = \emptyset$ **then**

3   **return** $''$;

4 **else**

5 $SL_{min} = min_{\varphi \in I} \, sl(\varphi, v, l)$;

6 $I' = \{\varphi \in I : sl(\varphi, v, l) = SL_{min}\}$;

7   **return** $(SL_{min}) + $ minString$(I', v, l + 1)$;

---

assigns a binary string to r so that the string encodes a visit of the entire configuration C starting from v. We remark that minString is executed concurrently by all robots on each possible p-leader. The p-leader that receives the smallest string is elected as the leader. The visit performed by minString exploits the embedding described in the previous section: in particular, it assumes that a hanging of $BF(d)$ has been computed with a call to Hang(d). The binary string is built in blocks, each block is in correspondence to a level composed of the vertices having the same distance from v. Each value of the string, 1 or 0, corresponds to the presence or the absence, respectively, of a robot on each vertex of the level (read from left to right).

The string associated to r is computed by calling minString$(\text{Aut}(C), v, 0)$. Since r is a p-leader, each automorphism $\varphi \in \text{Aut}(C)$ fixes v, that is $\varphi(v) = v$. During the execution, at lines 2–3, the algorithm checks if there are neighbors of v at level l if not, it returns the empty string. Otherwise, starting from line 5, the algorithm calls the function sl for each automorphism $\varphi \in I$. The function sl works as follows. It takes as input an automorphism φ, the root v of the embedding produced by Hang(d), and a level l of the embedding, and returns the string associated with that level. It is called for each automorphism φ in I, where I is the set of automorphisms that minimize the associated string up to level l. Hence, it finds the minimum string SL_{min} among the strings associated with level l which are returned by each call. At line 6, the algorithm computes the subset I' of I given by all the automorphisms associated with the minimum string. Finally, the algorithm outputs the concatenation of the minimum string and the string obtained by a recursive call on the subset of automorphisms I' and the level $l + 1$.

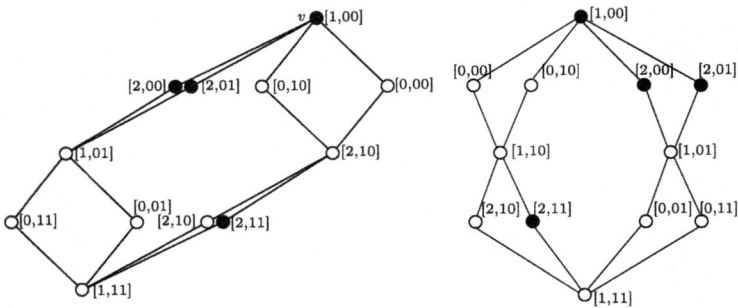

Fig. 3. *(left)* Embedding of a configuration defined on a $BF(2)$ and rooted at vertex $v = [1, 00]$. Black nodes are those occupied by robots. *(right)* An equivalent representation of the same configuration showing its automorphisms at a glance.

For example, in Fig. 3 a p-leader is located on vertex v. The string associated with such a p-leader is computed by calling minString$(I, v, 0)$, where I contains all possible automorphisms. During the execution, $N_l(v) = \{v\}$, $SL_{min} = (1)$, $I' = I$ and the algorithm is called on I', v, and $l = 1$. $N_l(v)$ now contains four vertices. Refer to Fig. 3 (right) to visualize better the automorphisms. The minimum string associated with level one is (0011) while now I' contains all the automorphisms that swap the vertices of the same levels but keep unchanged the minimum string associated with the vertices of level one. On level two, all the automorphisms in I' fix the vertices at that level, and the associated string is (00), while at level three the minimum string associated with that level is (0100). Indeed it is not possible to exchange the vertices of the graph to make the substring smaller at this level as the subset of the automorphisms prevents some exchanges between vertices. The string associated with level four is obviously (0). When the algorithm is called on level five, $N_l(v) = \emptyset$ it terminates returning the empty string. Hence, the minimum string associated to vertex v is $(1)(0011)(00)(0100)(0)$.

Lemma 1. *Let* $C = (BF(d), \lambda)$ *be an initial configuration. If* C *contains p-leaders then robots can elect a leader in* C.

4   Gathering Algorithm

In this section, we formalize an algorithm $\mathcal{A}$ that solves the Gathering for synchronous robots in leader configurations defined on $BF(d)$. We assume $d \geq 2$ since the case $d = 1$ is trivial.

Definition 2. Let $C = (BF(d), \lambda)$, $d \geq 2$, be a configuration. The *leader vertex* v of C is defined as follows:

- if C is initial and contains p-leaders, v corresponds to the vertex where the leader robot identified by Lemma 1 resides;
- if C contains multiplicities, v is leader if O_i is the first orbit of $BF(d)$ with a single occupied vertex, and v corresponds to such a vertex.

We first provide an overview of the strategy behind the provided algorithm. Given an initial configuration C admitting p-leaders, Lemma 1 is used to elect a leader robot r (assume located at O_k) and move all the robots so that (1) r moves to a lower orbit of $BF(d)$ (i.e. O_{k-1}) whereas the other robots leave O_{k-1}. This creates a new configuration C' where, by Observation 1, either r remains leader or r is located on a leader vertex (it depends on possible multiplicities created by robots different from r when they change the orbit and for which the final destination is chosen by the adversary). The same task is repeated until r reaches a boundary layer at O_0 (thus forming C''). Once the leader (vertex) is on O_0, the distance of all the robots from the leader is reduced. A special synchronous movement may be necessary when some robots must pass through O_0 (in particular, through the layer L complementary to the one in which the leader resides) to reach the leader. In such a case, the leader goes to O_1 while the other robots enter L. After that, r returns to O_0 (still preserving the property of occupying a leader vertex) while all the other continue reducing the distance from the leader. After this synchronization, all the robots can safely move toward the leader vertex along shortest paths, until the Gathering is accomplished.

To transform this strategy into an algorithm working under the LCM model, the Gathering problem on $BF(d)$ must be subdivided into a set of very simple tasks where each task can be formalized according to three elements: (a) a precondition that can be evaluated by all robots according to perceived configuration, (b) a subset of robots selected for the movements, and (c) a specification of the move to be performed by each selected robot. The move should be performed by the robots selected in the task whose precondition is evaluated as true. In the remainder of this section, we formalize each task required by $\mathcal{A}$. Clearly, according to the properties holding in the initial configuration, not necessarily all the tasks must be performed.

Task T_1 In this task, the leader robot is located in an internal layer. It moves so that its distance from a boundary layer is reduced and, at the same time, all the other robots move so that the leader remains unchanged.

- *Precondition* $\mathbf{pre}_1$: there exists a leader vertex (say v), v is located in O_k with $k > 0$, there is only one robot (say r) on v;

- *Selected robots:* all the elements of M_1, that is the set containing all robots in O_{k-1}, and the robot r;
- *Move m_1:* r moves to O_{k-1} and all robots in M_1 move to O_k.

Notice that at the end of the first execution of this task, r reduced its distance from a boundary layer. In case this distance is greater than zero, the precondition of this task remains true and hence the task is applied again until r reaches a boundary layer of O_0. When r reaches O_0, the subsequent task T_2 may start.

Task T_2 In this task, the leader vertex v is located in a boundary layer. Let r be the unique robot on v. In case O_0 contains other robots than r, all robots different from the leader (forming set M_2) will move to O_1.

- *Precondition* $\texttt{pre}_2$: there exists a leader vertex (say v), there is only one robot (say r) on v, r resides in O_0, and O_0 contains two or more robots;
- *Selected robots:* all robots in M_2, i.e., all robots in O_0 different from r;
- *Move m_2:* all robots in M_2 move to O_1.

At the end of this task, the leader vertex v results to be the only occupied vertex in O_0. This ensures that v remains the leader vertex (with just r on v) and task T_3 may start.

Task T_3 In this task, all the robots that must pass through O_0 to reach the leader vertex v are involved (cf. Proposition 1). The goal is to minimize their distance from the leader vertex, but without letting them enter O_0 (this is to keep the leader unchanged). To formalize precondition and move, consider the following assumptions and definitions:

- there exists only one occupied vertex (say v) in O_0, and let r be the only robot on v,
- let L'' be the layer of O_1 furthest from v,
- let M_3 be the set of all the robots that must pass through O_0 to reach v.

According to such notation, in this task, the algorithm works as follows:

- *Precondition* $\texttt{pre}_3$: v is the leader vertex occupied only by robot r, and M_3 is not empty;
- *Selected robots:* all the robots in M_3;
- *Move m_3:* each robot $r' \in M_3$ moves to any vertex belonging to a r', r-shortest path not on O_0.

Note that this task may be repeated several times (consecutive rounds during the execution of $\mathcal{A}$) and when the last round is concluded, all the robots in M_3 will be located on L''.

Task T_4 This task starts immediately after the conclusion of T_3 and operates on the same set of robots. The goal is to further reduce the distance of these robots from the leader vertex v, but now a special move is needed to keep always the same leader. Consider the following assumptions and definitions:

- there exists only one occupied vertex (say v) in O_0,
- let L' (L'', resp.) be the layer of O_1 closest to (furthest from, resp.) v,
- let L''' be the layer complementary to the one in which v resides.

According to this notation, in this task, the algorithm works as follows:

- *Precondition* $\mathtt{pre}_4$: v is the leader vertex, r is the only robot on v, all robots that must pass through L''' to reach v are on L'', and L'' is occupied.
- *Selected robots:* r and all robots in O_1.
- *Move* m_4: r moves to O_1, each robot on L' moves to O_2, each robot on L'' moves to L'''.

Task T_5 In this task, the leader robot moves on a vertex in O_0 and robots on O_0 move on vertices in O_1. Consider the following assumptions and definitions:

- there exists only one occupied vertex (say v) in O_1,
- let L' (L'', resp.) be the layer of O_0 closest to (furthest from, resp.) v,
- let M_5 be the set containing all robots on L''.

According to such notation, in this task, the algorithm works as follows:

- *Precondition* $\mathtt{pre}_5$: v is a leader vertex, there is only one robot (say r) on v, there are no robots on L', there are no robots that must pass through O_0 to reach v, and M_5 is not empty;
- *Selected robots:* r and all robots in M_5;
- *Move* m_5: r moves from O_1 to O_0, each robot $r' \in M_5$ moves toward a vertex in O_1 belonging to a r', r-shortest path.

When this task finishes, the leader v is the unique occupied vertex on a boundary level, r is the unique robot on v, and all the remaining robots are at a distance at most $d - 1$ from r.

Task T_6 The algorithm $\mathcal{A}$ finalizes the Gathering by moving all robots on the leader vertex v. Notice that, after the first round in which this task is executed, v may contain a multiplicity. $\mathcal{A}$ works as follows:

- *Precondition* $\mathtt{pre}_6$: there exists a leader vertex (say v) in O_0 and all robots reside on the same sub-Butterfly $BF(d - 1)$ which v belongs to;
- *Selected robots:* all the elements of M_6, i.e., the set containing all robots not located on v;
- *Move* m_6: each robot $r' \in M_6$ selects a vertex along a shortest path to v as destination.

Task T_7 It represents the case in which all robots recognize that the Gathering is accomplished and hence none of them moves.

- *Precondition* $\mathtt{pre}_7$: all the robots reside on a same vertex v, i.e., the configuration is final;
- *Selected robots:* none.

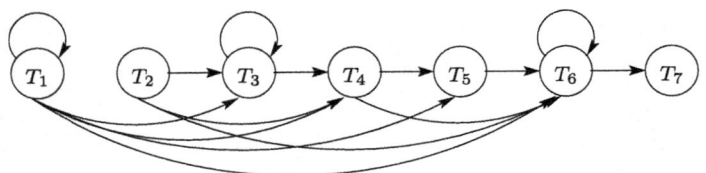

Fig. 4. Transition graph generated by $\mathcal{A}$.

4.1   Correctness and Complexity

The algorithm $\mathcal{A}$ has been designed according to the general methodology proposed in [7]. To detect which task must be accomplished in any configuration observed during an execution, a **predicate** P_i is assigned to task T_i, for each i. P_i is defined as follows: $P_i = \mathrm{pre}_i \wedge \neg(\mathrm{pre}_{i+1} \vee \mathrm{pre}_{i+2} \vee \ldots \vee \mathrm{pre}_7)$.

The predicate is evaluated in the Compute phase, based on the view acquired during the Look phase. As soon as the robots recognize that a task T_i must be accomplished, a subset of designed robots perform the move m_i associated with that task. In the Compute phase, each robot evaluates – for the perceived configuration and the provided input – the predicates starting from P_7 and proceeding in the reverse order until a true precondition is found. In case all predicates $P_7, \ldots, P_3$ are evaluated false, then either T_2 or T_1 is performed. Indeed, one of the two preconditions certainly holds because the algorithm is working on a configuration with p-leaders. Summarizing, the provided algorithm can be used by each robot r in the Compute phase as follows:

- *if r detects that P_i holds, then it applies move m_i associated with T_i.*

Consider now an execution of $\mathcal{A}$, and assume that a task T_i, $i < 7$, is performed for the current configuration C. If $\mathcal{A}$ transforms C into C' and this new configuration has to be assigned the task T_j, then we say that $\mathcal{A}$ can generate a transition from T_i to T_j. The set of all possible transitions of $\mathcal{A}$ determines a directed graph called *transition graph* (cf. Fig. 4). Of course, the transition graph should include a sink vertex corresponding to the scenario where robots recognize that the problem is solved and no move is required.

According to the proposed methodology, in [7] it is shown that the correctness of $\mathcal{A}$ can be obtained by proving that all the following properties hold:

- H_1: for each task T_i, the tasks reachable from T_i through transitions are exactly those represented in the transition graph (i.e., the transition graph is correct);
- H_2: possible cycles in the transition graph (including self-loops) must be performed a finite number of times;
- H_3: unsolvable configurations are not generated by $\mathcal{A}$ (concerning the addressed problem, this means that $\mathcal{A}$ does not generate configurations without a leader vertex).

Lemma 2. *Let* $C = (BF(d), \lambda)$, $d \geq 2$, *be any configuration of* $n \geq 2$ *synchronous robots with p-leaders. During any execution of* $\mathcal{A}$ *on* C, *properties* H_1, H_2 *(limited to self-loops), and* H_3 *hold in each task* T_i, $1 \leq i < 7$

Theorem 1. *If* $C = (BF(d), \lambda)$, $d \geq 2$ *is any initial configuration of* $n \geq 2$ *synchronous robots with p-leaders, then* $\mathcal{A}$ *optimally solves the Gathering problem on* C *in* $\Theta(d)$ *time.*

5 Concluding Remarks

We considered the Gathering problem in the $\mathcal{OBLOT}$ model for synchronized robots moving on Butterfly graphs. We have shown how to deal with leader configurations, i.e., those where it is possible to elect a leader robot. The difficulties provided by the isometries of the Butterfly topology have been addressed by suitably choosing an embedding of the Butterfly on the Euclidean plane. As future work, it is worth investigating what can be done for the configurations that are not leader nor partitive. Other challenges come from choosing different topologies, like Hypercubes (not oriented) or Cube Connected Cycle graphs.

References

1. Bose, K., Kundu, M.K., Adhikary, R., Sau, B.: Optimal Gathering by Asynchronous Oblivious Robots in Hypercubes. In: Gilbert, S., Hughes, D., Krishnamachari, B. (eds.) ALGOSENSORS 2018. LNCS, vol. 11410, pp. 102–117. Springer, Cham (2019). https://doi.org/10.1007/978-3-030-14094-6_7
2. Cicerone, S., Di Fonso, A., Di Stefano, G., Navarra, A.: Arbitrary pattern formation on infinite regular tessellation graphs. Theor. Comput. Sci. **942**, 1–20 (2023). https://doi.org/10.1016/j.tcs.2022.11.021
3. Cicerone, S., Di Stefano, G., Navarra, A.: Asynchronous arbitrary pattern formation: the effects of a rigorous approach. Distributed Comput. **32**(2), 91–132 (2019). https://doi.org/10.1007/S00446-018-0325-7
4. Cicerone, S., Di Stefano, G., Navarra, A.: Asynchronous robots on graphs: gathering. In: Flocchini, P., Prencipe, G., Santoro, N. (eds), Distributed Computing by Mobile Entities, Current Research in Moving and Computing, vol. 11340 of *LNCS*, pp. 184–217. Springer (2019). https://doi.org/10.1007/978-3-030-11072-7_8
5. Cicerone, S., Di Stefano, G., Navarra, A.: Gathering robots in graphs: The central role of synchronicity. Theor. Comput. Sci. **849**, 99–120 (2021). https://doi.org/10.1016/j.tcs.2020.10.011
6. Cicerone, S., Di Stefano, G., Navarra, A.: Solving the pattern formation by mobile robots with chirality. IEEE Access **9**, 88177–88204 (2021). https://doi.org/10.1109/ACCESS.2021.3089081
7. Cicerone, S., Di Stefano, G., Navarra, A.: A structured methodology for designing distributed algorithms for mobile entities. Inf. Sci. **574**, 111–132 (2021). https://doi.org/10.1016/j.ins.2021.05.043
8. Cicerone, S., Di Stefano, G., Navarra, A.: Embedded pattern formation by asynchronous robots without chirality. Distrib. Comput. **32**(4), 291–315 (2019). https://doi.org/10.1007/S00446-018-0333-7

9. Cieliebak, M., Flocchini, P., Prencipe, G., Santoro, N.: Distributed computing by mobile robots: Gathering. SIAM J. on Computing **41**(4), 829–879 (2012)
10. D'Angelo, G., Di Stefano, G., Klasing, R., Navarra, A.: Gathering of robots on anonymous grids and trees without multiplicity detection. Theor. Comput. Sci. **610**, 158–168 (2016)
11. D'Angelo, G., Di Stefano, G., Navarra, A.: Gathering asynchronous and oblivious robots on basic graph topologies under the look-compute-move model. In: Search Theory: A Game Theoretic Perspective, pp. 197–222. Springer, Berlin (2013)
12. D'Angelo, G., Di Stefano, G., Navarra, A.: Gathering on rings under the look-compute-move model. Distrib. Comput. **27**(4), 255–285 (2014)
13. D'Angelo, G., Di Stefano, G., Navarra, A.: Gathering six oblivious robots on anonymous symmetric rings. J. Discrete Algorithms **26**, 16–27 (2014)
14. D'Angelo, G., Di Stefano, G., Navarra, A., Nisse, N., Suchan, K.: Computing on rings by oblivious robots: a unified approach for different tasks. Algorithmica **72**(4), 1055–1096 (2015)
15. D'Angelo, G., Navarra, A., Nisse, N.: A unified approach for gathering and exclusive searching on rings under weak assumptions. Distrib. Comput. **30**(1), 17–48 (2017)
16. D'Emidio, M., Di Stefano, G., Frigioni, D., Navarra, A.: Characterizing the computational power of mobile robots on graphs and implications for the euclidean plane. Inf. Comput. **263**, 57–74 (2018)
17. Di Stefano, G., Navarra, A.: Gathering of oblivious robots on infinite grids with minimum traveled distance. Inf. Comput. **254**, 377–391 (2017)
18. Di Stefano, G., Navarra, A.: Optimal gathering of oblivious robots in anonymous graphs and its application on trees and rings. Distrib. Comput. **30**(2), 75–86 (2017)
19. Flocchini, P., Prencipe, G., Santoro, N. (eds.): Distributed Computing by Mobile Entities, Current Research in Moving and Computing, volume 11340 of Lecture Notes in Computer Science. Springer, Berlin (2019). https://doi.org/10.1007/978-3-030-11072-7
20. Flocchini, P., Prencipe, G., Santoro, N. (eds.): Distributed Computing by Oblivious Mobile Robots. Synthesis Lectures on Distributed Computing Theory. Morgan & Claypool Publishers (2012)
21. Guilbault, S., Pelc, A.: Gathering asynchronous oblivious agents with local vision in regular bipartite graphs. Theor. Comput. Sci. **509**, 86–96 (2013)
22. Kamei, S., Lamani, A., Ooshita, F., Tixeuil, S., Wada, K.: Asynchronous gathering in a torus. In: 25th International Conference on Principles of Distributed Systems (OPODIS), volume 217 of LIPIcs, pp. 9:1–9:17. Schloss Dagstuhl - Leibniz-Zentrum für Informatik (2021)
23. Klasing, R., Kosowski, A., Navarra, A.: Taking advantage of symmetries: gathering of many asynchronous oblivious robots on a ring. Theor. Comput. Sci. **411**, 3235–3246 (2010)
24. Klasing, R., Markou, E., Pelc, A.: Gathering asynchronous oblivious mobile robots in a ring. Theor. Comput. Sci. **390**, 27–39 (2008)
25. Manuel, P.D., Abd-El-Barr, M.I., Rajasingh, I., Rajan, B.: An efficient representation of Benes networks and its applications. J. Discrete Algorithms **6**(1), 11–19 (2008). https://doi.org/10.1016/j.jda.2006.08.003
26. Suzuki, I., Yamashita, M.: Distributed anonymous mobile robots: formation of geometric patterns. SIAM J. Comput. **28**(4), 1347–1363 (1999)

Brief Announcement: Pebble Guided Rendezvous Despite Fault

Ashish Saxena[1](✉), Barun Gorain[2], Subhrangsu Mandal[3],
and Kaushik Mondal[1]

[1] Indian Institute Of Technology Ropar, Rupnagar, Punjab, India
{ashish.21maz0004,kaushik.mondal}@iitrpr.ac.in
[2] Indian Institute Of Technology Bhilai, Raipur, Chattisgarh, India
barun@iitbhilai.ac.in
[3] Indian Institute Of Technology (ISM) Dhanbad, Dhanbad, Jharkhand, India

Abstract. We consider the rendezvous problem in an anonymous port-labelled connected simple graph. The objective is for two mobile agents to meet at some node of the graph without prior knowledge of the graph or the other agent's position. An oracle, that knows the graph and the starting positions of the agents, helps the agents by placing identical pebbles, at most one per node at some of the nodes. We introduce faults by considering the presence of a single faulty node that may remove a pebble that is kept on the node, or may add a pebble where there was no pebble placed by the oracle. The position of the faulty node is unknown to the agents as well as the oracle. Our goal is to find an efficient rendezvous algorithm regardless of the number of pebbles placed by the oracle in the presence of a faulty node. For trees, we present an algorithm that uses $O(D \log \Delta)$ pebbles and runs in time $O(D \log \Delta)$, where Δ is the maximum node degree and D is the shortest path distance between the initial agent positions. We prove that our algorithm for trees is optimal in terms of time. Additionally, we study the problem in general graphs with the constraint that the initial agent positions are no more than distance three apart. We propose an algorithm using $O(\log \Delta)$ pebbles with run time $O(\log^3 \Delta)$.

Keywords: Mobile agents · Anonymous graphs · Rendezvous · Faults · Deterministic algorithm

1 Introduction

The rendezvous is a well-studied problem in networks and related fields. It has many real-life applications. One such application is where a set of orbital manoeuvres during which two spacecraft, one of which is often a space station, must meet with other spacecraft, and it should be fast due to the limited fuel capacity in the spacecraft. In such scenarios, agents receive external help from an oracle in the form of a binary string called advice [6,7] to guide them for rendezvous. The oracle knows the network and agents' positions. Instead of giving

T. Masuzawa et al. (Eds.): SSS 2024, LNCS 14931, pp. 121–125, 2025.
https://doi.org/10.1007/978-3-031-74498-3_8

large strings to the agents, the objective is to achieve the same goal by providing just 1 bit of information to some nodes in the network. Agents collect these information by visiting nodes and use them for the rendezvous. The oracle may store this 1-bit information in any node by keeping (or not keeping) a pebble on the node.

In real life, faults are inevitable and can occur at some nodes, corrupting the information stored by the oracle i.e., such faulty nodes may remove a pebble that is kept on the node, or it may add a pebble where there was no pebble placed by the oracle. These nodes can be controlled by an adversary who knows the graph and the pebble placement. Therefore, studying the rendezvous problem in the presence of faulty nodes is crucial. Before we discuss the formal model and problem definition, let's review existing works on this topic.

1.1 Related Work

Rendezvous has been extensively studied in recent decades. If one of the agents does not move, then this problem is known as a treasure hunt problem. Treasure hunt by a mobile agent is a well-studied problem. The algorithm with advice paradigm has been primarily studied in networks, where the introduction of this model [4] improved the existing solution efficiency. The advice is in the form of either a string or identical pebbles. Gorain et al. [5] studied the treasure hunt problem in the graphs with pebbles and provided a lower bound on the run time using any number of pebbles. Later, Bhattacharya et al. [1] studied an efficient algorithm for a given number of pebbles, a general version of [5]. There is another work that studies the problem in the presence of faulty advice [3]. The model in [3] is the closest to our model. In [3], Dieudonné et al. studied the exploration in the presence of Byzantine tokens. These tokens are nothing but stationary pebbles. In the Byzantine token model of [3], if an agent visits a node when there is a token, an adversary may remove this token, however, if there is no token at some node, an adversary can not place a token. In our model, we provide more power to the adversary. It can place a pebble at a node where there is no pebble placed by the oracle, as well as remove an already placed pebble. In this work, we give an algorithm to solve a rendezvous problem using any number of pebbles in the presence of a faulty node.

1.2 Model and the Problem

The network is modelled as an undirected anonymous port-labelled simple graph, denoted as $G = (V, E)$, where V and E denote the sets of nodes and edges, respectively. The nodes of the graph are anonymous, and the edges at a node v of degree $deg(v)$ are labelled as $0, 1, \ldots, deg(v) - 1$ in an arbitrary order. More specifically, each undirected edge (u, v) has two labels, denoted as its port numbers at u and v. Port numbering is local, i.e., there is no relationship between the port numbers at u and v. Note that on an anonymous graph, without port numbers, edges belonging to a node would be indistinguishable to the agents, so rendezvous would often be impossible [2].

There are two agents present at some nodes in graph G. These agents are anonymous, equipped with infinite memory and computationally designed as Turing machines. The agents are situated at a distance of D. The agents have no prior information about the network G or the value of D. The agents move according to a deterministic algorithm at each node, choosing a port and moving to the next node using the chosen port. At the start, an agent only knows the degree of the initial node. From a node u, when the agent reaches node v using the port p at u, it learns the degree of the node v and the port q at v through which it reaches v. In a round, when two agents move via an edge but in the opposite direction, they can not see each other. Also, while being at a node, an agent can see if there is a pebble at that node or not.

The algorithm runs in synchronous rounds. In each round, an agent performs one look-compute-move cycle. During look, it gathers new information, then does computation based on the information it has, and finally moves to another node or stays on the same node based on the computation it did. The time complexity of the rendezvous is the number of rounds before they meet.

There is an oracle. The oracle knows the graph G and the starting positions of both agents. The oracle places identical pebbles at nodes. At most, one pebble can be placed at each node as the 1-bit information, and an agent sees a pebble only when it reaches a node (say v) and finds a pebble at v. In our model, we consider the presence of at most one faulty node, and its position is unknown to both the agents as well as the oracle.

1.3   Notations

Let $G = (V, E)$ be an anonymous port-labelled simple graph. For any node $v \in V$, by $deg(v)$, we denote the degree of the node v. By Δ, we denote the maximum degree of G. For any node $v \in V$, by $v(0), v(1), \ldots, v(deg(v)-1)$, we denote the neighbours of v that are connected through port numbers $0, 1, \ldots, deg(v) - 1$, at v respectively. For any two nodes $u, v \in G$, we denote the shortest distance between u and v in G by $dist(u, v)$. Let $s_i \in V$ be the starting point of the agent A_i where $i \in \{1, 2\}$. Let $P = (s_1 =)v_0, v_1, \ldots, v_{D-1}, v_D(= s_2)$ be a shortest path from s_1 to s_2, where $D = dist(s_1, s_2)$. Also let $p_0, p_1, \ldots, p_{D-1}$ be the sequence of port numbers corresponding to the path P such that $v_m(p_m) = v_{m+1}$ for all $m \in [0, D-1]$. Set L_i contains nodes which are at i distance from node s_1.

2   Rendezvous in Trees Despite Fault

Let G be an anonymous port-labelled tree. In our problem, both agents can move, and there is one unknown faulty node. This faulty node can disturb the encoding by placing a pebble on a node or removing an existing pebble. Therefore, the main challenge is to handle the fault.

High-level idea of our algorithm: We divide our algorithm into two phases. In the first phase, if one or more nodes of s_1, $s_1(0)$, and $s_1(1)$ are common with one or more nodes of s_2, $s_2(0)$, and $s_2(1)$, then they meet with each other within

17 rounds, with the help of the oracle, unless s_1, s_2 are the only nodes in the tree. Else both agents understand that no node of s_1, $s_1(0)$, $s_1(1)$ is common with any node of s_2, $s_2(0)$, $s_2(1)$ and start executing the second phase.

In the second phase, we convert the rendezvous problem into a treasure hunt problem. The oracle tells one agent to move and the other to stay put. The oracle encodes 111 in the neighbours of s_1, say, and encodes 000 in the neighbours of s_2. The agent A_i reads s_i, $s_i(0)$, $s_i(1)$, where $i = 1$ or 2. As per the encoding, A_1 finds a pebble on at least two nodes among s_1, $s_1(0)$, $s_1(1)$. Similarly, A_2 finds no pebble on at least two nodes among s_2, $s_2(0)$, $s_2(1)$. Using this pebble placement strategy, agent A_2 becomes a treasure and A_1 becomes the moving agent. Agent A_1 uses the following strategy to meet with A_2. If the degree of the node v_i on path P is at least 98 [1], it decodes the outgoing port to the next node on path P. Otherwise, it visits at most two length paths from this node to figure out the next node to visit by finding pebbles placed by the oracle. Using this idea, the agent A_1 meets with the agent A_2. The final results on trees are as follows.

Theorem 1. *Either A_1 and A_2 meet in the tree within $O(D \log \Delta)$ rounds using $O(D \log \Delta)$ pebbles, or understand in finite rounds that rendezvous is impossible.*

Theorem 2. *There exists a tree G of maximum degree Δ (≥ 2) and diameter D (≥ 3) such that any deterministic algorithm must require $\Omega(D \log \Delta)$-time for the rendezvous irrespective of the number of pebbles placed on the nodes of G.*

3   Rendezvous in General Graph with $D \leq 3$

In this section, we assume that the distance between s_1 and s_2 is at most 3 (i.e., $D \leq 3$). Again, the algorithm of [5] does not work due to the presence of the unknown fault. Our main challenge is to do an encoding overcoming the unknown fault such that the agents can meet using the encoding.

High-level idea of our algorithm: We divide the algorithm into two phases. In the first phase, the agents perform an algorithm of 44 rounds. As it was the case for trees, if one or more nodes of s_1, $s_1(0)$, and $s_1(1)$ are common with one or more nodes of s_2, $s_2(0)$, and $s_2(1)$, the agents meet with each other within these 44 rounds with the help of the oracle, unless s_1, s_2 are the only nodes in the graph or the agents are in a cycle of length three. Else, they perform the second phase with the understanding that no node of s_1, $s_1(0)$, and $s_1(1)$ is common with any node of s_2, $s_2(0)$, and $s_2(1)$. In the second phase, one of the agents (say A_2) stays at the starting position and waits for A_1 (like trees). We say that a node is heavy if its degree is at least $c_1 \log \Delta + c_2$, where c_1 and c_2 are a constant[2]. The agent A_1 searches for the node u and decodes the information

---

[1] To encode the correct encoding of the outgoing port of v_i, we need $14(1 + \lfloor \log deg(v_i) \rfloor)$ nodes. The inequality $14(1 + \lfloor \log deg(v_i) \rfloor) \leq deg(v_i)$ must hold to encode the correct encoding, which is not the case when $deg(v_i) < 98$. Therefore, the correct encoding is possible when $deg(v_i) \geq 98$. The detailed description of this is in the full version.

[2] The value of c_1 and c_2 is 486 and 1134, respectively. The computation of c_1 and c_2 with reasoning is in the full version

about the position of the first heavy node (say v_{heavy}) on P, i.e. v_{heavy} is present in the set L_i, $0 \leq i \leq 2$. If it finds the position of v_{heavy} in L_i, then agent A_1 visits the paths of length i from s_1 and selects the node with the largest degree as v_{heavy}. If there are more than one such nodes, it selects the node that lies on the lexicographically shortest path of s_1. After reaching v_{heavy}, it decodes the entire path P from the neighbours of v_{heavy}. If there is no heavy node between L_0 and L_2, it can find A_2 by visiting at most three length paths from s_1. Our final result on general graphs (when $D \leq 3$) is as follows.

Theorem 3. *Either A_1 and A_2 meet in G within $O(\log^3 \Delta)$ rounds using $O(\log \Delta)$ pebbles, or they understand in finite rounds that rendezvous is impossible.*

4   Conclusion

We propose an algorithm for rendezvous in trees in the presence of a faulty node, which takes $O(D \log \Delta)$ rounds using $O(D \log \Delta)$ pebbles. We also show that our algorithm is optimal w.r.t. time. Then, we solve the problem in general graphs, considering agents start within the shortest distance of 3. The algorithm requires $O(\log^3 \Delta)$ rounds and uses no more than $O(\log \Delta)$ pebbles. The immediate problem to solve is the 1-fault Rendezvous problem in general graphs, considering agents start from arbitrary positions in the graph.

References

1. Bhattacharya, A., Gorain, B., Mandal, P.S.: Treasure hunt in graph using pebbles. In: Stabilization, Safety, and Security of Distributed Systems, pp. 99–113. Springer International Publishing (2022)
2. Dessmark, A., Fraigniaud, P., Kowalski, D.R., Pelc, A.: Deterministic rendezvous in graphs. Algorithmica **46**(1), 69–96 (2006)
3. Dieudonné, Y., Pelc, A.: Deterministic network exploration by a single agent with byzantine tokens. Inf. Process. Lett. **112**(12), 467–470 (2012)
4. Fraigniaud, P., Ilcinkas, D., Pelc, A.: Tree exploration with advice. Inf. Comput. **206**(11), 1276–1287 (2008)
5. Gorain, B., Mondal, K., Nayak, H., Pandit, S.: Pebble guided optimal treasure hunt in anonymous graphs. Theoret. Comput. Sci. **922**, 61–80 (2022)
6. Gorain, B., Pelc, A.: Deterministic graph exploration with advice. ACM Trans. Algorithms (TALG) **15**(1), 1–17 (2018)
7. Miller, A., Pelc, A.: Tradeoffs between cost and information for rendezvous and treasure hunt. In: Principles of Distributed Systems, pp. 263–276. Springer International Publishing (2014)

Complete Graph Identification in Population Protocols

Haruki Kanaya[1] and Yuichi Sudo[2]($\boxtimes$)

[1] Nara Institute of Science and Technology, Nara, Japan
kanaya.haruki.kk3@naist.ac.jp
[2] Hosei University, Tokyo, Japan
sudo@hosei.ac.jp

Abstract. We consider the population protocol model where indistinguishable state machines, referred to as agents, communicate in pairs. The communication graph specifies potential interactions (i.e., communication) between agent pairs. This paper addresses the complete graph identification problem, requiring agents to determine if their communication graph is a clique or not. We evaluate various settings based on: (i) the fairness preserved by the adversarial scheduler—either global fairness or weak fairness, and (ii) the knowledge provided to agents beforehand—either the exact population size n, a common upper bound P on n, or no prior information. Positively, we show that $O(n^2)$ states per agent suffice to solve the complete graph identification problem under global fairness without prior knowledge. With prior knowledge of n, agents can solve the problem using only $O(n)$ states under weak fairness.

Keywords: Population protocols · Graph class identification

1 Introduction

The population protocol model, proposed by Angluin et al. [8], is a widely recognized computational model in distributed computing. Originally introduced to represent passively mobile sensor networks, this model is also applicable to chemical reactions systems, molecular computing, and similar systems. In this model, n state machines, called *agents*, form a population. These agents are indistinguishable, i.e., they lack unique identifiers. The states of agents are updated through pairwise communications, called *interactions*. The *communication graph* specifies which agent pairs may interact. Specifically, the communication graph $G = (V, E)$ is a simple and weakly-connected digraph. Each arc $(u, v) \in E$ implies that u and v can interact where u and v serve as the initiator and responder, respectively. An adversarial scheduler selects exactly one interactable ordered pair of agents at each time step. When two agents u and v are selected by the scheduler, they update their states according to their states and their roles. The

This work is supported by JST FOREST Program JPMJFR226U.

two roles, initiator and responder, may be helpful to break the symmetry: two agents with the same state may get different states after they interact. Throughout this paper, we denote the number of agents by $n = |V|$ and the set of non-negative integers by $\mathbb{N}$.

The adversarial scheduler must preserve some constraint, either *global fairness* or *weak fairness*. Global fairness, the most commonly assumed fairness in population protocol literature, ensures that if a configuration C appears infinitely often in an execution, every configuration reachable from C must also appear infinitely often. As we will discuss later, a protocol $\mathcal{A}$ solves a problem $\mathcal{P}$ under global fairness if and only if A solves $\mathcal{P}$ with probability 1 under the *uniformly random scheduler*, which selects any interactable ordered pair of agents uniformly at random at each time step. Thus, in terms of solvability, the globally fair scheduler can be considered equivalent to the uniformly random scheduler. Weak fairness is straightforward; it merely ensures that an interaction between initiator u and responder v occurs infinitely often if $(u, v) \in E$.

The graph class identification problem was first studied by Angluin, Aspnes, Chan, Fischer, Jiang, and Peralta [7]. This problem requires the agents to determine whether the communication graph belongs to a given graph class. The importance of this problem stems from the fact that many protocols in the literature are designed for specific graphs. Notably, most studies on population protocols assume a complete communication graph. Angluin et al. [7] showed that under the global fairness, the population can identify various graph classes, i.e., (i) directed lines, (ii) directed rings, (iii) directed stars, (iv) directed trees, (v) graphs with both in-degree and out-degree of every node bounded by a given integer k, (vi) graphs containing a given subgraph, (vii) graphs containing any directed cycle, and (viii) graphs containing any odd-length directed cycle.

Yasumi, Ooshita, and Inoue [42] studied graph class identification, assuming the communication graph $G = (V, E)$ is *undirected*, i.e., $(u, v) \in E$ if and only if $(v, u) \in E$ for all $u, v \in V$. They showed that the algorithms given by Angluin et al. [7] can be extended to identify (undirected) lines, rings, stars, and bipartite graphs under global fairness, by treating two arcs (u, v) and (v, u) as a single undirected edge $\{u, v\}$. Moreover, they gave two algorithms that identify trees and k-regular graphs for a given k, respectively, under global fairness. The latter requires all agents to know a common upper bound P on the population size $n = |V|$. They also explored the solvability under weak fairness, revealing that: (i) lines, rings, and bipartite graphs are unidentifiable even if all agents know the exact number of agents n; (ii) stars can be identified if all agents know the exact n, but not under the knowledge of a common upper bound P on n. The unidentifiability of rings implies that k-regular graphs cannot be identified under weak fairness when $k = 2$, even with the knowledge of exact n. The identifiability of k-regular graphs under weak fairness for $k \neq 2$, including when $k = n - 1$ (i.e., complete graphs), remains an open question.

Table 1. Complete Graph Identification Protocols. Time complexities are evaluated as the number of steps until the output of all agents stabilize under the uniformly random scheduler, bounded both in expectation and with high probability. A given upper bound on n is denoted by P, while k is a design parameter satisfying $k \in [1, n]$. See Sect. 1.1 for the definition of $\hat{G}$ and $\mathcal{H}_{\hat{G}}$.

	Fairness	Knowledge	States	Time complexity
[42]	Global	n	$O(n \log n)$	Exponential
CIG	Global	None	$O(n^2)$	$O((\mathcal{H}_{\hat{G}} + n^2)n \log n)$
CIW_n	Weak	n	$O(n)$	$O(n^3 \log n)$
$\text{CIW}_{n,k}$	Weak	n	$O(nk2^k)$	$O\left(\frac{n^3}{k} \log n\right)$
Impossible	Weak	P	–	–

1.1   Our Contributions

This paper focuses on identifying complete graphs, i.e., determining whether the communication graph forms a clique. As previously mentioned, most studies on population protocols assume that the communication graph is a complete graph. Thus, the identifiability of complete graphs is of significant importance.

We clarify the feasibility of identifying complete graphs based on fairness and prior knowledge of the population size. The summary of our results is listed in Table 1. First, we show that under global fairness, complete graphs can be identified without any prior knowledge of the population size. Specifically, we present an algorithm CIG, which identifies complete graphs using $O(n^2)$ states per agent. Next, we show that even under weak fairness, complete graphs can be identified with exact knowledge of n: we introduce an algorithm CIW_n that identifies complete graphs with $O(n)$ states per agent under weak fairness. The requirement for exact knowledge of the population size is justified by the fact that complete graphs cannot be identified under weak fairness even if the agents are aware of a common upper bound P on n. This impossibility directly follows from a lemma provided by Yasumi et al. [42] (Lemma 5 in [42]).

In addition to the number of states, we also evaluate the time complexity of our protocols. In line with the conventions of population protocols, we measure time complexity as the number of interactions required until all agents stabilize their outputs under the uniformly random scheduler. The time complexities of our two protocols, CIG and CIW_n, are $O((\mathcal{H}_{\hat{G}} + n^2)n \log n)$ and $O(n^3 \log n)$ interactions both in expectation and with high probability, respectively. Here, $\hat{G} = (V, \hat{E})$ is the (non-simple) undirected graph obtained by replacing each arc $(u, v) \in E$ with an undirected edge $\{u, v\}$, and $\mathcal{H}_{G'}$ denotes the maximum hitting time of an undirected graph G', i.e., the maximum expected number of moves required for a token performing the simple random walk on G' to move from any node s to any node t, where the maximum is taken over all possible pairs of s and t. Note that $\hat{E}$ is a multiset where $\{u, v\}$ appears at most twice for any pair $u, v \in V$. Since the hitting time of any connected undirected graph

$G' = (V', E')$ is known to be $O(|E'| \cdot D_{G'})$, where $D_{G'}$ is the diameter of G', the time complexity of CIG is $O(n^4 \log n)$ interactions both in expectation and with high probability. With the precise knowledge of n, CIW$_n$ achieves lower time complexity (i.e., $O(n^3 \log n)$ interactions) and fewer states per agent (i.e., $O(n)$ states). By introducing a design parameter $k \in [1, n]$, we show that CIW$_n$ can be generalized to a protocol CIW$_{n,k}$, which identifies complete graphs within $O((n^3 \log n)/k)$ interactions both in expectation and with high probability and requires $O(nk2^k)$ states per agent.

If we assume exact knowledge of the population size n, we can utilize Yasumi et al.'s protocol for identifying k-regular graphs [42] to identify complete graphs by setting $k = n - 1$. However, our protocol CIW$_n$ offers several advantages for identifying complete graphs: (i) CIW$_n$ requires only weak fairness, whereas Yasumi et al.'s protocol requires global fairness, (ii) CIW$_n$ uses fewer agent states, (iii) CIW$_n$ is faster (see Table 1). Although Yasumi et al. [42] focus only on solvability and the number of states, and do not address time complexity, their protocol requires exponential time when $k = \Omega(n)$. Specifically, it does not stabilize before every agent u experiences an event in which its k most recent interactions involve exactly k different neighbors. If the degree of u is exactly k, this event requires at least $\Omega(k^k/k!) = \Omega(e^k/k^{1/2})$ interactions in expectation due to Stirling's approximation $k! > \sqrt{2\pi k}(k/e)^k$. Therefore, for our purpose (i.e., $k = n - 1$), this protocol requires $\Omega(e^n/\sqrt{n})$ expected interactions. It may be worth mentioning that our protocol, designed for *directed communication graphs*, also functions under the *undirected* communication graphs assumed by Yasumi et al. By definition, undirected communication graphs are simply special cases of directed graphs.

In the original model of population protocols, agents are defined as *state machines*, and the set of agent states must be specified in advance when designing a protocol. Therefore, the number of agent states must remain constant if a protocol is *uniform*, i.e., it does not assume any prior knowledge of the population size. However, solving some problems and/or designing fast protocols require a super-constant number of states. For example, Doty and Soloveichik [26] proved that $\omega(1)$ states are necessary to solve the leader election problem within $o(n^2)$ expected interactions under the uniformly random scheduler. Thus, several studies [19, 24] have adopted a generalized model in which agents are defined as Turing machines, and the number of agent states is defined as $|\Sigma|^\ell$, where Σ is the tape alphabet and ℓ is the maximum number of tape cells written by any agent in any execution of the protocol.[1] We also adopt this generalized model for our uniform protocol CIG, which does not assume any prior knowledge, yet utilizes $O(n^2)$ states. If we want to adhere to the original model for our protocol CIG, we require knowledge of a common upper bound P on n. In this case, the protocol CIG$_P$, now depending on the knowledge of P, identifies complete graphs using $O(P^2)$ states. Here, we never use the knowledge P except for specifying the number of states, that is, bounding the domain of two variables.

---

[1] The *expected* number of states can also be evaluated under the uniformly random scheduler, e.g., in [19].

1.2   Related Works

Since Angluin, Aspnes, Diamadi, Fischer, and Peralta introduced this model in 2004 [8], various problems have been studied, including the leader election problem [2,3,8,11,19,20,26–28,34,37,39], the majority problem [1,2,5,10,17–19,21,25,30,35,36], predicate computation [8,9,11,12], and the counting problem [14,16]. Most studies focus only on complete graphs, while several explore other graph classes, including arbitrary graphs [4,6,8,31,42,43]. There are also numerous studies on *self-stabilizing* population protocols [15,22,23,38,40,44,45], initiated by Angluin, Aspnes, Fischer, and Jiang [13].

There are some variants of the population protocol model, such as the mediated population protocol model and the network construction model, which are closely related to the graph class identification problem. The mediated population protocol model is an extension of the population protocol model introduced by Michail, Chatzigiannakis, and Spirakis [32]. This model enables information to be stored not only in the states of agents but also in the communication links between them. The network construction model, an extension of the mediated protocol model, was introduced by Michail, and Spirakis [33]. This model aims to construct specific networks based on the states of agents and the information carried by links between agents.

2   Preliminaries

2.1   Population Protocols

The communication graph is represented by a simple and weakly connected digraph $G = (V, E)$, where V denotes the set of agents, and $E \subseteq \{(a, b) \in V \times V \mid a \neq b\}$ denotes the set of ordered pairs of agents that can interact: agents u and v can have interaction such that u is the initiator and v is the responder if $(u, v) \in E$. We say that v is an *out-neighbor* of u if $(u, v) \in E$. At each time step, an adversarial scheduler selects one ordered pair $(u, v) \in E$.

As mentioned in Sect. 1, we adopt the original model for non-uniform protocols (i.e., CIW_n and $\text{CIW}_{n,k}$) and adopt the generalized model given by [24] for uniform and $\omega(1)$-space protocols. In the original model, agents are defined as finite state machines. A protocol is defined as a 5-tuple $\mathcal{P} = (Q, \rho, Y, \pi, \delta)$, where Q is the set of agent states, $\rho \in Q$ the initial state, Y the set of output symbols, $\pi : Q \to Y$ the output function, and $\delta : Q \times Q \to Q \times Q$ the state transition function. An agent outputs the symbol $\pi(s) \in Y$ when in state $s \in Q$. Suppose agents u and v engage in an interaction, with u as the initiator and v as the responder, while in states p and q respectively. They then update their states to p' and q' respectively, where $(p', q') = \delta(p, q)$. The number of states of a protocol $\mathcal{P} = (Q, \rho, Y, \pi, \delta)$ is simply defined as $|Q|$. Roughly speaking, the generalized model, which accommodates both uniform and super-constant protocols, can be described as follows:

- The definition of a protocol $\mathcal{P} = (Q, \rho, Y, \pi, \delta)$ is the same as in the original model except for the set Q of agent states.

- An agent maintains a constant number of variables $x_1, x_2, \ldots, x_s$, and the combination of their values constitutes the state of the agent. The first element Q of the protocol is the set of all such states. The domain of each variable may not be bounded, thus Q may be an infinite set.
- For any agent $a \in V$, define a's amount of information at time step t as $f(a, t) = \sum_{i=1}^{s} \iota_i$, where each ι_i is the number of bits required to encode the value of variable x_i. The number of states of the protocol $\mathcal{P}$ is defined as $\max\{2^{f(a,t)} \mid a \in V, t = 0, 1, \ldots\}$ in any execution of protocol $\mathcal{P}$. Therefore, the number of states of $\mathcal{P}$ may depend on the population size n, even if $\mathcal{P}$ itself does not depend on n.

Refer to Sect. 2 in [24] for the formal definition of the generalized model, which is based on a Turing Machine.

A global state or a *configuration*, denoted $C : V \to Q$, specifies the state of all agents. The *initial configuration*, where all agents are in the initial state ρ, is denoted by $\mathcal{I}$. A configuration C *changes* to C' via an interaction $(u, v) \in E$, denoted by $C \overset{(u,v)}{\to} C'$, if $(C'(u), C'(v)) = \delta(C(u), C(v))$ and $C'(w) = C(w)$ for all $w \in V \setminus \{u, v\}$. When C changes to C' via any interaction, we say that C *can change* to C', denoted as $C \to C'$. An *execution* is defined as an infinite sequence of configurations $\Xi = C_0, C_1, \ldots$ that starts from the initial configuration $C_0 = \mathcal{I}$ and satisfies $C_i \to C_{i+1}$ for every $i \in \mathbb{N}$. A configuration C' is *reachable* from C if there exists a sequence of configurations $C_0, C_1, \ldots, C_t$ with $C = C_0$, $C' = C_t$, and $C_i \to C_{i+1}$ for each i from 0 to $t-1$. A configuration C is *stable* if, for any agent $a \in V$ and any configuration C' reachable from C, agent a outputs the same symbol in both C and C', i.e., $\pi(C(a)) = \pi(C'(a))$. We say that an execution *stabilizes* when it reaches a stable configuration.

Fairness is defined as a predicate on executions, i.e., it specifies the set of acceptable executions of a protocol $\mathcal{P}$. An execution $\Xi = C_0, C_1, \ldots$ is *globally fair* if for any configuration C that appears infinitely often in Ξ, every configuration reachable from C also appears infinitely often in Ξ. An execution $\Xi = C_0, C_1, \ldots$ is *weakly fair* if there exists a sequence of interactions $\gamma = \gamma_0, \gamma_1, \ldots$ such that every ordered pair $(u, v) \in E$ appears infinitely often in γ.

2.2 Complete Graph Identification

A communication graph $G = (V, E)$ is *complete* if $(u, v) \in E$ for any distinct agents $u, v \in V$. A protocol $\mathcal{P}$ *solves the complete graph identification* or *identify complete graphs* under global fairness (resp. weak fairness) if any globally fair (resp. weakly fair) execution of $\mathcal{P}$ eventually reaches a stable configuration where the outputs of all agents are **yes** if G is complete, and **no** otherwise.

2.3 Time Complexity

The time complexity is defined as the number of interactions required to reach a stable configuration, under the assumption of a uniformly random scheduler that selects an ordered pair of agents uniformly at random at each time step.[2] Given that we assume a uniformly random scheduler, the time complexity is considered a random variable; therefore, we evaluate it in terms of expectation and/or with high probability. In this paper, *with high probability* is defined as *with probability* $1 - O(1/n)$.

Introducing (asynchronous) *rounds* can be useful for analyzing the time complexities of protocols that work under weak fairness.

Definition 1. (rounds) Let $\gamma = \gamma_0, \gamma_1, \ldots$ be an infinite sequence of interactions. The first round of γ is the shortest prefix of γ such that every $(u, v) \in E$ appears at least once in this prefix. Let t_1 be the length of the first round of γ, i.e., the prefix is $\gamma_0, \gamma_1, \ldots, \gamma_{t_1-1}$. For any $i \geq 1$, the $(i + 1)$-th round of γ is defined as the shortest prefix of the infinite sequence $\gamma_{S_i}, \gamma_{S_i+1}, \ldots$, where each pair $(u, v) \in E$ appears at least once in the prefix, with $S_i = \sum_{k=1}^{i} t_k$ and t_k being the length of the k-th round of γ.

The standard analysis of the coupon collector's problem give the following lemma by considering each $(u, v) \in E$ as a type of coupon. Note that $|E| = \Omega(n)$ holds here by the weak connectivity of the communication graph.

Lemma 1. *Let* $\Gamma = \Gamma_0, \Gamma_1, \ldots$ *be the infinite sequence of interactions chosen by the uniformly random scheduler, where* $\Pr(\Gamma_i = (u, v)) = 1/|E|$ *for any* $(u, v) \in E$ *and* $i \geq 0$, *and this probability is independent of any other interaction* Γ_j $(j \neq i)$. *Then, for any* $i \geq 0$, *the i-th round of Γ has an expected length of* $O(|E| \log |E|) = O(|E| \log n)$. *Moreover, for any constant* $c \geq 1$, *the length is* $O(c \cdot |E| \log n)$ *with probability* $1 - O(n^{-c})$.

3   Complete Graph Identification Under Weak Fairness

This section presents the protocol $\mathtt{CIW}_n$, which identifies complete graphs using the exact knowledge of n. It also introduces a generalized version, $\mathtt{CIW}_{n,k}$, which incorporates a design parameter k where $1 \leq k \leq n$.

3.1   Protocol $\mathtt{CIW}_n$

In protocol $\mathtt{CIW}_n$, each agent $a \in V$ maintains four variables: $a.\mathtt{leader} \in \{F, L\}$, $a.\mathtt{phase} \in \{1, 2, 3, 4\}$, $a.\mathtt{mode} \in \{0, 1\}$, and $a.\mathtt{cnt} \in \{0, 1, \ldots, n\}$, resulting in exactly $16(n+1) = O(n)$ states per agent. Initially, $a.\mathtt{leader} = L$, $a.\mathtt{phase} = 1$, $a.\mathtt{mode} = 0$, and $a.\mathtt{cnt} = 1$. An agent a is considered a *leader* if $a.\mathtt{leader} = L$; otherwise, it is termed a *follower*. The set of leaders is denoted by $V_L \subseteq V$. For each $i \in \{1, 2, 3, 4\}$, an agent a is said to be in phase i if $a.\mathtt{phase} = i$, and the

---

[2] When focusing on complete graphs, time complexity is often expressed as *parallel time*, which refers to the number of interactions required for stabilization divided by the number of agents. In this paper, we do not use parallel time because we are addressing complete graph identification on arbitrary graphs.

Algorithm 1: CIW_n (a protocol for weak fairness with exact n)

---

when $a, b \in V$ *interacts with* a *as initiator and* b *as responder* **do**

1   **if** $a \in V_L \wedge b \in V_L$ **then**
2      $a.\text{cnt} \leftarrow a.\text{cnt} + b.\text{cnt}$
3      $(b.\text{leader}, b.\text{cnt}) \leftarrow (F, 0)$
4   **if** $a.\text{cnt} = n$ **then**
5        $(a.\text{phase}, a.\text{cnt}) \leftarrow (2, 0)$

6   **else if** $a \in V_L \cap V_2 \wedge a.\text{mode} = b.\text{mode}$ **then**
7      $a.\text{cnt} \leftarrow a.\text{cnt} + 1$
8      $b.\text{mode} \leftarrow 1 - b.\text{mode}$
9   **if** $a.\text{cnt} = n - 1$ **then**
10       $(a.\text{phase}, a.\text{cnt}, a.\text{mode}) \leftarrow (3, 1, 1 - a.\text{mode})$

11   **else if** $a \in V_L \cap V_3 \wedge b \in V_1$ **then**
12      $a.\text{leader} \leftarrow F$
13      $(b.\text{leader}, b.\text{phase}) \leftarrow (L, 2)$

14   **else if** $a \in V_3 \wedge b \in V_3 \wedge a.\text{cnt} > 0 \wedge b.\text{cnt} > 0$ **then**
15      $a.\text{cnt} \leftarrow a.\text{cnt} + b.\text{cnt}$
16      $b.\text{cnt} \leftarrow 0$
17   **if** $a.\text{cnt} = n$ **then**
18       $a.\text{phase} \leftarrow 4$

19   **else if** $a \in V_4$ **then**
20      $b.\text{phase} \leftarrow 4$

---

set of agents in phase i is denoted by $V_i \in V$. Agents output yes if and only if they are in phase 4. The objective is to ensure that all agents reach phase 4 if the communication graph G is complete (i.e., the out-degrees of all agents are $n - 1$), and that no agents reach phase 4 otherwise.

Our protocol CIW_n, presented in Algorithm refal:n, can be summarized as follows:

- Agents in phase 1 participate in leader election (lines 1–5). Once a unique leader is elected, it advances to phase 2.
- The unique leader in phase 2 counts the number of its out-neighbors (lines 6–8). It moves to phase 3 upon determining that the number of its out-neighbors equals $n - 1$ (lines 9–10). At the next interaction where this leader meets a follower in phase 1, the leadership status is transferred to the follower, who then moves to phase 2 and starts its own count (line 11–13).
- Agents in phase 3 count the number of agents in phase 3 (lines 14–16). If some agent observes that $|V_3| = n$, indicating that the out-degrees of all agents are $n - 1$, it transitions to phase 4 (line 17–18).
- Once an agent in phase 4 appears, all agents proceed to phase 4 via the well-known one-way epidemic protocol [11] (line 19–20): In each interaction where an agent $a \in V_4$ is the initiator and an agent $b \in V_3$ is the responder, b transitions to phase 4.

In what follows, we describe how to elect a leader, count the out-degree, and determine if $|V_3| = n$, using the variable cnt. Note that cnt is utilized in phases 1, 2, and 3 for different purposes. This *re-use* of cnt contributes to reducing the number of states of CIW_n from $O(n^3)$ to $O(n)$.

Leaders in phase 1 elect the unique leader. Initially, all agents are leaders. When two leaders meet, one becomes a follower. Given that all agents know the exact value of n, the completion of the leader election is easily detected as follows: (i) initially, all agents set cnt = 1, (ii) when two leaders interact, the surviving leader absorbs the cnt of the defeated leader, (iii) the unique leader recognizes its uniqueness when cnt = n. Upon this detection, the unique leader advances to phase 2 with resetting cnt value to zero.

A leader in phase 2 determines if its out-degree is $n - 1$. When a new leader $a \in V_2$ emerges, either from the leader election or the transfer of leadership status, a.cnt is 0, and all agents share the same mode value, either 0 or 1. Each time a, as the initiator, meets another agent b with the same mode value, it increments a.cnt by one and toggles b.mode from 0 to 1 or from 1 to 0. Thus, a.cnt reaches $n - 1$ if and only if a meets $n - 1$ different out-neighbors, implying that a's out-degree is $n - 1$. Then, it transitions to phase 3, setting cnt to 1 and toggling its mode value. Toggling a.mode ensures that all agents share the same mode value when a meets a follower b in phase 1 and transfers its leadership status to b, making b a new leader in phase 2.

Agents in phase 3 determine whether $|V_3| = n$. As mentioned previously, every agent sets its cnt value to 1 upon transitioning to phase 3. When two agents in phase 3 with non-zero cnt values meet, one agent absorbs the cnt value of the other. The weak fairness ensures that all agents will meet each other infinitely often, leading eventually to a configuration where one agent in phase 3, denoted as a, has a.cnt = $|V_3|$. If a.cnt = n, then a confirms that the out-degrees of all agents are exactly $n - 1$, indicating that the communication graph is complete. Subsequently, a advances to phase 4.

Lemma 2. $\sum_{a \in V_3 \cup V_4} a.\text{cnt} = |V_3 \cup V_4|$ *always holds in an execution of* CIW_n.

Proof. This lemma follows directly from the facts that: (i) each time an agent enters phase 3, it sets its cnt value to 1; (ii) an agent in phase 3 updates its cnt value only during an interaction with another agent in phase 3, which does not alter the sum of the cnt values of those two agents; and (iii) no agent in phase 4 updates its cnt value. □

Theorem 1. *Given the exact population size n, protocol* CIW_n *solves the complete graph identification problem under weak fairness, using* $O(n)$ *states.*

Proof. While all agents are in phase 1, the invariant $\sum_{a \in V_1} a.\text{cnt} = n$ always holds. Thus, only one leader may have a.cnt = n and transition to phase 2, with all agents having a zero cnt value. If the communication graph is complete, weak fairness guarantees that a unique leader is eventually elected. If not, the leader election may remain incomplete, but this does not matter to us. In both scenarios, once an agent transitions to phase 2, there is always exactly one leader.

Suppose that the communication graph is *not* complete. Then, there exists at least one agent, say b, with an out-degree of at most $n - 2$. Even if b transitions from phase 1 to phase 2, its cnt value is zero at that time. Each out-neighbor of b can increase b.cnt by at most one, ensuring that b.cnt never reaches $n - 1$. Therefore, b does not advance to phase 3, and consequently, $|V_3| < n$ always holds. By Lemma 2, this implies that no agent observes cnt $= n$, and thus no agent transitions to phase 4. Therefore, every agent always outputs no from the beginning of an execution if the communication graph is not complete.

Next, suppose that the communication graph is complete. As mentioned previously, the unique leader is eventually elected. When this leader, say ℓ, is elected and moves to phase 2, its cnt value is zero and the mode values of all agents are zero. Weak fairness ensures that ℓ meets its $n - 1$ out-neighbors as the initiator, changing the mode values of all agents including ℓ to 1, and then moves to phase 3. The leader ℓ gives the leadership status to another agent, say ℓ', in the next interaction. When ℓ' becomes a leader, ℓ'.cnt $= 0$ and the mode values of all agents are 1. Similarly, weak fairness ensures that ℓ'.cnt reaches $n - 1$, the mode values of all agents go back to 0, and ℓ' transitions to phase 3. Repeating this process, all agents eventually enter phase 3. By Lemma 2, weak fairness ensures that the event that some agent a in phase 3 observes a.cnt $= n$ eventually occurs, and a enters phase 4. After that, all agents enter phase 4 by meeting a or another agent in phase 4, by weak fairness. $\qquad\square$

Lemma 3. *If the communication graph is not complete, every execution of* $\mathtt{CIW}_n$ *stabilizes in zero time, i.e., the initial configuration* $\mathcal{I}$ *is stable.*

Proof. As shown in the proof of Theorem 1, no agent enters phase 4 if the communication graph is not complete, which gives the lemma. $\qquad\square$

Lemma 4. *If the communication graph is complete, every execution of* $\mathtt{CIW}_n$ *stabilizes in* $O(n)$ *rounds under weak fairness.*

Proof. The leader election will be completed before every pair of agents meets, thus finishing in one round. Each agent requires one round to recognize that its out–degree is $n - 1$ and advance to phase 3. Simultaneously, it takes one round for a leader in phase 3 to encounter an agent in phase 1 and transfer leadership status. Therefore, all agents reach phase 3 within $2n + 1$ rounds. An additional round is sufficient for some agent to detect that $|V_3| = n$ and move to phase 4, after which all agents enter phase 4 in one subsequent round. $\qquad\square$

The following theorem is derived from Lemmas 1, 3, and 4.

Theorem 2. *Any execution of* $\mathtt{CIW}_n$ *stabilizes in* $O(n^3 \log n)$ *steps, both in expectation and with high probability, under the uniformly random scheduler.*

3.2 Protocol $\mathtt{CIW}_{n,k}$

This section presents $\mathtt{CIW}_{n,k}$, a generalized version of $\mathtt{CIW}_n$ that incorporates a design parameter k, where $1 \leq k \leq n$. Due to space limitations, we omit detailed descriptions of this protocol. Instead, we provide the key strategies of $\mathtt{CIW}_{n,k}$.

The time complexity of $\mathtt{CIW}_n$ is primarily determined by the degree recognition process, where n agents sequentially determine whether their out-degrees are $n-1$, requiring $O(n)$ rounds. We reduce the duration of this period by dividing the population into k groups, nearly equally. Specifically, after the unique leader ℓ is elected, ℓ assigns group identifiers to all agents as follows:

- Each agent a maintains a variable $a.\mathtt{group} \in \{0, 1, \ldots, k\}$ to indicate the group to which it belongs. The initial value of this variable is k, which corresponds to the *null* group.
- When ℓ becomes the unique leader, $\ell.\mathtt{cnt} = n$ holds. Thereafter, whenever ℓ encounters an agent b with $b.\mathtt{group} = k$, ℓ decrements $\ell.\mathtt{cnt}$ by one and assigns ($\ell.\mathtt{cnt} \bmod k$) to $b.\mathtt{group}$. In parallel, if $\ell.\mathtt{cnt} < k$ after decrementing, ℓ reverts the opposing agent b to a leader, i.e., setting $b.\mathtt{leader}$ to L.
- When $\ell.\mathtt{cnt}$ reaches 1, it recognizes that it has already assigned group identifiers to all agents except itself. Then, ℓ assigns 0 to its $\mathtt{group}$ and completes the process of assigning group identifiers.

Through this process, the population is divided into k groups, each consisting of either $\lfloor n/k \rfloor$ or $\lceil n/k \rceil$ agents, with exactly one leader per group. The leader of each group $g \in \{0, 1, \ldots, k-1\}$ counts the number of its out-neighbors in the same manner as $\mathtt{CIW}_n$, but in parallel. Unlike $\mathtt{CIW}_n$, in protocol $\mathtt{CIW}_{n,k}$, the variable $a.\mathtt{mode}$ is an array of size k: $a.\mathtt{mode}[g] \in \{0, 1\}$ is used for counting the number of out-neighbors in group g. The remaining parts of the protocols $\mathtt{CIW}_n$ and $\mathtt{CIW}_{n,k}$ are identical.

Since $\mathtt{mode}$ is the binary array with size k, variable $a.\mathtt{mode}$ requires 2^k states, while $\mathtt{cnt}$ and $\mathtt{group}$ require $n+1$ and $k+1$ states, respectively. All other variables requires only $O(1)$ states. Thus, $\mathtt{CIW}_{n,k}$ requires $O(nk2^k)$ states per agent. At the cost of increased states, $\mathtt{CIW}_{n,k}$ reduces the time complexity from $O(n^3 \log n)$ to $O((n^3/k) \log n)$ steps both in expectation and with high probability because each group consists of at most $\lceil n/k \rceil$ agents, thus $O(n/k)$ rounds are sufficient to complete the out-degree counting process of all agents.

We have the following theorems. (See the full version for detailed implementation and proofs.)

Theorem 3. *Given the exact population size n and a design parameter k with $1 \leq k \leq n$, protocol $\mathtt{CIW}_{n,k}$ solves the complete graph identification problem under weak fairness, using $O(nk2^k)$ states.*

Theorem 4. *Any execution of $\mathtt{CIW}_{n,k}$ stabilizes in $O((n^3/k) \log n)$ steps, both in expectation and with high probability, under the uniformly random scheduler.*

4   Complete Graph Identification Under Global Fairness

This section proposes a protocol $\mathtt{CIG}$ for identifying complete graphs with no initial knowledge under global fairness using $32n(n+1) = O(n^2)$ states.

This protocol uses $\mathtt{CIW}_n$ as a submodule. However, we do not assume the knowledge of exact n here, whereas $\mathtt{CIW}_n$ requires knowledge of n in advance. In

CIG, the agents try to compute the number of agents n and store the estimated value of n on a variable size. As we will see later, for any agent $a \in V$, a.size is monotonically non-decreasing and eventually reaches n. Whenever two agents with the same size value meets, they execute CIW_{size}, i.e., updates the four variables leader, phase, cnt, and mode according to the transition function of CIW_{size}. Each time a.size is updated, all variables of CIW_n in agent a is reset by their initial values.

We compute the number of agents as follows:

- An agent a maintains two variables: a.token $\in \{\bot, \top\}$ and a.size $\in \mathbb{N}$. Initially, a.token $= \top$ and a.size $= 1$. An agent a is said to have a token if a.token $= \top$.
- When two agents a and b, each with a token, interact, the two tokens merge. This is done by setting b.token to $\bot$ and simultaneously increasing a.size by b.size. No other rules update the size value of an agent with a token. Therefore, the invariant $\sum_{a \in V, a.\text{token}=\top} a$.size $= n$ always holds. This token-merge process guarantees that the population eventually reaches a configuration where exactly one agent retains a token, denoted as a_T. Thereafter, a_T.size $= n$ always holds, thanks to the aforementioned invariant.
- Whenever an agent a without a token encounters another agent b with a.size $< b$.size, a.size is updated to b.size. Thus, the maximum value a_T.size $= n$ eventually propagates to the entire population. Consequently, all agents will have size $= n$, although they do not know whether this value equals to the actual number of nodes, meaning they cannot detect the termination of the size computation.

This protocol CIG requires only $O(n^2)$ states per agent because size and cnt never exceeds n.

We briefly analyze the stabilization time of CIG here. According to the analysis by Sudo, Shibata, Nakamura, Kim, and Masuzawa [41] (Lemma 4), the number of tokens becomes one within $O(\mathcal{H}_{\hat{G}} \cdot n \log n)$ steps both in expectation and with high probability. Here, the corresponding undirected graph $\hat{G}$ of the communication graph G and the hitting time $\mathcal{H}_{\hat{G}}$ were defined in Sect. 1. Although Sudo et al.[41] provided this upper bound only in expectation, we can easily show that this bound also holds with high probability by slightly modifying their proof. The propagation of the maximum value, often called *epidemic*, is well-studied in the field of population protocols. On arbitrary communication graphs, Alistarh, Rybicki, and Voitovych [6] proved that the epidemic completes within $O(m(\log n + D))$ steps in expectation (Theorem 3.2), and thus within $O(m(\log n + D) \log n)$ steps with high probability, where m is the number of arcs in the communication graph G, and D is the diameter of $\hat{G}$. Consequently, all agents will have size $= n$ within $O((\mathcal{H}_{\hat{G}} + n^2) \cdot n \log n)$ steps, both in expectation and with high probability. Since CIW_n completes in $O(n^3 \log n)$ steps, the overall time complexity is also $O((\mathcal{H}_{\hat{G}} + n^2) n \log n)$ steps, both in expectation and with high probability.

We have the following theorems. (See the full version [29] for detailed implementation and proofs.)

Theorem 5. *Protocol* CIG *solves the complete graph identification problem under global fairness, using* $O(n^2)$ *states.*

Theorem 6. *Any execution of* CIG *stabilizes in* $O((\mathcal{H}_{\vec{G}} + n^2)n \log n)$ *steps both in expectation and with high probability under the uniformly random scheduler.*

5   Impossibility

Yasumi et al. [42] introduced the transformed graph $f(G) = (V', E')$ for any communication graph $G = (V, E)$.

Definition 2. For any digraph $G = (V, E)$ with $V = \{v_1, v_2, ..., v_n\}$, the digraph $f(G) = (V', E')$ is defined as follows: $V' = \{v'_1, v'_2, ..., v'_{2n}\}$ and $E' = \{(v'_x, v'_y), (v'_{x+n}, v'_{y+n}) \in V' \times V' | (v_x, v_y) \in E\} \cup \{(v'_1, v'_{z+n}), (v'_{1+n}, v'_z) \in V' \times V' | (v_1, v_z) \in E\} \cup \{(v'_{z+n}, v'_1), (v'_z, v'_{1+n}) \in V' \times V' | (v_z, v_1) \in E\}$.

They proved that for any simple and connected directed graph G, no protocol can distinguish between G and $f(G)$ (Lemma 5 in [42]).[3] Specifically, they showed that for any protocol $\mathcal{P}$, there exists a weakly-fair execution Ξ' on $f(G)$ that stabilizes to output yes (respectively, no) if there is a weakly-fair execution Ξ of $\mathcal{P}$ on G that stabilizes to output yes (respectively, no). Here, the statement that an execution stabilizes to output x means that the execution reaches a stable configuration where all agents output x.

For any complete directed graph K_n with size n, where $n \geq 2$, $f(K_n)$ is not complete. (See Fig. 1 in the full version [29] for the case $n = 2$.) Consequently, the following theorem directly follows from the aforementioned lemma presented by Yasumi et al. [42].

Theorem 7. *There is no protocol that identifies complete graphs with initial knowledge of P under weak fairness.*

References

1. Alistarh, D., Aspnes, J., Eisenstat, D., Gelashvili, R., Rivest, R.L.: Time-space trade-offs in population protocols. In: SODA. pp. 2560–2579 (2017)
2. Alistarh, D., Aspnes, J., Gelashvili, R.: Space-optimal majority in population protocols. In: SODA, pp. 2221–2239 (2018)
3. Alistarh, D., Gelashvili, R.: Polylogarithmic-time leader election in population protocols. In: ICALP, pp. 479–491 (2015)
4. Alistarh, D., Gelashvili, R., Rybicki, J.: Fast graphical population protocols. In: OPODIS 2021, pp. 14:1–14:18 (2022)
5. Alistarh, D., Gelashvili, R., Vojnović, M.: Fast and exact majority in population protocols. In: PODC, pp. 47–56 (2015)

---

[3] They prove this fact for any *undirected graph* $G = (V, E)$, where "undirected" means that $(u, v) \in E$ holds if and only if $(v, u) \in E$ for any $u, v \in V$. However, their proof also applies to general directed graphs without any modifications.

6. Alistarh, D., Rybicki, J., Voitovych, S.: Near-optimal leader election in population protocols on graphs. In: PODC, pp. 246–256 (2022)
7. Angluin, D., Aspnes, J., Chan, M., Fischer, M.J., Jiang, H., Peralta, R.: Stably computable properties of network graphs. In: DCOSS, pp. 63–74 (2005)
8. Angluin, D., Aspnes, J., Diamadi, Z., Fischer, M.J., Peralta, R.: Computation in networks of passively mobile finite-state sensors. Distrib. Comput. **18**(4), 235–253 (2006)
9. Angluin, D., Aspnes, J., Eisenstat, D.: Stably computable predicates are semilinear. In: PODC, pp. 292–299 (2006)
10. Angluin, D., Aspnes, J., Eisenstat, D.: A simple population protocol for fast robust approximate majority. In: DISC, pp. 20–32 (2007)
11. Angluin, D., Aspnes, J., Eisenstat, D.: Fast computation by population protocols with a leader. Distrib. Comput. **21**(3), 183–199 (2008)
12. Angluin, D., Aspnes, J., Eisenstat, D., Ruppert, E.: The computational power of population protocols. Distrib. Comput. **20**(4), 279–304 (2007)
13. Angluin, D., Aspnes, J., Fischer, M.J., Jiang, H.: Self-stabilizing population protocols. ACM Trans. Auton. Adapt. Syst. **3**(4) (2008)
14. Aspnes, J., Beauquier, J., Burman, J., Sohier, D.: Time and space optimal counting in population protocols. In: OPODIS, pp. 13:1–13:17 (2017)
15. Beauquier, J., Blanchard, P., Burman, J.: Self-stabilizing leader election in population protocols over arbitrary communication graphs. In: OPODIS, pp. 38–52 (2013)
16. Beauquier, J., Burman, J., Clavière, S., Sohier, D.: Space-optimal counting in population protocols. In: DISC, pp. 631–646 (2015)
17. Ben-Nun, S., Kopelowitz, T., Kraus, M., Porat, E.: An o(log3/2 n) parallel time population protocol for majority with o(log n) states. In: PODC, pp. 191–199 (2020)
18. Berenbrink, P., Elsässer, R., Friedetzky, T., Kaaser, D., Kling, P., Radzik, T.: A population protocol for exact majority with O(log5/3 n) stabilization time and theta(log n) states. In: DISC, pp. 10:1–10:18 (2018)
19. Berenbrink, P., Elsässer, R., Friedetzky, T., Kaaser, D., Kling, P., Radzik, T.: Time-space trade-offs in population protocols for the majority problem. Distrib. Comput. **34**(2), 91–111 (2021)
20. Berenbrink, P., Giakkoupis, G., Kling, P.: Optimal time and space leader election in population protocols. In: STOC, pp. 119–129 (2020)
21. Bilke, A., Cooper, C., Elsässer, R., Radzik, T.: Brief announcement: Population protocols for leader election and exact majority with o(log2 n) states and o(log2 n) convergence time. In: PODC, pp. 451–453 (2017)
22. Chen, H.P., Chen, H.L.: Self-stabilizing leader election. In: PODC, pp. 53–59 (2019)
23. Chen, H.P., Chen, H.L.: Self-stabilizing leader election in regular graphs. In: PODC, pp. 210–217 (2020)
24. Doty, D., Eftekhari, M.: Efficient size estimation and impossibility of termination in uniform dense population protocols. In: PODC, pp. 34–42 (2019)
25. Doty, D., Eftekhari, M., Gasieniec, L., Severson, E., Uznanski, P., Stachowiak, G.: A time and space optimal stable population protocol solving exact majority. In: FOCS, pp. 1044–1055 (2022)
26. Doty, D., Soloveichik, D.: Stable leader election in population protocols requires linear time. Distrib. Comput. **31**(4), 257–271 (2018)
27. Gąsieniec, L., Stachowiak, G., Uznanski, P.: Almost logarithmic-time space optimal leader election in population protocols. In: SPAA, pp. 93–102 (2019)

28. Gąsieniec, L., Stachowiak, G.: Fast space optimal leader election in population protocols. In: SODA, pp. 265–266 (2018)
29. Kanaya, H., Sudo, Y.: Complete graph identification in population protocols (2024). https://arxiv.org/abs/2408.12862
30. Mertzios, G.B., Nikoletseas, S.E., Raptopoulos, C.L., Spirakis, P.G.: Determining majority in networks with local interactions and very small local memory. In: ICALP, pp. 871–882 (2014)
31. Mertzios, G.B., Nikoletseas, S.E., Raptopoulos, C.L., Spirakis, P.G.: Population protocols for majority in arbitrary networks. In: Extended Abstracts Summer 2015, pp. 77–82 (2017)
32. Michail, O., Chatzigiannakis, I., Spirakis, P.G.: Mediated population protocols. Theoret. Comput. Sci. **412**(22), 2434–2450 (2011)
33. Michail, O., Spirakis, P.G.: Simple and efficient local codes for distributed stable network construction. In: PODC, pp. 76–85 (2014)
34. Michail, O., Spirakis, P.G., Theofilatos, M.: Simple and fast approximate counting and leader election in populations. Inform. Comput. **285**(A), 104698 (2022)
35. Mocquard, Y., Anceaume, E., Aspnes, J., Busnel, Y., Sericola, B.: Counting with population protocols. In: NCA, pp. 35–42 (2015)
36. Mocquard, Y., Anceaume, E., Sericola, B.: Optimal proportion computation with population protocols. In: NCA, pp. 216–223 (2016)
37. Sudo, Y., Masuzawa, T.: Leader election requires logarithmic time in population protocols. Parallel Proces. Lett. **30**(01), 2050005 (2020)
38. Sudo, Y., Nakamura, J., Yamauchi, Y., Ooshita, F., Kakugawa, H., Masuzawa, T.: Loosely-stabilizing leader election in population protocol model. In: SIROCCO, pp. 295–308 (2010)
39. Sudo, Y., Ooshita, F., Izumi, T., Kakugawa, H., Masuzawa, T.: Time-optimal leader election in population protocols. IEEE Trans. Parallel Distrib. Syst. **31**(11), 2620–2632 (2020)
40. Sudo, Y., Ooshita, F., Kakugawa, H., Masuzawa, T.: Loosely stabilizing leader election on arbitrary graphs in population protocols without identifiers or random numbers. IEICE Trans. Inform. Syst. **E103.D**(3), 489–499 (2020)
41. Sudo, Y., Shibata, M., Nakamura, J., Kim, Y., Masuzawa, T.: Self-stabilizing population protocols with global knowledge. IEEE Trans. Parallel Distrib. Syst. **32**(12), 3011–3023 (2021)
42. Yasumi, H., Ooshita, F., Inoue, M.: Population protocols for graph class identification problems. In: OPODIS, pp. 13:1–13:19 (2021)
43. Yasumi, H., Ooshita, F., Inoue, M., Tixeuil, S.: Uniform bipartition in the population protocol model with arbitrary graphs. Theoret. Comput. Sci. **892**, 187–207 (2021)
44. Yokota, D., Sudo, Y., Masuzawa, T.: Time-optimal self-stabilizing leader election on rings in population protocols. IEICE Trans. Fundament. Electron. Commun. Comput. Sci. **E104.A**(12), 1675–1684 (2021)
45. Yokota, D., Sudo, Y., Ooshita, F., Masuzawa, T.: A near time-optimal population protocol for self-stabilizing leader election on rings with a poly-logarithmic number of states. In: PODC, pp. 2–12 (2023)

Efficient Self-stabilizing Simulations of Energy-Restricted Mobile Robots by Asynchronous Luminous Mobile Robots

Keita Nakajima[1]($\boxtimes$) (ID), Kaito Takase[2] (ID), and Koichi Wada[2] (ID)

[1] Tokyo Institute of Technology, Tokyo, Japan
nakajima.k.au@m.titech.ac.jp
[2] Hosei University, Tokyo, Japan
kaito.takase.6z@stu.hosei.ac.jp, wada@hosei.ac.jp

Abstract. In this study, we explore efficient simulation implementations to demonstrate computational equivalence across various models of autonomous mobile robot swarms. Our focus is on RSYNCH, a scheduler designed for energy-restricted robots, which falls between FSYNCH and SSYNCH. We propose efficient protocols for simulating $n(\geq 2)$ luminous ($\mathcal{LUMI}$) robots operating in RSYNCH using $\mathcal{LUMI}$ robots in SSYNCH or ASYNCH. Our contributions are twofold. (1) We introduce protocols that simulate $\mathcal{LUMI}$ robots in RSYNCH using $4k$ colors in SSYNCH and $5k$ colors in ASYNCH, for algorithms that employ k colors. This approach, based on the simulation mechanism in [9], notably reduces the number of colors needed for SSYNCH simulations of RSYNCH, compared to previous efforts. Meanwhile, the color requirement for ASYNCH simulations remains consistent with previous ASYNCH simulations of SSYNCH, facilitating the simulation of RSYNCH in ASYNCH. (2) We establish that for $n = 2$, RSYNCH can be optimally simulated in ASYNCH using a minimal number of colors. Additionally, we confirm that all of our proposed simulation protocols can become self-stabilizing, ensuring functionality from any initial configuration without adding colors.

Keywords: Autonomous mobile robots · Luminous robots · Simulation · Energy-restricted robots · Self-stabilization

This work was supported in part by JSPS KAKENHI Grant Number 20K11685, and 21K11748.

T. Masuzawa et al. (Eds.): SSS 2024, LNCS 14931, pp. 141–155, 2025.
https://doi.org/10.1007/978-3-031-74498-3_10

1   Introduction

1.1   Background and Motivation

The computational issues of autonomous mobile entities operating in a Euclidean space in *Look-Compute-Move* (*LCM*) cycles have been the subject of extensive research in distributed computing. In the *Look* phase, an entity, viewed as a point and usually called *robot*, obtains a snapshot of the space; in the *Compute* phase it executes its algorithm (the same for all robots) using the snapshot as input; it then moves towards the computed destination in the *Move* phase. Repetition of these cycles allows robots to collectively perform some tasks and solve some problems. Research interest has been in determining the impact that *internal* capabilities (e.g., memory, communication) and *external* conditions (e.g. synchrony, activation scheduler) have on the solvability of a problem.

In the most common model, $\mathcal{OBLOT}$, in addition to the standard assumptions of *anonymity* and *uniformity* (robots have no IDs and run identical algorithms), the robots are *oblivious* (no persistent memory to record information of previous cycles) and *silent* (without explicit means of communication). The computing in this model has been the object of intensive research since its introduction in [25]. Extensive investigations have been carried out to clarify the computational limitations and powers of these robots for basic coordination tasks such as *Gathering* (e.g., [1–3,6–8,12,18,25]), *Pattern Formation* (e.g., [13,16,25–27]), and *Flocking* (e.g., [5,17,24]); for a recent account of the state of the art on some of these problems, see [10] and the chapters therein.

A $\mathcal{LUMI}$ model, which provides a limited but persistent memory and means of communication, is available for the $\mathcal{OBLOT}$ model robot, formally defined and analyzed in [9], following a suggestion in [23]. In this model, each robot is equipped with a constant-sized memory (called *light*), whose value (called *color*) can be set during the *Compute* phase. The light is visible to all the robots and is persistent in the sense that it is not automatically reset at the end of a cycle. Hence, these luminous robots are capable of both remembering and communicating a constant number of bits in each cycle.

An important result is that, despite these limitations, the simultaneous presence of persistent memory and communication renders luminous robots strictly more powerful than oblivious robots [9]. This, in turns, has opened the question about the individual computational power of the two internal capabilities, memory and communication, and motivated the investigations on two sub-models of $\mathcal{LUMI}$: $\mathcal{FSTA}$ where the robots have a constant-size persistent memory but are silent, and $\mathcal{FCOM}$, where robots can communicate a constant number of bits but are oblivious (e.g., see [4,14,15,21,22]).

All these studies across various models have highlighted the crucial role played by two interrelated *external* factors: the level of synchronization and the activation schedule provided by the system. As in other types of distributed computing systems, there are two different settings; the synchronous and the asynchronous settings. In the setting *synchronous* (also called *semi-synchronous*) (SSYNCH) setting, introduced in [25], time is divided into discrete intervals, called *rounds*. In each round, an arbitrary but nonempty subset of the robots is activated, and they simultaneously perform exactly one atomic *Look-Comp-Move* cycle. The selection of which robots are activated at a given round is made by an adversarial scheduler, constrained only to be fair, i.e., every robot is activated infinitely often. Weaker form of synchronous adversaries have also been introduced and investigated. The most important and extensively studied is the *fully-synchronous* (FSYNCH) scheduler, which activates all the robots in every round.

In the *asynchronous* setting (ASYNCH), introduced in [11], there is no common notion of time and each robot is activated independently of the others. it allows for arbitrary but finite delays between the *Look*, *Comp* and *Move* phases, and each movement may take an arbitrary but finite amount of time. The duration of each robot's cycle, as well as the timing of robot's activation, are controlled by an adversarial scheduler, constrained only to be fair, i.e., every robot must be activated infinitely often.

In this paper, we focus on energy-restricted robots and their scheduler RSYNCH, where RSYNCH is a scheduler designed with the energy constraints of robots in mind [4]. In this framework, robots possess energy that is both consumable and renewable. Specifically, an activated robot completes one LCM cycle, using up all its energy, which is then restored during a subsequent period of inactivity. Research has explored the computational capabilities of these energy-constrained robots, revealing that RSYNCH serves as an intermediary between FSYNCH and SSYNCH, which are models for robots without energy restrictions. The distinct computational potentials under these three schedules have been established. RSYNCH represents a novel consideration in robotic models. It typically entails a sequence where all robots are activated (FSYNCE), followed by rounds so that the robots activated in any round and those activated in the next round are disjoint. Operating robots in energy-constrained environments is a critical challenge for ensuring the efficiency and sustainability of robotic systems. Additionally, enhancing compatibility between different synchronization models, such as synchronous and asynchronous models, allows for the development of more flexible and adaptive robotic systems. Motivated by these considerations, this paper considers the simulation of RSYNCH by SSYNCH or ASYNCH.

1.2   Contributions

Like in other types of distributed systems, understanding the computational difference between various levels of synchrony and asynchrony has been a primary research focus. In the robot model, to "separate" between the computational power of robots in two settings, we demonstrate that certain problems are solvable in one model but unsolvable in another. For example, in $\mathcal{OBLOT}$, *Rendezvous* is unsolvable in SSYNCH, but solvable in FSYNCH[25], indicating a separation between FSYNCH and SSYNCH in $\mathcal{OBLOT}$. Conversely, to show that a weaker model is equivalent to a stronger model, we devise a *simulation protocol* that allows the correct execution of any protocol from the stronger model in the weaker model. The first attempt in the robot model involves constructing a simulation protocol for any $\mathcal{LUMI}$ protocol in SSYNCH using $\mathcal{LUMI}$ robots in ASYNCH [9]. This protocol employs $5k$ colored lights in $\mathcal{LUMI}$ robots to simulate SSYNCH protocols using k colors.[1]

In this paper, we focus on making such simulations more efficient, specifically, considering simulations that involve $\mathcal{LUMI}$ robots operating in RSYNCH and $\mathcal{LUMI}$ robots operating in ASYNCH under the most unrestricted adversary. Though RSYNCH is introduced for modeling energy-restricted robots [4], it is interesting in its own right because $\mathcal{LUMI}$ robots in ASYNCH have the same power as those in RSYNCH [4,9], and $\mathcal{FCOM}$ robots in RSYNCH have the same power as $\mathcal{LUMI}$ robots in RSYNCH [4].

The simulator of $\mathcal{LUMI}$ robots with k colors in SSYNCH uses $\mathcal{LUMI}$ robots in ASYNCH and utilizes $5k$ colors [9], including 5 colors to control the simulation. Therefore, we aim to reduce the number of colors used to control the simulation.

Table 1 presents both the previous simulation results and our new results. Here for any model, $M \in \{\mathcal{FCOM}, \mathcal{LUMI}\}$ and any adversarial scheduler $A \in \{$ RSYNCH, SSYNCH, ASYNCH$\}$, M^A denotes the robot model M working in A.

The first simulation protocol was designed for $\mathcal{LUMI}$ robots in SSYNCH using $\mathcal{LUMI}$ robots under the strongest adversary, ASYNCH. This simulation ensures that in every round, the only selection of which robots are activated is made by an adversarial scheduler, constrained only to be fair [9]. The simulation employs 5 light colors to control the process. A unique property of this simulator is that not only does the simulated protocol function in SSYNCH, but it also ensures that any robot is activated exactly once during a certain duration, maintaining fairness in the scheduler. We will leverage this property to decrease the number of colors used when simulating $\mathcal{LUMI}$ robots in RSYNCH. When simulating a protocol involving k-color $\mathcal{LUMI}$ robots in an unfair SSYNCH, the simulator will use $3k$ colors for $\mathcal{LUMI}$ robots in ASYNCH [19].

In this paper, we first establish that by utilizing the property of the simulator detailed in [9],

(1) k-color $\mathcal{LUMI}$ robots in RSYNCH can be simulated by $4k$-color $\mathcal{LUMI}$ robots in SSYNCH.

(2) k-color $\mathcal{LUMI}$ robots in RSYNCH can be simulated by $5k$-color $\mathcal{LUMI}$ robots in ASYNCH.

Previously, case (1) required $36k$ colors [4]. In contrast, our simulator for case (2) also uses $5k$ colors, effectively simulating $\mathcal{LUMI}$ robots in RSYNCH using $\mathcal{LUMI}$ robots in ASYNCH. Additionally, we demonstrate that when the number of robots is limited to 2 ($n = 2$), the simulator in the case (1) can be implemented more efficiently, Specifically, we show that:

(3) k-color $\mathcal{LUMI}$ robots in RSYNCH can be simulated by $3k$-color $\mathcal{LUMI}$ robots in ASYNCH.

---

[1] In the case of $k = 1$, this protocol simulates $\mathcal{OBLOT}$ robots working in SSYNCH with $\mathcal{LUMI}$ robots with 5 colors in ASYNCH.

(4) In the case of $k = 1$, $\mathcal{OBLOT}$ robots in RSYNCH cannot be simulated by 2-color $\mathcal{LUMI}$ robots in ASYNCH. This demonstrates that the number of colors used in the simulator shown in (3) is optimal.

We also confirm that all our proposed simulation protocols are self-stabilizing, ensuring functionality from any initial configuration. These self-stabilization can be done without increasing the number of colors.

Table 1. The previous results and this paper's results.

# Robots	Simulating model	Simulated model	# Colors	Self-stabilized	References
$n \geq 2$	$\mathcal{LUMI}^A$	$\mathcal{LUMI}^S$	$5k$	Yes (modify)	[9]
	$\mathcal{LUMI}^S$	$\mathcal{LUMI}^{RS}$	$36k$	No	[4]
	$\mathcal{LUMI}^A$	$\mathcal{LUMI}^{S*}$	$3k$	Yes	[19]
	$\mathcal{FCOM}^F$	$\mathcal{LUMI}^F$	$2k^2$	No	[15]
	$\mathcal{FCOM}^{RS}$	$\mathcal{LUMI}^S$	$64k2^k$	No	[4]
	$\mathcal{LUMI}^S$	$\mathcal{LUMI}^{RS}$	$4k$	Yes	This paper
	$\mathcal{LUMI}^A$	$\mathcal{LUMI}^{RS}$	$5k$	Yes	This paper
$n = 2$	$\mathcal{LUMI}^A$	$\mathcal{LUMI}^{RS}$	$3k$ (optimal)	Yes	This paper

* Unfair SSYNCH

2 Preliminaries

2.1 Robots

The systems considered in this paper consist of a team $R = \{r_0, \ldots, r_{n-1}\}$ of computational entities moving and operating in the Euclidean plane $\mathbb{R}^2$. Viewed as points and called *robots*, the entities can move freely and continuously in the plane. Each robot has its own local coordinate system and it always perceives itself at its origin; there might not be consistency between the coordinate systems of the robots. A robot is equipped with sensorial devices that allow it to observe the positions of the other robots in its local coordinate system.

Robots are *identical*: they are indistinguishable by their appearance, and they execute the same protocol. Robots are *autonomous*, without central control.

At any time, a robot is *active* or *inactive*. Upon becoming active, a robot r executes a *Look-Compute-Move* (*LCM*) cycle performing the following three operations:

1. *Look:* The robot activates its sensors to obtain a snapshot of the positions occupied by the robots with respect to its own coordinate system.[2]
2. *Compute:* The robot executes its algorithm using the snapshot as input. The result of the computation is a destination point.

---

[2] This is called the *full visibility* (or unlimited visibility) setting; restricted forms of visibility have also been considered for these systems [12].

3. *Move:* The robot moves in a straight line towards the computed destination; if the destination is the current location, the robot stays still.

When inactive, a robot is idle. All robots are initially idle. The time it takes to complete a cycle is assumed to be finite and the operations *Look* and *Compute* are assumed to be instantaneous.

In the standard model, $\mathcal{OBLOT}$, the robots are *silent*: they have no explicit means of communication; furthermore, they are *oblivious*: at the start of a cycle, a robot has no memory of observations and computations performed in previous cycles.

In the other common model, $\mathcal{LUMI}$, each robot r is equipped with a persistent variable of visible state $Light[r]$, called *light*, whose values are taken from a finite set C of states called *colors* (including the color that represents the initial state when the light is off). The colors of the lights can be set in each cycle by r at its *Compute* operation. A light is *persistent* from one computational cycle to the next: the color is not automatically reset at the end of a cycle; the robot is otherwise oblivious, forgetting all other information from previous cycles. If any color is not set to some light, the color of the light remains unchanged. In $\mathcal{LUMI}$, the *Look* operation produces a colored snapshot; i.e., it returns the set of pairs (*position, color*) of the other robots.[3] Note that if $|C| = 1$, then the light is not used; thus, this case corresponds to the $\mathcal{OBLOT}$ model.

In all the above models, a *configuration* $\mathcal{C}(t)$ at time t is the multiset of the n pairs $(x_i(t), c_i(t))$, where $c_i(t)$ is the color of robot r_i at time t.

2.2   Schedulers, Events

With respect to the activation schedule of the robots, and the duration of their *LCM* cycles, the fundamental distinction is between the *synchronous* and *asynchronous* settings.

In the *synchronous* setting (SSYNCH), also called *semi-synchronous* and first studied in [25], time is divided into discrete intervals, called *rounds*; in each round, a non-empty set of robots is activated and they simultaneously perform a single *Look-Comp-Move* cycle in perfect synchronization. The selection of which robots are activated at a given round is made by an adversarial scheduler, constrained only to be fair (i.e., every robot is activated infinitely often). The particular synchronous setting, where every robot is activated in every round is called *fully-synchronous* (FSYNCH). In a synchronous setting, without loss of generality, the expressions "i-th round" and "time $t = i$" are used as synonyms.

In the *asynchronous* setting (ASYNCH), first studied in [11], there is no common notion of time, the duration of each phase is finite but unpredictable and might be different in different cycles, and each robot is activated independently of the others. The duration of the phases of each cycle as well as the decision of when a robot is activated is controlled by an adversarial scheduler, constrained only to be fair, i.e., every robot must be activated infinitely often.

---

[3] If (strong) multiplicity detection is assumed, the snapshot is a multi-set.

In the asynchronous settings, the execution by a robot of any of the operations *Look*, *Compute* and *Move* is called an *event*. We associate relevant time information to events: for the *Look* (resp., *Compute*) operation, which is instantaneous, the relevant time is t_L (resp., t_C) when the event occurs; for the *Move* operation, these are the times t_B and t_E when the event begins and ends, respectively. Let $\mathcal{T} = \{t_1, t_2, ...\}$ denote the infinite ordered set of all relevant times; i.e., $t_i < t_{i+1}, i \in \mathbb{N}$. In the following, to simplify the presentation and without any loss of generality, we will refer to $t_i \in \mathcal{T}$ simply by its index i; i.e., the expression "time t" will be used to mean "time t_t".

Consider now the synchronous scheduler, we shall call RSYNCH, obtained from SSYNCH by adding the following *restricted-repetition condition* to its activation sequences:

$$\left[\forall i \geq 1, e_i = R\right] \text{ or } \left[\exists p \geq 0 : \left(\left[\forall i \leq p, (e_i = R)\right] \text{ and}\right.\right.$$

$$\left.\left.\left[\forall i > p, (e_i \neq \emptyset \text{ and } e_i \neq R \text{ and } e_i \cap e_{i+1} = \emptyset)\right]\right)\right],$$

where an *activation sequence* of R is an infinite sequence $E = \langle e_1, e_2, \ldots, e_i, \ldots \rangle$, and $e_i \subseteq R$ denotes the set of robots activated in round i. That is, this scheduler is composed of sequences where the prefix is a (possibly empty) sequence of R and, if the prefix is finite, the rest are non-empty sets satisfying the constraint $(e_i \cap e_{i+1} = \emptyset)$.

2.3   Computational Relationships

Let $\mathcal{M} = \{\mathcal{LUMI}, \mathcal{OBLOT}\}$ be the set of models under investigation and $\mathcal{S} = \{$ RSYNCH, SSYNCH, ASYNCH$\}$ be the set of schedulers under consideration.

We denote by $\mathcal{R}$ the set of all robot teams that satisfy the core assumptions (i.e. they are identical, autonomous and operate in LCM cycles), and operate under rigidity of movements, chirality, and variable disorientation. By $\mathcal{R}_n \subset \mathcal{R}$ we denote the set of all teams of size n.

Given a model $M \in \mathcal{M}$, a scheduler $S \in \mathcal{S}$, and a team of robots $R \in \mathcal{R}$ we denote by $Task(M, k, S; R)$ the set of problems solvable by R in M with k colors under adversarial scheduler S.

For simplicity of notation, let $M_k^{RS}(R), M_k^S(R)$, and $M_k^A(R)$ denote $Task(M, k, \text{RSYNCH}; R), Task(M, k, \text{SSYNCH}; R)$, and $Task(M, k, \text{ASYNCH}; R)$, respectively.[4]

3   Simulation of RSYNCH on $\mathcal{LUMI}$

3.1   4-Color Simulation of RSYNCH by SSYNCH

In this section, we show that semi-synchronous systems equipped with a light with 4 colors are at least as powerful as *restricted-repetition* system (RSYNCH) without lights. More precisely, we have:

---

[4] Since $\mathcal{OBLOT}$ robots have no light (one color), the suffix $k = 1$ is omitted.

Theorem 1. $\forall R \in \mathcal{R}, \mathcal{OBLOT}^{RS}(R) \subseteq \mathcal{LUMI}_4^S(R).$

We show that every problem solvable by a set of $\mathcal{OBLOT}$ robots under RSYNCH can be also solved by a set of $\mathcal{LUMI}$ robots with 4 colors under SSYNCH. We do so constructively: we present a *simulation* protocol for $\mathcal{OBLOT}$ robots that allows them to correctly execute in RSYNCH any protocol $\mathcal{P}$ given in input.

We first simulate a restricted semi-synchronous scheduler called *multiple-slicing*, in which any robot is activated exactly once in some duration, where the duration is called *mega-cycle*. Then we modify the simulator working to attain the condition of RSYNCH.

A scheduler that a group of n robots, starting from time $t = 0$, after n successive activation rounds (slices), all robots in the system will have been activated exactly once. This is called centralized slicing SSYNCH. We extend the centralized slicing SSYNCH to $R_1^1, R_2^1, \ldots, R_{k_1}^1; R_1^2, R_2^2, \ldots, R_{k_2}^2; \ldots; R_1^i, R_2^i, \ldots, R_{k_i}^i; \ldots (1 \leq k_i \leq n)$, where for each $i \geq 1$, $R_1^i, R_2^i, \ldots, R_{k_i}^i$ are a partition of R. This scheduler is called *multiple-slicing*[5] SSYNCH. At this time, R_j^i is called the j-th *stage* in the i-th *mega-cycle*.

If the multiple slicing SSYNCH satisfies that $R_{k_i}^i \cap R_1^{i+1} = \emptyset$ for every $i \geq 1$, this scheduler works to satisfy the *disjoint* condition of RSYNCH.

Specifically, the robot should have one of the following colors:

- *T(rying)*: denotes not having executed $\mathcal{P}$ in a current mega-cycle yet.
- *M(oving)*: denotes having already executed $\mathcal{P}$ once.
- *S(topped)*: denotes after executing $\mathcal{P}$ except in the last stage of a mega-cycle.
- *S'(topped)*: denotes after executing $\mathcal{P}$ in the last stage of a mega-cycle.

When a robot with α state, it is called an α-robot. We also denote state set as a global configuration (e.g. $\forall T, S$ means each robot's state either T or S, and there is at least one robot with such state).

Considering the states in this way, the rule of protocol can be considered as follows.

Protocol Description

Figure 1-(a) shows the transition diagram representation of SIM_S^{RS}.[6] The protocol SIM_S^{RS} uses four colors: T, M, S, S'. Initially, all lights are set to T.

The protocol simulates a sequence of mega-cycles, each of which starts with some robots trying to execute protocol $\mathcal{P}$ and ends with all robots finishing the mega-cycle having executed $\mathcal{P}$ once. After this end configuration, it transits to start one, and a new mega-cycle starts.

During each mega-cycle, each robot gets the opportunity to execute one step of the protocol $\mathcal{P}$. A T-robot r, tries to execute protocol $\mathcal{P}$. However, the robot is allowed to execute $\mathcal{P}$ only if there are no M-robots (i.e. robots that executed $\mathcal{P}$ before this stage). If that is the case, r changes its color to M. Otherwise, it does nothing (i.e. it waits until no M-robots exist). M-robots, after executing $\mathcal{P}$, will change to S only when no S'-robot exists and T-robots exist (i.e. it is not in the

---

[5] It is centralized slicing if $|R_j^i| = 1$ for every $i(\geq 1)$ and $j(1 \leq j \leq k_i)$.

[6] The pseudo code of the protocol is presented in [20].

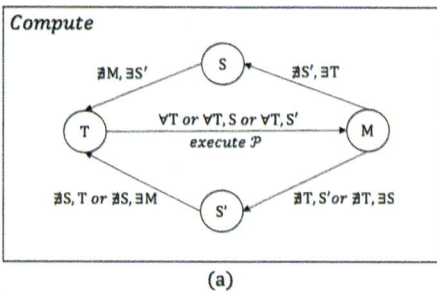
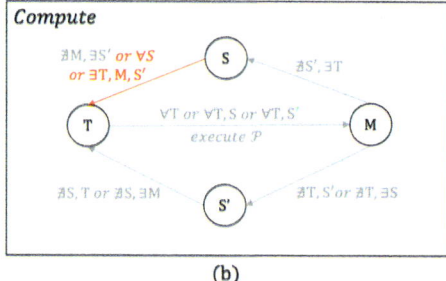

(a)   (b)

Fig. 1. Transition diagram representations of protocol SIM_S^{RS} (a), and self-stabilizing protocol ss-SIM_S^{RS} (b). The label in the nodes represents the color of the light of the executing robot. The label of an edge expresses the condition on the lights of all the other robots that must be satisfied for the transition to occur. The notation "$\forall A, B$" means: "$\{\mathrm{Light[r]} \mid \forall r \in R\} = \{A, B\}$", "$\exists A$" means: "$\exists r \in R, \mathrm{Light[r]} \in \{A\}$", "$\nexists A$" means: "$\{\mathrm{Light[r]} \mid \forall r \in R\} \cap \{A\} = \emptyset$". Conditions, colored red in (b) are newly added.

last stage). Changing to S' only when no $c (\in \{T, S'\})$-robot exists (i.e. all robots execute $\mathcal{P}$), or no T-robot exists and S-robots exist (i.e. it is in the last stage). If the robots change to S, after some time, each robot will be colored either S (i.e. executed $\mathcal{P}$) or T (i.e. not executed $\mathcal{P}$), else all robot will be colored S' (i.e. this happens when all robots execute $\mathcal{P}$ at the same stage). At this time, T-robots are given another opportunity to execute $\mathcal{P}$. Thus, a cycle of protocol SIM_S^{RS} consists of several stages such that, in each stage, at least one robot executes $\mathcal{P}$ while other robots wait. Eventually all robots will colored either S or S' (i.e. each robot has executed $\mathcal{P}$ once), and the cycle ends when S-robots change to T. At this point, the S'-robots do nothing; when this process is completed, a new cycle starts. Whenever a new mega-cycle begins, T-robots and S'-robots can exist, and some of the T-robots execute $\mathcal{P}$ in the first stage. Here, since the S'-robots have executed $\mathcal{P}$ in the last stage of the previous mega-cycle, the executing T-robots and the S'-robots are mutually disjoint and the condition of RSYNCH is satisfied. S'-robots change to T only when no $c (\in \{T, S\})$-robot exists (i.e. all robots execute $\mathcal{P}$ at the same stage), or no S-robot exists and M-robots exist (i.e. after the first stage of new mega-cycle).

Correctness The left part of Fig. 2 shows the transition diagram of configurations when performing SIM_S^{RS}. Each megacycle begins with $\forall T$ and ends with $\forall S'$ or $\forall S, S'$. If it ends with $\forall S'$, it means that all robots have executed the algorithm (FSYNCH-phase), while ending with $\forall S, S'$ represents all other cases (Disjoint-phase). The S'-robot signifies that it has executed at the end of this megacycle, and it is guaranteed not to execute at the beginning of the next megacycle. If a megacycle starts with $\forall T, S'$, it indicates the FSYNCH-phase has ended, and the Disjoint-phase is being executed, ending with $\forall S, S'$. Regarding stages, the FSYNCH-phase starts with $\forall T$ and ends with $\forall S'$ in one stage, while the Disjoint-phase begins with either $\forall T, S$ or $\forall T, S'$ and ends with either $\forall T, S$ or $\forall S, S'$. A stage that ends with $\forall S, S'$ transitions to $\forall T, S'$, and a new mega-cycle begins. The next robots to execute are chosen from all the non-S'-robots, operating in a way that satisfies the disjointness of RSYNCH. Due to space limitations in the paper, details of the correctness are relegated to [20], we obtain the following theorem.

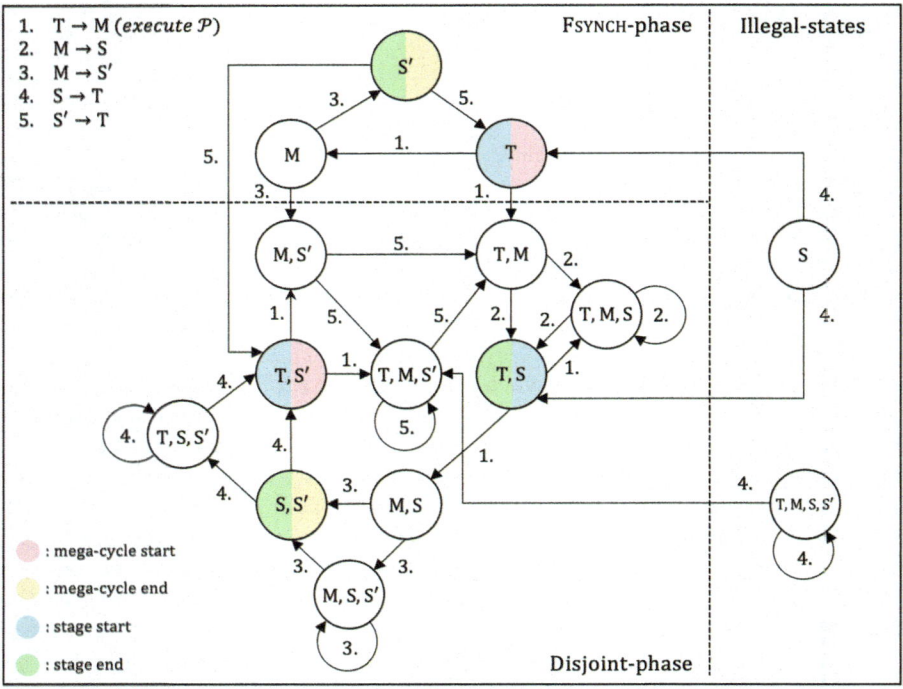

Fig. 2. Transition diagram of configurations (protocol SIM_S^{RS}(the left part of this figure) and self-stabilizing ss-SIM_S^{RS}).

Theorem 2. *Protocol* SIM_S^{RS} *is correct, i.e. any execution of protocol* SIM_S^{RS} *in* SSYNCH *corresponds to a possible execution of* $\mathcal{P}$ *in* RSYNCH.

Thus, Theorem 1 and its corollary hold.

Corollary 1. $\forall R \in \mathcal{R}, \mathcal{LUMI}_k^{RS}(R) \subseteq \mathcal{LUMI}_{4k}^S(R)$.

3.2   Making SIM_S^{RS} Self-stabilizing

An simulation protocol is *self-stabilizing* for protocol $\mathcal{P}$ if it satisfies the conditions of the scheduler under which $\mathcal{P}$ is executed from any initial configuration, stating with all robots in inactive.

We can make the protocol SIM_S^{RS} self-stabilizing (denoted as ss-SIM_S^{RS}). Figure 1 (b) shows ss-SIM_S^{RS}, and the red-labeled part was added to achieve self-stabilization.[7]

We can show that SIM_S^{RS} works correctly even if it starts from any configuration appearing on the left part of Fig. 2, and ss-SIM_S^{RS} works correctly from any configuration by adding the red-labeled part in Fig. 1 (b). The proofs will be shown in [20].

Theorem 3. *Protocol ss-SIM_S^{RS} is correct and self-stabilizing, i.e. from any initial global configuration, any execution of protocol SIM_S^{RS} in* SSYNCH *corresponds to a possible execution of* $\mathcal{P}$ *in* RSYNCH.

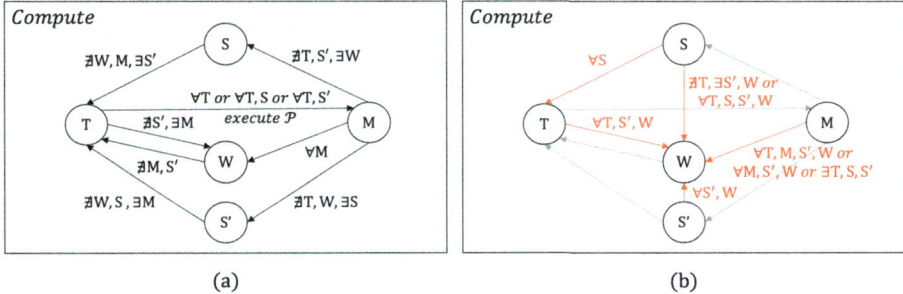

(a)   (b)

Fig. 3. Transition diagram representations of (a) protocol SIM_A^{RS} and (b) self-stabilizing protocol ss-SIM_A^{RS}. The condition colored red in (b) is newly added to SIM_A^{RS} to achieve self-stabilization, but not newly added in ss-SIM_A^{RS} are omitted.

3.3 5-Color Simulation of RSYNCH by ASYNCH

If we use one more color (that is, use 5 colors), protocol SIM_S^{RS} (resp. ss-SIM_S^{RS}) can be extended so that it works in ASYNCH to simulate RSYNCH from an initial configuration (resp. any initial configuration). They are called SIM_A^{RS} and ss-SIM_A^{RS}, and shown in Fig. 3 (a) and (b), respectively.[8]

In addition to T, M, S, S' which have the same meaning as the colors in Protocol SIM_S^{RS}, a color W (Waiting) is introduced. A robot r that executes the simulated algorithm changes color T to M and executes the algorithm. Unlike in SSYNCH, in ASYNCH, robots observing r before it changes to M get the same snapshot as r, but after changing to M, the snapshot differs, thus robots observing an M-robot cannot execute the algorithm. At this time, a robot observing an M-robot changes T to W and pend its execution until the next stage. Afterward, once all M-robots have completed their algorithm execution and changed to S,

---

[7] Also the pseudo code of the protocol is presented in [20].

[8] These pseudo codes are shown in [20].

the W-robots return to T, preparing for the execution of the next stage. These transitions, along with those involving the color of W, are the same as in Protocol SIM_S^{RS}, except for the transitions related to W, where M-robots change their color from M to S if W-robots exist and S'-robots and T-robots do not exist (i.e. it is not in the last stage). Changing M to S' if S-robots exist and T-robots and W-robots do not exist (i.e. it is in the last stage). $c(\in \{S, S'\})$-robots change to T if there do not exist W-robots.[9] Moreover, after all robots have executed the algorithm (FSYNCH-phase), they all become M-robots. In this case, any robot observing all M-robots changes their color to W. These W-robots then become capable of executing the algorithm in the next stage.

The transition diagrams of configurations when performing SIM_A^{RS} and ss-SIM_A^{RS}, correctness of the protocols can be shown in a way similar to the cases of SIM_S^{RS} and ss-SIM_S^{RS}, are shown in [20].

Theorem 4. *Protocols* SIM_A^{RS} *and* ss-SIM_A^{RS} *are correct and the latter is self-stabilizing.*

Thus, we have the following theorem and corollary.

Theorem 5. $\forall R \in \mathcal{R}, \mathcal{OBLOT}^{RS}(R) \subseteq \mathcal{LUMI}_5^A(R).$

Corollary 2. $\forall R \in \mathcal{R}, \mathcal{LUMI}_k^{RS}(R) \subseteq \mathcal{LUMI}_{5k}^A(R).$

4   Optimal Simulation of RSYNCH by ASYNCH with Two Robots

4.1   3-Color Simulation of RSYNCH by ASYNCH

In this subsection, we show the following theorem constructively.

Theorem 6. $\forall R \in \mathcal{R}_2, \mathcal{OBLOT}^{RS}(R) \subseteq \mathcal{LUMI}_3^A(R).$

To do so, we present a $\mathcal{LUMI}_3^A$ protocol SIM-2$_A^{RS}$ that produces RSYNCH execution of any $\mathcal{OBLOT}^{RS}$ protocol $\mathcal{P}$. We also show that the number of colors used in SIM-2$_A^{RS}$ is optimal.

The transition diagram representation is shown in Fig. 4-(a) and the pseudo code of the protocol is shown in [20]. It uses three colors $T, M,$ and S. The meaning of the colors is almost the same as those of Protocol SIM_S^{RS}. The transition of configurations is shown in Fig. 4-(b). If SIM-2$_A^{RS}$ works in SSYNCH, it is easily verified that it simulates $\mathcal{P}$ working in RSYNCH as follows; As long as the both robots continue to be activated simultaneously, since the transition repeats $(T,T) \rightarrow (M,M) \rightarrow (S,S) \rightarrow (T,T)$, and a and b have performed $\mathcal{P}$ simultaneously when changing T to M, SIM-2$_A^{RS}$ makes $\mathcal{P}$ work in FSYNCH. Once

---

[9] Based on the meanings of S and S', the transition from S to T occurs when there are no W-robots and M-robots, but there are S'-robots present. Conversely, the transition from S' to T occurs when there are no W-robots and S-robots, but M-robots are present.

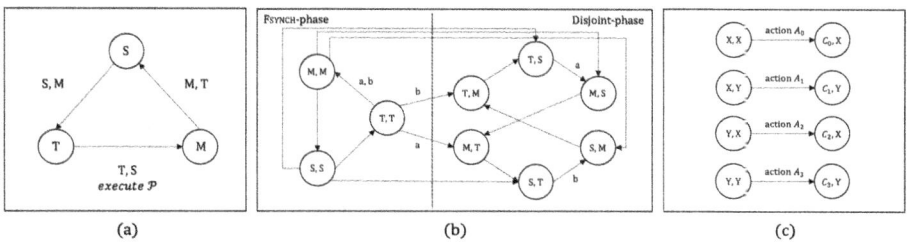

Fig. 4. (a) Transition diagram of protocol SIM-2_A^{RS}. (b) Transition diagram of state transitions (protocol SIM-2_A^{RS}). (c) Transition diagram of protocol two colors SIM-2_S^{RS}.

only one robot, say a, is activated at (T, T),[10] the configuration becomes (M, T) and a has executed $\mathcal{P}$. After that the transition of the configurations repeats the loop of $(M, T) \to (S, T) \to (S, M) \to (T, M) \to (T, S) \to (M, S) \to (M, T)$ shown in Fig. 4-(b) for any activation schedule in SSYNCH, where b performs $\mathcal{P}$ and then a performs $\mathcal{P}$ in the loop. For example, when the configuration is (M, T), since a changes M to S and b does not change its color in SIM-2_A^{RS}, if a is activated once regardless of activation of b, the configuration becomes (S, T). Other transitions are similar. Thus these activation satisfies RSYNCH adding the preceding simultaneous activation of the both robots. In [20], we show that SIM-2_A^{RS} can work correctly in ASYNCH, although it is very complicated due to asynchronicity. Also it is verified that SIM-2_A^{RS} works correctly from any initial configuration shown in 4-(b), that is, SIM-2_A^{RS} is self-stabilizing.

Theorem 7. *Protocol SIM-2_A^{RS} is correct and self-stabilizing.*

Thus, we have Theorem 6 and the following corollary.

Corollary 3. $\forall R \in \mathcal{R}_2, \mathcal{LUMI}_k^{RS}(R) \subseteq \mathcal{LUMI}_{3k}^A(R)$.

4.2   The Impossibility of the Simulation with 2 Colors

In this subsection, we show that any simulation of two $\mathcal{OBLOT}^{RS}$ robots by two $\mathcal{LUMI}^S$ robots with two colors is impossible. Thus, the three-color simulation in the preceding subsection is optimal with respect to the number of colors.

Let a and b be two $\mathcal{LUMI}^S$ robots with two colors X and Y. Since any action in any simulation protocol depends on only own color and the other's color, when α-robot observes β-robot, determining action $\mathcal{A} \in \{$ "Execute $\mathcal{P}$", "no action"$\}$ and the next color γ defines a simulating protocol, where $\alpha, \beta, \gamma \in \{X, Y\}$.[11] For example, in algorithm two colors SIM-2_S^{RS} setting $\mathcal{A}_0 = $ execute $\mathcal{P}$, $\mathcal{A}_i = $ no action $(i = 1, 2, 3)$ and $C_0 = Y$, $C_i = X (i = 1, 2, 3)$, one simulating protocol is defined. However, this protocol cannot simulate RSYNCH by $\mathcal{LUMI}^S$ robots,

---

[10] When it is activated at (M, M) or (S, S), we can show similarly noting that at any time, when only one robot is activated, two robots have executed $\mathcal{P}$ simultaneously.

[11] The pseudo code of the protocol is presented in [20].

because considering a schedule that only a robot is activated, a performs $\mathcal{P}$ consecutively and so it violates RSYNCH.

There are 2^8 simulating protocols including meaningless and these are all protocols which simulate two $\mathcal{LUMI}^{RS}$ robots by two $\mathcal{LUMI}^S$ robots with two colors, and we can verify that schedules all the simulating protocols produce violate RSYNCH. Thus, we obtain the following theorem.

Theorem 8. *SIM-2_A^{RS} is an optimal simulating protocol with respect to the number of colors.*

5   Concluding Remarks

In this paper, we discuss efficient protocols for simulating RSYNCH under ASYNCH or SSYNCH for $\mathcal{LUMI}$ robots. In particular, for the simulation of RSYNCH under SSYNCH we have significantly reduced the number of colors previously required. Also, in the simulation of RSYNCH under ASYNCH, we have achieved the simulation with the same number of colors as used in previous work's simulations of SSYNCH under ASYNCH. Furthermore, for the case of $n = 2$, we have realized the simulation of RSYNCH under ASYNCH with an optimal number of colors. We have also shown that all our proposed protocols are self-stabilizing and their self-stabilization can be done without increasing the number of colors. An outstanding issue is the reduction of the number of colors needed for simulating $\mathcal{LUMI}$ robots under RSYNCH for $\mathcal{FCOM}$ robots.

References

1. Agmon, N., Peleg, D.: Fault-tolerant gathering algorithms for autonomous mobile robots. SIAM J. Comput. **36**(1), 56–82 (2006)
2. Ando, H., Osawa, Y., Suzuki, I., Yamashita, M.: A distributed memoryless point convergence algorithm for mobile robots with limited visivility. IEEE Trans. Robot. Autom. **15**(5), 818–828 (1999)
3. Bouzid, Z., Das, S., Tixeuil, S.: Gathering of mobile robots tolerating multiple crash faults. In: The 33rd International Conference on Distributed Computing Systems, pp. 334–346 (2013)
4. Buchin, K., Flocchini, P., Kostitsyna, I., Peters, T., Santoro, N., Wada, K.: On the computational power of energy-constrained mobile robots: algorithms and cross-model analysis. In: Proceedings of 29th International Colloquium on Structural Information and Communication Complexity (SIROCCO), pp. 42–61 (2022)
5. Canepa, D., Potop-Butucaru, M.: Stabilizing flocking via leader election in robot networks. In: Proceedings 10th International Symposium on Stabilization, Safety, and Security of Distributed Systems (SSS), pp. 52–66 (2007)
6. Cicerone, S., Di Stefano, G., Navarra, A.: Gathering of robots on meeting-points. Distrib. Comput. **31**(1), 1–50 (2018)
7. Cieliebak, M., Flocchini, P., Prencipe, G., Santoro, N.: Distributed computing by mobile robots: Gathering. SIAM J. Comput. **41**(4), 829–879 (2012)
8. Cohen, R., Peleg, D.: Convergence properties of the gravitational algorithms in asynchronous robot systems. SIAM J. Comput. **34**(15), 1516–1528 (2005)

9. Das, S., Flocchini, P., Prencipe, G., Santoro, N., Yamashita, M.: Autonomous mobile robots with lights. Theoret. Comput. Sci. **609**, 171–184 (2016)
10. Flocchini, P., Prencipe, G., Santoro (eds).: Distributed Computing by Mobile Entities. Springer, Berlin (2019)
11. Flocchini, P., Prencipe, G., Santoro, N., Widmayer, P.: Hard tasks for weak robots: the role of common knowledge in pattern formation by autonomous mobile robots. In: 10th International Symposium on Algorithms and Computation (ISAAC), pp. 93–102 (1999)
12. Flocchini, P., Prencipe, G., Santoro, N., Widmayer, P.: Gathering of asynchronous robots with limited visibility. Theoret. Comput. Sci. **337**(1–3), 147–169 (2005)
13. Flocchini, P., Prencipe, G., Santoro, N., Widmayer, P.: Arbitrary pattern formation by asynchronous oblivious robots. Theoret. Comput. Sci. **407**, 412–447 (2008)
14. Flocchini, P., Santoro, N., Viglietta, G., Yamashita, M.: Rendezvous with constant memory. Theoret. Comput. Sci. **621**, 57–72 (2016)
15. Flocchini, P., Santoro, N., Wada, K.: On memory, communication, and synchronous schedulers when moving and computing. In: Proceedings of 23rd International Conference on Principles of Distributed Systems (OPODIS), pp. 25:1–25:17 (2019)
16. Fujinaga, N., Yamauchi, Y., Ono, H., Kijima, S., Yamashita, M.: Pattern formation by oblivious asynchronous mobile robots. SIAM J. Comput. **44**(3), 740–785 (2015)
17. Gervasi, V., Prencipe, G.: Coordination without communication: The case of the flocking problem. Discret. Appl. Math. **144**(3), 324–344 (2004)
18. Izumi, T., Souissi, S., Katayama, Y., Inuzuka, N., Défago, X., Wada, K., Yamashita, M.: The gathering problem for two oblivious robots with unreliable compasses. SIAM J. Comput. **41**(1), 26–46 (2012)
19. Nakai, R., Sudo, Y., Wada, K.: Asynchronous gathering algorithms for autonomous mobile robots with lights. In: Proceedings of 23rd International Symposium (SSS), pp. 410–424 (2021)
20. Nakajima, K., Takase, K., Wada, K.: Efficient self-stabilizing simulations of energy-restricted mobile robots by asynchronous luminous mobile robots (2024). arXiv.org, cs(ArXiv:2403.05542)
21. Okumura, T., Wada, K., Défago, X.: Optimal rendezvous $\mathcal{L}$-algorithms for asynchronous mobile robots with external-lights. In: Proceedings of 22nd International Conference on Principles of Distributed Systems (OPODIS), pp. 24:1–24:16 (2018)
22. Okumura, T., Wada, K., Katayama, Y.: Brief announcement: Optimal asynchronous rendezvous for mobile robots with lights. In: Proceedings 19th International Symposium on Stabilization, Safety, and Security of Distributed Systems (SSS), pp. 484–488 (2017)
23. Peleg, D.: Distributed Coordination Algorithms for Mobile Robot Swarms: New Directions and Challenges. In: Pal, A., Kshemkalyani, A.D., Kumar, R., Gupta, A. (eds.) IWDC 2005. LNCS, vol. 3741, pp. 1–12. Springer, Heidelberg (2005). https://doi.org/10.1007/11603771_1
24. Souissi, S., Izumi, T., Wada, K.: Oracle-based flocking of mobile robots in crash-recovery model. In: Proceedings of 11th International Symposium on Stabilization, Safety, and Security of Distributed Systems (SSS), pp. 683–697 (2009)
25. Suzuki, I., Yamashita, M.: Distributed anonymous mobile robots: Formation of geometric patterns. SIAM J. Comput. **28**, 1347–1363 (1999)
26. Yamashita, M., Suzuki, I.: Characterizing geometric patterns formable by oblivious anonymous mobile robots. Theoret. Comput. Sci. **411**(26–28), 2433–2453 (2010)
27. Yamauchi, Y., Uehara, T., Kijima, S., Yamashita, M.: Plane formation by synchronous mobile robots in the three-dimensional euclidean space. J. ACM **64**(3), 16:1–16:43 (2017)

Brief Announcement: Perpetual Exploration of Triangular Grid by Myopic Oblivious Robots Without Chirality

Raja Das[(✉)] [ID], Pritam Goswami [ID], and Buddhadeb Sau [ID]

Jadavpur University, Kolkata, India
{rajad.math.rs,buddhadeb.sau}@jadavpuruniversity.in,
pgoswami.academic@gmail.com

Abstract. Exploration of different network topologies is one of the fundamental problems of distributed systems. The problem has been studied on networks like lines, rings, tori, rectangular grids, etc. In this work, we have considered a *rectangle enclosed triangular grid* (RETG). A RETG is a part of an infinite triangular grid and the part is enclosed by a rectangle whose one pair of parallel sides aligns with a family of parallel straight lines of the infinite triangular grid. We have studied the problem of perpetual exploration on a RETG using oblivious robots. We have considered the robots with limited visibility i.e. the robots are myopic. Infinite visibility becomes impractical for a very large network. Limited visibility is more practical than infinite visibility. The robots have neither any chirality nor any axis agreement. An algorithm is provided to explore the RETG perpetually without any collision. The algorithm works under a synchronous scheduler. The algorithm requires three robots with two hop visibility.

Keywords: Myopic robot · Autonomous robots · Oblivious robot · Triangular grid · Perpetual exploration · Distributed algorithms

1 Introduction

There has been extensive research on swarm robotics under distributed systems for the past two decades. A swarm of robots is a collection of simple robots. Each robot executes the same algorithm and they together perform a particular task. Exploration, gathering, pattern formation, dispersion etc. can be done using a swarm of robots. There are several applications of swarm robotics in the real world. Patrolling areas inaccessible to humans, cleaning large surfaces and maintaining networks are some applications of swarm robotics. The robots work autonomously i.e., without any central control. They are identical i.e. one robot can not be distinguished from another robot. They are anonymous i.e., they are without any identifier. In some cases, robots can have persistent memory and can communicate with other robots. There are four types of robot models based on persistent memory and communication. In oblivious robot ($\mathcal{OBLOT}$) model

T. Masuzawa et al. (Eds.): SSS 2024, LNCS 14931, pp. 156–160, 2025.
https://doi.org/10.1007/978-3-031-74498-3_11

a robot does not have persistent memory and can not communicate with other robots. In finite state ($\mathcal{FSTA}$) model a robot has persistent memory but can not communicate with other robots. In finite communication ($\mathcal{FCOM}$) model a robot does not have persistent memory but can communicate with other robots. In luminous ($\mathcal{LUMI}$) model a robot has persistent memory and can communicate with other robots. This can be achieved using finite bits or lights. There are generally two types of visibility. In infinite visibility, a robot can see up to an infinite distance. But finite visibility is more practical. In that case, a robot can see up to a finite distance in the Euclidean plane and up to a finite hop in discrete domain. Robots with finite visibility are called *myopic*. Robots may have consistent chirality i.e., the clockwise direction sense remains the same in each round. If the clockwise direction sense of any robot matches with the same of the remaining robots then we say that the robots have a common chirality. A robot has two states i.e., idle state and non idle state. Activation of a robot means the transformation from idle state to non idle state. The non idle state consists of three phases i.e., Look phase, Compute phase, and Move phase. These three phases form the LCM cycle. In the look phase, a robot takes a snapshot of its surroundings and gets the position and states of other robots. In the compute phase, a robot runs the algorithm to get an output. In the move phase, a robot stays idle or moves to a node at one hop distance, depending on the output. After these phases, a robot enters into idle state until the next activation. Then the robot runs the LCM cycle again. Activation plays an important role in swarm robotics. It is considered that the scheduler is responsible for activating the robots and determining the time slot for the phases. There are generally three types of scheduler. These are fully synchronous (FSYNC), semi synchronous (SSYNC) and asynchronous (ASYNC). Under FSYNC scheduler each robot is activated at the same time and they perform the look-compute-move cycle (LCM cycle) in each round. Time, given for Look, Compute, Move is same for each robot. Under SSYNC scheduler everything happens like FSYNC scheduler except the fact that some robots may be inactivated in each round. But a robot can not be inactivated for infinitely many rounds. Under ASYNC scheduler there is no common notion of time. Each robot performs its LCM cycle independently.

2   Related Works and Our Contributions

Exploration in a graph network means each node of the graph has to be visited by at least one robot within finite time. There are two types of exploration, terminating and perpetual. In terminating exploration the robots terminate after each node is visited by at least one robot. In perpetual exploration each node has to be visited infinitely often. In [1] perpetual exploration of a finite rectangular grid was studied using myopic luminous and non luminous robots, assuming common chirality. The same problem was then studied in [2] assuming no common chirality. In this work, we have considered the problem of perpetual exploration of a *rectangle enclosed triangular grid* (RETG) by myopic oblivious robots having no chirality. Informally a RETG is a part of an infinite triangular grid and the part is enclosed by a rectangle whose one pair of parallel sides aligns with a family of parallel straight lines of the infinite triangular grid. Note that the algorithms of finite rectangular grids can not be applied to RETG directly. The boundary nodes of a finite rectangular grid are of degree three but the boundary nodes of a RETG are of degree three, four and five. The corner nodes of a finite rectangular grid are of degree two but the corner nodes of a RETG are of degree two or three. We have overcome these challenges in this work. We have provided an algorithm to explore a RETG perpetually using three oblivious robots without chirality. The robots can see up to two hop distance and they do not have any axis agreement.

3   Model, Definitions and Preliminaries

An infinite triangular grid is a graph with infinite nodes (N be the set of all nodes) where the degree of each node is six and each face of the graph is an equilateral triangle of unit length sides. If we consider the infinite straight lines within the infinite triangular grid then there exist three families of parallel straight lines. Now we place a rectangle on the infinite triangular grid in such a way that a pair of parallel sides of the rectangle aligns with a pair of parallel straight lines of the infinite triangular grid. Let N' (a subset of N) be the set of all nodes which lie on or inside the rectangle. The subgraph induced by the set of nodes N' is said to be a *rectangle enclosed triangular grid* (RETG). In this work, we have considered the underlying network to be a RETG. Depending on the degrees of the corner nodes, different types of RETGs can be formed. In this work, we have considered all possible RETGs (see Fig 1).

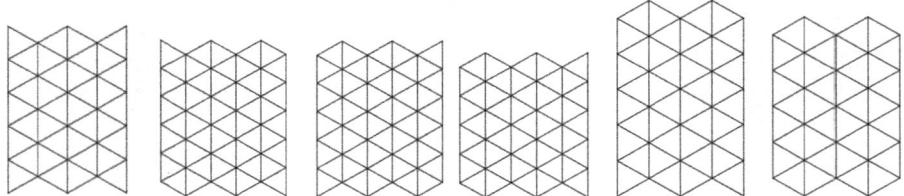

Fig. 1. Different type of RETGs

In this work, we have considered $\mathcal{OBLOT}$ robot model which means the robots have no persistent memory and can not communicate with each other. The robots work under a fully synchronous scheduler. The robots can see up to two hop distance only. The robots have neither any chirality nor any axis agreement.

The straight lines of a *rectangle enclosed triangular grid* (RETG) are called columns if they are parallel to a side of the rectangle. The columns of a RETG are denoted as C_i, where $1 \leq i \leq k$. C_i being the i-th column from "left". This "left" is from a global perspective, required only for understanding the behaviour of the algorithm. In this work, we have considered the RETGs with at least four columns and each column with at least four nodes. The straight lines of a *rectangle enclosed triangular grid* (RETG) are called diagonal lines if they are not parallel to any side of the rectangle.

Our aim is to provide an algorithm so that the robots present in the RETG will execute the algorithm and each node of the RETG will be visited by the robots infinitely many times.

4   Algorithm A_{231}^{VRL}

In this section, we introduce the algorithm A_{231}^{VRL} . At first, we give some definitions and preliminaries. Then we give the description of the algorithm A_{231}^{VRL} for three oblivious robots with two hop visibility.

4.1   Definitions and Preliminaries

Definition 1 ($InitA_{231}^{VRL}$). *A configuration (r_1, r_2, r_3) with two robots that are on adjacent nodes of a RETG and another robot at a node such that it is at two hop distance from both the robots, is called an $InitA_{231}^{VRL}$ configuration.*

In $InitA_{231}^{VRL}$ configuration, the robot which is at two hop distance from the other two robots is called tail. The robot which is on the same column or diagonal line of the tail is called head. The robot which is adjacent to head is called arm.

An $InitA_{231}^{VRL}$ configuration with the head robot and the tail robot on the same column is called an E-type configuration. E-type configuration can be of four different types. These are $E^<$ configuration, $E_<$ configuration, $E^>$ configuration, and $E_>$ configuration.

An E-type configuration is said to be an $E^<$ configuration on C_i if the head robot is on C_i where $2 \leq i \leq k$, and the tail robot is two hop below the head robot and the arm robot is on C_{i-1}. An E-type configuration is said to be an $E_<$ configuration on C_i if the head robot is on C_i where $2 \leq i \leq k$, and the tail robot is two hop above the head robot and the arm robot is on C_{i-1}. An E-type configuration is said to be an $E^>$ configuration on C_i if the head robot is on C_i where $1 \leq i \leq k-1$, and the tail robot is two hop below the head robot and the arm robot is on C_{i+1}. An E-type configuration is said to be an $E_>$ configuration on C_i if the head robot is on C_i where $1 \leq i \leq k-1$, and the tail robot is two hop above the head robot and the arm robot is on C_{i+1}. An $InitA_{231}^{VRL}$ configuration, which is not an E-type configuration is called a D-type configuration.

4.2   Description of A_{231}^{VRL}

At the beginning of A_{231}^{VRL}, three robots can be anywhere on a RETG provided they form $InitA_{231}^{VRL}$ configuration. $InitA_{231}^{VRL}$ configuration looks like an arrow. The robots move in the direction from the tail robot to the head robot. The position of the arm robot indicates a right turn or left turn of the arrow after hitting boundary.

Now we present a high level idea of the algorithm A_{231}^{VRL}. If (r_1, r_2, r_3) is not an E-type configuration then (r_1, r_2, r_3) must be a D-type configuration. If $InitA_{231}^{VRL}$ is a D-type configuration then the configuration will become E-type configuration first. In E-type configuration the head robot and the tail robot belong to the same column and the arm robot belongs to its left or right column. Then the robots move along their column until r_1 reaches the boundary maintaining the E-type configuration. After that the robots go to the next column with E-type configuration and start moving in the opposite direction i.e. if the robots were going downwards through the previous column then they will move upwards through the current column or if the robots were going upwards through the previous column then they will move downwards through the current column.

Suppose the robots move upwards with $E^>$ configuration through C_i where $1 \le i \le k - 2$ then after reaching the top most node of C_i the robots will enter C_{i+1} and start moving downwards with $E_>$ configuration. After reaching the bottom most node of C_{i+1} the robots will enter C_{i+2} and start moving upwards with $E^>$ configuration. In this way, the robots explore each column from C_{i+1} to C_{k-1} and will reach C_{k-1}. After exploring C_k the robots will become $E^<$ configuration or $E_<$ configuration on C_k. Now the robots explore columns from higher indices to lower indices. We can observe when C_1 or C_k is explored twice, all columns will be explored. Thus after finite rounds, all the nodes of RETG will be explored.

Acknowledgement. The first author is supported by Council of Scientific & Industrial Research (CSIR), Govt. of India. The second author is supported by University Grants Commission (UGC), Govt. of India. The third author is supported by Science and Engineering Research Board (SERB), Govt. of India.

References

1. Bramas, Q., Lafourcade, P., Devismes, S.: Optimal exclusive perpetual grid exploration by luminous myopic opaque robots with common chirality. Theor. Comput. Sci. **977**, 114162 (2023). https://doi.org/10.1016/J.TCS.2023.114162
2. Rauch, A., Bramas, Q., Devismes, S., Lafourcade, P., Lamani, A.: Optimal exclusive perpetual grid exploration by luminous myopic robots without common chirality. In: Echihabi, K., Meyer, R. (eds.) Networked Systems—9th International Conference, NETYS 2021, Virtual Event, May 19-21, 2021, Proceedings. Lecture Notes in Computer Science, vol. 12754, pp. 95–110. Springer (2021). https://doi.org/10.1007/978-3-030-91014-3_7

An Optimal Algorithm for Geodesic Mutual Visibility on Hexagonal Grids

Sahar Badri, Serafino Cicerone$^{(\boxtimes)}$, Alessia Di Fonso, and Gabriele Di Stefano

Department of Information Engineering, Computer Science and Mathematics (DISIM), University of L'Aquila, Italy
sahar.badri@graduate.univaq.it,
{serafino.cicerone,alessia.difonso,gabriele.distefano}@univaq.it

Abstract. For a set of robots (or agents) moving in a graph, two properties are highly desirable: confidentiality (i.e., a message between two agents must not pass through any intermediate agent) and efficiency (i.e., messages are delivered through shortest paths). These properties can be obtained if the GEODESIC MUTUAL VISIBILITY (GMV) problem is solved: oblivious robots move along the edges of the graph, without collisions, to occupy some vertices that guarantee they become pairwise geodesic mutually visible. This means there is a shortest path (i.e., a "geodesic") between each pair of robots along which no other robots reside. In this work, we optimally solve GMV on finite hexagonal grids G_k. This, in turn, requires first solving a graph combinatorial problem, i.e. determining the maximum number of mutually visible vertices in G_k.

Keywords: Mutual visibility · Grid graphs · Oblivious robots · Synchronous robots · Oblot model

1 Introduction

Problems about sets of points in the Euclidean plane and their mutual visibility have been investigated for a long time. For example, in [16] Dudeney posed the famous *no-three-in-line* problem: finding the maximum number of points that can be placed in an $n \times n$ grid such that there are no three points on a line.

Mutual visibility in graphs for a set of vertices has been recently introduced and studied in [15] in terms of the existence of a shortest path between two vertices without a third vertex from the set. The visibility property is then understood as the absence of "obstacles" between the two vertices along the shortest path, which makes them "visible" to each other. For example, in communication networks, mutually visible agents can communicate both efficiently

The work has been partially supported by the European Union - NextGenerationEU under the Italian Ministry of University and Research (MUR) National Innovation Ecosystem grant ECS00000041-VITALITY—CUP E13C22001060006, SICURA—CUP C19C200005200004 and by the Italian National Group for Scientific Computation (GNCS-INdAM).

T. Masuzawa et al. (Eds.): SSS 2024, LNCS 14931, pp. 161–176, 2025.
https://doi.org/10.1007/978-3-031-74498-3_12

(i.e., through shortest paths) and confidentially (i.e., the message does not pass through intermediate agents). Formally, let G be a connected and undirected graph, and $X \subseteq V(G)$ a subset of the vertices of G. Two vertices $x, y \in V(G)$ are X-*visible* if there exists a shortest x, y-path where no internal vertex belongs to X, and X is a *mutual-visibility set* if its vertices are pairwise X-visible. Any largest mutual-visibility set of G is called μ-set and its cardinality is the *mutual-visibility number* of G (denoted as $\mu(G)$). In [15], it is shown that computing $\mu(G)$ is an NP-complete problem. Still, exact formulae exist for the mutual-visibility number of special graph classes like paths, cycles, blocks, cographs, grids and distance-hereditary graphs [9].

Ever since its introduction, this graph-based mutual-visibility concept has garnered significant interest within the research community (e.g., see references within recent works [8,12,20]). These contributions provided new structural and computability results and identified several connections between the mutual-visibility problem and some classical combinatorics topics. For example, there exist relationships with the Zarankiewicz problem [10], the Turán problem [12], and the classical Bollobás-Wessel theorem [3]. There is also a close relationship between the mutual-visibility problem and the general position problem [19] is a distance-related topic that has attracted great interest recently.

The graph-based notion of mutual-visibility has already been applied in the context of mobile robots operating on discrete environments modeled by graphs where visibility is verified along shortest paths [5–7]. In particular, starting from a configuration composed of any number of robots located on distinct vertices of an arbitrary graph, within finite time the robots must reach (if possible), a configuration where they are all mutually visible. This problem is called GEODESIC MUTUAL VISIBILITY problem (GMV, for short) to distinguish it from the well-known MUTUAL VISIBILITY problem on the Euclidean plane, where two robots are visible if there is not a third robot on the straight line segment between them (e.g., see [1,14]).

RESULTS. In this paper, we are interested in solving GMV for synchronous robots moving on finite hexagonal grids embedded in the plane. Such grids are denoted as G_k, $k \geq 1$, and correspond to the $k \times k \times k$ grid graph defined in [21]. Note that, regular grid graphs are a typical discretization of the plane used in swarm robotics (e.g., see [2,4] for OBLOT and [13] for the programmable matter).

It is worth noting that any algorithm solving GMV on a graph G for any possible number of robots must know not only $\mu(G)$ but also some μ-set of G to use it as a pattern defining the final positions that the robots must reach. For this reason, our contribution is twofold. First, from a graph-theoretical point of view, we provide an exact formula for $\mu(G_k)$ and determine a μ-set for G_k. Then, we design a distributed algorithm $\mathcal{A}$ that solves optimally GMV in each configuration defined on G_k and composed of synchronous robots endowed with chirality. Algorithm $\mathcal{A}$ uses any μ-set of G_k as a pattern. We remark that the general PATTERN FORMATION problem (where robots belonging to an initial configuration C are asked to move and form a configuration F – i.e., the pattern

– that is provided as input to robots) has been already studied on infinite grid graphs, but all the algorithms proposed so far consider only asymmetric configurations in input, while $\mathcal{A}$ must solve GMV even when the initial configuration is symmetric (e.g., see [2,4,18]).

2 Robot Model and the Addressed Problem

An OBLOT system (cf. [17]) comprises a set of robots that live and operate in a graph. Robots are **identical** (indistinguishable from their appearance), **anonymous** (they do not have distinct ids), **autonomous** (they operate without a central control or external supervision), **homogeneous** (they all execute the same algorithm), **silent** (they have no means of direct communication of information to other robots), and **disoriented** (each robot has its own local coordinate system - LCS) but they agree on a cyclic orientation (e.g., clockwise) of the plane, i.e., a common sense of **chirality** is assumed. A robot can observe the positions (expressed in its LCS) of all the robots. We consider **synchronous** robots that operate according to the Look-Compute-Move (LCM) computational cycle [17]:

– Look. The robot obtains a snapshot expressed in its own LCS of the positions of all the other robots.
– Compute. The robot performs a local computation according to a deterministic algorithm $\mathcal{A}$ (i.e., the robot executes $\mathcal{A}$), which is the same for all robots, and the output is a vertex among its neighbors or the one where it resides.
– Move. The robot performs a *nil* movement if the destination is at its current location otherwise it **instantaneously moves** to the computed neighbor.

Robots are **oblivious** (they have no memory of past events), thus the Compute phase depends only on the information of the current Look phase. A data structure containing the information elaborated from the current snapshot represents what is called the **view** of a robot. Since each robot refers to its own LCS, the view cannot exploit absolute orienteering and it is based on the relative positions of robots. Hence, if symmetries occur (see Sect. 4), then symmetric robots have the same view. In turn, (i) the algorithm cannot distinguish between symmetric robots – even when placed in different positions, and (ii) symmetric robots perform the same movements.

Robots are placed in a simple, undirected, and connected graph $G = (V, E)$. A function $\lambda : V \to \mathbb{N}$ represents the number of robots on each vertex of G, and we call $C = (G, \lambda)$ a **configuration** whenever $\sum_{v \in V} \lambda(v)$ is bounded and greater than zero. A vertex $v \in V$ such that $\lambda(v) > 0$ is said *occupied*, *unoccupied* otherwise. We say that a collision occurs on a vertex v if $\lambda(v) > 1$.

Definition 1. *(GMV problem)* Let $C = (G, \lambda)$ *be any configuration with* $\lambda(v) \leq 1$ *for each* $v \in G$. *Design a deterministic distributed algorithm working under the* LCM *model that, starting from* C *and in a finite number of computational cycles, moves the robots without collisions until they form a configuration* $C' = (G, \lambda')$ *where the occupied vertices are in mutual visibility.*

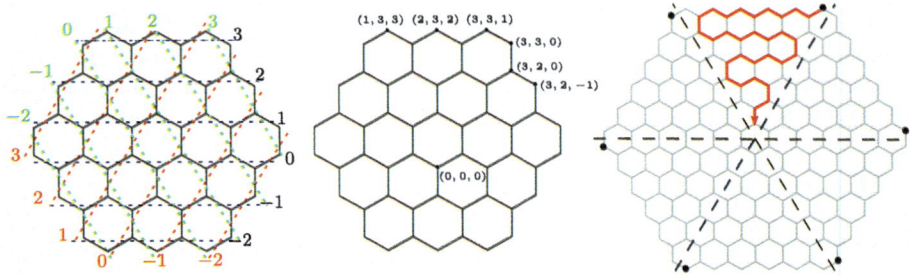

Fig. 1. *(left)* Visualization of G_3 with the three types of oriented lines used for the vertex labeling: h-lines (in blue), l-lines (in green), and r-lines (in red); *(middle)* Visualization of some vertex labels; *(right)* The 6 sectors of G_3, the corner in each sector, and a special-path within one sector.

Here we study GMV in finite hexagonal grids. Note that, given a configuration $C = (G, \lambda)$ with n robots, any algorithm designed to solve GMV on C must be provided with a strategy that, in a finite time, guides robots to occupy vertices that are in mutual-visibility. This can be obtained by providing the algorithm with a mutual-visible set X of G such that $|X| \geq n$. To our knowledge, computing $\mu(G)$ when G is a finite hexagonal grid is an open problem.

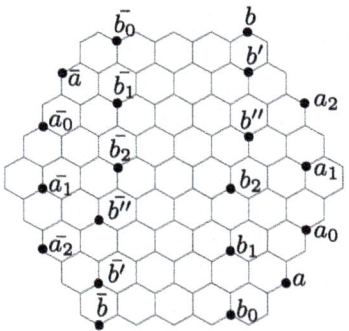

Fig. 2. Visualization of the μ-set X_k when $k = 5$.

3   Mutual Visibility on Hexagonal Grid Graphs

We consider finite subgraphs of **hexagonal grid graphs**: G_1 is the grid graph of just one hexagon, G_2 is obtained by surrounding G_1 with a "crown" of 6 additional hexagons – one per side, and, in general, G_k with $k \geq 2$ can be informally built by surrounding G_{k-1} with a crown of $6(k-1)$ additional hexagons (see

Fig. 1). Formally, G_k corresponds to the $k \times k \times k$ grid graph defined in [21] and results in a finite graph with $6k^2$ vertices. It can be divided into 6 **sectors** by using the three lines passing through the middle points of parallel edges of the initial sub-grid G_1. The **perimeter** of G_k is formed by all the external vertices and edges of the grid, and it consists of 6 **sides** formed by a path of length 1 when $k = 1$ and of length $2k$ when $k \geq 2$ (in these grids, two consecutive sides share one edge). Vertices can be labeled by extending the classical Cartesian coordinates used in the Euclidean plane to three directions. To this end, consider the three kinds of straight lines parallel to the sides of G_k: according to the orientation, we call them r-lines (right-oriented lines), l-lines (left-oriented lines) and h-lines (horizontal lines); sometimes, h-lines are also called "levels". These lines can be numbered according to the distance from the lowest vertex of the central hexagon, starting from 0 (cf. Fig. 1). A vertex at the intersection of three lines is labeled by a triple consisting of the numbers assigned to those lines in the order (l-line, h-line, r-line).

We now define a μ-set X_k for G_k for each $k \geq 2$. To this end, consider the following subsets (cf Fig. 2):

- $A_k = \{a\} \cup \{a_i | i = 0, \ldots, k - 3\}$, where $a = (2, -k + 2, -k + 1)$ and $a_i = (3 + i, -k + 4 + 2i, -k + 1 + i)$;
- $B_k = \{b, b', b''\} \cup \{b_i | i = 0, \ldots, k - 3\}$, where $b = (k, k, 1)$, $b' = (k - 1, k - 1, 0)$, $b'' = (k - 2, k - 3, -1)$ and $b_i = (i, -k + 1 + 2i, -k + 2 + i)$;
- $\bar{A}_k$ ($\bar{B}_k$, resp.) contains the vertices of G_k obtained through a 180° rotation of the elements of A_k (B_k, resp.).

Finally, let $X_k = A_k \cup B_k \cup \bar{A}_k \cup \bar{B}_k$. Note that $|A_k| = k - 1$, $|B_k| = k + 1$, and hence $|X_k| = 4k$. The next result states that $\mu(G_k) = 4k$.

Theorem 1. X_k is a μ-set of G_k, for each $k \geq 4$.

4 Notation and Concepts for GMV

Given $C = (G_k, \lambda)$, $R = \{r_1, r_2, \ldots, r_n\}$ denotes the set of robots located on C. The distance between vertices u and v is denoted $d(u, v)$, and, given $r_i, r_j \in R$, $d(r_i, r_j)$ represents the distance between the vertices in which the robots reside. Finally, $D(r) = \sum_{r_i \in R} d(r, r_i)$.

SYMMETRIC CONFIGURATIONS. As chirality is assumed, the possible symmetries that can occur in G_k are only rotations. A rotation is defined by the center c of G_k and a minimum angle of rotation $\alpha \in \{60, 120, 180, 360\}$ working as follows: if the configuration is rotated around c by an angle α, then a configuration coincident with itself is obtained. The **symmetricity** of a configuration C is denoted as $\rho(C)$ and corresponds to $360/\alpha$. C is **rotational** if $\rho(C) > 1$, and is **asymmetric** if $\rho(C) = 1$.

VIEW OF ROBOTS. In Sect. 3 we have defined the perimeter of G_k as composed of 6 sides, with two consecutive sides sharing one edge. If we consider two clockwise consecutive sides, we assume that the shared edge belongs to the second side.

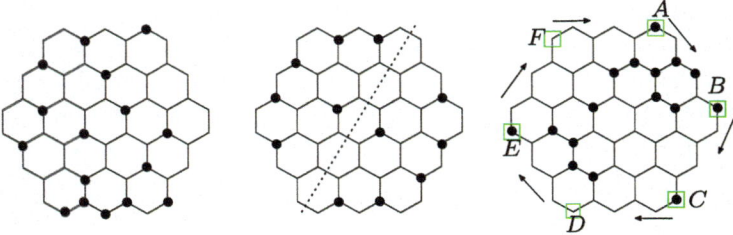

Fig. 3. *(left)* A configuration C with $\rho(C) = 1$; *(middle)*: a configuration divided into two sectors; *(right)* Computing the view of robots: corners are highlighted by squares, the reading starts at each corner and proceeds along the side of G_3, $LSS(C) = 0000010$ 000011111 00001001101 1011000000 001100000 0000001 and it is obtained from F.

Hence, each edge on the perimeter belongs to a unique side. A **corner** of G_k is defined as the rightmost vertex on a side of G_k. Robots encode the G_k grid starting from a corner of G_k and proceeding clockwise along the side of the grid; then, all the lines parallel to that side are analyzed one by one proceeding along the same direction. When one vertex is encountered, it is encoded as 1 or 0 according to the presence or the absence, respectively, of a robot. This produces a binary string for each starting corner, and then six strings in total are generated. Among such strings, the **lexicographically smallest string** is denoted as $LSS(C)$. If two strings obtained from different corners are equal, then the configuration is rotational, otherwise it is asymmetric. Hence, the number of generated strings equal to $LSS(C)$ corresponds to $\rho(C)$. The robot(s) with **minimum view** is the one with the minimum position in $LSS(C)$. Figure 3.right shows how the view is computed by robots.

SPECIAL-PATHS. We have already remarked that G_k can be divided into six sectors. In each sector, we define a **special-path** that starts from each corner of G_k and proceeds as indicated in Fig. 1.right. Regarding the number of vertices in a sector and in a special-path, consider the topmost sector in Fig. 1.right and analyze the vertices level by level, starting from level 1: there are $1 + 3 + 5 + \ldots + 2k - 1 = k^2$ vertices in total, while the defined special-path leaves one vertex per level (except level 1) untouched, thus having $k^2 - (k - 1)$ vertices in total.

5   A Resolving Algorithm for GMV on Hexagonal Grids

According to Theorem 1, at most $4k$ robots can be arranged in mutual visibility on the vertices of G_k. As a consequence, GMV defined on a given $C = (G_k, \lambda)$ can be restricted to $n \leq 4k$ robots only. As already remarked, a final configuration C' must be provided to the algorithm designed to solve GMV. For instance, if C has $4k$ robots, this can be done as follows: let $C' = (G_k, \lambda')$ be the configuration such that $\lambda'(v) = 1$ iff $v \in X_k$, where X_k is the μ-set of G_k as defined in Sect. 3.

It is worth noting that solving GMV on G_k by using X_k to define C' resembles the **pattern formation problem**, where robots belonging to an initial configuration C are required to arrange themselves to form a configuration F (i.e., the pattern) which is provided as input to robots. In this context, it is well-known that "$\rho(C)$ divides $\rho(F)$" is a necessary condition for solving the pattern formation problem (e.g., see [4]). Since $\rho(C') = 2$, the existence of only X_k as a μ-set for G_k would restrict the analysis of the problem to input instances C such that $\rho(C) \in \{1, 2\}$.

OUR APPROACH. We assume that each input configuration $C = (G_k, \lambda)$ contains n robots, with $k \geq 4$ and $12 \leq n \leq 4k$ (these limitations for k and n are imposed to avoid cases defined by small instances which, as is often the case, would require specific approaches). Given C, we then assume there exists a configuration $F = (G_{k'}, \lambda')$ where λ' indicates the presence of **target vertices** instead of robots, and such that all the following conditions hold:

1. $k' = \lceil n/4 \rceil$;
2. λ' with at most two targets per line, and at least one target on the perimeter;
3. $\rho(C)$ divides $\rho(F)$.

According to these assumptions, *we design $\mathcal{A}$ such that, in a finite number of LCM cycles, it transforms C into a configuration C' having robots disposed as in F.* By using this general approach, if $\mathcal{A}$ take as input $C = (G_k, \lambda)$ with $n = 4k$ robots and $\rho(C) \leq 2$, then X_k can be used for defining F as described above; otherwise, if C contains $n \in \{4k - 1, 4k - 2, 4k - 3\}$ robots, a corresponding subset of X_k with n elements can be used. So far it is still an open problem to find a μ-set for G_k with symmetricity greater than two. But if this problem will be solved in the future, $\mathcal{A}$ can be used to solve configurations with $\rho(C) > 2$.

5.1   Description of the Algorithm

We present here $\mathcal{A}$, the algorithm for solving the GMV problem for fully synchronous robots endowed with chirality and moving on a finite grid G_k. By using the methodology provided in [11], the problem GMV is divided into a set of sub-problems that are simple enough to be thought of as "tasks" to be performed by (a subset of) robots. Algorithm $\mathcal{A}$ detects the task T_i to be accomplished, by means of a **predicate** P_i assigned to T_i, for each $i = 1 \ldots n$, where n is the number of tasks. Each predicate P_i assumes the form $P_i = \mathtt{pre}_i \wedge \neg(\mathtt{pre}_{i+1} \vee \mathtt{pre}_{i+2} \vee \ldots \vee \mathtt{pre}_n)$, where $\mathtt{pre}_i$ is a condition that must be true for task T_i and, at the same time, all the preconditions $\mathtt{pre}_j$, $j > i$ must be false. In the Compute phase, each robot evaluates the predicates starting from the last predicate and proceeding in the reverse order until a true predicate is found. The provided algorithm $\mathcal{A}$ can be used by each robot in the Compute phase as follows: – *if a robot r executing $\mathcal{A}$ detects that predicate P_i holds, and if r is involved in the task, then r simply performs move m_i associated with T_i.*

What follows represents a high-level description of $\mathcal{A}$. The first sub-problem is the "Placement of guards", in which $\mathcal{A}$ selects $\rho(C)$ robots and places each of them on different and symmetric corners of G_k. As robots are disoriented (only sharing chirality), the positioning of these robots allows the creation of a common reference system used by robots in the successive stages of the algorithm. For this reason, these special robots are called **guards**, and each of them will be usually denoted as r_g. By exploiting chirality, the position of the guards allows robots to identify and enumerate two main directions, hereafter called rows and columns, among the three available in G_k. Then, each guard does not move until the final stage of the algorithm. The "Placement of guards" needs to be decomposed into three simple tasks named T_{1a}, T_{1b}, and T_{1c}.

Given the common reference system formed by the guards, all robots can agree on the embedding of the pattern F in G_k: robots identify the center of F with c and place the $\rho(F)$ corners of F with the maximum view on the column in which the guards r_g reside and closest to r_g. Note that since robots in C are synchronous, irrespective of the algorithm operating on C, the center c of the configuration is invariant.

Task T_2 solves the "Distribute robots on rows" sub-problem. In particular, $\mathcal{A}$ first exploits the common reference system formed by the guards to let robots agree on the embedding of the pattern F in G_k. Then, $\mathcal{A}$ moves the robots in each sector along columns to obtain the suitable number of robots for each row according to the targets defined by F. During this task, the algorithm avoids occupying any corner of G_k equivalent to the one on which a guard resides (thus preserving the reference system). These vertices are hereafter called **forbidden**.

Task T_3 solves the "Moving toward targets" sub-problem, where all robots but the guards move along rows toward their final target. The moved robots either reach the target or stop one step away if their final destination is at a forbidden vertex (there are at most five of such robots as there are at most five forbidden vertices).

Finally, in task T_4, all non-guard robots not on target reach their final destination, and all the guards move away from the corner of G_k toward their final target. In this way, the "Pattern finalization" sub-problem is solved.

In what follows, we detail each of the above tasks.

TASK. T_1 During this task, $\rho(C)$ robots called guards move to occupy a corner of the grid G_k. This task is divided into three sub-tasks based on the number of robots occupying the corners of G_k. Let RP be the number of robots on the perimeter, and let RC be the number of robots on the corners of G_k. The precondition associated with Task T_{1a} is the following: $\texttt{pre}_{1a} \equiv [RP = 0]$. All robots r such that $D(r)$ is maximum, and of minimum view in case of ties, are selected for moving (for symmetrical reasons, exactly $\rho(C)$ robots are selected). The planned move is the following: $m_{1a} \equiv$ "*each selected robot moves toward a closest vertex belonging to a side of G_k, chosen clockwise in case of ties*". At the end of T_{1a}, $\rho(C)$ robots are on the perimeter of G_k. Notice that, when $\rho(C) > 1$ each moved robot will be on a distinct side of the grid.

In Task T_{1b} precondition $\texttt{pre}_{1b} \equiv [RP \geq \rho(C) \wedge RC = 0]$ holds. In this case, there are at least $\rho(C)$ robots on the perimeter of G_k but none on corners. Then, on each side, the algorithm selects the robot closest to a corner, those with the minimum view in case of ties. Notice that $\rho(C)$ robots are selected. The planned

---

Algorithm 1 MoveAlongSpecial-Path

Require: a configuration $C = (G_k, \lambda)$
1: **if** $fos = 0$ **then**
2:    Let S be the set of occupied special-paths whose first robot has the minimum view.
3:    **move:** all the robots on a special-subpath and not on a special-path of S move toward the neighbor vertex along the special-path.
4: **if** $fos = 1$ **then**
5:    Let I be the fully-occupied special-path
6:    **move:** all the robots on a special-subpath of an occupied special-path different from I move toward the neighbor vertex along the special-path

---

move is $m_{1b} \equiv$ "*each selected robot moves toward the closest corner of G_k, chosen clockwise in case of ties*". Finished T_{1b}, exactly $\rho(C)$ robots occupy a corner.

To activate Task T_{1c} precondition $\text{pre}_{1c} \equiv [RC > \rho(C)]$ must hold. In this case, $RC - \rho(C)$ robots are moved away from the corners along special-paths. The movements are provided by Procedure MoveAlongSpecial-Path (cf. Algorithm 1). This algorithm uses some additional definitions: a special-path is said **occupied** if its corner (i.e., the first vertex of the path) is occupied; a special-path is said to be **fully-occupied** if each of its vertices is occupied; if P is an occupied special-path, but not fully-occupied, then the **special-subpath** of P is the maximal subpath that starts at the corner of P and is composed of occupied vertices only. Finally, the number of fully-occupied special-paths is denoted by fos. Note that, due to the maximum number of robots, there cannot be more than one fully-occupied special-path.

At line 1, the algorithm checks if there are no fully-occupied special-paths. The $\rho(C)$ robots occupying the corners and having the minimum view are elected as guards. The move is designed to empty all the other corners of G_k except those elected as guards. In each occupied special-paths P, except those occupied by guards, the robots on the special-subpath of P move forward along the special-path. At line 4, there is precisely one fully-occupied special-path, say I. Therefore, robots on that fully-occupied special-path are kept still. Concerning any other occupied special-paths P different from I, the robots on the special-subpath of P move forward along the special-path. A configuration where exactly one corner of G_k in each sector is occupied is obtained in a single LCM cycle.

TASK. T_2 This task moves the suitable number of robots for each row according to the pattern F. This is the most complex task and it needs several auxiliary concepts. When this task starts, we are sure that T_1 is concluded and hence the initial configuration C has been transformed so that there are exactly $\rho(C)$ guards positioned on corners. The position of guards allows robots to agree on how to subdivide G_k into $\rho(C)$ equivalent sub-configurations hereafter called **configuration-sectors** (c-sectors, for short). For defining c-sectors, robots use the lines passing through the center of G_k and cutting in half the edge connecting the vertex occupied by the guard robot and the first vertex of the next side (see Fig. 4). There are $\rho(C)$ c-sectors.

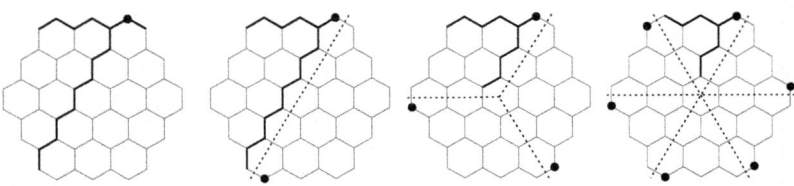

Fig. 4. Visualization of the subdivision into c-sectors when $\rho(C) = 1, 2, 3, 6$, respectively. In each case, the guard and the row/column with index 0 are emphasized.

Robots can now agree on a **global reference system** valid in each c-sector. To this end, they can identify two lines among the three available to be used as rows and columns: a **row** is any line parallel to the side of G_k in which the guard robot resides (and these lines are assumed as h-lines), and a **column** is any line obtained by a 60° counter-clockwise rotation of a h-line (these lines are assumed as r-lines, see Fig. 4). Moreover, rows and columns can be enumerated within each c-sector as follows: the guard occupies the vertex having coordinates $(0, 0, 0)$ while all other vertices are labeled as explained in Sect. 4 by using the coordinates of the guard and the h-lines, r-lines and, as a consequence, l-lines defined in each c-sector. Given the common reference system formed by the guards, all robots agree on the embedding of the pattern $F = (G_{k'}, \lambda')$ into G_k, as follows:

Definition 2. *Let $C = (G_k, \lambda)$ be the configuration created by $\mathcal{A}$ at the end of Task T_1, and let $F = (G_{k'}, \lambda')$ be the target configuration. The **embedding** of F into C is given by identifying the center of F with the center of C and by superimposing the edges of F into the edges of C. Moreover, given a guard, r_g on a side L of C, the side L' of F parallel to L and closest to it is the one such that its rightmost target minimizes the distance with r_g and is of minimum view.*

This embedding allows the algorithm to detect the number of robots to be moved on each row as required by F. In particular, $\mathcal{A}$ identifies M as the number of rows in a c-sector, t_h as the number of targets on row h, $(t_1, t_2, \ldots, t_M)$ as the vector of the number of targets, and $(\overline{r_1}, \overline{r_2}, \ldots, \overline{r_M})$ as the number of robots on each of the rows. For each row h, the algorithm computes the number of exceeding robots above and below h to the number of targets, to determine the number of robots that need to leave row h. Given a row h, let $R_{up}(h) = \sum_{i=1}^{h-1} \overline{r_i}$, and let $R_{down}(h) = \sum_{i=h+1}^{M} \overline{r_i}$. Accordingly, let $T_{up}(h) = \sum_{i=1}^{h-1} t_i$ and $T_{down}(h) = \sum_{i=h+1}^{M} t_i$. Concerning the number of targets, given a row h, let $ER_{up}(h)$ be the number of exceeding robots above h, h included, and let $ER_{down}(h)$ be the number of exceeding robots below h, h included. Formally, $ER_{up}(h) = (R_{up}(h) + \overline{r_h}) \dot{-} (T_{up}(h) + t_h)$, and $ER_{down}(h) = (R_{down}(h) + \overline{r_h}) \dot{-} (T_{down}(h) + t_h)$ where for a, b integers the operator $a \dot{-} b$ is defined as $a \dot{-} b = 0$ if $a < b$ and $a \dot{-} b = a - b$, otherwise.

Let $MR_{down}(h) = \overline{r_h} - (\overline{r_h} \dot{-} ER_{up}(h))$ be the number of robots that must move downward from row h and let $MR_{up}(h) = \overline{r_h} - (\overline{r_h} \dot{-} ER_{down}(h))$ be the number of robots that must move upward from row h. Task T_2 can be activated only when the following precondition holds:

$$\text{pre}_2 \equiv [RC = \rho(C) \wedge \exists\ h \in \{1, 2, \ldots, M\} : ER_{down}(h) \neq 0 \vee ER_{up}(h) \neq 0].$$

Informally, the precondition identifies all the configurations with $\rho(C)$ guards on corners of G_k and at least one row having an excess of robots.

The move planned in this task is defined in Procedure MoveAlongColumns (cf. Algorithm 2), which is based on the following rationale. The procedure moves non-guard robots along columns in each c-sector, where they operate independently and concurrently. Since columns have different lengths in G_k, in some cases, the robot might be not able to proceed along a column. Another reason why a robot should not proceed is when it could occupy a **forbidden vertex**, that is any corner of G_k equivalent to the one on which a guard resides (to keep the global reference system). Let FV be the set containing all the forbidden vertices, then $|FV| = 6 - \rho(C)$. Therefore there are $\frac{6-\rho(C)}{\rho(C)}$ in each c-sector. Let h be a row of a c-sector: robots can distinguish the **leftmost** (**rightmost**, resp.) vertex on h as the vertex $v = (l, h, r)$ with minimum (maximum, resp.) r-coordinate (and therefore distinguish the left and right side of h). To identify the robots to be moved, let U_h be the set of $MR_{up}(h)$ robots of row h selected from right and let D_h be the set of $MR_{down}(h)$ robots of row h selected from the left. U_h, D_h, and forbidden vertices are used to define the concept of blocked robot.

Definition 3. *Let $r \in U_h$ ($r \in D_h$, resp.) and belonging to a column c, then it is **blocked-up** (**blocked-down**, resp.) when one of the following holds:*

- *at row $h - 1$ ($h + 1$, resp.) and column c there is a forbidden vertex,*
- *h is the first (last, resp.) row of column c and $h \neq 1$ ($h \neq M$, resp.),*
- *there is a blocked robot at distance one when moving up (down, resp.),*
- *$\rho(C) = 1$, r is located at $h(r) = M - 1$ and $c(r) = 1$, and there exists a robot r' at $h(r') = M$ and $c(r') = 1$.*

*Finally, r is **blocked** when it is blocked-up or blocked-down.*

Let $H = \bigcup_h (U_h \cup D_h)$. If h is a row with $D_h \neq \emptyset$, we call *right-shifting path* the longest path on h starting from the leftmost occupied vertex of h and composed of occupied robots only. Symmetrically, if h is a row with $U_h \neq \emptyset$, we call *left-shifting path* the longest path on h starting from the rightmost occupied vertex of h and composed of only occupied robots. $\mathcal{R}$ ($\mathcal{L}$, resp.) contains all the right-shifting (left-shifting, resp.) paths.

We can now describe Procedure MoveAlongColumns. In the initial block of lines 1–5, the robot r executing the code computes all the data necessary for the subsequent move: the view, guards, and c-sectors, the vertex labeling local to each c-sector, the embedding of F into C, the sets U_h or D_h for each row h, and the sets H, $\mathcal{R}$, $\mathcal{L}$. Then, at line 7 the algorithm checks if all the robots selected to move are all blocked according to Definition 3. In the affirmative case, the

---

Algorithm 2 MoveAlongColumns

Require: $C = (G_k, \lambda)$ and $F = (G_{k'}, \lambda')$
  1: compute the view
  2: detect guard robots and the corresponding c-sectors
  3: assign the c-sector vertex labeling by using rows and columns
  4: perform the embedding of F into C
  5: compute U_h and D_h for each row h; compute H, $\mathcal{R}$, and $\mathcal{L}$
  6: let $h(r)$ and $c(r)$ be the row and column in which r resides
  7: **if** all robots in H are blocked **then**
  8:     **if** r is not a guard and is on a path of $\mathcal{R}$ ($\mathcal{L}$, resp.) **then**
  9:         r moves to the right (left, resp.) neighbor on $h(r)$
 10: **if** r is not a guard and $r \in H$ **then**
 11:     **if** r not blocked **then**
 12:         r move along $c(r)$ according to the membership to $U_{h(r)}$ or $D_{h(r)}$
 13: **else**
 14:     **if** r is not a guard and is *AlignedUp* or *AlignedDown* with r' **then**
 15:         r moves to a neighbour vertex on $h(r)$ that is not forbidden

---

robot r (along with all the other robots belonging to the same r-shifting or l-shifting path) performs one horizontal movement toward the assigned direction, see Line 8. This implies that a fraction of blocked robots will change columns and will be able to move in successive cycles. At line 10, the robot r checks whether it is selected to move. If it is not blocked, at Line 12, it moves along the column $c(r)$ in which it resides either upward or downward, depending if it belongs to the set U_h or D_h. Even if r does not belong to $U_h \cup D_h$, it is possible that r must move because it is "aligned" along the column with another robot r' that is involved in the move at Line 12. Formally, we say that a robot r is *AlignedUp* with r' when r resides on row l, $U_{l+1} = \{r'\}$ and $c(r) = c(r')$. Similarly, r is *AlignedDown* with r' when r resides on row l, $D_{l-1} = \{r'\}$ and $c(r) = c(r')$. At line 14, r checks if it is *AlignedUp* or *AlignedDown* with another robot r'. If so, to avoid a collision with r', r moves to a non-forbidden neighbor on l.

TASK. T_3 Non-guard robots move along rows toward their final target. Note that, if a robot encounters a forbidden vertex during its movement, then it will stop at a distance of one from it. During the task, each guard robot will stay in place. This task is activated only when task T_2 is over, therefore **pre$_3$** holds:

$$\textbf{pre}_3 \equiv [RC = \rho(C) \wedge \forall \text{ row } h \in \{1, 2, \ldots, M\} : (ER_{down}(h) = 0 \wedge ER_{up}(h) = 0)].$$

According to the conclusion of Task T_2, in each row h there are at most two embedded targets of F (i.e., $t_h \leq 2$) and at most two robots (i.e., $\overline{r_h} \leq 2$). Moreover $\overline{r_h} = t_h$. Each non-guard robot r determines its **target vertex** $t(r)$ as follows: the leftmost robot on h is assigned to the leftmost target in h, and the rightmost robot on row h (if any) is assigned to the rightmost target in h. Consider the shortest path of r toward $t(r)$, and let $v(r)$ be the next vertex toward $t(r)$ along such a path. Then, move m_3 is defined as follows: *if $v(r) \notin FV$ then move to $v(r)$.*

TASK. T_4 This task is executed only when T_3 is concluded, hence all robots matched their final target except for the guards and those robots having a forbidden along the path to reach their final target. Note that, when the guard robots start moving from the corners of G_k, the common reference system is lost, hence concepts like guards, c-sectors, rows, and columns cannot be used.

To verify that T_4 must be performed, robots try all the possible embeddings (at most 6) to check whether there exist special conditions. Given $C = (G_k, \lambda)$ and $F = (G_{k'}, \lambda')$, we say that an embedding of F into C is **conclusive** if all the following conditions hold:

1. if U denotes the set of all the unmatched robots of C, then $|U| \leq 6$,
2. U can be partitioned into U_g (set of guards) and U_f (set of robots to be moved on forbidden vertices), where $|U_g| = \rho(C)$ and $0 \leq |U_f| \leq |U| - \rho(C)$,
3. if $r \in U_g$, then r is either outside F or on its boundary, and the target of F closest to r is unmatched and located on the boundary of F,
4. if $r \in U_f$, then r is adjacent to a corner of G_k, and this corner corresponds to an unmatched target of F.

As a consequence, this task can be activated only when the precondition $\mathtt{pre}_4 \equiv$ [there exists a conclusive embedding of F into C] holds. The corresponding move is called m_4 and is defined as follows: *each robot in U moves toward the unmatched target of F specified in the conclusive embedding of F into C.*

TASK. T_5 Each robot recognizes that there exists an embedding of F into C such that all robots are matched (i.e., "F is formed") and no more movements are required. The precondition is $\mathtt{pre}_5 \equiv [F$ is formed]. When this precondition is verified, each robot performs the nil move keeping the current position.

5.2   Formalization and Correctness

The algorithm has been designed according to the methodology proposed in [11]. The decomposition into tasks of $\mathcal{A}$ is summarized in Table 1.

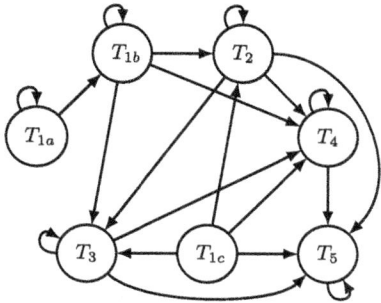

Fig. 5. Transition graph.

Table 1. Phases of the algorithm. Columns from left to right report: a summary of the task's goal, the task's name, the precondition to enter each task, and the transitions among tasks.

Sub-problems	Task	Precondition	Transitions
Placement of guards	T_{1a}	$RP = 0$	T_{1b}
	T_{1b}	$RP \geq \rho(C) \wedge RC = 0$	T_{1b}, T_2, T_3, T_4
	T_{1c}	$RC > \rho(C)$	T_2, T_3, T_4, T_5
Distribute robots on rows	T_2	$RC = \rho(C) \wedge \exists h \in \{1, 2, \ldots, M\}:$ $(ER_{down}(h) \neq 0 \vee ER_{up}(h) \neq 0)$	T_2, T_3, T_4, T_5
Moving toward targets	T_3	$RC = \rho(C) \wedge \forall h \in \{1, 2, \ldots, M\}:$ $(ER_{down}(h) = 0 \wedge ER_{up}(h) = 0)$	T_3, T_4, T_5
Pattern finalization	T_4	there exists a conclusive embedding of F into C	T_4, T_5
Termination	T_5	F formed	T_5

Assume T_i is performed on a current configuration C, $\mathcal{A}$ transforms C into C', and C' is assigned the task T_j: we say that $\mathcal{A}$ generates a transition from T_i to T_j. The set of all possible transitions of $\mathcal{A}$ determines a directed graph called *transition graph* (cf. Fig. 5). The correctness of $\mathcal{A}$ can be provided by showing that all the following properties hold: (1) for each task T_i, the transitions from T_i are exactly those reported in the transition graph, (2) each transition occurs within finite time, (3) possible cycles among transitions are traversed a finite number of times, (4) there exists a sink node assigned to configurations where GMV is solved, (5) the algorithm is collision-free.

Concerning the complexity, the execution time of $\mathcal{A}$ is determined by counting the required LCM cycles. $\mathcal{A}$ is time-optimal since it can be observed that (1) any algorithm needs at least $\Omega(k)$ cycles to solve GMV, and (2) each task of $\mathcal{A}$ requires $O(k)$ cycles. Proving the correctness and complexity of $\mathcal{A}$ leads to the following results:

Theorem 2. *If there exists a mutual-visibility set X per G_k, then GMV can be optimally solved in each configuration $C = (G_k, \lambda)$, $k \geq 4$, composed of $|X|$ synchronous robots with chirality and such that $\rho(C)$ divides $\rho(X)$.*

Corollary 1. *There exists a time-optimal algorithm that solves GMV in each $C = (G_k, \lambda)$, $k \geq 4$, of synchronous robots with chirality and such that $\rho(C) \leq 2$.*

6   Final Remarks

The full characterization of GMV on hexagonal grids requires investigating the existence of μ-sets with symmetricity six. So far, we were able to find μ-sets for G_{3k}, with $1 \leq k \leq 7$. Open problems include studying GMV for G_k under different schedulers (semi-synchronous or asynchronous) and/or with completely disoriented robots (no common chirality) and whether is possible to solve it optimally under these settings. Finally, extending the analysis to general $k \times m \times n$ hexagonal grids, with generic integers k, m, and n, would also be challenging.

References

1. Adhikary, R., Bose, K., Kundu, M.K., Sau, B.: Mutual visibility by asynchronous robots on infinite grid. In: Algorithms for Sensor Systems—ALGOSENSORS 2018. LNCS, vol. 11410, pp. 83–101. Springer (2018). https://doi.org/10.1007/978-3-030-14094-6_6
2. Bose, K., Adhikary, R., Kundu, M.K., Sau, B.: Arbitrary pattern formation on infinite grid by asynchronous oblivious robots. Theor. Comput. Sci. **815**, 213–227 (202https://doi.org/10.1016/J.TCS.2020.02.016
3. Brešar, B., Yero, I.G.: Lower (total) mutual-visibility number in graphs. Appl. Math. Comput. **465**, 128411 (2024). https://doi.org/10.1016/j.amc.2023.128411
4. Cicerone, S., Di Fonso, A., Di Stefano, G., Navarra, A.: Arbitrary pattern formation on infinite regular tessellation graphs. Theor. Comput. Sci. **942**, 1–20 (2023https://doi.org/10.1016/j.tcs.2022.11.021
5. Cicerone, S., Di Fonso, A., Di Stefano, G., Navarra, A.: The geodesic mutual visibility problem for oblivious robots: the case of trees. In: 24th International Conference on Distributed Computing and Networking, ICDCN 2023, pp. 150–159. ACM (2023). https://doi.org/10.1145/3571306.3571401
6. Cicerone, S., Di Fonso, A., Di Stefano, G., Navarra, A.: The geodesic mutual visibility problem: Oblivious robots on grids and trees. Pervasive Mob. Comput. **95**, 101842 (2023). https://doi.org/10.1016/j.pmcj.2023.101842
7. Cicerone, S., Di Fonso, A., Di Stefano, G., Navarra, A.: Time-optimal geodesic mutual visibility of robots on grids within minimum area. In: Proceedings of the Stabilization, Safety, and Security of Distributed Systems, pp. 385–399. Springer (2023). https://doi.org/10.1007/978-3-031-44274-2_29
8. Cicerone, S., Di Fonso, A., Di Stefano, G., Navarra, A., Piselli, F.: Mutual Visibility in Hypercube-Like Graphs. In: Structural Information and Communication Complexity—SIROCCO 2024. LNCS, vol. 14662, pp. 192–207. Springer (2024). https://doi.org/10.1007/978-3-031-60603-8_11
9. Cicerone, S., Di Stefano, G.: Mutual-visibility in distance-hereditary graphs: a linear-time algorithm. Procedia Comput. Sci. **223**, 104–111 (2023). https://doi.org/10.1016/J.PROCS.2023.08.219
10. Cicerone, S., Di Stefano, G., Klavžar, S.: On the mutual visibility in cartesian products and triangle-free graphs. Appl. Math. Comput. **438**, 127619 (2023). https://doi.org/10.1016/j.amc.2022.127619
11. Cicerone, S., Di Stefano, G., Navarra, A.: A structured methodology for designing distributed algorithms for mobile entities. Inform. Sci. **574**, 111–132 (2021). https://doi.org/10.1016/j.ins.2021.05.043
12. Cicerone, S., Di Stefano, G., Klavžar, S., Yero, I.G.: Mutual-visibility problems on graphs of diameter two. Eur. J. Combinatorics **120** (2024). https://doi.org/10.1016/j.ejc.2024.103995
13. Daymude, J.J., Hinnenthal, K., Richa, A.W., Scheideler, C.: Computing by programmable particles. In: Distributed Computing by Mobile Entities, Current Research in Moving and Computing, LNCS, vol. 11340, pp. 615–681. Springer (2019). https://doi.org/10.1007/978-3-030-11072-7_22

14. Di Luna, G.A., Flocchini, P., Chaudhuri, S.G., Poloni, F., Santoro, N., Viglietta, G.: Mutual visibility by luminous robots without collisions. Inf. Comput. **254**, 392–418 (2017)
15. Di Stefano, G.: Mutual visibility in graphs. Appl. Math. Comput. **419**, 126850 (2022). https://doi.org/10.1016/j.amc.2021.126850
16. Dudeney, H.E.: Amusements in Mathematics. Nelson, Edinburgh (1917)
17. Flocchini, P., Prencipe, G., Santoro, N.: Moving and computing models: robots. In: Distributed Computing by Mobile Entities, Current Research in Moving and Computing, LNCS, vol. 11340, pp. 3–14. Springer (2019).https://doi.org/10.1007/978-3-030-11072-7_1
18. Ghosh, S., Goswami, P., Sharma, A., Sau, B.: Move optimal and time optimal arbitrary pattern formations by asynchronous robots on infinite grid. Int. J. Parallel Emerg. Distrib. Syst. **38**(1), 35–57 (2023). https://doi.org/10.1080/17445760.2022.2124411
19. Manuel, P., Klavžar, S.: A general position problem in graph theory. Bull. Aust. Math. Soc. **98**(2), 177–187 (2018). https://doi.org/10.1017/S0004972718000473
20. Roy, D., Klavžar, S., Lakshmanan, A.: Mutual-visibility and general position in double graphs and in mycielskians. CoRR abs/2403.05120 (2024).https://doi.org/10.48550/arXiv.2403.05120
21. Wolfram MathWorld: Hexagonal Grid Graph. https://mathworld.wolfram.com/HexagonalGridGraph.html (2023) [Online; Entries Last Updated: Sat Jul 29 2023]

Coating in **SILBOT** with One Axis Agreement

Alfredo Navarra[iD] and Francesco Piselli[(✉)][iD]

Department of Mathematics and Computer Science, University of Perugia,
Perugia, Italy
`alfredo.navarra@unipg.it`, `francesco.piselli@unifi.it`

Abstract. In the context of *Programmable Matter* (PM), we consider
the *Coating* problem. A swarm of weak and self-organizing computational
entities, called *particles*, are required to move so as to ensure the closed
surrounding of an object. As a model for PM, we consider the SILBOT,
where asynchronous particles are modeled as finite state automata, living and operating on a triangular grid embedded in the plane. So far,
within SILBOT, the Coating problem has been investigated for n particles sharing a common handedness, i.e., chirality.
Here we investigate the case where particles share the direction of one
axis of the coordinate system instead of chirality. We present a time
optimal deterministic distributed algorithm – along with the correctness
proof, that in $\Theta(n^2)$ rounds solves the Coating problem, where a round
concerns the minimal time window within which each particle is activated
at least once.

Keywords: Programmable Matter · Coating · Asynchrony ·
Stigmergy

1 Introduction

In the realm of Distributed Computing, main attention has been devoted in
the last years to the so-called Programmable Matter (PM). This is intended as
a system of computational entities, called *particles*, that can be programmed
via distributed algorithms to collectively achieve global tasks. Particles have the
ability to change their physical properties (e.g., shape, density, optical properties,
etc.) in a programmable way. Various models have been introduced in order to
capture the essence of such systems [4, 6, 7]. In fact, the weaker a model, the
wider the range of applications. One model that is certainly gaining more and
more interest due to its minimal assumptions is SILBOT [3]. Within the SILBOT
model, particles are modeled as finite state automata, living and operating on
a triangular grid embedded in the plane. They are all identical, executing the

The work has been supported in part by the Italian National Group for Scientific
Computation GNCS-INdAM.

T. Masuzawa et al. (Eds.): SSS 2024, LNCS 14931, pp. 177–192, 2025.
https://doi.org/10.1007/978-3-031-74498-3_13

same distributed algorithm based on local observation of their surroundings, up to two-hops distance.

In SILBOT, particles operate independently, in a fully asynchronous way. The movement of a particle occurs from one node to another by alternating between a CONTRACTED state (a particle occupies one node) and an EXPANDED state (a particle occupies one node and an incident edge).

Apart for their visible state, particles do not admit any means of persistent memory with respect to past computations. Relative positioning among particles is their only (implicit) way of communicating, i.e., stigmergy is exploited.

Fundamental tasks have been approached within SILBOT, demonstrating the power of such a system. In [3], for instance, *Leader Election* has been solved. In [16], the *Scattering* problem has been approached. The requirement was to reach a configuration where no two particles are at one-hop distance. The problem has been solved also by generalizing the model with just one-hop visibility. The task to reach a configuration where it is guaranteed that all particles are disposed along one straight path (*Line Formation*) has been studied in [13,14]. The resolution of Line Formation has required the particles to share a common orientation. Finally, in [15], the COATING problem has been approached. The requirement is to reach a configuration where the closed surrounding of a convex object is ensured. The problem has been solved when particles share a common handedness, i.e., *chirality* is assumed.

1.1 Our Results

In this paper, we still investigate the COATING problem in SILBOT. The main question left open in [15] was whether one can get rid of chirality. Even though it is still unknown whether COATING can be solved without any additional assumption if chirality is removed, we investigate whether it is enough to consider a common knowledge of the direction of just one axis of the coordinate system. In fact, the two assumptions, i.e., chirality and one-axis agreement, are somehow incomparable in terms of additional power for entities in distributed computing. Hence, it is worth investigating the COATING problem in such a setting.

More precisely, given a convex object occupying some connected nodes of the grid and given a sufficiently large set of particles, the problem asks for a distributed algorithm that moves particles so as to surround the object by occupying all the nodes adjacent to it. As in [15], for the ease of the discussion, we prefer to deal with convex objects. In fact, when dealing with concave objects, still particular cases must be excluded, see e.g., the so-called *tunnels* as in [5,10].

We provide a deterministic and distributed algorithm that optimally solves COATING within $\Theta(n^2)$ rounds, with n being the number of particles, and a round being the minimal time window within which each particle is activated at least once. Similarly to the algorithm proposed in [15], while solving COATING, our algorithm also solves the *Hole Compaction*, where a hole is a connected subset of unoccupied nodes, enclosed by particles and possibly the object, provided that particles are able to recognize holes. An example of hole composed of just two nodes can be seen in the bottom-right of Fig. 1, where two empty nodes are surrounded by eight particles. In fact, we exploit part of the algorithm proposed in [15] as a subroutine, but then the technique to end up in a configuration where COATING is solved is completely different. It is mainly based on some geometric properties and it certainly deserves main attention.

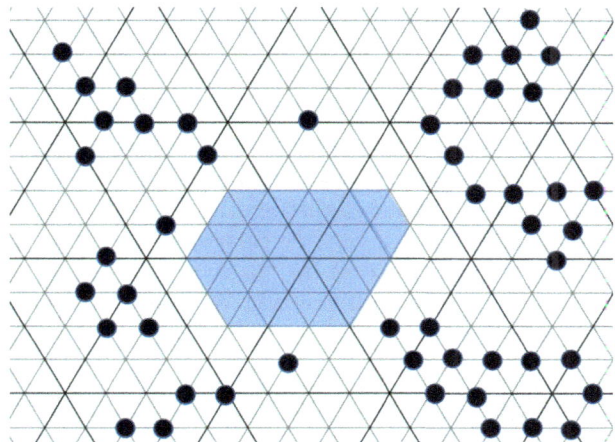

Fig. 1. Example of initial configuration where the black circles represent CONTRACTED particles and the shaded area includes all the nodes occupied by a convex object.

1.2   Outline

In the next section, we review the SILBOT model in detail and we formalize the COATING problem. In Sect. 3, we describe and formally present our algorithm for COATING. In Sect. 4, the correctness proof and the complexity for the proposed algorithm are presented. Finally, in Sect. 5 we provide concluding remarks and pose some interesting research directions.

2   SILBOT Model and COATING Problem

In this section, we review the SILBOT model for PM introduced in [3], and we formalize the COATING problem.

- Operating Environment. Particles operate on an infinite triangular grid embedded in the plane, where each node has six incident edges. Each node can be occupied by at most one particle. There are n particles and there is an object *obj* occupying a connected and convex set of nodes, see e.g., Fig. 1.

- Particles and Configurations. Each particle is an automaton admitting two states, CONTRACTED or EXPANDED (it does not have any other form of persistent memory). In the former state, a particle occupies a single node of the grid while in the latter, a particle occupies one single node and one of the adjacent edges. Hence, a particle always occupies one node, at any time.

Each particle can sense its surroundings up to two-hops distance, i.e., if a particle occupies a node u, then it can see the neighbors of u and the neighbors of the neighbors of u. Specifically, a particle can determine (i.e., sense) if a node is empty or occupied by a CONTRACTED particle, or occupied by an EXPANDED particle, for each node in its two-hops visibility range. Any positioning of CONTRACTED or EXPANDED particles that includes all n particles composing the system plus obj is referred to as a *configuration*.

Furthermore, particles have the so called *exterior awareness*, that informally is the power of detecting what is "outside" the configuration and what is "inside" (that is, holes). Formally, a hole is a subset of empty nodes enclosed by particles and possibly by obj. A particle can distinguish whether a node v within its visibility range is CONT, EXP, IN, OUT or OBJ where: a CONT node is a node occupied by a CONTRACTED particle; an EXP node is a node occupied by an EXPANDED particle; an IN node is an empty node that is part of a hole; an OUT node is an exterior empty node (an empty node that is not part of any hole); an OBJ node is a node occupied by obj.

- Connectivity and Initial Configurations. About the configurations from which COATING can be solved, a relevant property concerns *connectivity*.

Definition 1. *A configuration is said to be* connected *if the set of nodes that are occupied or semi-occupied or containing the object or holes form a connected subgraph of the grid.*

In [15], a configuration was said to be *initial* if connected (in the meaning of Def. 1) and containing only CONTRACTED particles. Here we try to slightly generalize such a definition with the aim to enlarge the set of initial configurations admitted. We first need the following:

Definition 2. *The* External Perimeters EP, EP^{+1} and EP^{+2} *of an object obj are composed by all the nodes at distance 1, 2 and 3, respectively, from obj.*

See Fig. 2 for a visualization of the defined External Perimeters. A consequence of what was intended as initial configuration in [15], is that either the object is immersed in a hole, or particles can be partitioned into maximal subsets such that each subset is connected and occupies at least one node in EP (i.e., particles plus obj plus holes form a connected subgraph of the grid). Furthermore, it follows that if a subset P contains more than one node in EP, then each node in EP within the most extreme ones of P either belongs to P or it is an IN node.

In this paper, for the initial configurations we require that the above property is true but with respect to EP^{+1} rather than EP. Hence,

Definition 3. *A configuration is* initial *if it is composed of* CONTRACTED *particles where either the object is immersed in a hole, or particles can be partitioned into connected (according to Def. 1) maximal subsets such that each subset P*

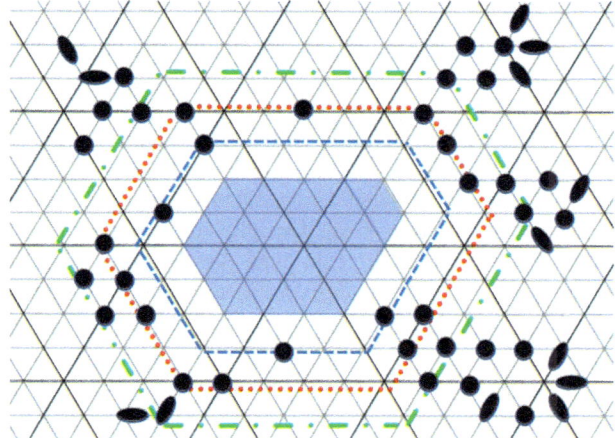

Fig. 2. A possible evolution of the configuration from Fig. 1 where the black ellipses occupying one node and one neighboring edge represent EXPANDED particles; the dashed line represents EP, the dotted line represents EP^{+1}; the dash-dot line represents EP^{+2}.

contains at least a node in EP^{+1} and if P contains more than one node in EP^{+1}, then each node in EP^{+1} within the most extreme ones of P either belongs to P or it is an IN *node.*

Note that, by the above definition, an initial configuration is not necessarily connected as it was in [15], see, e.g., Fig. 1. A configuration is said *not-truly disconnected* if it admits all the properties of the initial configurations but neglecting on the status of the particles that can also be EXPANDED.

- Movement and States. As described above, each particle p can occupy only one node u at a time. In order to move to a neighboring node v, p expands on the edge (u, v). Thus, in the EXPANDED state, p occupies node u and edge (u, v) (see, e.g., in Fig. 2 how EXPANDED particles look like).

Note that, node v may still be occupied by another particle p'. If p' leaves node v in the future, then p, the EXPANDED particle, will contract onto node v during its next activation. There might be arbitrary but finite delays between the actions of these two particles. If p' moves to another node, edge (u, v) is still occupied by p, the originally EXPANDED particle. In this case, we say that node v is *semi-occupied*. We denote by SO the set of semi-occupied nodes.

A very strong constraint of **SILBOT**, called *Commitment Property*, is that: *A particle commits itself into moving to node v by expanding in that direction and, at the next activation of the same particle, it is constrained to move to node v, if v is empty. A particle cannot revoke its expansion once committed.*

- Asynchrony and Rounds. There might be possible delays between observations and movements performed by the particles. This reminds the so-called ASYNC model designed for mobile and oblivious robots (see, e.g., [2,11,12]).

All operations performed by the particles are non-atomic: there can be arbitrarily finite delays between the actions of sensing the surroundings, computing the next decision (e.g., expansion or contraction), executing the decision.

There are no assumptions nor restrictions on the scheduling of these events: any possible execution of a physical system can be captured by the model.

A *round* is the time within which all particles have been activated and concluded their activation time at least once. Clearly, the duration of a round is finite but unknown and may vary from time to time. The well-established fairness assumption is also included, that is, each particle must be activated within finite time, infinitely often in any execution of the particles system. Due to the asynchronous nature of the system, it may happen that a particle decides (or is forced, in case of contraction) at time t to take an action, and that this action will actually be executed at time $t' > t$, when other particles might have changed their state; in other words, the action executed at time t' might be based on the obsolete observation of the surroundings taken at time t. The time required to accomplish an action is finite but arbitrary. Hence a round is the shortest time period during which each particle has performed an action (where an action could be the *nil* one if a CONTRACTED particle decides to remain as such, or if an EXPANDED particle finds the target node still occupied).

- **Orientation.** Differently from [15], we remove the assumption on chirality but we consider all particles sharing the knowledge of the direction of one axis, say *South − North* (S-N). Various papers on COATING assume chirality (e.g., [1,8,9,15]) but it is worth investigating the one-axis assumption as there is no hierarchy among such assumptions. For the sake of readability, we assume that the embedding of the triangular grid is such that one of its three directions is parallel to the S-N axis.

- **Randomness.** Particles take deterministic decisions and do not have access to random bits. Each particle may be activated at any time independently from the others. Once activated, a particle decides (deterministically) its next action.

The activation of each particle is intended to be decided by an 'adversarial' scheduler. If two CONTRACTED particles decide to expand on the same edge simultaneously, exactly one of them (arbitrarily chosen by the scheduler) succeeds. If two particles are EXPANDED along two distinct edges incident to the same node v, and both particles are activated simultaneously, exactly one of them (again, chosen arbitrarily by the scheduler) contracts to node v, while the other particle does not change its EXPANDED state.

- **Coating and Holes Compaction Problem.**

Definition 4. *[COATING and Holes Compaction Problem (CHCP)] Given an initial configuration of $n \geq |EP|$ particles, an algorithm solves CHCP if there exists a time t after which no expansion occurs and the following conditions hold:*

i) the configuration is connected, does not contain holes, $SO = \emptyset$;
ii) EP is fully occupied by particles.

Algorithm 1 Procedure EROSION

Require: A node $p \notin \{EP \cup EP^{+1}\}$ occupied by a CONTRACTED particle.
Ensure: Action to be taken by the particle in p.
1: **if** $G(N(p, IC))$ is connected $\wedge |N(p) \cap SO| = 0$ **then**
2:    **if** $N(p, IC) = \{q\}$ **then** Expand along (p, q);
3:    **else if** $N(p, IC) = \{q, r\}$ **then**
4:        **if** $q \in N(p, I)$ **then** Expand along (p, q)
5:        **else** Expand along (p, r);
6:    **else if** $N(p, IC) = \{r, q, s\}$, with q being between r and s **then**
7:        **if** $\{r, q, s\} \cap N(p, I) = \{x\}$ **then** Expand along (p, x)
8:        **else if** $|N(r, C)| > 2 \wedge |N(q, C)| > 3 \wedge |N(s, C)| > 2$ **then**
            Expand along (p, q)
9:        **else if** $\{r, q\} \subseteq N(p, C) \cap EP^{+1} \wedge |N(r, C)| = 2 \wedge |N(q, C)| > 3 \wedge$
            $|N(s, C)| > 2$ **then** Expand along (p, q);

Algorithm 2 FUNNEL

Require: A configuration with node p occupied by a CONTRACTED particle.
Ensure: COATING plus Hole Compaction.
1: **if** $p \notin \{EP \cup EP^{+1}\}$ **then** EROSION(p);
2: **else if** $p \in EP^{+1}$ **then** let r, p and s be sequential along EP^{+1} with $\alpha_r \leq \beta_s$;
3:    **if** $|N(p, I) \cap EP^{+2}| = 0 \wedge |N(p) \cap EP^{+2} \cap SO| = 0$ **then**
4:        **if** $|N(p, C) \cap EP^{+2}| = 0$ **then**
5:            **if** $\alpha_r = \beta_s$ **then**
6:                **if** $\alpha_r + \beta_s > 120°$ **then** Expand along (p, s)
7:                **else** Expand along (p, q), with $q \in EP$;
8:            **else** Expand along (p, s);
9: **else if** $p \in EP$ **then** let r, p and s be sequential along EP;
10:    **if** $\exists x \in EP^{+1}$: x is EXPANDED toward p **then** Expand along (p, r)
11:    **else if** r (resp. s) $\in EP$ is EXPANDED toward p **then**
            Expand along (p, s) (resp. (p, r));

Of course, it must be $n \geq |EP|$ as otherwise there are not enough particles to accomplish the COATING task.

3 Algorithm FUNNEL

In this section, we propose an algorithm to solve CHCP within SILBOT, with the additional assumption that the particles are endowed with the knowledge of the direction of one axis, e.g., the S-N axis. Starting from any initial configuration with an object *obj*, the algorithm leads the particles to surround *obj* and to compact holes, if any. The algorithm is divided into two parts. The first part represented by Procedure EROSION is reported in Algorithm 1, whereas Algorithm 2 shows the pseudo-code of the main Algorithm FUNNEL.

Before going into the details of the algorithms, we first need to introduce the next additional notation. Given a particle p, we denote by $N(p)$ the set of six nodes adjacent to p, by $N(p,C)$ and $N(p,I)$ the set of CONT and IN nodes, respectively, adjacent to p, and by $N(p,IC)$ the set $N(p,I) \cup N(p,C)$.

We also add the concept of *siblings*: the siblings of a node/particle p are its two neighboring nodes at the same distance from *obj* (Fig. 3.a), e.g., the siblings of a node in EP^{+1} are also in EP^{+1}. Thanks to the two-hops visibility, if $p \in EP^{+1}$, then p is able to determine its neighbors in EP, then the neighbors shared between p itself and the ones in EP are its two siblings (in EP^{+1}) whereas all its remaining neighbors are in EP^{+2}. Moreover, because of the triangular grid, a particle p has only two siblings r and s and since it knows the direction of the S-N axis, it can calculate the angles α_r and β_s between the known axis and the two siblings. To do so, for each sibling, p computes the smallest angle necessary to rotate the S-N axis centered on p so as to have such a sibling in its north (Fig. 3.b). Whenever $\alpha_r \neq \beta_s$, without loss of generality, we always consider $\alpha_r < \beta_s$.

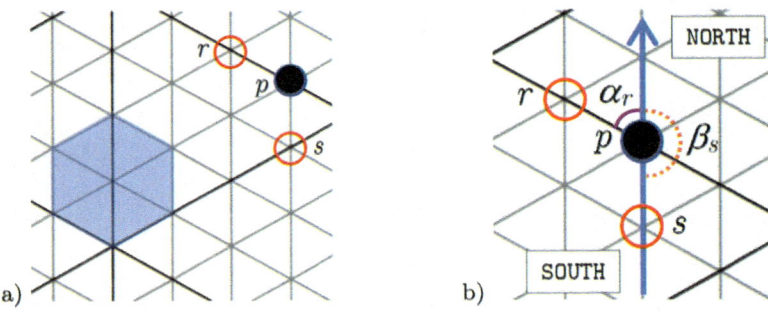

Fig. 3. *a*) a particle p in EP^{+1} with its two siblings r and s; *b*) the angles α_r and β_s computed by p.

Definition 5. *The* pivot *is a node/particle with its two siblings positioned at $60°$ from the known axis.*

Lemma 1. *For each positive distance d from obj, the pivot is unique.*

Proof. Recall that the embedding of the triangular grid is such that one of its three directions is parallel to the S-N axis. The proof proceeds by considering $d = 1$. Then, the proof can be easily generalized to the case of any $d > 1$.

Since *obj* is convex (it might be even just a single node or a segment) lying over a triangular grid, by considering the convex hull CH_{EP} of EP, we have that each internal angle γ of CH_{EP} measures exactly 120°. In fact, any wider angle would imply that *obj* is concave, whereas $\gamma = 60°$ would require three mutual neighboring nodes belonging to EP which cannot occur. Moreover, the southernmost angle of CH_{EP} is unique and subtends a node p. In fact, in order to have more than one of such angles, again obj should be concave. It follows that there does not exists another node of EP with the same properties of p. As already observed, $\gamma = 120°$ means that the siblings of p, say r and s, are not neighboring to each other. Considering the $\mathbb{S}$-$\mathbb{N}$ axis passing through p, since γ is the southernmost angle of EP, then $\alpha_r = \beta_s = 60°$, i.e., p is a pivot.

Increasing the distance from obj does not change the provided arguments. In fact, for each given distance $d > 1$, the only pivot is the southernmost node with respect to the $\mathbb{S}$-$\mathbb{N}$ axis, at distance d from obj. □

Remark 1. By the proof of Lemma 1, we can observe some useful properties holding for each positive distance d from the object. In fact, not only the pivot is unique but also the node p such that its siblings r and s provide $\alpha_r = \beta_s = 120°$ is unique. Furthermore, any other angle of the convex hull with respect to the considered distance d lies on a node whose siblings define two different angles, with the bigger (denoted by β) determining the southern (with respect to the known axis) direction.

The strategy adopted by FUNNEL is ideally based on three phases. First it makes the most external particles of the configuration EXPANDED toward the ones closer to obj, until pushing those in EP^{+1}. Afterwards, particles in EP^{+1} enter EP only from EP's pivot, by means of movements toward south (cf. Remark 1). Finally, EP is surrounded by movements starting from the pivot, possibly in both directions.

The first phase is realized by means of Procedure EROSION, which is directly imported from [15]. The only difference is that the erosion process is adapted so as to stop with the particles in EP^{+1}, instead of reaching EP as it was originally. Moreover, in the first phase every IN node not in EP is occupied.

For the second phase, a particle p in EP^{+1} first checks if it is neighboring to any IN or SO node in EP^{+2}. If not, it checks if it has neighboring particles in EP^{+2} and with respect to that decides whether and where to expand. The proposed algorithm allows particles in EP^{+1} to expand toward nodes in EP only from a specific node, i.e., the pivot of EP^{+1}, hence the name FUNNEL.

Regarding a particle in EP not on the pivot, it is allowed to move along EP only if another particle from EP is EXPANDED toward it. A particle p lying on the EP's pivot can only receive a push from a particle on the pivot of EP^{+1} EXPANDED toward it, hence p can expand toward one of its siblings indiscriminately.

To better specify the aforementioned strategy, FUNNEL determines three different behaviors for the particles, according to whether they belong to a set not in $EP \cup EP^{+1}$, to EP^{+1} or to EP.

- For particles not in $EP \cup EP^{+1}$, FUNNEL calls Procedure EROSION, where a particle p is allowed to expand if $G(N(p, IC))$ is connected, there are no semi-occupied nodes in $N(p)$ and $|N(p, IC)| \leq 3$.

If p has only one neighbor, i.e., $|N(p, IC)| = 1$, then it expands toward that one neighbor once awakened (see Line 2 of Procedure EROSION).

If p has two neighbors, $|N(p, IC)| = 2$, then it needs to check if one of those is an IN node. If yes, then p expands toward that one, otherwise it expands toward one of its two neighbors, indiscriminately (Lines 3–5). It is worth to remark that the expansion priority is always toward IN nodes, if any.

If p has three neighbors, $|N(p, IC)| = 3$, then we can have the following:

(i) if there is only one IN node, p expands toward that one (Line 7);

(ii) if the condition at Line 8 is satisfied, p expands toward the central CONTRACTED neighbor. This is necessary to avoid that particles with degree three expand toward opposite directions possibly causing a disconnection.

Moreover, the condition at Line 9 is a special case where a particle $p \in EP^{+2}$ has three neighbors r, q and s, with $\{r, q\} \in EP^{+1}$. If r has only two CONTRACTED neighbors, then the condition is satisfied and p expands toward the central neighbor q. Without this exception, the system may incur in a deadlock;

- For particles in EP^{+1}, the behaviour dictated by FUNNEL is the following. A particle $p \in EP^{+1}$ checks if it is neighboring to any IN or SO node in EP^{+2} (Line 3 of Algorithm FUNNEL). If yes, it waits for those nodes to be occupied by EROSION. If not, it checks if it has neighboring CONTRACTED particles in EP^{+2}:

(i) if yes, condition at Line 4 is not satisfied and it will wait for those particles to be EXPANDED;

(ii) if not (Lines 5–8), first of all p waits for its neighbors in EP^{+2} to be all EXPANDED. This is necessary to guarantee that there is no disconnection between particles in EP^{+1} and EP^{+2}. Once p is allowed to expand, then it checks the angles of its siblings and, if those angles are equal (Line 5), another check needs to be done: if the sum of the angles is greater than $120°$ then p expands toward one of its siblings, say s (Line 6), otherwise p expands toward EP since it recognizes that it is a pivot (Line 7). If the siblings' angles are different, p directly expands toward s (Line 8), its sibling with greatest angle, i.e., the southernmost one;

- For particles in EP, FUNNEL allows a particle p to expand only if there exists another particle EXPANDED toward p. Particle $p \in EP$ can receive such a push in two different ways: the pushing comes from a particle in EP^{+1} or from a particle in EP (i.e., from a sibling of p). In the former case p is the pivot and can expand toward one of its siblings (Line 10). In the latter case, p expands toward the other sibling (Line 11). This implies that particles in EP are allowed to move only inside EP.

If we have $|EP| = n$, where n is the number of CONTRACTED particles in the initial configuration, then the last movement of the particles will complete the COATING of *obj* with all CONTRACTED particles. Instead, when $|EP| > n$, after finite time, all the particles will be EXPANDED and COATING solved.

4   Correctness and Complexity of FUNNEL

In this section, we show the correctness of FUNNEL, i.e., it terminates in a finite number of rounds in a connected configuration where the object is totally surrounded by particles and all the holes from the initial configuration, if any, have been compacted. Moreover, we show the complexity of the algorithm, proving it to be optimal in terms of rounds.

Similarly to [15], we model each execution of FUNNEL as a path in a directed graph $H = (V, E)$, where, given an object obj, the vertices in V correspond to any not-truly disconnected configuration with $n \geq |EP|$ particles, and the edges in E correspond to transitions among configurations determined by FUNNEL. Recall that for any particle $p \notin \{EP \cup EP^{+1}\}$, we exploit Procedure EROSION which has been basically imported from [15]. Hence, the correctness of that procedure is implicitly derived from [15]. We can then focus on particles in EP^{+1} and in EP. We first show that from any not-truly disconnected configuration from which Procedure EROSION does not make any particle expand, there always exists at least a particle $p \in EP^{+1}$ that can expand and decides to do so once activated. Then, we show that if a particle $p \in EP^{+1}$ has neighbors in EP^{+2}, then the movements from p never cause a disconnection between EP^{+1} and EP^{+2}. On the other hand, if a particle $p \in EP^{+1}$ has no neighbors (occupied by CONTRACTED particles, belonging to IN or to SO) in EP^{+2} then it can move along EP^{+1} or toward EP. Finally, we show that since particles in EP can expand only after a 'push' and they only move along EP, each new configuration generated is not-truly disconnected and the algorithm leads to the resolution of CHCP.

Theorem 1. *Given an initial configuration with a convex object obj and $n \geq |EP|$ CONTRACTED particles, Algorithm FUNNEL terminates within $\Theta(n^2)$ rounds in a connected configuration where EP is totally occupied and there are no holes. Moreover, any configuration generated during the execution of FUNNEL is not-truly disconnected.*

Proof. Initially there are n CONTRACTED particles forming a not-truly disconnected configuration with obj and possible holes. We prove the correctness of FUNNEL by showing the three following properties:

P1 Evolution. Each vertex in H, excluding those corresponding to final configurations where COATING is accomplished, has at least one outgoing edge;
P2 (Dis-)Connectivity. Any configuration C obtained by an action dictated by FUNNEL is not-truly disconnected;
P3 Acyclicity. Graph H is acyclic.

The finite number of particles and the type of movements allowed by the algorithm, imply that the number of configurations obtainable is finite. Hence, those three properties guarantee that a final configuration where CHCP is solved will be reached, eventually. More in details, P1 guarantees that from any configuration there exists at least a particle able to expand or contract, generating a new configuration (i.e., a new vertex in graph H is reached); P2 guarantees that each new configuration is not-truly disconnected, meaning that each subset of particles never 'loses sight' of obj; P3 guarantees that no configuration is generated more than once, hence the execution of the algorithm always leads to a final configuration.

Proof of property P1 (Evolution). First we show that in any non-final con-figuration there always exists at least a CONTRACTED particle p that is allowed by FUNNEL to move. As long as there are CONTRACTED particles not in $EP \cup EP^{+1}$, then from [15] we know that P1 is true. If none of the particles are moved by EROSION, we show that at least a particle (i) in EP^{+1} or (ii) in EP is moved by FUNNEL.

If the configuration is non-final, we are assured by [15] that the particles not in $EP \cup EP^{+1}$ are all EXPANDED, if any, and all the IN nodes at distance greater than one from obj have already been occupied. We now show that if there are CONTRACTED particles in EP^{+1}, then there always exists at least one able to expand. A particle $p \in EP^{+1}$ can expand in two possible scenarios: (a) all its neighbors in EP^{+2} are EXPANDED and it has no IN or SO neighbors in EP^{+2} or, (b) p has no neighbors (occupied by CONTRACTED particles, belonging to IN or to SO) in EP^{+2}. We already know that both (a) and (b) will eventually occur thanks to Procedure EROSION. This condition implies that a particle in EP^{+1} will never move as long as there is at least one of its neighbor in EP^{+2} still CONTRACTED. Moreover, since we know that the movements of particles in EP^{+2} only go toward particles in EP^{+2}, EP^{+1} or IN nodes, this makes possible that each subset of particles never disconnects from EP^{+1}. Hence we can say that a particle in EP^{+1} that decides to expand following Lines 2–8 of FUNNEL eventually exists.

According to FUNNEL, particles in EP^{+1} only expand toward nodes in EP^{+1} or toward the pivot in EP. A particle p has only two siblings and they can be positioned, with respect to p and the direction of the known axis, at the same angle or at two different angles: if they are at the same angle and the sum of those two is greater than $120°$ (Lines 5 and 6) then the expansion of p can be toward any of the two siblings; if the sum of the two equal angles is exactly $120°$, then it must be $\alpha_r = \beta_s = 60°$ because of the convexity of obj. Hence, p is on a pivot and it will expand toward EP's pivot (Line 7). On the other hand, if those siblings' angles are different, p expands toward its sibling with greater angle (Line 8), i.e., the southern one (cf. Remark 1). Let us consider two sides of obj with respect to the direction of the known axis. From 'one side' of obj, p expands in clockwise direction, and from the 'other side' it expands in counter-clockwise direction. This means that the expansions from the two sides of obj eventually reach the pivot. From that node, according to Line 7, p will expand toward its only neighbor in EP.

Now we can finally consider the particles in EP. According to Lines 9–11, particles in EP are able to expand only when pushed from particles in EP^{+1} or from a neighboring particle in EP. We have already proven that the first case will occur, eventually, and that it happens always on the pivot. A particle p on EP's pivot, receiving a push from a particle $x \in EP^{+1}$ at time t, will be able to expand toward one of its two siblings, r or s, at time $t' > t$. If at time t', particle p is EXPANDED toward r (s, resp.) and r (s, resp.) is empty, then p is able to occupy r (s, resp.) at time $t'' > t'$. Particle x, will then be able to contract on the node previously occupied by p, once activated again. Otherwise, if r (s, resp.) is occupied by a particle q at time t', then q will receive a push from p and, once activated, expands toward its other sibling, at time $t'' > t'$. Therefore, a particle in EP will surely be able to expand and it will do so always remaining in EP, occupying the nodes necessary to solve COATING.

Proof of property P2 ((Dis-)Connectivity). Since particles have only two-hops visibility, maintaining the configuration not-truly disconnected is crucial to the correct execution of the algorithm.

Algorithm FUNNEL has been designed to mimic an erosion process where the particles from the outer part of the configuration expand toward their inner neighbors (or holes, if any), closer to obj. Then we have an oriented movement along EP^{+1} toward the pivot node and, finally, a movement along EP dictated only by a push from a particle to another.

More in details, a particle $p \notin \{EP \cup EP^{+1}\}$ can expand toward a node occupied by another particle, or toward an IN node. From [15] we are assured that none of those movements can cause a disconnection.

If $p \in EP^{+1}$, then p is allowed to expand toward another node in EP^{+1} or toward EP if it is positioned on the pivot. These movements can occur when p has no neighbors (occupied by CONTRACTED particles, belonging to IN or to SO) in EP^{+2} or, if any, when they are all EXPANDED and no neighboring node in EP^{+2} is IN or SO. In both cases, a disconnection between p and other neighbors in EP^{+2} doesn't happen. Instead, a disconnection between $p \in EP^{+1}$ and a neighbor $p' \in EP$ or in EP^{+1} could happen. However, for the property of a configuration to be not-truly disconnected, p's neighbors in EP are not considered, and disconnections between particles in EP^{+1} still maintain a not-truly disconnected configuration. Therefore, these disconnections do not hinder the execution of the algorithm.

Finally, if $p \in EP$, it can only expand when pushed from another particle. This implies that once p moves at time t, the node previously occupied by p is now semi-occupied and at a time $t' > t$ will become occupied. Hence the configuration always maintains its original connectivity.

Proof of property P3 (Acyclicity). Consider configuration C_u represented in the directed graph H by vertex u. Let us associate to u a pair (δ_u, ζ_u) where δ_u represents the sum of the number of CONT nodes, plus the number of IN nodes, plus $|SO|$, plus the number of empty nodes in EP, plus the number of empty nodes in EP^{+1}; ζ_u represents the sum of the number of IN nodes, plus the number of empty nodes in EP, plus the number of empty nodes in EP^{+1}.

We observe that for each edge (u, v) of H, (δ_v, ζ_v) is lexicographically smaller than (δ_u, ζ_u). In fact, after an action is performed, the following can happen:

– At least a CONTRACTED particle in C_u is EXPANDED in C_v, making the number of CONT nodes decrease;
– At least an EXPANDED particle in C_u is CONTRACTED in C_v, making $|SO|$ or the number of IN nodes or the number of empty nodes in EP or in EP^{+1} decrease by at least $k \geq 1$. However, the number of CONTRACTED nodes increases of the same amount k.

By the above arguments, it is possible to define a topological ordering of the nodes of H as a linear extension of the partial ordering given by the pair (δ_i, ζ_i) of the corresponding configurations C_i. Hence, any execution of FUNNEL can be seen as a path along 'decreasing' configurations, i.e., no configuration in H is generated twice during the execution of FUNNEL.

Time complexity of FUNNEL. About the complexity of the algorithm, each path in H that starts from a node corresponding to an initial configuration and ends in a node corresponding to a final configuration, has a length bounded by the sum of the initial number of CONTRACTED particles, plus the number of movements necessary to reach the pivot in EP^{+1}, plus the number of movements required to occupy EP, plus the number of movements necessary to 'fill' the initial IN nodes. The first factor is exactly n. For the second factor, we can think about particles disposed in a path-shape configuration having one particle in EP^{+1} with both siblings at $120°$. This is the worst possible position to have the particles, for the resolution of the problem, since they are at the opposite side of obj from the pivot. The first movement in EP^{+1} requires the movement of all the particles. The scheduler may activate all the particles concurrently and, in so doing, a whole round would be consumed to make only one particle expand, which is the last one of the path (i.e., the furthest from obj). Moreover, the adversary will make the particles occupy at least half EP^{+1} before starting to expand toward EP. Hence, the second factor is given by $n \cdot \frac{1}{2}|EP^{+1}|$, where $|EP^{+1}|$ roughly equals $|EP| \leq n$. It follows that the number of rounds necessary to occupy all EP^{+1} is at most $O(n^2)$.

For the third factor, when $|EP| = n$, we need at most $O(n^2)$ rounds to completely occupy EP. Finally, the number of movements necessary to fill the IN nodes is upper bounded by $O(n^2)$ since with n particles, the maximum dimension of a hole is of such an order.

By the arguments given above, we can observe that the case of particles disposed in a path-shape configuration turns out to be rather general, regardless of the strategy adopted to solve CHCP within SILBOT. In particular, with all the particles aligned touching obj in just one place, they are forced to proceed one by one moving along EP. Any other possible movement would possibly disconnect them, make them 'lose track' of obj or waste rounds for unnecessary movements.

We can then state that Algorithm FUNNEL requires $\Theta(n^2)$ rounds to solve the COATING and Holes Compaction problem.   □

5   Conclusion

Within the SILBOT model, by adding the knowledge of one axis direction to the particles, we have been able to solve the COATING and Holes Compaction problem. The requirement is to make particles move so as to suitably surround a given static object while obtaining a connected configuration. We proposed an optimal algorithm that starting from any initial configuration of n particles, eventually leads to solve the problem within $\Theta(n^2)$ rounds.

As future work, it still remains to investigate whether the assumption on the direction of one axis (or chirality) is necessary to solve the problem Furthermore, investigating on other basic tasks within SILBOT is certainly of main interest. A variant of the COATING problem requires that all the particles contribute to surround the object in a multi-layer fashion. In the current setting, such a variant seems rather unfeasible. Perhaps, enlarging the visibility range of the particles might be enough.

References

1. Bazzi, R.A., Briones, J.L.: Deterministic leader election in self-organizing particle systems. In: Proceedings of the 21st International Symposium on Stabilization, Safety, and Security of Distributed Systems (SSS), vol. 11914. Springer (2019)
2. Cicerone, S., Di Stefano, G., Navarra, A.: Solving the pattern formation by mobile robots with chirality. IEEE Access **9**, 88177–88204 (2021). https://doi.org/10.1109/ACCESS.2021.3089081
3. D'Angelo, G., D'Emidio, M., Das, S., Navarra, A., Prencipe, G.: Asynchronous silent programmable matter achieves leader election and compaction. IEEE Access **8**, 207619–207634 (2020)
4. D'Angelo, G., D'Emidio, M., Das, S., Navarra, A., Prencipe, G.: Leader election and compaction for asynchronous silent programmable matter. In: Proceedings of 19th International Conference on Autonomous Agents and Multiagent Systems (AAMAS), pp. 276–284. International Foundation for Autonomous Agents and Multiagent Systems (2020)
5. Daymude, J.J., Derakhshandeh, Z., Gmyr, R., Porter, A., Richa, A.W., Scheideler, C., Strothmann, T.: On the runtime of universal coating for programmable matter. Nat. Comput. **17**(1), 81–96 (2018)
6. Daymude, J.J., Richa, A.W., Scheideler, C.: The canonical Amoebot model: algorithms and concurrency control. Distrib. Comput. **36**(2), 159–192 (2023)
7. Derakhshandeh, Z., Dolev, S., Gmyr, R., Richa, A.W., Scheideler, C., Strothmann, T.: Brief announcement: amoebot—a new model for programmable matter. In: Proceedings of 26th ACM Symposium on Parallelism in Algorithms and Architectures (SPAA), pp. 220–222. ACM (2014)
8. Derakhshandeh, Z., Gmyr, R., Richa, A.W., Scheideler, C., Strothmann, T.: Universal coating for programmable matter. Theor. Comput. Sci. **671**, 56–68 (2017)
9. Derakhshandeh, Z., Gmyr, R., Strothmann, T., Bazzi, R.A., Richa, A.W., Scheideler, C.: Leader election and shape formation with self-organizing programmable matter. In: Proceedings of 21st International Conference on DNA Computing and Molecular Programming (DNA), vol. 9211, pp. 117–132. Springer (2015)
10. Kim, Y., Katayama, Y., Wada, K.: Pairbot: A novel model for autonomous mobile robot systems consisting of paired robots. CoRR **abs/2009.14426** (2020), https://arxiv.org/abs/2009.14426
11. Kirkpatrick, D., Kostitsyna, I., Navarra, A., Prencipe, G., Santoro, N.: On the power of bounded asynchrony: convergence by autonomous robots with limited visibility. Distrib. Comput. **37**, 279–308 (2024). https://doi.org/10.1007/s00446-024-00463-7
12. Klasing, R., Kosowski, A., Navarra, A.: Taking advantage of symmetries: gathering of many asynchronous oblivious robots on a ring. Theor. Comput. Sci. **411**, 3235–3246 (2010)

13. Navarra, A., Piselli, F.: Asynchronous silent programmable matter: line formation. In: Proceedings of 25th International Symposium on Stabilization, Safety, and Security of Distributed Systems (SSS). LNCS, vol. 14310, pp. 598–612 (2023)
14. Navarra, A., Piselli, F.: Brief announcement: line formation in silent programmable matter. In: Proceedings of 37th International Symposium on Distributed Computing (DISC). LIPIcs, vol. 281, pp. 45:1–45:8 (2023)
15. Navarra, A., Piselli, F.: Silent programmable matter: coating. In: 27th International Conference on Principles of Distributed Systems, (OPODIS). LIPIcs, vol. 286, pp. 25:1–25:17. Schloss Dagstuhl—Leibniz-Zentrum für Informatik (2023). https://doi.org/10.4230/LIPICS.OPODIS.2023.25
16. Navarra, A., Prencipe, G., Bonini, S., Tracolli, M.: Scattering with programmable matter. In: Proceedings of 37th International Conference on Advanced Information Networking and Applications (AINA), pp. 236–247. Advances in Intelligent Systems and Computing. Springer (2023)

Rendezvous and Merging for Two Metamorphic Robotic Systems Without Global Compass

Ryonosuke Yamada[1], Tomoyuki Usami[1], and Yukiko Yamauchi[2(✉)] (iD)

[1] Graduate School of ISEE, Kyushu University, 744 Motooka, Nishi-ku, Fukuoka 819-0395, Japan
[2] Faculty of ISEE, Kyushu University, 744 Motooka, Nishi-ku, Fukuoka 819-0395, Japan
yamauchi@inf.kyushu-u.ac.jp

Abstract. A *metamorphic robotic system (MRS)* consists of anonymous *modules*, each of which autonomously moves in the 2D square grid by sliding and rotation with keeping connectivity among the modules. Existing literature considers distributed coordination among modules so that they collectively form a single MRS. In this paper, we consider distributed coordination for two MRSs. We first present a rendezvous algorithm that makes the two MRSs gather so that each module can observe all the other modules. Then, we present a merge algorithm that makes the two MRSs assemble and establish connectivity after rendezvous is finished. These two algorithms assume that each MRS consists of five modules, that do not have a common coordinate system. Finally, we show that five modules for each MRS is necessary to solve the rendezvous problem. To the best of our knowledge, our result is the first result on distributed coordination of multiple MRSs.

Keywords: Metamorphic robotic system · rendezvous · merging

1 Introduction

The rendezvous problem is the simplest agreement problem in distributed coordination of autonomous mobile computing entities. The problem requires two mobile computing entities to gather at some position, which is not given in advance. The problem has been considered for mobile agents [2,6] and mobile robots [8] in a graph and the 2D Euclidean space. Most existing literature presents rendezvous algorithms together with impossibility results, which conversely show necessary assumptions and equipment of the mobile computing entities. For example, two mobile robots can solve the rendezvous problem in the fully-synchronous model, while they cannot in the semi-synchronous and asynchronous models [8]. However, when each mobile robot is equipped with a light, that can take a constant number of colors, the rendezvous problem is solvable in the semi-synchronous and asynchronous models [1].

This work was supported by JSPS KAKENHI Grant Number JP18H03202.

T. Masuzawa et al. (Eds.): SSS 2024, LNCS 14931, pp. 193–209, 2025.
https://doi.org/10.1007/978-3-031-74498-3_14

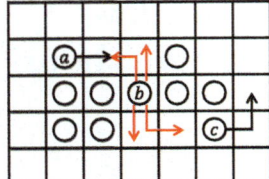

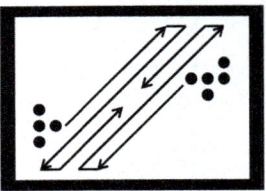

Fig. 1. Movements and connectivity   **Fig. 2.** Rendezvous of two MRSs

In this paper, we investigate the rendezvous problem for the *metamorphic robotic systems (MRSs)*. An MRS consists of a collection of autonomous mobile computing entities called *modules*, each of which autonomously moves in the 2D square gird. Each cell of the square grid can accept at most one module at each time instance. Modules are *anonymous* (indistinguishable) and each module can observe other modules in the constant-size visibility range, and perform a *rotation* or a *sliding* at one time. At each time instance, more than one module can perform rotations and slidings as long as there is no collision and the modules keep their connectivity. We consider a graph over the modules, where the vertex set is the set of modules and two modules are connected by an undirected edge when they are placed at side-adjacent cells. Then, an MRS is *connected* if the graph over the modules is connected. Figure 1 shows an example, where module a can perform a sliding and module c can perform a rotation. However, module b cannot perform the rotations or slidings shown by the red arrows because these movements break the connectivity of the MRS.

A variety of problems have been considered for a single MRS. The *reconfiguration problem* requires an MRS to change its initial shape to a given target shape [5]. Dumitrescu et al. showed that any horizontally convex shape can be transformed to another horizontally convex shape via a canonical chain shape. The *locomotion problem* requires an MRS to move toward a given direction. Dumitrescu et al. presented a shape that achieves fastest locomotion to the horizontal direction and another shape for the orthogonal direction [4]. The *search problem* requires an MRS to find a target in a field, which is a rectangular field surrounded by walls. Doi et al. pointed out that the necessary and sufficient number of modules to solve a problem represents the complexity of the problem [3]. That is, as the number of modules increases, the number of shapes (i.e., states) of an MRS increases and the MRS is expected to be able to solve more complicated problems. The authors also investigated the effect of common knowledge of the coordinate systems. The modules are equipped with the *global compass* when they agree on the north, south, east, and west in the 2D square grid. They showed that three modules are necessary and sufficient for the search problem if the modules are equipped with global compass, otherwise five modules are necessary and sufficient. The *evacuation problem* requires an MRS to get out of a field through an exit placed on the boundary of the field. Nakamura et al. showed that two modules are necessary and sufficient if the modules are equipped with global compass, otherwise four modules are necessary and sufficient for a rectangular field [7]. Existing results consider distributed coordination among the modules so that the modules collectively form a single MRS.

In this paper, we consider distributed coordination among MRSs. We first consider the rendezvous problem for two MRSs placed in a rectangular field. The goal of the rendezvous problem is to gather the two MRSs so that each module can observe all modules of the two MRSs in its constant visibility. When the modules are equipped with the global compass, the two MRSs can solve the rendezvous problem in the same approach as the evacuation problem [7]. That is, the two MRSs first move to some direction until it reaches a wall, and then move toward the south west corner. However, this algorithm cannot solve the rendezvous problem when the modules are not equipped with the global compass. In the proposed algorithm, we adopt an orthogonal "path" in the field and make the two MRSs move back and forth on the path. Thus, they can finish rendezvous in finite time. See Fig. 2 as an example. However, an MRS may follow a mirror image of the path due to the lack of the global compass. We will show that an MRS can return to the correct path if they are not moving on it. The proposed rendezvous algorithm is designed for two MRSs each of which consists of five modules not equipped with the global compass and with visibility range of 7. We also show that five modules are necessary to solve the rendezvous problem based on the results in [3].

We then consider the *merge problem* that requires the modules of the two MRSs to assemble and establish entire connectivity from a configuration where each module of the two MRSs can observe all the other modules.[1] A simple solution for the merge problem is to make one MRS wait the other MRS to come closer and merge. The proposed merge algorithm is based on a labeling of all 18 shapes of five modules and it makes the MRS R with the smaller label move toward the other MRS R' with the larger label. When the two MRS are initially in the same shape, each module first computes the *view* of each MRS, which is a configuration of the modules observed in the local coordinate system defined for each shape of the MRS. Then the MRS with a smaller view moves toward the MRS with a larger view. Finally, we present exceptional movements for initial configurations, where the two MRS cannot break their tie by their shapes or views. The proposed algorithm is designed for two MRSs each of which consists of five modules not equipped with the global compass and with visibility range of 9. To the best of our knowledge, this is the first result on distributed coordination for two MRSs. We believe our results open up new vistas for the distributed coordination theory of the MRS model.

---

[1] The rendezvous problem does not require connectivity among the modules of the two MRSs.

2   Preliminary

A *metamorphic robotic system* consists of a set of anonymous *modules* that autonomously moves in a 2D square grid. Let $R = \{m_1, m_2, \ldots, m_n\}$ be an MRS that consists of n modules. We call n the *size* of R. We use m_i just for notation. When we consider more than one MRS $R_1, R_2, \ldots$, each MRS is represented by $R_i = \{m_{i,1}, m_{i,2}, \ldots, m_{i,n_i}\}$, where n_i is the size of R_i. In this case, we also use R_i and $m_{i,j}$ just for notation.

We consider MRSs moving in a finite *field* in the 2D square grid. The field is a rectangle of width w and height h with $w \neq h$. Without loss of generality, we assume $w > h$. We consider the global coordinate system, whose origin is a corner of the field. Then, we call the positive y direction *north* and the negative y direction *south*. Thus, the positive x direction is *east* and the negative x direction is *west*. A cell whose bottom left coordinate is (x, y) is denoted by $c_{x,y}$. Then, the field consists of cells $c_{x,y}$ for $x \in \{0, 1, \ldots, w-1\}$ and $y \in \{0, 1, \ldots, h-1\}$. The field is surrounded by four walls, i.e., the north wall $W_N = \{c_{x,y} \mid x \in \{-1, 0, 1, \ldots, w\}, y = h\}$, the south wall $W_S = \{c_{x,y} \mid x \in \{-1, 0, 1, \ldots, w\}, y = -1\}$, the east wall $W_E = \{c_{x,y} \mid x = w, y \in \{-1, 0, 1, \ldots, h\}\}$, and the west wall $W_W = \{c_{x,y} \mid x = -1, y \in \{-1, 0, 1, \ldots, h\}\}$. See Fig. 3, where walls are represented by black cells.

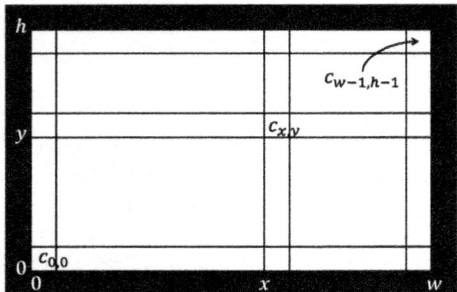

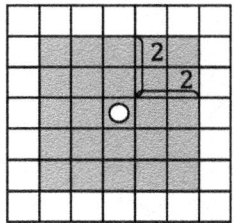

Fig. 4. Visibility when $k = 2$

Fig. 3. Field and walls

Cell $c_{x,y}$ is *side-adjacent* to four cells, $c_{x-1,y}$, $c_{x,y-1}$, $c_{x+1,y}$, and $c_{x,y+1}$. We also say that module m_j is side-adjacent to module m_i if the two cells occupied by m_i and m_j are side-adjacent. Each cell can accept at most one module and we say a cell is *empty* when there is no module in the cell. A *configuration* C_t at time t is the set of cells occupied by the modules at time t. We now define the connectivity requirement for a single MRS $R = \{m_1, m_2, \ldots, m_n\}$. We consider the *side-adjacency* graph $G_t = (R, E_t)$, where $\{m_i, m_j\} \in E_t$ if m_i is side-adjacent to m_j in C_t. We say R is *connected* at time t if G_t is connected.

The modules have limited visibility. When a module occupies a cell $c_{x,y}$, it can observe whether each cell in its k-*neighborhood* $\{c_{x+i,y+j} \mid i,j \in [-k,k]\}$ is occupied or not. We call the value of k the *visibility range* of the module. See Fig. 4 as an example. Each module uses its own *local coordinate system* when they observe the cells in its visibility. We assume that each local coordinate system is right-handed, its origin is the center of the cell that the module occupies, and its axes are parallel to the rows and columns of the 2D grid. When all local coordinate systems have the same directions and orientation, we say the modules are equipped with the *global compass*. When the modules are not equipped with the global compass, they can still agree on the clockwise direction. A *state* of an MRS is its local shape. When the modules are not equipped with the global compass, a state of an MRS does not contain directions. For example, consider the following two configurations of an MRS of two modules: $C = \{c_{x,y}, c_{x+1,y}\}$ and $C' = \{c_{x,y}, c_{x,y+1}\}$. If the modules are equipped with the global compass, the state of the MRS in C is different from that in C'. Otherwise, the state of the MRS in C is the same as that in C'.

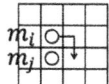

Fig. 5. Rotation of m_i around m_j.

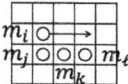

Fig. 6.   2-sliding   of m_i along m_j, m_k, m_ℓ.

Each module can perform a *rotation* or a *sliding* at each time instance. Each movement requires other modules that guide the movement and do not move during the movement, and it must go through empty cells. The destination cell also must be empty. In addition, the modules cannot go through nor enter the cells of the four walls surrounding the field.

- A rotation is a rotation by $\pi/2$ around a side-adjacent module that does not move simultaneously. See Fig. 5 as an example, where module m_i performs a rotation around module m_j, which is side-adjacent to m_i and does not move simultaneously.
- An ℓ-*sliding* is a vertical or horizontal movement by ℓ-cells along a line of modules that do not move simultaneously. See Fig. 6 as an example, where module m_i performs 2-sliding while the three modules m_j, m_k, m_ℓ along the moving track does not perform any movement simultaneously. In this paper, we use rotations, 1-slidings, and 2-slidings.

We consider discrete time $t = 0, 1, 2, \ldots$. The modules are *synchronous*, i.e., at each time instance, each module observes all cells in its visibility, computes its movement, and performs the computed movement. The modules are *uniform*, i.e., a module computes it movement by a common deterministic algorithm.

The modules are *oblivious*, i.e., when a module computes its next movement, the input to the common algorithm is its preceding observation. That is, the modules cannot use past observations nor computation. Let B_t be the set of *backbone modules* that do not move at time t. At each time instance t, the modules of a single MRS R must keep the following conditions:

1. The resulting state of R is connected.
2. B_t is connected.
3. Moving trajectories of the modules do not overlap.

Hence, more than one module can move at each time instance as long as the above conditions are satisfied.

When we consider multiple MRSs $R_1, R_2, \ldots$, each MRS must satisfy the first two conditions on connectivity, and moving trajectories of all modules must not overlap.

The movements of modules generate the evolution of the MRS(s). An *execution* from an initial configuration C_0 is a sequence of configurations $C_0, C_1, C_2, \ldots$, where C_{t+1} is obtained from C_t by the movements of modules at time t. We say an algorithm solves a problem from an initial configuration C_0 if the execution starting from C_0 reaches configuration C_t in a finite time and C_t satisfies the problem requirement and all modules do not move after t.

The *rendezvous problem* requires two MRSs initially placed in the field to gather so that each module of the two MRSs observe all the modules.

We say two MRSs R_1 and R_2 *merge* into a single MRS when the side adjacency graph with the vertex set $R_1 \cup R_2$ is connected. The *merge problem* requires the two MRSs to merge into a single MRS. In this paper, for the merge problem, we consider initial configurations where each module can observe all the modules of the two MRSs.

3   Rendezvous Algorithm

In this section, we present a rendezvous algorithm for two MRSs each of which consists of five modules with visibility range 7 and not equipped with the global compass. The proposed algorithm makes the two MRSs go back and forth on a path uniquely fixed in the rectangular field. The path starts from the southwest corner and goes at a $\pi/4$ angle to the north wall. An MRS follows this path by a zigzag move, which will be presented later. Then, it turns on the north wall and goes at a $\pi/4$ angle to the south wall. The path is shifted to east during these moves. By repeating these moves, the path ends at the northeast corner. We call this path "*path A*" and its mirror image is called "*path B*." See Fig. 7 (*a*) and (*b*).

Due to limited visibility and the lack of the global compass, the modules of an MRS cannot recognize which path they are following when they move in the middle of the field. The difference between path A and path B is the angle formed by the path and the north wall (or the south wall). When an MRS approaches the north wall along path A, if we consider the MRS's moving direction is its positive y direction, the angle formed by its x-axis and the wall is $\pi/4$. When an MRS moves along path B, the angle is $-\pi/4$. Based on this fact, the proposed algorithm makes the MRS return to path A when it is not moving on path A. That is, each module of an MRS does not need the global compass to recognize which path it is following. We say an MRS is moving with a *correct angle* when its moving direction forms angle of $\pi/4$ with the longer wall (i.e., the north or the south wall) lying in front of it. Otherwise, we say the MRS is moving with a *wrong angle*. We call a parallelogram filled by path A the *area of path A*. We abuse the word "acute angle" of the area of path A when we refer to the corners of the field that path A touches.

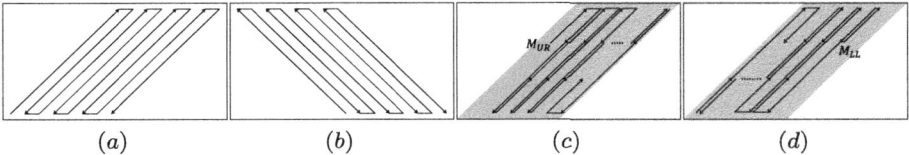

(a) (b) (c) (d)

Fig. 7. Paths on the field. (a) Path A. (b) Path B. (c) Forward move on path A. (d) Backward move on path B.

We first show the nine unit moves of the proposed algorithm. Each of the nine moves consists of a short sequence of states and movements in each state as shown in Fig. 8 to 16. The states used in a move are not used in the other moves, thus each module can recognize which move the MRS is performing. Each module execute the proposed algorithm as follows: At each time instance, each module checks the state of the MRS to which it belongs, identifies the ongoing move, and performs the movement defined for the current state. The following is the list of the nine moves.

- Move to upper right denoted by M_{UR} and shown in Fig. 8. The MRS moves diagonally to the right. In Fig. 8, the module with X is a landmark of the MRS. After the sequence of M_{UR}, the landmark moves up by one cell and right by one cell.
- Move to lower left denoted by M_{LL} and shown in Fig. 9. The MRS moves diagonally to the left. In Fig. 9, the cell with X is a landmark of the MRS. After the sequence of M_{LL}, the landmark moves down by one cell and left by one cell.
- Turn on the top wall denoted by T_{TW} and shown in Fig. 10. The MRS approaching to a wall by M_{UR} turns $180°$ and leaves the wall by M_{LL}. The gray cells show that the trajectory of the MRS shift by one cell after T_{TW}.
- Turn on the bottom wall denoted by T_{BW} and shown in Fig. 11. The MRS approaching to a wall by M_{LL} turns $180°$ and leaves the wall by M_{UR}. The gray cells show that the trajectory of the MRS does not shift after T_{BW}.
- Turn at a top corner denoted by T_{TC} and shown in Fig. 12. The MRS approaching to a corner by M_{UR} turns $180°$ and leaves the corner by M_{UR}. The gray cells show that the trajectory of the MRS does not shift after T_{TC}.

- Turn on the left wall denoted by T_{LW} and shown in Fig. 13. The MRS approaching a wall with M_{LL} turns 90° and leaves the wall by M_{UR}. The gray cells show that the trajectory of the MRS shift after T_{LW}.
- Turn on the right wall denoted by T_{RW} and shown in Fig. 14. The MRS approaching a right wall by M_{UR} turns and starts M_{RW}. The gray cells show the direction of M_{UR} of the MRS.
- Move down along the right wall denoted by M_{RW} and shown in Fig. 15. The MRS on a right wall moves down along the wall.
- Turn at a bottom corner denoted by T_{BC} and shown in Fig. 16. The MRS approaching a corner by M_{RW} turns and leaves the corner by M_{UR}. The gray cells show the direction of M_{UR} of the MRS.

As explained above, there are two types of moves, i.e., straight moves and turns. A turn changes a straight move to another straight move, and Fig. 10 to 16 show the changes by gray cells, that show the first state of the straight moves.

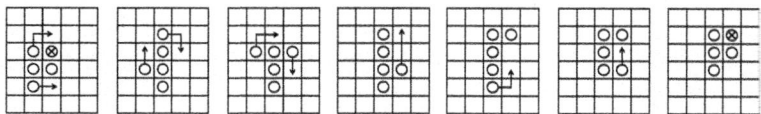

Fig. 8. M_{UR}. The module with X is a reference point, which shows that the MRS moves to upper right by M_{UR}.

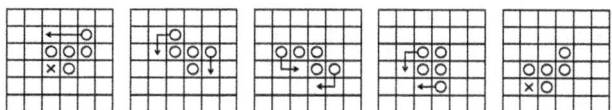

Fig. 9. M_{LL}. The cell with X is a reference point.

These nine moves yield two types of trajectories on path A as shown in Fig. 7(c) and (d). Because the modules are not equipped with the global compass, an MRS may approach the top wall by executing M_{LL} with a correct angle. In this case, the MRS changes its direction by T_{BW} and moves toward the bottom wall by M_{UR}. When the MRS reaches the bottom wall, it changes its direction

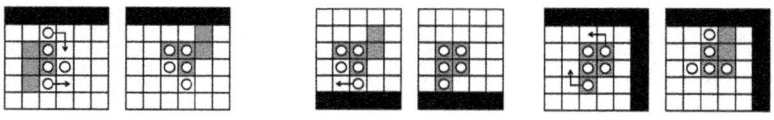

Fig. 10. T_{TW} **Fig. 11.** T_{BW} **Fig. 12.** T_{TC}

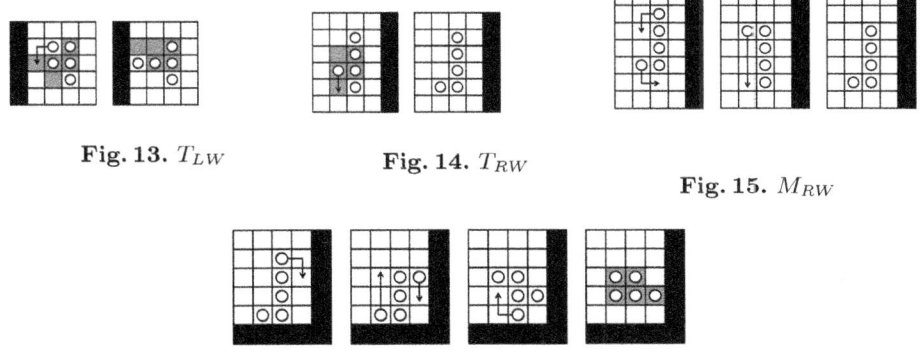

Fig. 13. T_{LW} **Fig. 14.** T_{RW}

Fig. 15. M_{RW}

Fig. 16. T_{BC}

by T_{TW} with sliding its move by one and again moves toward the top wall by executing M_{LL}. This "backward" trajectory is shown in Fig. 7(d). When it reaches the bottom corner by executing M_{UR}, it turns by T_{TC} and moves toward the top wall by executing M_{UR}. The trajectory is now switched to the "forward" trajectory shown in Fig. 7(c), which reaches the top wall by M_{UR}. It turns on the top wall by T_{TW} with sliding its move by one. Then it moves toward the bottom wall by M_{LL}. When it reaches the bottom wall, it changes its direction by T_{BW} and moves toward the top wall by M_{UR}. When it reaches the top corner, it turns by T_{TC} and it moves toward the bottom wall by M_{UR}. Now, the trajectory is again switched to the backward trajectory.

When an MRS is moving with a wrong angle or moving outside of the area of path A, it can return to path A. For example, when an MRS approaches a top or bottom wall by M_{UR} with a wrong angle, it turns by T_{RW} on the wall and moves along the wall by M_{RW} until it reaches a corner. Then, it turns by T_{BC} and starts M_{UR} with a right angle. The corner is an acute corner of path A, thus the MRS now moves along path A. We can show that the proposed algorithm makes an MRS return to path A for the remaining seven cases, but we omit the detail due to page limitation.

As shown in Sect. 4, five modules not equipped with the global compass have 18 states. The proposed rendezvous algorithm does not use the states shown in Figs. 17 and 18. In each state in Fig. 17, there exists a pair of modules that are placed at symmetric positions. Thus, the two modules have the same observation in the worst case, and move with keeping their symmetry forever. The proposed algorithm does not accept these four symmetric states as an initial state and it does not have these states during execution. Figure 18 shows the remaining three states together with the movements that translate the states into the first state of M_{UR}.

In the proposed algorithm, we assumed visibility range of 7. Figure 19 shows a collision when the visibility range is 6. In the figure, the MRS in the top, say R_1, is performing M_{RW} while the MRS in the bottom, say R_2, is performing

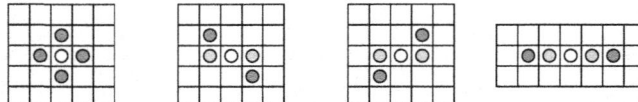

Fig. 17. Symmetric states. In each state, a pair of modules that may have a same observation is painted with a same color.

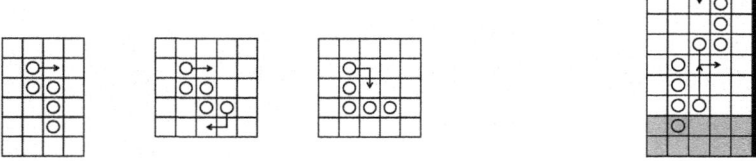

Fig. 19. Collision with visibility range $k = 6$

Fig. 18. Exceptional states and transformation to M_{UR}

M_{UR}. The gray module cannot observe the gray cells when its visibility range is 6, hence it cannot determine whether R_2 is performing M_{UR} or the first exceptional move in Fig. 18. If R_2 is performing M_{UR}, R_1 and R_2 will have a collision as shown in Fig. 19. To avoid this case, we allow the modules to have visibility range of 7.

Consequently, we have the following theorem.

Theorem 1. *Two MRSs each of which consists of five modules not equipped with the global compass can solve the rendezvous problem from an arbitrary initial configuration, where the state of each MRS is not symmetric in a field of size $w \times h$ ($w \neq h$).*

4   Merge Algorithm

In this section, we present a merge algorithm for an initial configuration, where all modules of the two MRSs are in a square area of size 8×8. The initial configurations includes all the terminal configurations of the rendezvous algorithm in Sect. 3. In the proposed merge algorithm, we adopt visibility range of 9. Other assumptions on the MRSs are the same as Sect. 3, i.e., the size of each MRS is five and the modules are not equipped with the global compass.

The main trick of the proposed algorithm is a labeling of all the 18 states of an MRS that consists of five modules not equipped with the global compass. Figure 20 shows all these states with their labeling in the form of S_i^5. We say label S_i^5 is smaller than label S_j^5 if $i < j$. Figure 20 also shows the local coordinate system in each state, which is used to break the ties when the two MRSs are in the same state.

When the initial states of the two MRSs are different, the proposed algorithm makes the MRS with the smaller label move to the other MRS that stays in its initial position. Otherwise, the proposed algorithm makes the two MRSs to compare their observations translated by their local coordinate systems shown in Fig. 20. The *view* of an MRS in state S_i^5 is a sequence $((x_1, y_1), (x_2, y_2), \ldots)$ of coordinates of the centers of all modules in its visibility that satisfies (i) $x_i \leq x_{i+1}$ or (ii) $x_i = x_{i+1}$ and $y_i < y_j$ for all $i = 0, 1, 2, \ldots$. In states $S_{15}^5, S_{16}^5, S_{17}^5, S_{18}^5$, there are multiple choices for the local coordinate systems due to the symmetry of the states. In this case, we select the local coordinate system that minimizes the view in the lexicographic ordering. When the views of the two MRSs are different, the proposed algorithm makes the MRS with the smaller view in the lexicographic ordering move to the other MRS, that stays in its initial position. Finally, if the two MRSs are in the same state with the same view, (i.e., their states and coordinate systems are symmetric), they approach each other until they merge into one MRS.

To make the MRSs remember their initial roles, the proposed algorithm first makes the MRS with the larger label or larger view change its state to S_{18}^5 by increasing its label. Figure 21 shows the transitions to S_{18}^5.

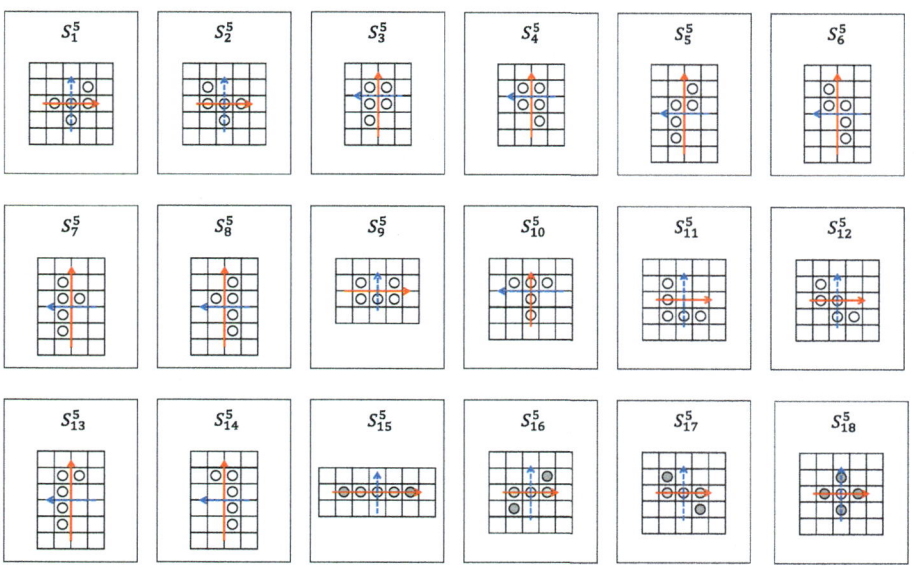

Fig. 20. 18 states of an MRS of size five. For each state, its local coordinate system is also shown by the red arrow (x-axis) and the blue arrow (y-axis).

After that, the MRS with the smaller label or smaller view changes its state to S_1^5 or S_2^5 by decreasing its label. Figure 22 shows the transitions to S_1^5 and S_2^5.

The transitions contain asymmetric movements for S_5^{16}, S_5^{17}, and S_5^{18}, because the local coordinate system of an MRS is uniquely fixed by the other MRS. Then, the MRS approaches to the other MRS by repeating transformations between S_1^5 and S_2^5.

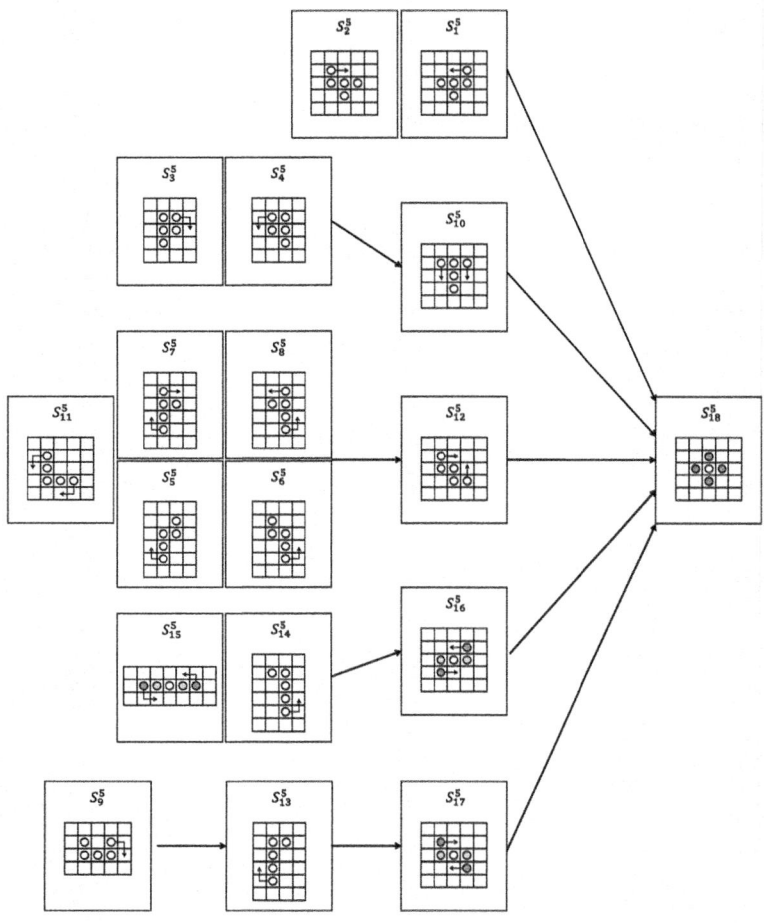

Fig. 21. Transformation to S_{18}^5 with increasing the label of an MRS

Figure 23 shows an execution of the proposed algorithm. Consider an initial configuration of two MRSs R_1 in S_3^5 and R_2 in S_{10}^5. The proposed algorithm first makes R_2 change its state to S_{18}^5 by monotonically increasing its label. After that, R_1 changes its state to S_1^5. Then, R_1 approaches R_2 by horizontal moves with respect to its local coordinate system. The horizontal movement in the positive direction is denoted by M_{x+} and shown in Fig. 24 and that in the negative direction is denoted by M_{x-} and shown in Fig. 25. When a module

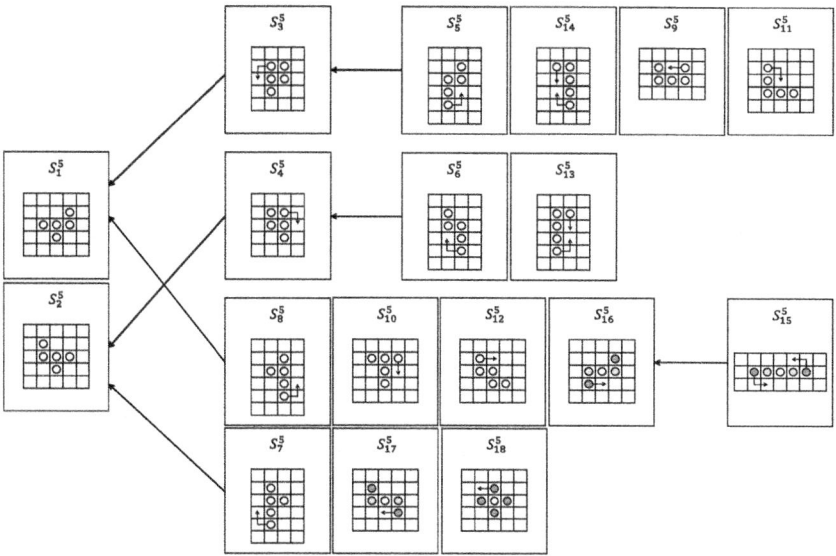

Fig. 22. Transformation to S_1^5 or S_2^5 with decreasing the label of an MRS

of R_1 computes its next move, it considers smallest enclosing rectangles of R_1 and R_2. For example, the smallest enclosing rectangle of S_{18}^5 is a square of size 3×3 enclosing the five modules. Then, it considers the projections of the smallest enclosing rectangles of R_1 and R_2 to the x-axis of R_1. See Fig. 23(b) as an example. We say R_1 is x-*overlapping* with R_2, if there is some overlap of the projections of the smallest enclosing rectangles. If R_1 is not x-overlapping with R_2, R_1 moves horizontally by M_{x+} or M_{x-} until it becomes x-overlapping with R_2 (Fig. 23(c)). Then, R_1 changes its moving direction by a turn (Fig. 23(d)). There are two turns; one is denoted by $T_{S_1^5}$ shown in Fig. 26 and the other is denoted by $T_{S_2^5}$ shown in Fig. 27. After the turn, if R_2 is not x-overlapping with R_1, R_1 moves horizontally until it merges with R_2 (Fig. 23(e)). During the move of R_1, no module cause a collision with the modules of R_2. For example, consider the case where the upper right module in S_1^5 of M_{x+} cause a collision by its rotation. In this case, the colliding module of R_2 is already side-adjacent to some module of R_1. In the same way, we can check that all movements of M_{x+} and M_{x-} do not yield any collision.

We then consider symmetric initial configurations, where the two MRSs have the same view. In this case, the proposed algorithm makes the two MRSs to perform the transformation to S_1^5 or S_2^5. By symmetry, the two MRSs are in the same state in the resulting configuration. Then, the proposed algorithm makes the two MRSs follow the same procedure as R_1 in the asymmetric initial configurations. They check the overlap of their smallest enclosing rectangles and perform M_{x+} or M_{x-} until they become x-overlapping. Then, they turn by $T_{S_1^5}$ or $T_{S_2^5}$ and perform M_{x+} or M_{x-} until they merge. All these moves are symmetric

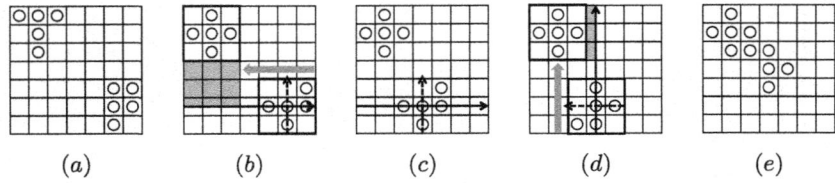

(a)   (b)   (c)   (d)   (e)

Fig. 23. Progress of the merge algorithm. (*a*) An initial configuration. The MRS in S_{10}^5 changes its state to S_{18}^5. (*b*) The MRS in S_1^5 moves by M_{x-}. (*c*) The MRSs are now *x*-overlapping and the MRS in S_1^5 performs $T_{S_1^5}$. (*d*) The MRS in S_2^5 moves by M_{x+}. (*e*) Two MRSs establish connectivity.

Fig. 24. M_{x+} **Fig. 25.** M_{x-}

with respect to the center of symmetry in the initial configuration. However, in the final step, the two MRS may caught in a deadlock due to collisions of modules. We incorporate exceptional movements for these cases to the proposed algorithm but we omit the detail due to page limitation.

In the proposed merge algorithm, we assume visibility range of 9. Consider a configuration, in which two MRSs R_1 in state S_{11}^5 and R_2 in state S_8^5 are placed at the diagonal corners of their smallest enclosing rectangles of a regular square of size 8. Hence, each module can observe all the ten modules when the visibility range is 7. Then, R_1 performs the transformation to S_{18}^5 and R_2 performs the transformation to S_1^5. However, during the translations, some modules of R_1 and R_2 go outside of the regular square of size 8. Thus, if the visibility range is 7, there exists some module that cannot observe some of the ten modules. These modules cannot continue the merge algorithm. In the proposed algorithm, when an MRS executes transformations to S_1^5, S_2^5 and S_{18}^5, it can go outside of its initial smallest enclosing rectangle at most once and at most by one. In the transformations to S_{18}^5 shown in Fig. 21, these cases are the moves starting from $S_3^5, S_4^5, S_5^5, S_6^5, S_7^5, S_8^5, S_9^5, S_{11}^5, S_{13}^5, S_{14}^5, S_{15}^5$, and in the transformations to S_1^5 or

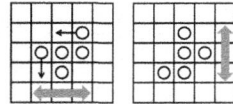

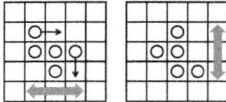

Fig. 26. $T_{S_1^5}$. Gray arrows show the moving direction by M_{x+} and M_{x-}.   **Fig. 27.** $T_{S_2^5}$.

S_2^5 shown in Fig. 22, these cases are the moves starting from $S_3^5, S_4^5, S_7^5, S_8^5, S_{15}^5$. Then, in the worst case, visibility size of 9 is sufficient to keep the modules visible to each other and continue the proposed merge algorithm.

Theorem 2. *Two MRSs each of which consists of five modules not equipped with the global compass can solve the merge problem from an arbitrary initial configuration, where each module of the two MRSs can observe all the modules in its visibility.*

We finally show how to synthesize the rendezvous algorithm and the merge algorithm. Each module starts the rendezvous algorithm and switches to the merge algorithm as soon as the smallest enclosing rectangle of the two MRSs becomes a regular square of size 8×8 or a smaller rectangle. The required visibility range is 9.

5   Impossibility

We finally show necessity of the five modules to solve the rendezvous problem. In [3], Doi et al. showed that an MRS that consists of four modules not equipped with the global compass can move horizontally or vertically. By considering movements on walls, an MRS can move on a cycle formed by two horizontal (or vertical) lines connected by two constant-length lines on the walls. When these cycles for the two MRSs do not overlap and the modules of one MRS cannot observe the other MRS, the two MRSs never meet. The authors also showed that when each MRS consists of less than four modules, the MRS cannot move. Thus, we have the following theorem.

Theorem 3. *Two MRSs each of which consists of less than five modules not equipped with the global compass cannot solve the rendezvous problem.*

6   Conclusion

In this paper we presented a rendezvous algorithm and a merge algorithm for two MRSs, each of which consists of five modules not equipped with the global compass. We then showed that five modules for each MRS are necessary to solve the rendezvous problem. Due to the page limitation, we omit the fact that the merge problem can be solved by two MRSs of size four.

While existing results consider distributed coordination among the modules so that they collectively form a single MRS, we considered distributed coordination among multiple MRSs. We believe the results open up new vistas for distributed coordination theory for MRSs. There are many future directions including conventional distributed coordination problems such as leader election, synchronization, and fault tolerance, and mobile computing problems such as gathering, separation, and patrolling. It is also important to develop basic building blocks for distributed algorithms for MRSs instead of tailor-made algorithms.

References

1. Das, S., Flocchini, P., Prencipe, G., Santoro, N., Yamashita, M.: Autonomous mobile robots with lights. Theor. Comput. Sci. **609**, 171–184 (2016)
2. Dieudonné, Y., Pelc, A., Petit, F.: Almost universal anonymous rendezvous in the plane. Algorithmica **85**(10), 3110–3143 (2023)
3. Doi, K., Yamauchi, Y., Kijima, S., Masafumi, Y.: Search by a metamorphic robotic system in a finite 2D square grid. Inf. Comput. **285**, 104695 (2022)
4. Dumitrescu, A., Suzuki, I., Yamashita, M.: Formations for fast locomotion of metamorphic robotic systems. Int. J. Robot. Res. **23**(6), 583–593 (2004)
5. Dumitrescu, A., Suzuki, I., Yamashita, M.: Motion planning for metamorphic systems: feasibility, decidability, and distributed reconfiguration. IEEE Trans. Robot. Autom. **20**, 409–418 (2004)
6. Kranakis, E., Santoro, N., Sawchuk, C., Krizanc, D.: Mobile agent rendezvous in a ring. In: Proceedings of the IEEE 33rd International Conference on Distributed Computing Systems, p. 592 (2003)
7. Nakamura, J., Kamei, S., Yamauchi, Y.: Evacuation from various types of finite two-dimensional square grid fields by a metamorphic robotic system. Concurr. Comput. Pract. Exp. **35**, e6628 (2023)
8. Suzuki, I., Yamashita, M.: Distributed anonymous mobile robots: formation of geometric patterns. SIAM J. Comput. **28**(4), 1347–1363 (1999)

Gathering Semi-Synchronously Scheduled Two-State Robots

Kohei Otaka[1]([✉])(iD), Fabian Frei[2,3]([✉])(iD), and Koichi Wada[1]([✉])(iD)

[1] Hosei University, Tokyo, Japan
kohei.otaka.8n@stu.hosei.ac.jp, wada@hosei.ac.jp
[2] Department of Computer Science, ETH Zurich, Zürich, Switzerland
fabian.frei@inf.ethz.ch
[3] CISPA Helmholtz Center for Information Security, Saarbrücken, Germany
fabian.frei@cispa.de

Abstract. We study the problem *Gathering* for n autonomous mobile robots in synchronous settings with a persistent memory called *light*. It is well known that Gathering is impossible in the basic model ($\mathcal{OBLOT}$) where robots have no lights, even if the system is semi-synchronous (called Ssynch). Gathering becomes possible, however, if each robot has a light of some type that can be set to a constant number of colors. In the $\mathcal{FCOM}$ model, the robots can only see the lights of other robots. In the $\mathcal{FSTA}$ model, each robot can only observe its own light. In the $\mathcal{LUMI}$ model, all robots can see all lights. This paper focuses on $\mathcal{FSTA}$ robots with 2-colored lights in synchronous settings. We show that 2-color $\mathcal{FSTA}$ and $\mathcal{FCOM}$ robots cannot solve Gathering in Ssynch without additional conditions, even with rigid movement and agreement of chirality and the minimum moving distance. We also improve the condition of the previous Gathering algorithm for $\mathcal{FSTA}$ robots with 2-color working in Ssynch.

1 Introduction

Background and Motivation. The computational power of autonomously acting, simple, mobile robots has been the object of intense research in the field of distributed computing. Ever since Suzuki and Yamashita's seminal work [26], a large amount of work has been dedicated to the research of theoretical models of such autonomous mobile robots [1,3,6,10,18,21,25]. In the default setting, a robot is modeled as a point in the two-dimensional plane, and its capabilities are rather weak. In particular, the robots are assumed to be *oblivious* (have no memory to record past history), *anonymous* (have no IDs), and *uniform* (run identical algorithms) [11].

Robots operate in synchronous *Look-Compute-Move* (*LCM*) cycles. In each *round*, a nonempty set of (possibly all) robots is activated, and all activated

---

This work was supported in part by JSPS KAKENHI Grant Number 20K11685, and 21K11748.

T. Masuzawa et al. (Eds.): SSS 2024, LNCS 14931, pp. 209–224, 2025.
https://doi.org/10.1007/978-3-031-74498-3_15

Table 1. Previous gathering algorithms for robots with lights.

Scheduler	Movement	$\mathcal{LUMI}$	$\mathcal{FCOM}$	$\mathcal{FSTA}$	$\mathcal{OBLOT}$
Fsynch	Non-Rigid	→	→	→	○
Round-Robin	Rigid	↓	↓	2 [27]	× [9]
Cent	Non-Rigid	↓	2 [27]	↓	×* [9]
Ssynch	Rigid	↓	3,2** [27]	?	× [11]
	Non-Rigid($+\delta=$)	↓	?	2*** [27]	↑
	Non-Rigid	2 [27]	?	?	↑
Asynch	Rigid	↓	?	↓	↑
	Non-Rigid	3 [23]	?	∞**** [5]	

The symbols mean the following. *: Distinct gathering. **: Local-awareness.
: 2δ-distant. *: unlimited number of colors. ○: solvable. ×: unsolvable.
?: unknown. →, ↓, ↑: The same as what the arrow is pointing to. (Here, the possibility and impossibility are derived from the stronger and weaker model, respectively.)

robots then simultaneously perform an *LCM* cycle. The round ends as soon as all activated robots have performed their cycle. Each cycle is composed of three phases: in the *Look* phase, a robot obtains a snapshot of the plane showing the positions of the other robots; in the *Compute* phase, it executes its algorithm (which is identical for all robots) using the snapshot as input; then it moves towards the computed destination in the *Move* phase. Repetition of these cycles allows robots to collectively perform some tasks and solve some problems.

The selection of which robots are activated in a round is made by an adversarial scheduler. This general setting is usually called *semi-synchronous* (Ssynch). The special restricted setting where every robot is activated in every round is called *fully-synchronous* (Fsynch) [11].

These systems have been extensively investigated within distributed computing. The focus of the research has been on understanding the nature and extent of the impact of crucial factors, such as *memory persistence* and *communication capability*, have on the solvability of a problem and thus on the computational power of the system. To this end, four robot models with light have been identified and investigated: $\mathcal{OBLOT}$, $\mathcal{FSTA}$, $\mathcal{FCOM}$, and $\mathcal{LUMI}$. The most common (and weakest) model $\mathcal{OBLOT}$ [26] (which stand for *obli*vious *rob*ots) assumes basic robots without light. In the strongest model $\mathcal{LUMI}$ [8](which stands for *lumi*nous) robots can see their own lights as well as those of the other robots, whereas in $\mathcal{FCOM}$ and $\mathcal{FSTA}$ [14] (which stand for *f*inite *com*munication and *f*inite *sta*tes, respectively), they can see, respectively, only the lights of the other robots (granting some communication capabilities) or only their own lights (which translates to persistent internal memory).

Gathering and Previous Results. *Gathering* is one of the most fundamental tasks for autonomous mobile robots. Gathering is the process where n mobile robots, initially located in arbitrary positions, meet within finite time at an arbitrary single location. When there are two robots (that is, for $n = 2$), the task of Gathering is usually called *Rendezvous*. Since Gathering is a simple but essential problem, it has been intensively studied and a number of possibility and impossibility results have been shown under different assumptions [1–4, 6–9, 12, 15, 19, 20, 22, 24–26].

Table 1 summarizes the previous results for Gathering by robots with lights. For all of them, it is assumed that different robots may have different local coordinate systems with different length units (i.e., no consistency between robots is guaranteed), but each local coordinate system remains the same throughout all rounds (i.e., the robots are all self-consistent); this is referred to as *fixed disorientation*. Moreover, no capabilities for detecting multiplicity are assumed.

In the basic $\mathcal{OBLOT}$ model, Gathering is trivially solvable in FSYNCH but remains unsolvable in SSYNCH, even with assumptions such as rigidity of movement (i.e., robots always reach their target) or consistent chirality [11]. However, for robots equipped with lights, the problem becomes solvable in SSYNCH for various models like $\mathcal{LUMI}$, $\mathcal{FCOM}$, and $\mathcal{FSTA}$. Solvability in these settings results from different combinations of factors such as the number of available colors, algorithmic constraints, and movement restrictions. These differences highlight the complexity and versatility of solutions in the presence of different additional capabilities. Table 1 also includes results for the asynchronous scheduler (ASYNCH) for comparison purposes. In ASYNCH, there is no common notion of time; the robots may be activated independently of the others, letting them perform their *Look*, *Compute*, and *Move* operations at arbitrary times [13].

Regarding movement restriction, *Rigid* means that robots always reach the computed destination during the movement operation. *Non-Rigid* means that a robot x may be stopped before reaching the calculated destination but is be stopped before having moved some distance $\delta_x > 0$ unknown to the robots, guaranteeing that any destination located within a radius of δ_x can be reached. *Non-Rigid*$(+\delta=)$ is the same as Non-Rigid, except that now δ_x is the same for all robots and known to them. In the following, we assume Non-Rigid$(+\delta =)$.

We introduce some possibility and impossibility results for Gathering. It is known [9, 11] that Gathering for $\mathcal{OBLOT}$ is not solvable in SSYNCH. In particular, Gathering for $\mathcal{OBLOT}$ is deterministically unsolvable in a restricted subclass of SSYNCH, where exactly one robot is activated in each round and they are always activated in the same order (called ROUND-ROBIN). Moreover, if all robots are initially located in different positions (called Distinct Gathering), it is deterministically unsolvable under a 2-bounded CENT scheduler, where a scheduler is 2-bounded if between any two consecutive activations of any robot, any other robot is activated at most 2 times and CENT means that exactly one robot is activated in each round [9]. This impossibility holds even if we assume chirality and rigidity.

Multiplicity detection is a strong assumption when it comes to solving Gathering. We know [11] that, if strong multiplicity is assumed, Gathering for n robots is solvable in SSYNCH if and only if n is odd, and that distinct Gathering for n robots is solvable for $n \geq 3$ even in ASYNCH. Non-oblivious robots have persistent memory. This is also true for robots with internal lights, but the amount of memory with internal lights is restricted to a constant. Gathering was already shown to be solvable with non-oblivious robots [5]. However, the known algorithm stores the locations of other robots exactly, and the amount of memory exceeds any constant. It has remained unknown whether Gathering is solvable by robots with internal lights with a constant number of colors.

In the $\mathcal{LUMI}$ model, there is a 2-color algorithm with chirality and non-rigid movement in SSYNCH [27] and 3-color one with the same assumption in ASYNCH [23]. In the $\mathcal{FCOM}$ model, there is an algorithm with 3 colors assuming rigidity. Assuming local awareness (i.e., robots recognize other robots sharing the same location) reduces the number of used colors down to 2 [27]. In the $\mathcal{FSTA}$ model, there is an algorithm assuming Non-Rigid($+\delta=$) with lights of only 2 colors if the initial configuration of robots is 2δ-distant, where 2δ-distant means that the largest distance between two robots in the configuration is at least 2δ. There also exist 2-color Gathering algorithms in CENT and ROUND-ROBIN schedulers for $\mathcal{FCOM}$ and $\mathcal{FSTA}$ robots, respectively [27].

Our Contributions. We prove the impossibility of Gathering on $\mathcal{FCOM}$ or $\mathcal{FSTA}$ for robots with 2 colors. Specifically, we show that $\mathcal{FCOM}$ and $\mathcal{FSTA}$ robots with 2-colored lights cannot solve Gathering in SSYNCH, even under the assumptions of rigid movement, consistent chirality, and a shared unit of length. This result demonstrates that the assumptions of previous Gathering algorithms for $\mathcal{FCOM}$ and $\mathcal{FSTA}$ robots shown in Table 1 are optimal in the sense that Gathering becomes impossible without assuming them. For example, for the 2-color algorithm for $\mathcal{FCOM}$ robots working with rigid movement, the condition of *local-awareness* cannot be removed, and for the 2-color algorithm for $\mathcal{FSTA}$ robots working in Non-Rigid($+\delta=$), the condition of 2δ-*distant* cannot be removed either. The conditions that cannot be removed are minimal and should be as weak as possible. In this paper, we demonstrate that a Gathering algorithm for $\mathcal{FSTA}$ robots with 2 colors exists under conditions weaker than a 2δ-distant initial configuration. Specifically, we show that if only two forbidden patterns are excluded in the initial configuration, a Gathering algorithm for $\mathcal{FSTA}$ robots with 2 colors can be achieved.

2   Preliminaries

We consider a set of anonymous mobile robots $\mathcal{R} = \{r_1, ..., r_n\}$ located in $\mathbb{R}^2$. Each robot r_i has a persistent state l_i called its light, which may be taken from a finite set L of colors. We denote by $l_i(t) \in L$ the color that the light of robot r_i has at time t and by $p_i(t) \in \mathbb{R}^2$ the position occupied by r_i at time t represented in some global coordinate system. A configuration $C(t)$ at time t is a multiset of n pairs $(l_i(t), p_i(t))$, each defining the color of light and the position of the robot r_i at time t. When no confusion arises, $C(t)$ is simply denoted by C.

For a subset S of $L \times \mathbb{R}^2$, $\mathcal{L}(S)$ and $\mathcal{P}(S)$ denote the projections to L and $\mathbb{R}^2$ from S, respectively.

Each robot r_i has its own coordinate system where r_i is located at its origin at any time. These coordinate systems do not necessarily agree with those of other robots. This means that there is no guarantee of a common unit distance, nor for the directions of coordinate axes, nor for a clockwise orientation (chirality). However, each local coordinate system remains the same throughout all rounds. This is called the *fixed disorientation*.

At any time, any robot can be active or inactive. When a robot r_i is activated, it executes the *Look*, *Compute*, and *Move* cycles:

- *Look*: The robot r_i activates its sensors to obtain a snapshot which consists of a pair of light and position for every robot with respect to the coordinate system of r_i. Let $SS_i(t)$ denote the snapshot of r_i at time t. We assume that robots can observe all other robots (unlimited visibility). Note that $SS_i(t)$ represents a sub-multi-set of $C(t)$ according to imposed assumptions in the local coordinate system of r_i, where r_i is at the origin.
- *Compute*: The robot r_i executes its algorithm using the snapshot and (if visible) the color of its own light and returns a destination point des_i expressed in its own coordinate system and a light $l_i \in L$. The robot r_i sets its own light to the color l_i.
- *Move*: The robot r_i moves to the computed destination des_i. If the robot may be stopped by an adversary before reaching the computed destination, the movement is said to be *non-rigid*. Otherwise, it is said to be *rigid*. If stopped before reaching its destination, we assume that a robot has moved at least a minimum distance $\delta > 0$. Note that without this assumption an adversary could make it impossible for any robot to ever reach its destination. If the distance to the destination is at most δ, the robot can thus reach it. If the movement is non-rigid and robots know the value of δ, this is called Non-Rigid($+\delta =$).

In the *Look* operation, the snapshot SS_i of r_i should contain the positions of all robots, including r_i. However, if robots located on p_i and r_i can recognize the other robots, the robots have multiplicity detection at this point. Thus, we separately classify the observation of other robots located on p_i for robot r_i. If any robot r_i can observe the other robots located on p_i, it is said to be *local-aware*. Otherwise, it is said to be local-unaware. Note that if we assume local awareness, r_i recognizes whether other robots occupy location p_i or not. In the following, we usually use the assumption that the system is local-aware.

A scheduler decides which subset of robots is activated for every configuration. The scheduler we consider is semi-synchronous. Moreover, it is always assumed that schedulers are fair, that is, each robot is activated infinitely often.

- SSYNCH: The semi-synchronous scheduler (SSYNCH) activates a subset of all robots synchronously and their *Look-Compute-Move* cycles are performed at the same time. We can assume that activated robots at the same time obtain the same snapshot (adjusted to their local coordinate system) and their *Compute* and *Move* are executed instantaneously. In SSYNCH, we can assume that any activation happens in a discrete-time round and the *Look-Compute-Move* cycle is performed instantaneously in each round. In the following, since we consider SSYNCH and its subsets, we use round and time interchangeably.

As a special case of SSYNCH, if all robots are activated in each round, the scheduler is called fully-synchronous (FSYNCH).

Let $C(t)$ be a configuration in round t. When $C(t)$ reaches $C(t+1)$ by executing the cycle at t, this is denoted as $C(t) \rightarrow C(t+1)$, where $C(t+1)$ is obtained by activating the robots once at time t to execute the algorithm on $C(t)$. The reflective and transitive closure of $\rightarrow$ is denoted as $\rightarrow^*$. That is, a configuration transition $C(t) \rightarrow C(t') \rightarrow \cdots \rightarrow C(t'')$ is denoted by $C(t) \rightarrow^* C(t'')$.

Snapshots may be different by using assumptions even if these configurations are the same, and they depend on the multiplicity detection and on how robots can see lights of other robots when robots are equipped with lights. Robots are said to be capable of (weak) multiplicity detection if they can distinguish whether a point is occupied by at least two robots. The multiplicity detection is strong if the robots can detect the exact number of robots at any given point.

In our settings, robots have persistent lights and can change their color after *Compute* operation. With regard to the visibility of the lights, we consider the following robot model.

- $\mathcal{LUMI}$: The robot can recognize not only colors of lights of other robots but also its own color of light.
- $\mathcal{FCOM}$: The robot can recognize only colors of lights of other robots but cannot see its own color of light. Note that a robot can still set its own color in each round.
- $\mathcal{FSTA}$: The robot can recognize only the color of its own light but not the lights of other robots.

When a robot performs the *Look* operation in $\mathcal{FSTA}$, its snapshot is the same as in the case of robots without lights.

Given a snapshot SS_i of a robot r_i and a point $p_j (j \neq i)$ included in $\mathcal{P}(SS_i)$, a view $V_i[p_j]$ of p_j in SS_i is a subset of $AL_i[p_j] = \{l | (l, p_j) \in SS_i, r_j \neq r_i\}$, where $AL_i[p_j]$ is a multi-set of colors of other robots that r_i can see at point p_j. For any robot r_i and any point p in the snapshot of r_i, if $V_i[p] = AL_i[p]$, the view of the robots is called the *multiset view*. If $V_i[p]$ regards $AL_i[p]$ as just a set, it is called *set view*. If $V_i[p]$ is a set of any single element taken from $AL_i[p]$, it is called *arbitrary view*. Let V_i denote $\bigcup_{(l,p) \in SS_i} V_i[p]$.

Multiset view is a strong assumption because robots without lights (i.e., with one color) can have strong multiplicity detection if multiset view is assumed. In fact, we can solve the Gathering problem by using robots without lights and multiset view [16]. On the other hand, set view and arbitrary view do not imply multiplicity detection. In the following, we assume set view.

The n-Gathering task is defined as follows: given $n(\geq 2)$ robots initially placed at arbitrary positions in $\mathbb{R}^2$, let them congregate in finite time at a single location which is not predefined. In the following, the case 2-Gathering problem is called *Rendezvous* and the n-Gathering problem for $n \geq 3$ is simply called Gathering. Gathering is said to be distinct if all robots are initially placed in different positions. An algorithm solving Gathering is said to be *self-stabilizing* if the robots initially have their lights set to arbitrary colors and start their execution from the *Look* operation.

Given two points $p, q \in \mathbb{R}^2$, we indicate the line segment by $\overline{pq}$ and its length by $|\overline{pq}|$. Let SS be a configuration or a snapshot. Given SS, $SEC(SS)$ denotes the smallest enclosing circle containing $\mathcal{P}(SS)$, and the length of its diameter and center are denoted by $Diam(SS)$ and $CTR(SS)$, respectively. A longest distance segment (LDS, for short) in SS is a line segment $\overline{pq}$ such that $p, q \in \mathcal{P}(SS)$ and $|\overline{pq}| = max_{x,y \in \mathcal{P}(SS)}|\overline{xy}|$ and the set of the longest distance segments in SS is denoted by $LDS(SS)$. If $|LDS(SS)| = 1$, LDS in SS is denoted by $\overline{pq}_{SS}$ and $O(LDS(SS))$ denotes the set of points that are not within $\overline{pq}_{SS}$. If $|O(LDS(SS))| = 0$, SS is called $OnLDS$.

Formally we define *color configurations* as follows. Let $C(t)$ be a configuration at time t, and let p and q be the endpoints of $\overline{pq}_{C(t)}$. The configuration $C(t)$ has a color configuration.

1. $\alpha\beta$, if all robots at p have color α, all robots at q have color β ($\alpha, \beta \in \{A, B\}$) and there are no robots inside $\overline{pq}_{C(t)}$.
2. $\alpha\gamma\beta$, if all robots at p have color α, all robots at q have color β, all robots at the mid-point of the $\overline{pq}_{C(t)}$ have color γ ($\alpha, \beta \gamma in \{A, B\}$) and no robots are in other locations.
3. $\alpha_{\zeta}^{\gamma}\beta$, if all robots at p have color α, all robots at q have color β, all robots at the mid-point of the $\overline{pq}_{C(t)}$ have color γ or ζ ($\alpha, \beta, \gamma, \zeta \in \{A, B\}$) and no robots are in other locations.

3   Minimum Number of Lights for Gathering

The following Theorem 1 provides the complementing lower bound to the known upper bounds for both the case of $\mathcal{F}COM$ and $\mathcal{F}ST\mathcal{A}$ robots by showing that 2 colors of lights do not suffice. Note that this contrasts with the full-light model, where two lights are sufficient for Gathering in SSYNCH [27]. Moreover, note that the restriction to $\mathcal{F}ST\mathcal{A}$ in the last part of the theorem is inevitable because Rendezvous is possible with 3 colors even in ASYNCH with non-rigid movement [28, Thm. 4].

Theorem 1. (3 colors of $\mathcal{F}COM$ and $\mathcal{F}ST\mathcal{A}$ robots are necessary) *Consider the $\mathcal{F}COM$ or $\mathcal{F}ST\mathcal{A}$ model working in* SSYNCH *with rigid movement, consistent chirality, and a shared unit. With only 2 colors of lights, Gathering is impossible for any $n \geq 2$. Moreover, for the $\mathcal{F}ST\mathcal{A}$ model, it is impossible even with an unlimited number of colors if the algorithm is assumed to be self-stabilizing.*

Proof. The case of $\mathcal{F}ST\mathcal{A}$ robots and the case of $\mathcal{F}COM$ robots can be proved similarly. For the sake of simplicity, we assume $n = 2$. The proof is easily generalized to the case $n > 2$ by considering a 2-point configuration in which the robots of each point are always activated together.

We show that no algorithm can achieve a Gathering for every SSYNCH schedule. Fix an algorithm for the two robots. We assume that they have the same snapshot except that one is rotated by 180° with respect to the other. We construct a schedule for which the robots with this algorithm cannot gather, round by round. For any round that starts with a configuration where the robots do not gather if they are both activated, we activate them both. If activating them both leads to a Gathering, we distinguish two cases. If both robots move during their cycle, then we change the schedule such that only one robot is activated to prevent the Gathering. The robot to be activated can be chosen arbitrarily; we can therefore continue to use this strategy without violating fairness to prevent a Gathering as long as this case occurs.

The remaining case is a round that achieves Gathering with only one of the two robots moving, even if both are activated. In this case, the two robots behave differently—one is standing still while the other is moving towards it—and must thus see different internal lights. We first keep activating the non-moving robot, which might change its color, until it either decides to move or has cycled back to a previous color without any movement. (There is no third possibility since the number of colors is assumed to be finite. Indeed, Gathering is always possible with unlimited internal memory [5].) In the former case of movement, we continue the strategy as described above to prolong the schedule that prevents the Gathering. The latter case remains, where one of the robots cycles through a set of colors without ever moving.

Assume for this paragraph the case of $\mathcal{FSTA}$ and a self-stabilizing algorithm, that is, that we can impose an initial configuration that is identical except for both robots seeing the same color as the internal light of the non-moving robot. Then we can activate one of the robots arbitrarily often without any movement, until it loops back to a previously used color. We can do the same with other roobt. Thus both robots are in a non-moving loop and they never meet. This shows that Gathering is impossible for $\mathcal{FSTA}$ robots if we require the algorithm to work for any initial configuration where all lights are set to the same color, possibly different from A. Thus Gathering is impossible for $\mathcal{FSTA}$ robots even with an unlimited number of colors if the algorithm is assumed to be self-stabilizing.

We now drop the last paragraph's restriction to $\mathcal{FSTA}$ and the requirement of self-stabilization, allowing instead only initial configurations in which the robots are all set to light A. In exchange for weakening the adversarial scheduler, we also weaken the robots by granting them only the 2 colors A and B. In the final round as constructed before, the moving robot sees one of the 2 colors and the non-moving robot the other color. In this case they cannot achieve a Gathering from the initial configuration that is identical except for both robots having and thus seeing light A: If we keep activating both of them forever, then there are four subcases. The first one is that they both see the color that lets them move and that they keep the color while doing so. In this case, they will swap their positions forever. The second subcase is that they both see the color that lets them stay where they are and that they keep the color when doing this. In this case, they will stay where they are forever. The third subcase is that they both see the color that lets them stay, but they change the color while doing so. After one such swap we are in the second or the fourth subcase. The fourth and last subcase is that they both see the color that lets them stay where they are, but they change the color when doing so. This leads us back to the first or third subcase. The robots will thus keep synchronously swapping positions or staying where they are forever, preventing a Gathering.   □

Using this theorem, it is shown that all the conditions of previous Gathering algorithms for $\mathcal{FCOM}$ and $\mathcal{FSTA}$ robots shown in Table 1 are necessary.

Theorem 2. *(1) For the 3-color algorithm [27, Algorithm 3] for $\mathcal{FCOM}$ robots with rigid movement in* SSYNCH, *the number of used colors is optimal.*
(2) For the 2-color algorithm [27, Algorithm 4] for $\mathcal{FCOM}$ robots with rigid movement in SSYNCH, *the condition of local-awareness cannot be removed.*
(3) For the 2-color algorithm [27, Algorithm 5] for $\mathcal{FSTA}$ robots with non-rigid movement in SSYNCH, *the condition of 2δ-distant cannot be removed completely.*

In the next section, we weaken the condition of 2δ-distant in Theorem 2 (3) instead of removing it completely.

4   2-Color Gathering Algorithm for $\mathcal{FSTA}$ Robots

In this section, we show a Gathering algorithm for $\mathcal{FSTA}$ robots with 2 colors in Non-rigid($+\delta =$) with agreements of chirality if we exclude 2 patterns stated below from the initial configurations.

The views of the robots in $\mathcal{FSTA}$ are the same as those of the robots in $\mathcal{OBLOT}$, so the robots must determine their behavior using these views without colors and their own colors of lights. Thus Gathering algorithms in $\mathcal{FSTA}$ cannot seem to be constructed without additional knowledge such as distance information. In fact, known Rendezvous algorithms use the minimum distance of moving δ and/or the unit distance [17].

In our Gathering algorithm for $\mathcal{FSTA}$ robots, we assume the agreement of $d(< \frac{\delta}{4})$ and $\epsilon(< d)$. The prohibited initial configurations are as follows: (1) there are two points a and b such that $d - \frac{\epsilon}{2} \leq |\overline{ab}| < d$, or (2) there are three points a, b and c such that $2d - \epsilon \leq |\overline{ac}| < 2d$, and $|\overline{ab}| = |\overline{bc}|$ (Fig. 1). These patterns appear in the final phase of the Gathering algorithm and are used to achieve the Gathering using color. However, when these patterns appear as the initial configuration, the $\mathcal{FSTA}$ robots cannot distinguish them from the patterns of this final phase. If initial configurations do not include the two prohibited patterns, we can construct a Gathering algorithm for $\mathcal{FSTA}$ robots in Non-rigid($+\delta =$) and SSYNCH with 2 colors of light.

Our Gathering algorithm for $\mathcal{FSTA}$ robots (Algorithm 1) consists of three parts.

1. From any configuration, we make an $OnLDS$ configuration C such that $|\overline{pq}_C| \geq 2d$ (Algorithm 2).

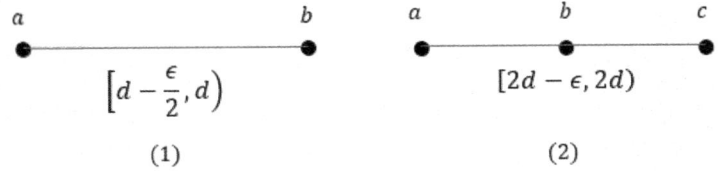

Fig. 1. The prohibited initial configurations

2. From any $OnLDS$ configuration C such that $|\overline{pq}_C| \geq 2d$, we make a 2-point configuration C' such that $2d - \epsilon < |\overline{pq}_{C'}| < 2d$ (Algorithm 3).
3. From any 2-point configuration C such that $2d - \epsilon \leq |\overline{pq}_C| < 2d$, we make a Gathering configuration (Algorithm 4).

Algorithm 1 Gathering-$\mathcal{FSTA}$-Robots(r_i)

Input: Any configuration except the 2 prohibited patterns, all robots have color A.
Output: Gathering configuration.

1: **if** $\neg OnLDS$ **or** $(OnLDS$ **and** $2d > |\overline{pq}_{SS_i}|)$ **then**
2:     ElectLDS(r_i)
3: **else if** $OnLDS$ **and** $2d \leq |\overline{pq}_{SS_i}|$ **then**
4:     Adjustment-LDS(r_i)
5: **else if** $|\mathcal{P}(SS_i)| = 2$ **and** $2d - \epsilon \leq |\overline{pq}_{SS_i}| < 2d$ **then**
6:     Gather(r_i)

Note that we do not use colors to solve Cases 1 and 2 and we only use two colors to solve Case 3. The output of i is the input of $i + 1$ for $i \in \{1, 2\}$ and the explanation of the algorithms is listed in order from 1 to 3. The configurations transitions of Algorithm 1 are shown in Fig. 2, where nodes denote configurations and a directed edge denotes the transition from a configuration to a configuration.

The outline of the behavior of Algorithm 1 is explained as follows. From any initial configuration C_0 of $\neg OnLDS$ or $OnLDS$, robots move radially outward by $4d < \delta$ from $CTR(C_0)$ if $Diam(C_0) < 4d$, making the configuration $Diam(C_1) \geq 4d$, and then robots make the $OnLDS$ configuration C_2 such that $Diam(C_2) \geq 2d$ (lines 1–2) by using ElectLDS-Preserving-Distance(r_i) [27], making $OnLDS$ preserving the diameter. When the configuration C_2 is obtained, robots located not at endpoints move to endpoints and robots located at endpoints stay and then a 2-point configuration C_3 such that $2d \leq Diam(C_3)$ results through special patterns $A3P$ or $A4P$ (lines 3–4). Then the robots reduce the diameter and make a 2-point configuration C_4 such that $2d - \epsilon \leq Diam(C_4) < 2d$ through special patterns $A3P$ or $A4P$. When a 2-point configuration C_4 is obtained, robots with color A change its color to B and move to the midpoint and robots with B color stay, Gathering is achieved (lines 5–6).

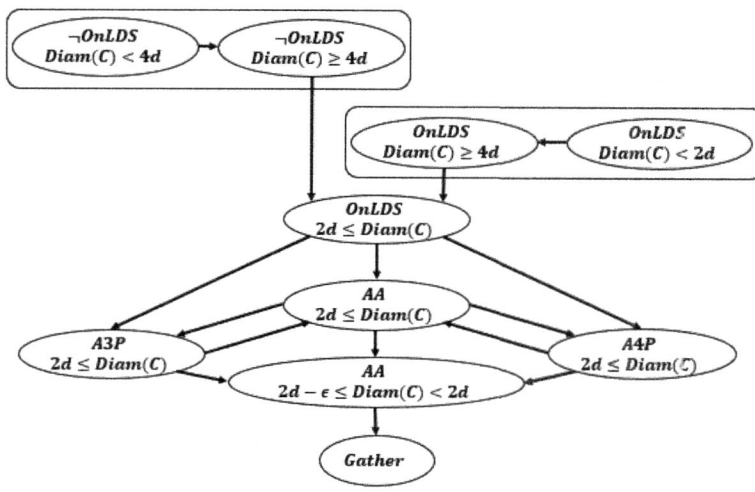

Fig. 2. The transition of configurations in Gathering-$\mathcal{FSTA}$-Robots(r_i).

Algorithm 2 is based on ElectLDS-Preserving-Distance(r_i) [27, Algorithm 8] and produces the unique LDS with its length at least $2d$ unless it produces a Gathering configuration.

Lemma 1. *[27] Using ElectLDS-Preserving-Distance, if $C(t)$ is any configuration, then there exists a time $t' > t$ such that $C(t) \rightarrow^* C(t')$, $C(t')$ is a Gathering configuration or an $OnLDS$ configuration with $Diam(C(t')) \geq Diam(C(t))/2$.*

In the following algorithms, the expression $[p, q] + \alpha - \beta$ denotes the point at distance $\alpha - \beta$ from point p on the line segment $\overline{pq}$.

Algorithm 2 ElectLDS(r_i)

Input: $\neg OnLDS$ or ($OnLDS$ and $2d > |\overline{pq}_{SS_i}|$)
Output: $OnLDS$ and $2d \leq |\overline{pq}_{SS_i}|$

1: **if** $Diam(SS_i) < 4d$ **then**
2:    $des_i \leftarrow [p_i, CTR(SS_i)] + 4d$ //move $4d$ outward from $CTR(SS_i)$
3: **else if** $Diam(SS_i) \geq 4d$ **then**
4:    ElectLDS-Preserving-Distance(r_i)   //[27, Algorithm 8]

Lemma 2 ensures that an $OnLDS$ configuration $C(t)$ with $|\overline{pq}_{C(t)}| \geq 2d$ is reached.

Lemma 2. *If $|O(LDS(C(t)))| \geq 0$, then there exists a time $t' > t$ such that $C(t) \rightarrow^* C(t')$, $C(t')$ is a Gathering configuration or an $OnLDS$ configuration with $|\overline{p'q'}_{C(t')}| \geq 2d$.*

Proof. Whenever the robots are not configured $OnLDS$, they consider $SEC(C(t))$ and $LDS(C(t))$. If $Diam(C(t))$ is below $4d$, then it is increased by letting any activated robot move radially outwards by $4d < \delta$ from $CTR(C(t))$. This increases $Diam(C(t))$ by at least $4d$ in one step for the following reason. For any point on or in the smallest enclosing circle $SEC(C(t))$, there is another point at an angle of at least $90°$ when viewed from the $CTR(C(t))$ because all points would lie on one side of some diameter of the $SEC(C(t))$, contradicting its minimality. The outward movements thus cannot cancel each other out; they will add up to an increase of at least $4d$. We now consider the second case where $Diam(C(t))$ is at least $4d$. Then the mentioned algorithm is used to reach a Gathering or a $OnLDS$ configuration $C(t'))$ with $Diam(C(t')) \geq 2d$ according to Lemma 1.   □

Next, we make a 2-point configuration C satisfying $2d - \epsilon \leq |\overline{pq}_C| < 2d$ from any $OnLDS$ configuration. This adjustment task is performed using Algorithm 3. In the algorithm for robot r_i, the endpoints of LDS are denoted by p_n and p_f for position p_i of r_i, where p_n is the nearest endpoint from p_i and p_f is the farthest one from p_i. Note that if p_i is one of the endpoints, then $p_i = p_n$ and p_f is the other endpoint.

In Algorithm 3, $A3P$ and $A4P$ denote the following predicates (Fig. 3). When configuration $C(t)$ is 2-point configuration, if robots move on one endpoint only, $C(t+1)$ satisfies $A3P$ and if robots move on both endpoints, $C(t+1)$ satisfies $A4P$.

$A3P$: $|\mathcal{P}(SS_i)| = 3, \mathcal{P}(SS_i) = \{p_n, p_{m_1}, p_f\}, |\overline{p_n p_{m_1}}| \neq |\overline{p_{m_1} p_f}|, |\overline{p_n p_{m_1}}| = \frac{\epsilon}{2}, 2d - \epsilon \leq |\overline{p_{m_1} p_f}|$,

$A4P$: $|\mathcal{P}(SS_i)| = 4, \mathcal{P}(SS_i) = \{p_n, p_{m_1}, p_{m_2}, p_f\}, |\overline{p_n p_{m_1}}| = |\overline{p_{m_2} p_f}| = \frac{\epsilon}{2}, 2d - 2\epsilon \leq |\overline{p_{m_1} p_{m_2}}|$

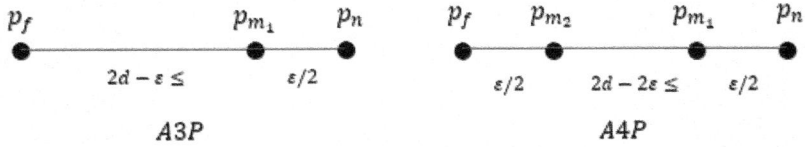

Fig. 3. Configurations in $A3P$ and $A4P$.

Algorithm 3 Adjustment-LDS(r_i)

Input: $OnLDS$ and $2d \leq |\overline{pq}_{SS_i}|$

Output: $|\mathcal{P}(SS_i)| = 2$ and $2d - \epsilon \leq |\overline{pq}_{SS_i}| < 2d$.

1: **if** $((((|\mathcal{P}(SS_i)| = 2)$ **and** $(2d \leq |\overline{pq}_{SS_i}|))$ **or** $A3P$ **or** $A4P)$ **and** $(p_i = p_n)$ **then**
2:     $des_i \leftarrow [p_n, p_f] - \frac{\epsilon}{2}$ $//\frac{\epsilon}{2}$ inward
3: **else if** $((|\mathcal{P}(SS_i)| \geq 3)$ **then**
4:     $des_i \leftarrow p_n$

Lemma 3. *If $C(t)$ is an OnLDS configuration with $|\overline{pq}_{(C(t))}| \geq 2d$, then by Algorithm 3, there are times t' and t'' $(t'' > t' > t)$ such that $C(t) \rightarrow^* C(t')$, $C(t')$ satisfies one of the following, and $C(t') \rightarrow^* C(t'')$, $C(t'')$ is a 2-point configuration with $2d - \epsilon \leq |\overline{p''q''}_{C(t'')}| < 2d$.*

1. $|\mathcal{P}(C(t'))| = 2$ and $|\overline{p'q'}_{C(t')}| = |\overline{pq}_{C(t)}|$
2. $|\mathcal{P}(C(t'))| = 2$ and $|\overline{p'q'}_{C(t')}| = |\overline{pq}_{C(t)}| - \frac{\epsilon}{2}$
3. $|\mathcal{P}(C(t'))| = 2$ and $|\overline{p'q'}_{C(t')}| = |\overline{pq}_{C(t)}| - \epsilon$

Proof. When $|\mathcal{P}(C(t))| \geq 3$, an activated robot r_i not positioned at the endpoints($p_i \neq p_n$), moves toward p_n (line 6).

1. When all robots reach p_n, $|\overline{p'q'}_{C(t')}|$ remains unchanged and becomes a 2-point configuration.
2. Before all robots reach p_n, $C(t)$ may become $A3P$. When $C(t)$ is $A3P$, the robot at p_n closest to p_{m_1} moves toward p_{m_1} (line 2), resulting in a 2-point configuration with $|\overline{p'q'}_{C(t')}|$ reduced by $\frac{\epsilon}{2}$.
3. Before all robots reach p_n, $C(t)$ may become $A4P$. When $C(t)$ is $A4P$, the robot at p_n moves toward p_{m_1} (line 2), resulting in a 2-point configuration with $|\overline{p'q'}_{C(t')}|$ reduced by ϵ.

After $t' + 1$, the activated robot moves $\frac{\epsilon}{2}$ inward (line 2). By iterating the above steps, a 2-point configuration $C(t'')$ with $2d - \epsilon \leq |\overline{p''q''}_{C(t'')}| < 2d$ is reached.□

By Lemmas 2–3, it is guaranteed that there is a time t such that a 2-point configuration $C(t)$ with $2d - \epsilon \leq |\overline{pq}_{C(t)}| < 2d$ is obtained.

Lastly, Algorithm 4 solves Gathering in $\mathcal{FSTA}$ if the initial configurations satisfy $|\mathcal{P}(\mathcal{SS}_i)| = 2$ and $2d - \epsilon \leq |\overline{pq}_{\mathcal{SS}_i}| < 2d$. Since $d \leq \frac{\delta}{4}$, every movement in Algorithm 4 is the same as the rigid one. The following lemma is easily verified for Algorithm 4.

Algorithm 4 Gather(r_i)

Input: $|\mathcal{P}(\mathcal{SS}_i)| = 2$ and $2d - \epsilon \leq |\overline{pq}_{\mathcal{SS}_i}| < 2d$, all robots have color A.
Output: Gathering configuration.

1: **if** $(2d - \epsilon \leq |\overline{pq}_{\mathcal{SS}_i}| < 2d)$ **and** $((|\mathcal{P}(\mathcal{SS}_i)| = 2)$ **or** $((|\mathcal{P}(\mathcal{SS}_i)| = 3)$ **and** $(|\overline{p_n p_m}| = |\overline{p_m p_f}|)))$ **and** $(l_i = A)$ **then**
2:     $l_i \leftarrow B$
3:     $des_i \leftarrow p_m$
4: **else if** $(d - \frac{\epsilon}{2} \leq |\overline{pq}_{\mathcal{SS}_i}| < d)$ **and** $(|\mathcal{P}(\mathcal{SS}_i)| = 2)$ **and** $(l_i = A)$ **then**
5:     $l_i \leftarrow B$
6:     $des_i \leftarrow p_f$
7: **else if** $l_i = B$ **then**
8:     $l_i \leftarrow B$
9:     $des_i \leftarrow p_i$ //stay

Lemma 4. *If $C(t)$ is a 2-point configuration with $|\overline{pq}_{C(t)}|$ satisfying $2d - \epsilon \leq |\overline{pq}_{C(t)}| < 2d$, there is a time $t' > t$ such that $C(t) \rightarrow^* C(t')$ and $C(t')$ is a Gathering configuration.*

Proof. From such a 2-point configuration, the activated robots move to the midpoint while changing light from A to B, and stay there afterwards. So when all the robots at the endpoints are activated at the same time, Gathering is achieved. When some of the robots at the endpoints are activated, $C(t_1)$ reaches a 3-point configuration which has the the color configuration ABA. In $C(t_1)$, the robots with color A move the same as before, so when the number of robots at the endpoints decreases and there are no more robots with color A at one endpoint, we have a 2-point configuration $C(t_2)$ with the color configuration $AB(BA)$. Then, since the 2-point configuration satisfies $d - \frac{\epsilon}{2} \leq |\overline{pq}_{C(t_2)}| < d$, the robots with color A move to the other endpoint while changing their light from A to B, and there is a time t' such that $C(t')$ is a Gathering configuration. $\square$

The prohibited patterns appear in this case. If the initial configuration is a 2-point configuration C such that $C = \{a, b\}$ and $d - \frac{\epsilon}{2} \leq |\overline{ab}| < d$ with the color configuration AA, the robots with color A move to the other endpoint while changing their lights from A to B, and stay there thereafter, and C_1 reaches a 2-point configuration with the color configuration BB, and thereafter the algorithm terminates and Gathering is not achieved. If the initial configuration is a 3-point configuration C such that $C = \{a, b, c\}$ and $2d - \epsilon \leq |\overline{ac}| < 2d$ and $|\overline{ab}| = |\overline{bc}|$ with the color configuration AAA, the activated robots located at the endpoints move to the midpoint while changing their lights from A to B, and stay there thereafter, the activated robots not located at the endpoints move to the endpoints without changing their lights, C_2 reaches a 3-point configuration with the color configuration $A \,^A_B A$ or a 2-point configuration with the color configuration $A \,^A_B (\,^A_B A)$. When C_2 is the 3-point configuration with the color configuration $A \,^A_B A$, the robots with color A move the same as before, so when the number of robots at the endpoints decreases and there are no more robots with color A at one endpoint, C_2' reaches the 2-point configuration with the color configuration $A \,^A_B (\,^A_B A)$. When C_2' is a 2-point configuration with the color configuration $A \,^A_B (\,^A_B A)$, for the same reason as in a 2-point configuration with $d - \frac{\epsilon}{2} \leq |\overline{ab}| < d$ with the color configuration AA, C_2'' reaches a 2-point configuration with the color-configuration BB, and thereafter the algorithm terminates and Gathering is not achieved.

We obtain the following theorem by Lemmas 2–4.

Theorem 3. *Let $d < \frac{\delta}{4}$ and $\epsilon < d$ be agreed upon by the robots. Gathering is solvable in $\mathcal{FSTA}$, Non-Rigid($+\delta$ =), and SSYNCHif robots have 2 colors, set view and agreement on chirality, and the following initial configuration excepted.*

1. *2-point configuration $C = \{a, b\}$ such that $d - \frac{\epsilon}{2} \leq |\overline{ab}| < d$*
2. *3-point configuration $C = \{a, b, c\}$ such that $2d - \epsilon \leq |\overline{ac}| < 2d$ and $|\overline{ab}| = |\overline{bc}|$*

5   Concluding Remarks

In this paper, we have shown that Gathering is impossible for $\mathcal{FCOM}$ and $\mathcal{FSTA}$ robots with 2-color lights in SSYNCH, even assuming agreement on chirality and the minimum distance δ, and consequently, that the conditions imposed on previously developed Gathering algorithms for $\mathcal{FCOM}$ and $\mathcal{FSTA}$ robots are necessary. We have also improved the condition imposed on the Gathering algorithm for $\mathcal{FSTA}$ robots with 2-color lights.

Interesting open questions are developing unconditional Gathering algorithms for $\mathcal{FSTA}$ robots with more than two colors in SSYNCH, and/or Gathering algorithms for $\mathcal{FCOM}$ or $\mathcal{FSTA}$ robots in ASYNCH.

References

1. Agmon, N., Peleg, D.: Fault-tolerant gathering algorithms for autonomous mobile robots. SIAM J. Comput. **36**(1), 56–82 (2006)
2. Ando, H., Osawa, Y., Suzuki, I., Yamashita, M.: A distributed memoryless point convergence algorithm for mobile robots with limited visibility. IEEE Trans. Robot. Autom. **15**(5), 818–828 (1999)
3. Bouzid, Z., Das, S., Tixeuil, S.: Gathering of mobile robots tolerating multiple crash faults. In The 33rd International Conference on Distributed Computing Systems, pp. 334–346 (2013)
4. Cicerone, S., Stefano, D., Navarra, A.: Gathering of robots on meeting-points. Distrib. Comput. **31**(1), 1–50 (2018)
5. Cieliebak, M.: Gathering non-oblivious robots. In: LATIN 2004: Theoretical Informatics, pp. 577–588 (2004)
6. Cieliebak, M., Flocchini, P., Prencipe, G., Santoro, N.: Distributed computing by mobile robots: gathering. SIAM J. Comput. **41**(4), 829–879 (2012)
7. Cohen, R., Peleg, D.: Convergence properties of the gravitational algorithms in asynchronous robot systems. SIAM J. Comput. **34**(15), 1516–1528 (2005)
8. Das, S., Flocchini, P., Prencipe, G., Santoro, N., Yamashita, M.: Autonomous mobile robots with lights. Theor. Comput. Sci. **609**, 171–184 (2016)
9. Défago, X., Potop-Butucaru, M., Raipin-Parvédy, P.: Self-stabilizing gathering of mobile robots under crash or byzantine faults. Distrib. Comput. **33**(5), 393–421 (2020)
10. Degener, B., Kempkes, B., Langner, T., Meyer auf der Heide, F , Pietrzyk, P., Wattenhofer, R.: A tight run-time bound for synchronous gathering of autonomous robots with limited visibility. In 23rd ACM SPAA, pp. 139–148 (2011)
11. Flocchini, P., Prencipe, G., Santoro, N.: Distributed Computing by Oblivious Mobile Robots. Morgan & Claypool (2012)
12. Flocchini, P., Prencipe, G., Santoro, N., Widmayer, P.: Gathering of asynchronous robots with limited visibility. Theor. Comput. Sci. **337**(1–3), 147–169 (2005)
13. Flocchini, P., Prencipe, G., Santoro, N., Widmayer, P.: Arbitrary pattern formation by asynchronous oblivious robots. Theor. Comput. Sci. **407**, 412–447 (2008)
14. Flocchini, P., Santoro, N., Viglietta, G., Yamashita, M.: Rendezvous with constant memory. Theor. Comput. Sci. **621**, 57–72 (2016)
15. Flocchini, P., Santoro, N., Wada, K.: On memory, communication, and synchronous schedulers when moving and computing. In: Proceedings of 23rd International Conference on Principles of Distributed Systems (OPODIS), pp. 25:1–25:17 (2019)

16. Flocchini, P., Prencipe, G., Santoro, N.: Distributed Computing by Oblivious Mobile Robots. Springer Nature (2022)
17. Flocchini, P., Santoro, N., Viglietta, G., Yamashita, M.: Rendezvous with constant memory. Theor. Comput. Sci. **621**, 57–72 (2016)
18. Izumi, T., Bouzid, Z., Tixeuil, S., Wada, K.: Brief announcement: the BG-simulation for byzantine mobile robots. In: 25th DISC, pp. 330–331 (2011)
19. Izumi, T., Katayama, Y., Inuzuka, N., Wada, K.: Gathering autonomous mobile robots with dynamic compasses: an optimal result. In: 21st DISC, pp. 298–312 (2007)
20. Izumi, T., Souissi, S., Katayama, Y., Inuzuka, N., Défago, X., Wada, K., Yamashita, M.: The gathering problem for two oblivious robots with unreliable compasses. SIAM J. Comput. **41**(1), 26–46 (2012)
21. Kamei, S., Lamani, A., Ooshita, F., Tixeuil, S.: Asynchronous mobile robot gathering from symmetric configurations without global multiplicity detection. In: 18th SIROCCO, pp. 150–161 (2011)
22. Lin, J., Morse, A.S., Anderson, B.D.O.: The multi-agent rendezvous problem. Parts 1 and 2. SIAM J. Comput. **46**(6), 2096–2147 (2007)
23. Nakai, R., Sudo, Y., Wada, K.: Asynchronous gathering algorithms for autonomous mobile robots with lights. In: Proceedings of 23rd International Symposium (SSS), pp. 410–424 (2021)
24. Okumura, T., Wada, K., Défago, X.: Optimal rendezvous $\mathcal{L}$-algorithms for asynchronous mobile robots with external-lights. Theor. Comput. Sci. **979**, 114198 (2023)
25. Souissi, S., Défago, X., Yamashita, M.: Using eventually consistent compasses to gather memory-less mobile robots with limited visibility. ACM Trans. Autonom. Adapt. Syst. **4**(1), 1–27 (2009)
26. Suzuki, I., Yamashita, M.: Distributed anonymous mobile robots: formation of geometric patterns. SIAM J. Comput. **28**, 1347–1363 (1999)
27. Terai, S., Wada, K., Katayama, Y.: Gathering problems for autonomous mobile robots with lights. Theor. Comput. Sci. **941**, 241–261 (2023)
28. Viglietta, G.: Rendezvous of two robots with visible bits. In: 10th International Symposium on Algorithms and Experiments for Sensor Systems, Wireless Networks and Distributed Robotics (ALGOSENSORS), pp. 291–306 (2013)

Selective Population Protocols

Adam Gańczorz[1], Leszek Gąsieniec[2(✉)], Tomasz Jurdziński[1],
Jakub Kowalski[1], and Grzegorz Stachowiak[1]

[1] University of Wrocław, ul. Joliot-Curie 15, 50-383 Wrocław, Poland
{adam.ganczorz,tju,jko,gst}@cs.uni.wroc.pl
[2] University of Liverpool, Liverpool, UK
l.a.gasieniec@liverpool.ac.uk

Abstract. The model of population protocols provides a universal platform to study distributed processes driven by pairwise interactions of anonymous agents. While population protocols present an elegant and robust model for randomized distributed computation, their efficiency wanes when tackling issues that require more focused communication or the execution of multiple processes. To address this issue, we propose a new, selective variant of population protocols by introducing a partition of the state space and the corresponding conditional selection of responders. We demonstrate on several examples that the new model offers a natural environment, complete with tools and a high-level description, to facilitate more efficient solutions. In particular, we provide fixed-state stable and efficient solutions to two central problems: leader election and majority computation, both with confirmation. This constitutes a separation result, as achieving stable and efficient majority computation requires $\Omega(\log n)$ states in standard population protocols, even when the leader is already determined. Additionally, we explore the computation of the median using the comparison model, where the operational state space of agents is fixed, and the transition function determines the order between (arbitrarily large) hidden keys associated with interacting agents. Our findings reveal that the computation of the median of n numbers requires $\Omega(n)$ time. Moreover, we demonstrate that the problem can be solved in $O(n \log n)$ time, both in expectation and with high probability, in standard population protocols. In contrast, we establish that a feasible solution in selective population protocols can be achieved in $O(\log^4 n)$ time.

Keywords: Population Protocol · Stability · Conditional Interactions · Median Computation

Adam Gańczorz, Tomasz Jurdziński and Grzegorz Stachowiak—Supported by the National Science Centre, Poland under project number 2020/39/B/ST6/03288.
Jakub Kowalski—Supported in part by the National Science Centre, Poland under project number 2021/41/B/ST6/03691.

T. Masuzawa et al. (Eds.): SSS 2024, LNCS 14931, pp. 225–239, 2025.
https://doi.org/10.1007/978-3-031-74498-3_16

1   Introduction

The standard model of population protocols originates from the seminal paper [5], providing tools suitable for the formal analysis of *pairwise interactions* between simple, indistinguishable entities referred to as *agents*. These agents are equipped with limited storage, communication, and computation capabilities. When two agents engage in a direct interaction, their states change according to the predefined *transition function*, which is an integral part of the population protocol. The weakest possible assumptions in population protocols pertain to the fixed (constant size) operational *state space* of agents, and the size of the population n is neither known to the agents nor hard-coded in the transition function. It is assumed that a protocol starts in the predefined *initial configuration* of agents' states representing the input, and it *stabilizes* in one of the *final configurations* of states representing the solution to the considered problem. In the *probabilistic variant* of population protocols adopted here, in each step of a protocol, the *random scheduler* selects an ordered pair of agents: the *initiator* and the *responder*, which are drawn from the whole population uniformly at random. The lack of symmetry in this pair is a powerful source of random bits utilized by population protocols. In the probabilistic variant, in addition to efficient *state utilization*, one is also interested in the *time complexity*, where the *sequential time* refers to the number of interactions leading to the stabilization of a protocol in a final configuration. More recently, the focus has shifted to the *parallel time*, or simply the *time*, defined as the sequential time divided by the size nn of the whole population. The (parallel) time reflects on the parallelism of simultaneous independent interactions of agents utilized in *efficient population protocols* that stabilize in time $O(\text{poly} \log n)$. All protocols presented in this paper are *stable* (always correct) and guarantee stabilization time with high probability (whp) defined as $1 - n^{-\eta}$, for a constant $\eta > 0$.

There are already several efficient protocols known for solving central problems in distributed computing, including *leader election* [1,8,13], *majority computation* [2,12], and the *plurality problem* [7]. While these protocols are efficient in terms of time, they rely on non-constant state space utilization and operate indefinitely. That is, they are not able to declare stabilization with probability 1. Moreover, the most efficient protocols are often non-trivial and hard to analyze. One can circumvent some of these deficiencies by relaxing probabilistic expectations, e.g., by dropping assumptions about the necessity of stabilization in protocols with predefined input [19], as well as in self-stabilizing protocols [9]. While such relaxation is beneficial, it does not solve some major deficiencies of the standard model, including depleting in time the number of meaningful interactions, limited computational power, and inefficient space-time trade-offs.

In order to circumvent some of these deficiencies, we propose a new *selective* variant of population protocols by imposing a simple *group (partition) structure* on the state space together with a conditional choice of the responder during random interacting pair selection. This model provides a natural extension of *passive mobile* sensor networks adopted in [5], where the focus is on single channel communication. In the new model the agents communicate over multiple communication channels, where each channel corresponds to one of a fixed number of groups (partitions) of the state space. Specifically, only agents currently listening on some specific communication channel C (their states belong to the corresponding group of states $\mathcal{G}_C$) are able to receive and respond to messages transmitted over this channel by agents with state indicating $\mathcal{G}_C$ as the *target group*. Alternative models with biased communication were previously used in the context of stochastic chemical reaction networks in [23] and data collection with non-uniform schedulers in [9]. The adopted selective model also refers to biased choices in nature studied earlier, in the context of small-world phenomena, where closer location in space results in a more likely interaction [18], or social preference, where agents with a greater array of similar attributes are more likely to know one another and, in turn, to interact [17]. A different motivation to study selective population protocols refers to more structural variant of population protocols known as *network constructors*, in which agents are allowed to be connected. As the expected parallel time to manipulate a specific edge is $\Theta(n)$, see, e.g., [15,20], the design of truly efficient protocols in this model is not currently feasible. Utilizing the concept of selective population protocols, one can give a higher probabilistic bias to interactions along existing edges, enabling more efficient computation, comparable to graphical population protocols [3,4].

Our contribution In this paper, we present initial studies on the (parallel) efficiency and stability of selective population protocols. We begin by discussing fundamental properties of this new promising variant, introducing the notion of *fragmented parallel time* as an equivalent measure to parallel time in the standard population protocol model. Additionally, we highlight that selective protocols offer a natural mechanism for deterministic emptiness (zero) testing. It is known, as indicated in [6], that such a test enables efficient simulation of $O(\log n)$-space Turing Machines with high probability. In contrast, we highlight that such simulations in selective protocols are not only efficient but also stable. Selective protocols can be utilized to design algorithms within this class that are both efficient and stable. Furthermore, we present fixed-state efficient and stable solutions to two central problems: leader election and majority computation (with confirmation, i.e., all agents stabilize while being aware of the conclusion of the process). This result is noteworthy as stable efficient majority computation requires $\Omega(\log n)$ states in standard population protocols [2], even when the leader is given. We also introduce the first non-trivial study on median computation in population protocols. We adopt a comparison model in which the operational state space of agents is constant, and the transition function determines the order between (arbitrarily large) hidden keys associated with the interacting agents. We demonstrate that computing the median of n numbers requires $\Omega(n)$ parallel time and the problem can be solved in $O(n \log n)$ parallel time in expectation and with high probability (whp) in standard population protocols. In contrast, we present an efficient median computation in selective population protocols, achieving $O(\log^4 n)$ parallel time. Furthermore, we delve into suitability of selective protocols for the high-level design of algorithms.

Please note that due to space restrictions the proofs of Lemmas 3, 4, 5, 6, 7, 8, 9, and Proposition 1 can be found in the full version of this paper [14].

2 Selective Population Protocols

As discussed in Sect. 1, in the standard population protocol model, the random scheduler draws consecutive pairs of interacting agents uniformly at random from the entire population. This is done irrespective of whether the states of interacting agents match some rule of the transition function or not. Consequently, many random pairwise interactions do not result in a transition and, in turn, do not bring the population closer to a final configuration.

In *selective population protocols* random interactions are scheduled differently. Specifically, the fixed-state space of agents S is partitioned into l groups of states $\mathcal{G}_1, \mathcal{G}_2, \ldots, \mathcal{G}_l$. In addition, any state s is mapped onto its *target group* $\mathcal{G}_{i(s)}$. We say that an interaction is *internal* if $s \in \mathcal{G}_{i(s)}$, and is *external* otherwise. This mapping is used during an attempt to form a random pair of interacting agents. The random scheduler first draws the initiator in state s uniformly from the whole population. This is followed by drawing the responder uniformly from all agents (different to the initiator) currently being in any state t belonging to the target group $\mathcal{G}_{i(s)}$. Such interaction is denoted by $s + \mathcal{G}_{i(s)}|t$, where $t = null$ when no responder is currently available in any state of $\mathcal{G}_{i(s)}$. The rules of the transition function refer to two types of outcomes of interaction attempts:

1. *Biased communication*
   Meaningful interaction (successful biased interaction attempt)
       $s + \mathcal{G}_{i(s)}|t \rightarrow s' + t'$.
   The purpose of meaningful interactions is to advance and in turn to maintain efficiency of the computing process.

2. *Interaction availability test*
   Emptiness (zero) test (unsuccessful external interaction attempt)
       $s + \mathcal{G}_{i(s)}|null \rightarrow s'$.
   Singleton test (unsuccessful internal interaction attempt)
       $s + \mathcal{G}_{i(s)}|null \rightarrow s'$.
   The two tests primarily confirm completion of computation processes.

One-way epidemic Consider a communication primitive known as *one-way epidemic* [6] in which state 1 of the source agent is propagated to all other agents initially being in state 0. The transition function has only one rule in the standard population protocol model $1 + 0 \rightarrow 1 + 1$. It is known that such epidemic process is stable and efficient, i.e., one-way epidemic stabilises with the correct answer in parallel time $O(\log n)$ whp, however during the final stages of the epidemic process the expected fraction of *meaningful interactions* decreases dramatically. Assume, that the state space $S = \{0, 1\}$ is partitioned into two singleton groups $\mathcal{G}_0 = \{0\}$ and $\mathcal{G}_1 = \{1\}$ in the new variant, and we have two transition rules instead:

$$(1)\ 1 + \mathcal{G}_0|0 \longrightarrow 1 + 1 \qquad\qquad (2)\ 1 + \mathcal{G}_0|null \longrightarrow Stop$$

Now every interaction initiated by an agent in state 1 is either meaningful, when there are still uninformed agents, or changes the initiator's state to *Stop*, which indicates the end of the epidemic process, and in turn the next stage of computation not requiring group $\mathcal{G}_0$.

2.1   Beyond Presburger Arithmetic

We find in [6] that the emptiness test, also known as *zero test*, is a powerful tool enabling efficient simulation of $O(\log n)$-space Turing Machine. The two-stage randomised simulation from [6] is based on simulation of *Register Machines* known to be equivalent with $O(\log n)$-space Turing Machines [21]. This approach hinges on the presence of a unique leader, crucial for achieving efficient and stable computations, which efficient computation requires at least $\Omega(\log \log n)$ states [1]. An alternative randomized two-stage simulation of Turing Machines, detailed in [23] within the context of a related *stochastic chemical reaction network* model, utilises the concept of *clockwise Turing Machines* [22]. Both simulations rely on zero tests, the correctness of which can be assured only with high probability in the adopted models, rendering them unsuitable for deployment in stable protocols. In contrast, selective protocols equipped with deterministic emptiness test provide a suitable platform for the design of efficient and stable *fixed-state* solutions in $O(\log n)$-space complexity class.

As the primary focus of this paper centers on the (parallel) *efficiency* of selective population protocols, and the computational power of such protocols is inherited from standard population protocols, we direct the reader to [6] for full simulation details. Instead, our current study delves into the parallelism of selective protocols, presenting several separation results. This includes *majority computation*, where any efficient stable algorithm in standard protocols requires $\Omega(\log n)$ states while a fixed-state space allows for an $O(\log n)$-time stable solution, as demonstrated in Sect. 2.4.

Several efficient algorithms presented in this paper follow a more direct approach, relying on a single stabilization process. Examples include the efficient and stable leader election and majority computation discussed in Sects. 2.3 and 2.4, respectively. However, in more complex solutions, the need arises for a leader to act as the *"program counter,"* overseeing the proper execution of potentially numerous individual stabilization processes encoded in the transition function in the correct order. This encompasses the preparation of input for each individual process, ensuring its proper termination, and further interpreting the output. It's worth noting that, due to state partitioning in selective protocols, each individual stabilization process can be executed on a distinct partition of states. This allows several processes to run efficiently at the same time, as demonstrated in the efficient ranking protocol presented in Sect. 4, where multiple leaders are employed. The leader is also responsible for translating the output from one process to the input of its successor. This is achieved by rewriting states (from one partition to another) via one-way epidemic. Ultimately, the termination of any process, including rewriting, is recognized through either an emptiness or singleton test.

2.2   Parallelism of Selective Protocols

Recall that in population protocols, the (parallel) time of a sequence I of interactions is defined as $|I|/n$. This definition is motivated by the observation that in a sequence of xn interactions, each agent has, on average, x interactions. However, it is noteworthy that only for $x = \Omega(\log n)$ does each agent engage in $\Theta(x)$ interactions whp. An interesting finding presented in [11] demonstrates that in this latter case, the sequence I can be simulated in time $\Theta(x)$ on a parallel computer whp. In the new variant, where the choice of the responder is likely biased, we must adopt a more nuanced definition of parallelism.

Fragmented Parallel Time In the novel selective variant of population protocols, the initiator is uniformly chosen at random. Consequently, in a sufficiently long sequence of interactions, any agent serves as the initiator with the same frequency, aligning with the pattern observed in the standard model. This stands in contrast to the selection of responders, where certain agents are more likely to be chosen than others. For example, consider an epidemic process with the state space $S = \{0, 1, 1^*\}$ partitioned into groups: $\mathcal{G}_0 = \{0\}$ with uninformed agents, $\mathcal{G}_1 = \{1\}$ with active informers, and $\mathcal{G}_{1^*} = \{1^*\}$ with informed and already rested agents, governed by two transition rules:

$$(1)\ 0 + \mathcal{G}_1 | 1 \longrightarrow 1^* + 1 \qquad\qquad (2)\ 1 + \mathcal{G}_0 | null \longrightarrow 1^*$$

If in the initial configuration there is exactly one informed agent in state 1, all other agents in state 0 contact this agent to get informed and rest, see rule (1). In the last meaningful interaction rule (2) rests the unique informer. While the number of interactions before stabilisation with all agents resting in state 1^* is $O(n \log n)$ whp, the parallelism of this epidemic process is very poor as only one agent informs others as the responder.

This potential imbalance in the workload of individual agents can be captured by tracking the frequency at which agents act as responders. To handle this imbalance, we propose a more subtle definition of (parallel) time. This new definition is whp asymptotically equivalent to the definition and the properties of time used in the standard model, see Lemma 1.

Definition 1 (Fragmented parallel time). Consider ways to divide the sequence of interactions I into subsequent disjoint *chunks*, where each chunk is a sequence of consecutive interactions in which any agent has at most $10 \ln n$ interactions as the responder. If the minimum number of chunks for such divisions is k, then we say that the *fragmented parallel time*, or in short the *fragmented time*, is $T_F = k \ln n$.

Recall that η is the quality parameter in the definition of high probability.

Lemma 1. *Consider a sequence of interactions I in the standard population protocol model executed in time $T = |I|/n$. If the fragmented time $T_F = k \ln n$, for $k > 11\eta/35$ and large enough n, then $T/10 \le T_F \le 2T$ whp.*

Proof. The total number of interactions during fragmented time T_F does not exceed $10kn \ln n$, ensuring $\frac{T}{10} \le T_F$. It remains to show that $T_F \le 2T$ whp.

Let us first estimate the probability that a given chunk corresponds to time smaller than $\ln n$. This probability is not greater than the probability that in time $\ln n$ (starting at the beginning of the chunk) some agent has the responder type interactions greater than $10 \ln n$. By Chernoff bound[4], the probability that in time $\ln n$ a given agent experiences $X > 10 \ln n = 10 \mathbb{E} X$ interactions as the responder can be estimated by

$$\Pr(X > 10 \ln n) = \Pr \left(X > (1+9) \mathbb{E} X \right)$$

$$< \exp \left(-\frac{9^2}{2+9} \mathbb{E} X \right) = \exp \left(-\frac{81}{11} \ln n \right) = n^{-81/11}.$$

By the union bound the probability that in time $\ln n$ some agent interacts as the responder more than $10 \ln n$ times is smaller than $n^{-70/11}$. Thus, for n large enough and $11\eta/35 < k$, the probability that at least half of k chunks correspond to time smaller than $\ln n$ does not exceed

$$\binom{k}{k/2} \left(n^{-70/11} \right)^{k/2} < 2^k n^{-35k/11} < (2^k n^{-k/5}) n^{-\eta} < n^{-\eta}.$$

In turn, whp we obtain time at least $\frac{k}{2} \ln n$, and $T_F \le 2T$. $\qquad\square$

We introduce a lemma for analyzing fragmented time in the new model, crucial for the examination of leader election and majority computation protocols.

Lemma 2. *Consider an interval of interactions I, s.t., $|I| > \frac{110}{35} \eta n \ln n$. If every agent acts as the responder in an external interaction in I at most once, then the fragmented parallel time of I is $\Theta(|I|/n)$ whp.*

Proof. The total number of fragmented time chunks is at least $k \ge \frac{|I|}{10n \log n}$, thus $T_F \ge \frac{|I|}{10n}$. We demonstrate that $T_F \le \frac{2|I|}{n}$ with high probability.

Consider a fixed agent in a given interaction. We first observe that the probability of an event A, that an interaction is internal and this agent acts as the responder is at most $1/n$. For an agent belonging to a group of size 1, when this agent can be counted as both the initiator and the responder, the probability of event A does not exceed $1/n$. For an agent belonging to a group of size $g > 1$ this probability is at most $\frac{g-1}{n} \cdot \frac{1}{g-1} = \frac{1}{n}$.

Let us divide all interactions into maximal subseries such that for any fixed agent there are at most $10 \ln n - 1$ events A involving this agent. Note that these subseries are simultaneously chunks of interactions in which any agent acts as the responder at most $10 \ln n$ times since any agent is a responder in I in an external interaction at most once. Let us first estimate the probability that a given subseries has less than $n \ln n$ interactions. This probability is not greater

---

[4] We utilise the Chernoff bound variant: $\Pr \left(X > (1+\delta) \mathbb{E} X \right) < \exp \left(-\delta^2 \mathbb{E} X / (1+\delta) \right)$ for $\delta > 0$.

than the probability that during $n \ln n$ interactions (counting from the beginning of the subseries) event A happens for some agent at least $10 \ln n$ times. Using calculations from Lemma 1, one can estimate that this probability for a specific agent is smaller than $n^{-81/11}$.

By the union bound the probability that in a given subseries event A occurs for some agent at least $10 \ln n$ times does not exceed $n^{-70/11}$. Analogously to the proof of Lemma 1, for sufficiently large n and $k \geq \frac{|I|}{10n \ln n} > \frac{11\eta}{35}$, the probability that at least half of the k subseries correspond to times smaller than $\ln n$ is $n^{-\eta}$.

As this occurs with negligible probability, we get time at least $\frac{2|I|}{n}$ whp.   □

Recall that if a group is a singleton, an attempt to execute pairwise interaction within this group fails. This is observed by the initiator via singleton test. Note also that such failed interactions do not affect parallelism as each failed interaction is attributed to the initiator. The next lemma enables the analysis of more complex protocols using the leader.

Lemma 3. *Consider an interval of interactions I, s.t., $|I| > \frac{60}{13}\eta n \ln n$. Assume also that every agent acts as the responder in an external interaction, which is not initiated by the leader in $|I|$ at most once, then the fragmented time of I is $\Theta(|I|/n)$ whp.*

2.3   Leader Election

In *leader election* (with confirmation) at least one agent from the initial configuration is a candidate to become the *unique leader*, and all other agents start as followers. The main goal in leader election is to distinguish and report selection of the unique leader, and to declare all other agents as followers. The state space of the leader election protocol presented below is $S = \{L, L^*, F, F^*\}$, where all initial leader and follower candidates are in states L and F, respectively. The remaining states include L^* referring to the confirmed unique leader, and F^* utilised by confirmed followers. The state space is partitioned into two groups $\mathcal{G}_0 = \{L, L^*, F^*\}$ and $\mathcal{G}_1 = \{F\}$.

LE-protocol: As at least one agent starts in state L, these agents target group $\mathcal{G}_0$ using a double rule (1) and (2) and when state L^* is eventually reached, with exactly one agent being in this state, the epidemic process defined by the transition rules (3)-(4) informs all followers about successful leader election.

$$
\begin{array}{ll}
(1)\ L + \mathcal{G}_0 | L \longrightarrow L + F & \quad (3)\ L^* + \mathcal{G}_1 | F \longrightarrow L^* + F^* \\
(2)\ L + \mathcal{G}_0 | null \longrightarrow L^* & \quad (4)\ F^* + \mathcal{G}_1 | F \longrightarrow F^* + F^*
\end{array}
$$

Lemma 4. *The fragmented time of fixed-state LE-protocol is $O(\log n)$ whp.*

2.4   Majority Computation

The state space of the majority protocol is $S = \{G, G^*, R, R^*, N\}$, and in the initial configuration each agent is either in state G or R. The main goal is to decide which subpopulation of agents in state G or R is greater than the other. If the subpopulation in state G is greater, all agents are expected to stabilise in state G^*. Otherwise, they must stabilise in state R^*. The majority protocol **M-protocol** described below uses also neutral state N. The state space is partitioned into three groups $\mathcal{G}_R = \{R, R^*\}$, $\mathcal{G}_N = \{N\}$, and $\mathcal{G}_G = \{G, G^*\}$.

M-protocol: The protocol has the following transition rules:

(1) $R + \mathcal{G}_G\|G \;\longrightarrow N + N$	(4) $G + \mathcal{G}_R\|null \longrightarrow G^*$
(2) $R + \mathcal{G}_G\|null \longrightarrow R^*$	(5) $R^* + \mathcal{G}_N\|N \;\longrightarrow R^* + R^*$
(3) $G + \mathcal{G}_R\|R \;\longrightarrow N + N$	(6) $G^* + \mathcal{G}_N\|N \;\longrightarrow G^* + G^*$

Transition rules (1) and (3) instruct agents in states R and G to become neutral for as long as pairs $R + G$ and $G + R$ can be formed. As soon as one of these states is no longer present in the population either rule (2) or (4) is used to change state R to R^* or G to G^*, respectively. In addition, either rule (5) or (6) is used to change neutral state to R^* or G^*, respectively. Alternatively, if all states G and R disappear after application of rules (1) and (3), the population stabilises in the neutral state N.

Lemma 5. *The fragmented time of fixed-state M-protocol is $O(\log n)$ whp.*

3   Computing the Median

In this section, we consider computing the median of n distinct keys, each of which is held by one of the n agents. For agents a, b belonging to the set S of agents, the relation $b < a$ denotes $\mathrm{key}(a) < \mathrm{key}(b)$. We adopt here a *comparison model* in which the transition function depends not only on the states of the agents, but also on the order of their keys. The keys are hidden and there is no other way to access them. The number of states remains fixed. A similar limited use of large keys can be found in *community protocols* in [16] to handle Byzantine failures.

For any agent $c \in S$, let $\mathbb{A}_c$ and $\mathbb{B}_c$ be the set of all agents above and below c respectively. The agent m is the unique *median* if $|\mathbb{B}_m| - |\mathbb{A}_m| = 0$, for odd n, or one of the two medians if $||\mathbb{B}_m| - |\mathbb{A}_m|| = 1$, for even n. In this version we assume that all keys are different and n is odd. The arbitrary case requires minor amendments, as the answer may refer to two agents. Before we consider selective protocols, we first consider median computation in comparison model with a standard random scheduler.

3.1   Median Computation in Standard Model

Theorem 1. *Finding the median in the comparison model requires $\Omega(n)$ time in expectation.*

Proof. Assume that n is odd and agents a_1 and a_2 share between themselves the median and the key immediately succeeding it (in the total order of keys). One can observe that before the first interaction between a_1 and a_2, all consecutive configurations of states of all agents are independent from whether a_1 or a_2 is associated with the median. Thus before the first interaction between a_1 and a_2 no algorithm can declare either of them as the median. And since the expected number of interactions preceding the first interaction between a_1 and a_2 is $\Omega(n^2)$, the thesis of the theorem follows.   □

Now we formulate an almost optimal median population protocol in the adopted model. All agents start this protocol in neutral state N. In due course, agents change their states, s.t., eventually all agents associated with keys smaller than the median end up in state B, those with keys greater than the median in state A, and the median conclude in state N. The median protocol uses the following symmetric transition function:

(1) $N + N \xrightarrow{\leq} B + A \triangleright$ initialisation	(3) $A + N \xrightarrow{\leq} N + A \triangleright$ fix order
(2) $A + B \xrightarrow{\leq} B + A \quad \triangleright$ fix order	(4) $N + B \xrightarrow{\leq} B + N \triangleright$ fix order

Note that there is always the same number of agents in states B and A and one agent will remain in state N as n is an odd number.

Theorem 2. *The fixed-state median protocol operates in $O(n \log n)$ time both in expectation and whp.*

Proof. We say that a pair of agents a and b is *disordered* if an interaction between a and b is meaningful. We define the *disorder* $d(C)$ of a configuration C as the total number of disordered pairs of agents in this configuration. Since all agents start in state N, any initial interaction is meaningful, via application of rule (1). And in turn the disorder of the initial configuration is $d(C_0) = \binom{n}{2}$. In the final configuration C_∞, when the agent with the median key is in state N, and all agents with smaller and larger (than the median) keys are in states B and A respectively, the disorder $d(C_\infty) = 0$, as none of the rules can be applied.

Proposition 1. *Any meaningful interaction reduces the disorder of a configuration.*

The probability of making a meaningful interaction in a configuration C, s.t., $d(C) = i$ is $p_i = i/\binom{n}{2}$. Let the random variable T_i be the number of interactions needed to observe a meaningful interaction for a configuration C when $d(C) = i$. We have $\mathbb{E}[T_i] = \binom{n}{2}/i$. The expected number of interactions $\mathbb{E}[T]$ to transition from C_0 to C_∞ is

$$\mathbb{E}[T] \leq \mathbb{E}[T_1] + \mathbb{E}[T_2] + \cdots + \mathbb{E}\left[T_{\binom{n}{2}}\right] = \binom{n}{2} H_{\binom{n}{2}} = O(n^2 \log n).$$

One can also prove that $T = O(n^2 \log n)$ whp applying Janson's bound. However, we show here an alternative proof by a potential function argument.

In particular, we show that for two subsequent configurations C and C' we have $\mathbb{E}[d(C')] \leq \left(1 - \frac{2}{n^2}\right) d(C)$. This inequality becomes trivial when $d(C) = 0$. Otherwise, when $d(C) \neq 0$,

$$\mathbb{E}[d(C')] \leq \left(1 - \frac{2d(C)}{n^2}\right) d(C) + \frac{2d(C)}{n^2} \left(d(C) - 1\right) = \left(1 - \frac{2}{n^2}\right) d(C).$$

After t interactions beyond configuration C_0 we get $\mathbb{E}[d(C_t)] \leq \left(1 - \frac{2}{n^2}\right)^t d(C_0) \leq \exp\left(-\frac{2t}{n^2}\right) \binom{n}{2}$. We obtain $\mathbb{E}[d(C_t)] < n^{-\eta}$ for $t \geq \binom{n}{2} \left(\ln \frac{n^2}{2} + \ln \eta\right)$. Finally, when $\mathbb{E}[d(C_t)] < n^{-\eta}$, by Markov's inequality, we get $\Pr(d(C_t) \geq 1) < n^{-\eta}$. This is equivalent to $\Pr(C_t \neq C_\infty) < n^{-\eta}$. □

3.2 Fast Median Computation

We present and analyse here efficient median computation in selective population protocols. The proposed solution is done by breaking up the full protocol into smaller blocks implemented as independent stabilisation processes with clearly defined inputs and outputs, as well as efficient and stable solutions. Each of these independent processes (see Fast-median algorithm below) including leader (pivot) election, partitioning agents wrt the key of the pivot, and majority computation, is executed on distinct partitions of states. Recall from Sect. 2.1 that the leader elected in the beginning of the computation process executes the code of the solution embedded in the transition function, and manages all input/output operations.

Algorithm 1 Fast-median.

Input: S – all agents set, $C = S$ – median candidate set
1: Select randomly leader (as pivot) $p \in S$ ▷ leader election
2: **repeat**
3:      Partition S to $\mathbb{B}_p = \{x \in S : x < p\}, \mathbb{A}_p = \{x \in S : x > p\}, \{p\}$;
4:      **switch**   ▷ majority computation
5:          **case** $(|\mathbb{B}_p| > |\mathbb{A}_p|) \longrightarrow C = C \cap \mathbb{B}_p$
6:          **case** $(|\mathbb{B}_p| < |\mathbb{A}_p|) \longrightarrow C = C \cap \mathbb{A}_p$
7:          **case** $(|\mathbb{B}_p| = |\mathbb{A}_p|) \longrightarrow C = \{p\}$
8:      $p \leftarrow$ randomly chosen (by current p) agent in C ▷ leader hand over
9: **until** $(|\mathbb{B}_p| = |\mathbb{A}_p|)$
10: **return** p ▷ result announcement

Recall that leader election and majority computation, were discussed in Sects. 2.3 and 2.4, respectively. Thus the focus in this section is on efficient partitioning of all agents in S to $\mathbb{B}_p = \{x \in S : x < p\}, \mathbb{A}_p = \{x \in S : x > p\}$, and $\{p\}$. Note that such partitioning is not trivial as due to the restrictions in the model the pivot p cannot distribute the value of its key to all agents in the population. Instead, the agents gradually learn their relationship with respect to the pivot by comparing their keys with other agents.

Theorem 3. *Fixed-state* Fast-median *protocol stabilises in fragmented parallel time* $O(\log^4 n)$ *whp.*

Proof. The proof is located at the end of Sect. 3.2.   □

Partitioning via Coloring We commence by providing an overview of the argument. The partitioning of all agents occurs in multiple phases, represented by consecutive stabilization processes. The objective in each phase is to correctly partition a constant fraction of agents that have not been partitioned yet. Each phase has a time complexity of $O(\log^2 n)$. Since partitioning requires $O(\log n)$ phases, the overall time complexity becomes $O(\log^3 n)$.

In the median protocol, we execute $O(\log n)$ partitioning steps, which constitute the majority of the computation time. Consequently, the total time complexity for computing the median is $O(\log^4 n)$. To analyze a single phase of partitioning, we categorize the set of uncolored agents above the pivot into $2 \log n$ buckets, and similarly for those below the pivot. We then demonstrate that within $O(\log^2 n)$ time, the algorithm successfully colors $\log n$ agents from the first bucket. In each successive time period indexed by $i = 2, 3, \ldots$, with a duration of $O(\log n)$, the algorithm colors $2^{i-1} \log n$ agents in the i-th bucket. Consequently, after $O(\log^2 n)$ time from the initiation of a phase, a constant fraction of uncolored agents acquires colors.

The input for the partitioning process consists of the leader agent in the pivot state P and all other agents in the state N_{in}. We utilize two groups of states: $\mathcal{G} = \{P, B_0, \ldots, B_{21}, A_0, \ldots, A_{21}, N\}$ and $\mathcal{G}_{in} = \{N_{in}\}$. We interpret states A_t, B_t, and N as above, below, and neutral colors, respectively. An agent adopts state A_t (B_t) as soon as it learns that its key is above (below) the key of the pivot.

During each phase, we color approximately a fraction of $1/22$ of yet uncolored agents. Upon the conclusion of the phase, these agents are moved to a different group, i.e., they do not participate in the partitioning of uncolored agents in the remaining phases.

In order to limit the activity of colored agents and, in turn, the duration of each phase, we introduce the concept of *tickets*. While the pivot has an unlimited number of tickets, any newly colored agent receives a fixed pool of 21 tickets. For as long as any colored agent has tickets, it targets agents in group $\mathcal{G}_{in}$ trying to color them. During such an interaction, a colored agent loses one ticket and moves one agent from $\mathcal{G}_{in}$ to $\mathcal{G}$. Once a colored agent loses all its tickets, it starts targeting group $\mathcal{G}$. The (partial) coloring phase concludes when the group $\mathcal{G}_{in}$ becomes empty. The set of relevant rules is given below.

(1) $P + \mathcal{G}_{in}\|N_{in} \xrightarrow{<} P + A_{21}$	(3) $A_{t>0} + \mathcal{G}_{in}\|N_{in} \xrightarrow{<} A_{t-1} + A_{21}$	
(1) $P + \mathcal{G}_{in}\|N_{in} \xrightarrow{>} P + B_{21}$	(3) $A_0 + \mathcal{G}\|N \xrightarrow{<} A_0 + A_{21}$	
(2) $B_{t>0} + \mathcal{G}_{in}\|N_{in} \xrightarrow{<} B_{t-1} + N$	(4) $N + \mathcal{G}\|P \xrightarrow{<} B_{21} + P$	
(2) $B_{t>0} + \mathcal{G}_{in}\|N_{in} \xrightarrow{>} B_{t-1} + B_{21}$	(4) $N + \mathcal{G}\|P \xrightarrow{>} A_{21} + P$	
(2) $B_0 + \mathcal{G}\|N \xrightarrow{>} B_0 + B_{21}$	(4) $N + \mathcal{G}\|B_t \xrightarrow{<} B_{21} + B_t$	
(3) $A_{t>0} + \mathcal{G}_{in}\|N_{in} \xrightarrow{>} A_{t-1} + N$	(4) $N + \mathcal{G}\|A_t \xrightarrow{>} A_{21} + A_t$	

We formulate here a tail bound that works for hypergeometric sequences for the case of small fraction p of black balls. We need this more sensitive bound in some of our proofs.

Lemma 6. *Assume we have an urn with n balls where pn of them are black. Let X_i be a binary random variable equal to one iff in the ith draw without replacement the drawn ball was black, and let $X = \sum_{i=1}^{\kappa} X_i$. Then, for $\kappa \leq n$ and $0 < \delta < 1$, we get $\Pr\left(X < (1-\delta)p\kappa - p\right) < \kappa \exp\left(\frac{-\delta^2 p\kappa}{2}\right)$.*

Denote the number of agents participating (not previously colored) in the phase by m. For any interaction t, let $k(t)$ be a number of agents in group $\mathcal{G}$ in t and $Inf_x(t)$ be a number of informed agents which are in states A or B in bucket x. The sequence of useful technical lemmas leading to the thesis of Theorem 4 follows.

Lemma 7. *If during interaction t the number of colored agents is $r \geq \log^2 n$, then $k(t + 500n) \geq \min\{20r, m\}$ whp.*

Lemma 8. *The fragmented time of one phase of partitioning by coloring is $O(\log^2 n)$ whp.*

Lemma 9. *There is a constant $c > 0$, such that if $|[t_0, t_1)| = cn \log^2 n$ and $|[t_{i-1}, t_i)| = cn \log n$ for all $i > 1$, then whp*

- $k(t_i) \geq \min\left\{20 \cdot 2^{i-1} \log^2 n, m\right\}$,
- $Inf_i(t_i) > \min\left\{2^{i-1} \log n, \frac{m}{10 \log n}\right\}$.

The phase ends when group $\mathcal{G}_{in}$ becomes empty. Each agent is relocated from $\mathcal{G}_{in}$ to $\mathcal{G}$ only if it either gets properly colored or it was given a ticket. Since each colored agent has only 21 tickets to utilise, we can formulate the following fact.

Fact 1. *After single coloring phase a fraction of $\frac{1}{22}$ uncolored agents gets colored.*

Thus, after $O(\log n)$ iterations of the coloring phase all agents are properly colored. This leads to the following theorem.

Theorem 4. *Partitioning by coloring stabilises in $O(\log^3 n)$ fragmented time.*

We conclude with the proof of Theorem 3.

Proof (of Theorem 3). The structure of the solution replicates the logic of a standard median computation protocol. Thus, the correctness of the solution follows from the correctness of the individual routines including leader election, majority computation and partitioning.

Concerning the time complexity, leader election and majority computation are implemented in fragmented parallel time $O(\log n)$ whp, see Lemmas 4 and 5. By Theorem 4, each partitioning stage takes $O(\log^3 n)$ time whp, and with probability $\frac{1}{2}$ at most $\frac{3}{4}$ candidates remain in C. Thus, with high probability after at most $O(\log n)$ iterations of this routine set C is reduced to a singleton containing the median.

Finally, Fast-median protocol stabilises in $O(\log^4 n)$ parallel time. □

4   Further Discussion

Suitable Programming Environment It is a natural solution to articulate solutions in any computational model using pseudocode. Such representation enhances the readability and understanding of the proposed solution within the context of the main features of the underlying computational model. Subsequently, this aids in conducting rigorous mathematical analysis. Various pseudocodes have been explored in the past to address challenges in population protocols, encompassing simple protocols [19], separation bounds [10], and leader-based computation [6]. The latter work forms the basis for our approach, which here focuses on the development of efficient parallel protocols. Our objective is to champion selective population protocols, enabling the development of simpler and more structured efficient solutions presented at higher programming level. The primary reasons for advocating this approach stem from the partitioning of the state space. Each partition represents the local variables of an independent process, supported by conditional interactions that also facilitate independent interaction availability (zero) tests. This, in turn, eliminates the necessity for a global clocking mechanism through the application of event-based distributed computation. For further detail consult the full version of this paper [14].

Final Comment We would like to postulate that the efficiency of selective protocols stand out when tackling problems that demand more extensive memory utilization and yield intricate outputs. A good example is the *ranking problem*, requiring assignment of unique labels from the range 1 to n to all agents, examined recently in the context of leader election in self-stabilizing protocols [9], and related sorting problem in the constructors model [15]. For these two problems, currently, no efficient solutions based on a polynomial number of states are known in standard population protocols. We would like to assert the following.

Conjecture 1. *Any efficient solution to the sorting problem necessitates exponential state space in standard population protocols.*

On the contrary, the evidence presented in the full version [14] demonstrates that selective protocols can efficiently solve sorting by ranking using a much smaller number of states. Specifically, we present transition rules of an efficient quick-sort-like, selective sorting by ranking. This algorithm has polynomial in n state space, utilises $O(n)$ partitions and stabilises in time $O(\log^2 n)$.

References

1. Alistarh, D., Aspnes, J., Eisenstat, D., Gelashvili, R., Rivest, R.: Time-space trade-offs in population protocols. In: Proceedings of SODA, pp. 2560–2579 (2017)
2. Alistarh, D., Aspnes, J., Gelashvili, R.: Space-optimal majority in population protocols. In: Proceedings of SODA, pp. 2221–2239 (2018)
3. Alistarh, D., Gelashvili, R., Rybicki, J.: Fast graphical population protocols. In: Proceedings of OPODIS, vol. 217, pp. 14:1–14:18 (2021)

4. Alistarh, D., Rybicki, J., Voitovych, S.: Near-optimal leader election in population protocols on graphs. In: PODC, pp 246–256 (2022)
5. Angluin, D., Aspnes, J., Diamadi, Z., Fischer, M., Peralta, R.: Computation in networks of passively mobile finite-state sensors. In: Proceedings of PODC, pp. 290–299 (2004)
6. Angluin, D., Aspnes, J., Eisenstat, D.: Fast computation by population protocols with a leader. Distributed Comput. 21(3), 183–199 (2008)
7. Bankhamer, G., Berenbrink, P., Biermeier, F., Elsässer, R., Hosseinpour, H., Kaaser, D., Kling, P.: Population protocols for exact plurality consensus: how a small chance of failure helps to eliminate insignificant opinions. In: PODC, pp. 224–234 (2022)
8. Berenbrink, P., Giakkoupis, G., Kling, P.: Optimal time and space leader election in population protocols. In: STOC, pp. 119–129 (2020)
9. Burman, J., Chen, H., Chen, H., Doty, D., Nowak, T., Severson, E., Xu, C.: Time-optimal self-stabilizing leader election in population protocols. In: Proceedings of PODC, pp. 33–44 (2021)
10. Czerner, P.: Brief announcement: population protocols decide double-exponential thresholds. In: Proceedings of PODC, pp. 28–31 (2023)
11. Czumaj, A., Lingas, A.: On parallel time in population protocols. Inf. Process. Lett. 179, 106314 (2023)
12. Doty, D., Eftekhari, M., Gąsieniec, L., Severson, E., Uznanski, P., Stachowiak, G.: A time and space optimal stable population protocol solving exact majority. In: FOCS, pp. 1044–1055 (2021)
13. Gąsieniec, L., Stachowiak, G.: Enhanced phase clocks, population protocols, and fast space optimal leader election. J. ACM 68(1), 2:1–2:21 (2021)
14. Gańczorz, A., Gąsieniec, L., Jurdziński, T., Kowalski, J., Stachowiak, G.: Selective Population Protocols. arXiv:abs/1906.04238 (2024)
15. Gąsieniec, L., Spirakis, P., Stachowiak, G.: New clocks, optimal line formation and self-replication population protocols. In: Proceedings of STACS, pp. 33:1–33:22 (2023)
16. Guerraoui, R., Ruppert, E.: Names trump malice: tiny mobile agents can tolerate byzantine failures. In: Proceedings of ICALP'09, pp. 484–495 (2009)
17. Kennedy, J.: Thinking is social: experiments with the adaptive culture model. J. Conflict Resolut. 42(1), 56–76 (1998)
18. Kleinberg, J.: The small-world phenomenon: an algorithmic perspective. In: Proceedings of STOC, pp. 163–170 (2000)
19. Kosowski, A., Uznański, P.: Brief announcement: population protocols are fast. In: Proceedings of PODC, pp. 475–477 (2018)
20. Michail, O., Spirakis, P.: Network constructors: a model for programmable matter. In: Proceedings of SOFSEM, pp. 15–34 (2017)
21. Minsky, M.: Computation: Finite and Infinite Machines. Prentice-Hall (1967)
22. Neary, T., Woods, D.: Four small universal Turing machines. Fundam. Informaticae 91(1), 123–144 (2009)
23. Soloveichik, D., Cook, M., Winfree, E., Bruck, J.: Computation with finite stochastic chemical reaction networks. Nat. Comput. 7(4), 615–633 (2008)

Partially Disjoint Shortest Paths and Near-Shortest Paths Trees

Yefim Dinitz[1], Shlomi Dolev[1]($\boxtimes$), Manish Kumar[2]($\boxtimes$), and Baruch Schieber[2]($\boxtimes$)

[1] Ben-Gurion University of the Negev, Beer Sheva, Israel
dinitz@cs.bgu.ac.il, dolev@cs.bgu.ac.il
[2] New Jersey Institute of Technology, Newark, NJ, USA
manish.kumar@njit.edu, sbar@njit.edu

Abstract. One of the ways to increase communication reliability is by sending k duplicate messages along different routes. This gives rise to the problem of finding k shortest paths between a given source and destination. An unconstrained solution of the k shortest paths problem may output paths that overlap in almost all edges. Clearly, using such paths will have an adverse impact on the communication reliability. On the other extreme, a solution of k independent shortest paths, which are paths that share neither an edge nor an intermediate node may not be realistic for several reasons: such paths may not exist, if they exist they may be very long compared to the shortest path, and the computational effort of finding such paths may be prohibitive. This motivated us to investigate the intermediate case in which the number of edges that are not shared among any two paths in the output k paths is parameterized. We explore both *exactly shortest paths* and *near-shortest paths*. Our results are also generalized to the case of multi-criteria prioritized weights.

Next, we consider the related albeit different problem of computing the k shortest paths trees, which are the k spanning trees with minimum total path length. This problem was introduced by Sedeño-Noda and González-Martín (2010). They solved it using a greedy algorithm and proved its correctness using linear programming theory. We provide an alternative, combinatorial and simpler proof of the correctness of the same greedy algorithm. We believe that the combinatorial approach can lead to a better understanding and possible extensions of the related results.

1 Introduction

Optimizing the cost of paths is a fundamental task in Computer Science and Operations Research. In many scenarios, there is a need to compute the second best or in general, the k best alternatives to the optimal (shortest) path. The variety in the obtained k-set of best solutions can facilitate a choice of solutions due to other considerations (e.g., preferences of geographic locations or communication channels) among the close to optimal (or allowed budget) solutions.

T. Masuzawa et al. (Eds.): SSS 2024, LNCS 14931, pp. 240–254, 2025.
https://doi.org/10.1007/978-3-031-74498-3_17

In some cases, the usage of all (or several) solutions from the set is preferable in order to allow the diversity of routing patterns while still being close to the optimal solution.

One of the applications of finding k shortest paths is in communication networks. Communication reliability can be obtained by sending k duplicate messages along different routes. This gives rise to the problem of finding k shortest paths between a given source and destination. An unconstrained solution of the k shortest paths problem may output paths that overlap in almost all edges. Clearly, using such paths will have an adverse impact on the communication reliability. On the other extreme, a solution of k independent shortest paths, which are paths that share neither an edge nor an intermediate node may not be realistic for several reasons: such paths may not exist, if they exist they may be very long compared to the shortest path, and the computational effort of finding such paths may be prohibitive as detailed below. One way to trade off the two extremes is to parameterize the number of edges that are not shared among any two paths in the output k paths. This is one variant of a class of problems we call *partially independent shortest paths*.

Another example of the applicability of partially independent shortest paths is in genome exploration evaluation (e.g., for viruses variants). There, the distance relation is defined as the probability of possible changes in the genome. The k near-optimal solutions enable the tracing of the most probable paths (routes) from one virus variant to another or from one virus variant to many others.

Partially independent paths are applicable in a scenario where multiple objects or robots want to move (almost) collision-free [10,14]. More generally, partially independent paths are applicable in multi-agent systems such as object transportation [17,21], search and rescue [15], robot path reconfiguration [11,18], tasks spanning assemble [13], evacuation [20], and formation control [23].

The complexity of computing partially independent paths is an interesting theoretical question as it considers the relaxation of the independence constraint. On the one end, finding k independent shortest paths [9] is a challenging computational task, where polynomial algorithms exist only for very limited cases. On the other end, k 1-edge independent shortest paths (i.e., paths that differ by at least one edge) can be computed in polynomial time. The increase from 1-edge independence to the entire path independence changes the complexity of the problem dramatically, and thus it is interesting to analyze this transition.

1.1   Preliminaries

We study variations of the k *partially independent shortest paths*. The input to our problem consists of either directed or undirected weighted graph $G(V, E, w)$ with a length $w(e)$ associated with each edge $e \in E$, and an integer k. We assume that the graph has no parallel edges and no self-loops. A (simple) path p connecting two terminals $s, t \in V$ is a sequence of vertices: $s = v_0, v_1, \ldots, v_r = t$ such that all v_i's are distinct and $(v_i, v_{i+1}) \in E$, for $i \in [0, 1, \ldots, r - 1]$. Let $E(p)$ denote the set of edges in the path p, and $w(p)$ denote the length (weight) of path p, that is, $w(p) = \sum_{e \in E(p)} w(e)$. We are also given some independence measure as discussed below. Given this measure, the goal is to find k partially independent shortest paths connecting s to t. (This can be extended naturally to the multiple terminal pairs case.)

The independence measure of shortest paths can be defined in several ways. In this paper, we view a path as a set of edges (or nodes) and consider the size of either the set difference or the symmetric difference of two paths as their independence measure.

As a warm-up, consider the 2 partially independent shortest paths problem and let the size of the set difference between the edge sets of the two paths be the independence measure of those paths. By the pigeonhole principle, any collection of $m + 1$ paths must include a pair of paths whose set difference is at least 2 (even assuming parallel edges). Such $m + 1$ near-shortest paths can be found using Yen's algorithm [24] that computes paths in the order of their length. This yields a solution of the 2 partially independent shortest-path problem for this variant of the independence measure. (Below, we give additional solutions for this independence measure.)

1.2   Our Results

In this paper, we consider two variants of the partially independent paths problem. Section 2 considers the case in which only strictly shortest paths qualify (and thus all qualified paths should be equal in length). Given a subset of the edges, that we call *sensitive edges*, we show how to compute the set of shortest paths of maximum cardinality in which no pair of paths share a sensitive edge. Then, we relax the disjointness constraint and show how to find a set of r shortest paths that cause a load of at most 2 on the sensitive edges. Lastly, we show how to find a set of r shortest paths that minimizes the maximum load over all sensitive edges. All these results can be extended to *sensitive nodes*. Section 3 considers the case in which the output may include near-shortest paths and the independence measure is the size of the symmetric difference between either the edges or the nodes of any pair of paths. We show that for small values of independence measure (up to 6 in case of edges and 4 in case of vertices) it is enough to consider a small number of shortest paths (in order of their lengths) to find two near-shortest paths with the required independence measure. On the other hand, we show that such approach cannot scale by proving that for any natural number $\bar{q}$, there are graphs where for any $q \leq \bar{q}$ there are exponential (in q) number of near-shortest paths whose pairwise distance measure is q. In Sect. 4 we consider the related albeit different problem of computing the k shortest paths trees, which are the k spanning trees with minimum total path length. This problem was introduced by Sedeño-Noda and González-Martín [22]. They solved the problem using a greedy algorithm and proved its correctness using linear programming theory. We provide an alternative, combinatorial and simpler proof of the correctness of the same greedy algorithm. We believe that this alternative approach can lead to a better understanding and possible extensions of the related results.

Note that all our results can be generalized to the multi-criteria prioritized weights case. That is, in the case where there are several weights per edge and any arbitrarily small amount of weight i is more important than an arbitrarily big amount of weight j, for any $i < j$. This is possible by the reduction of this case to the case of a single weight provided in [8].

1.3   Related Work

One of the related directions in the literature is studying optimal sets of paths, where a bounded number of *shared* vertices or edges is allowed There, a set of paths is called optimal (shortest) if the *sum of the path lengths* is minimum possible, where the length of a path is defined as the sum of the weights of all its contained edges. Observe that this objective does not imply any guarantee on the quality of each path. Guo et al. [12] studied the problem of finding the shortest set of k paths that are edge-disjoint and partially vertex disjoint. They considered the *δ-vertex k edge-disjoint shortest path ($\delta V - kEDSP$)* problem, where at most δ vertices (besides s and t) are shared by at least any two paths. As mentioned there, the $0V - kEDSP$ problem can be solved simply via min-cost max-flow. For $k = 2$ and any positive δ, they solve the $\delta V - kEDSP$ problem in time $O(\delta m + n \log n)$, for a graph $G(V, E, w)$ with $n = |V|$ and $m = |E|$. For general k, the problem is still open. Similarly, the k partially edge-disjoint path problem ($kPEDSP$) computes the shortest set of k paths connecting s and t such that at most δ edges are shared by any pair of paths. For $k = 2$, Yunyun et al. [6] introduced an exact algorithm with a runtime $O(mn \log_{(1+m/n)} n + \delta n^2)$. In the above works, $\delta > 0$ was referred to as the *disjointness* factor.

Chondrogiannis et al. [4] introduced the k Shortest Paths with Limited Overlap ($kSPwLO$) problem seeking to find k alternative paths which are (a) as short as possible and (b) sufficiently dissimilar based on a user-controlled similarity threshold. Given a set of simple paths P from a source s to destination t in an edge weighted graph $G(V, E, w)$, Chondrogiannis et al. called a path $p(s \rightarrow t)$ an *alternative path* to P if p is sufficiently dissimilar to every path $p' \in P$. The similarity of two simple paths p and p' is determined by their overlap ratio:

$$Sim(p, p') = \frac{\sum_{(x,y) \in p \cap p'} w_{xy}}{w(p')},$$

where $w(p)$ is the length of a path p, and $p_i \cap p'$ denotes the set of edges shared by p and p. The range of overlap ratio is $0 \leq Sim(p, p') \leq 1$, where $Sim(p, p') = 0$ holds if p shares no edge with p' and $Sim(p, p') = 1$ if $p = p'$. Chondrogiannis et al. introduced two algorithms. The first is a baseline algorithm based on Yen's algorithm [24]. The second algorithm, *OnePass algorithm*, considers the overlap constraint in each expansion step while traversing the network.

In another work by Chondrogiannis et al. [5] they considered the same similarity constraint and introduced the *MultiPass (exact) algorithm* that traverses the network $k - 1$ times and employs pruning criteria to reduce the number of potential alternative paths; the criteria are same as in the *OnePass algorithm*.

The MultiPass algorithm is based on the following observation. Let P be the given set of paths from source s to destination t, and p_i, p_j be two paths from source s to some intermediate node t'. If $w(p_i) < w(p_j)$ and $\forall p \in P\ Sim(p_i, p) \leq Sim(p_j, p)$ holds, then the path p_j cannot be the prefix of any of the shortest alternative paths to P. It takes $O(m + K \cdot n \cdot \log n)$ time, where K ($K >> k$) is the number of shortest paths that have to be computed in order to cover the k results of the kSPwLO query. In comparison to the OnePass algorithm, the MultiPass algorithm may have to construct all paths from s to t, which results in higher time complexity, but experimental evaluation showed that MultiPass is much faster than OnePass.

Below, we list a sample of previous results on the disjoint paths problem. Given a set of k pairs of terminals in graph, the existence of k vertex-disjoint paths connecting each pair of terminals can be determined in $O(|V|^3)$ time, for any fixed k, as shown by Robertson and Seymour [19] in their graph minor project. The problem of two disjoint shortest paths was first considered by Eilam-Tzoreff [9]. Eilam-Tzoreff provided a polynomial-time algorithm for $k = 2$, based on a dynamic programming approach for the weighted undirected vertex-disjoint case. This algorithm has a running time of $O(|V|^8)$. Later, Akhmedov [1] improved the algorithm of Eilam-Tzoreff, and achieved an algorithm whose running time is $O(|V|^6)$ for the unit-length case of the 2-Disjoint Shortest Path and $O(|V|^7)$ for the weighted case of the 2-disjoint shortest path. In both cases, Akhmedov [1] considered the undirected vertex disjoint shortest path. In recent past, Bentert et al. [2] improved the result of Akhmedov [1]. In other work of Bérczi et al. [3] they showed that the undirected k-DSPP (disjoint shortest paths problem) and the vertex-disjoint version of the directed k-DSPP can be solved in polynomial time if the input graph is planar and k is a fixed constant. Lochet [16] shows that for any fixed k, the disjoint shortest paths problem admits a slicewise polynomial time algorithm.

2 Partially Edge-Disjoint Exact Shortest Paths

In this section we study variants of finding partially edge-disjoint paths among the (exact) *shortest* paths. The restricted scope allows for achieving new interesting results. Our main tool is the *subgraph of shortest paths* $\tilde{G}$ introduced in [8, Sect. 4].[1] For any graph $G = (V, E, w)$ and its nodes s and t, its subgraph $\tilde{G} = \tilde{G}(s,t)$ is composed of the nodes and edges of all shortest paths from s to t, keeping their weights. If G is undirected, then its edges are directed along the shortest path(s) going through them, so the subgraph of shortest paths $\tilde{G}$ is always directed. Its main properties are as follows. We denote by $d(u,v)$ the distance (=the length of the shortest path) from node u to node v.

---

[1] This subgraph may be considered as a generalization of the layered network introduced in [7]. A layered network $L = L(G)$ is a subgraph of the given unweighted graph G, such that the set of *all shortest paths* from s to t in G coincides with the set of *all paths* from s to t in L.

- Graph $\tilde{G}$ is acyclic. For any node u of $\tilde{G}$, $d(s,u) + d(u,t) = d(s,t)$. For any edge (u,v) of $\tilde{G}$, $d(s,u) + w(u,v) + d(v,t) = d(s,t)$.
- A path from s to t in G is shortest if and only if it belongs to $\tilde{G}$.
- Any path in $\tilde{G}$ is shortest between its end-nodes in G.
- If v is reachable from u in $\tilde{G}$, then all shortest paths from u to v in G are contained in $\tilde{G}$.

Since we need only the subgraph of shortest paths for studying the shortest paths from s to t, we *assume* $G = \tilde{G}$ in the follows, for simplicity of notation.

First, assume that our goal is to find the maximal number of disjoint shortest paths. Let us build flow network $N_1 = (G, s, t, u_1)$ by assigning capacity $u_1(e) = 1$ to each edge e of G. The following statement is classic.

Theorem 1. *The maximal number of disjoint shortest paths is equal to the size of the maximal flow f_{max} in N_1. The set of such paths can then be found by the flow decomposition of f_{max}.*

Our new results generalize this theorem to the setting where a set of *sensitive edges* $S \subseteq E$ is distinguished in G, and the independence requirement applies only to the sensitive edges. Let us define flow network $N_2 = (G, s, t, u_2)$ by assigning capacity $u_2(e) = 1$ to each edge $e \in S$ and $u_2(e) = \infty$ to each other edge of G. The proof of the following theorem is straightforward.

Theorem 2. *The maximal number of shortest paths disjoint at the edges in S is equal to the size of the maximal flow f_{max} in N_2. The set of such paths can then be found by the flow decomposition of f_{max}.*

Recall that the case of node capacities can be easily reduced to that of edge capacities. Therefore, this and the following statements related to S can be extended to the case of subset $S \subseteq V$ of *sensitive nodes* in G.

Next, we relax the independence constraint that requires that any edge of S may be included in at most *one path*, and instead require that any edge of S may be included in at most *two paths*. We say that the *load* of an edge e caused by a set of (simple) paths is the number of paths in the set that include the edge e. Using this terminology the relaxed constrained is that the load of any edge of S caused by the set of shortest paths is at most 2.

The objective is to find a set of r shortest paths that cause a load of at most 2 on all edges of S and minimizes the number of edges of S with load 2. Note that the problem may be infeasible in case there is no set of r shortest paths that cause a load of at most 2 on all edges of S. Let us define a flow network with edge-costs $N_3 = (G, s, t, u_3, c_3)$ by taking N_2, assigning edge cost zero to all its edges, and for any edge $(u,v) \in S$, adding a new node x_2 and a pair of edges (u,x_2) and (x_2,v) of capacity 1 and of cost 1. The following theorem follows from the definition of N_3.

Theorem 3. *A set of r shortest paths that cause a load of at most 2 on all edges of S and minimizes the number of edges of S with load 2 can be computed by finding the min-cost flow $f_{mincost}$ of size r in N_3 and applying to it the flow decomposition. If there is no flow of size r, then the problem is infeasible.*

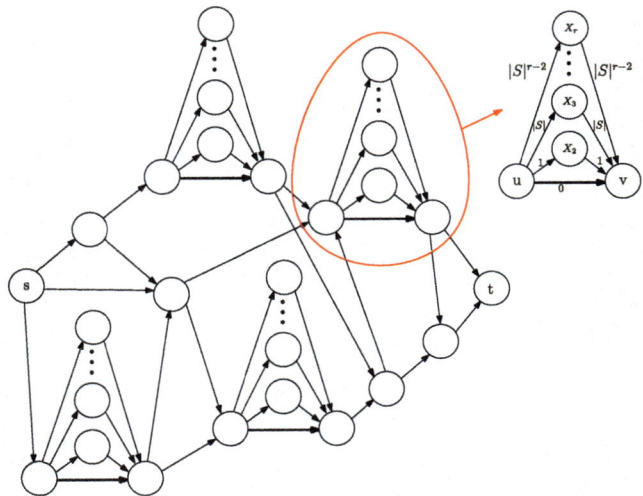

Fig. 1. The construction of Theorem 4. The thick edges belong to S. In each gadget, all edge capacities are 1, and edge costs are as shown in the zoomed copy.

Next, we remove the constraint on the maximum load of any edge of S, and instead consider the objective of minimizing the maximum load caused by the set of shortest oaths over all sensitive edges. Note that the problem may be infeasible in case there is no set of r shortest paths.

Let us define flow network $N_4 = (G, s, t, u_4, c_4)$ by taking N_3, and for any edge $(u, v) \in S$ and any $i : 3 \leq i \leq r$, adding a new node x_i and a pair of edges (u, x_i) and (x_i, v) of capacity 1 and of cost $|S|^{i-2}$. See Fig. 1 for illustration.

Theorem 4. *The set of r shortest paths that minimizes the maximum load caused by it over all sensitive edges can be computed by finding the min-cost flow $f_{mincost}$ of size r in N_4 and applying to it the flow decomposition. If there is no flow of size r, then the problem is infeasible.*

Note that all basic flow problems mentioned in this section are known as solvable in a polynomial time. So, all the above results can also be implemented in a polynomial time.

3   Partial Independence Among Near-Shortest Paths

In this section we study variants of finding partially edge-disjoint paths among the *near-shortest* paths. We consider the following approach: *generating several near-shortest paths in the order from best to worst (e.g., using Yen's algorithm [24]), and finding a pair of paths with the highest independence measure among them.* The independence measure that we consider is the size of the symmetric difference between either the edges or the nodes of the two paths. Our positive results are for small values of independence, and show that in this case the number of near-shortest paths that need to be generated is small. The proof of the following theorem is postponed to the full version due to space constraints.

Theorem 5. *For any graph with nodes s and t:*

1. *The first and the second near-shortest paths from s to t have at least three different edges and at least one different node.*
2. *Among the first three near-shortest paths from s to t, there are two paths with at least four different edges and at least two different nodes.*
3. *(conjectured) Among the first $O(n)$ near-shortest paths, there are two paths with at least six different edges and at least four different nodes.*

Proof (sketch). We assume that there are no loops and multiple edges in the graph. Note that any two distinct paths from s to t coincide up to a node u and continue by two disjoint paths p_1 and p_2 up to a common node v. As a minimum, w.l.o.g., $|p_1| = 1$ and $|p_2| = 2$; we denote this by *1,2-case*. This proves item 1.

In any non-minimum case, either $|p_1|, |p_2| \geq 2$ or $|p_1| \geq 1$ and $|p_2| \geq 3$. That is, we are done with item 2 unless the first three near-shortest paths, P_1, P_2, P_3, form 1,2-cases in all pairs. In the remaining situation, if both P_1, P_2 and P_1, P_3 are 1,2-pairs, item 2 holds either if the same edge or different edges in P_1 are replaced by two-edge paths. Otherwise, w.l.o.g., both P_3, P_1 and P_1, P_2 are 1,2-pairs. Then, simple counting shows that item 2 holds if the single replaced edge in P_1 w.r.t. pair P_1, P_2 either lies or does not lie on the replacing two-edge path in P_1 w.r.t. pair P_3, P_1.

For the last item, we still do not have a full proof.   □

On the other hand, we show that the effectiveness of the considered approach is bounded since its worst-case time complexity is exponential in the prescribed independence measure between the two paths. This holds even for planar graphs with all generated near-shortest paths of different lengths.

Theorem 6. *For any natural number $\bar{q}$, there is a wide class of planar graphs where for any $q \leq \bar{q}$, the maximal distance between any pair of paths among the first $\Omega(2^q)$ near-shortest paths is $O(q)$ edges and nodes. This holds even if the lengths of all involved near-shortest paths are required to be different.*

Proof. We construct two graph examples that have the required property. For any $\bar{q}$, Example 1 is a planar graph with $n = \bar{q}(r + 1) + 1$ nodes, where for any $q \leq \bar{q}$, among the first r^q near-shortest paths, all of different lengths, the maximal distance between two paths is $4q$ edges and $2q$ nodes. There, the edge weights are integers, and the maximal edge weight is less than $(r + 1)^{\bar{q}}$.

For the case $r = 9$, such a graph is shown at Fig. 2. There are $9^{\bar{q}}$ paths from s to t. The shortest path is horizontal, the next 8 near-shortest paths are different at the rightmost "tower" only, the next $8 \cdot 9 = 72$ ones are different at the two rightmost "towers", and so on. The transformation to any r is straightforward.

We now show that this is not the case of singular bad examples. The preliminary variant of Example 2, also parameterized by $\bar{q}$, shows how an *arbitrary graph in a wide graph class* can be "spoiled" locally so that among the

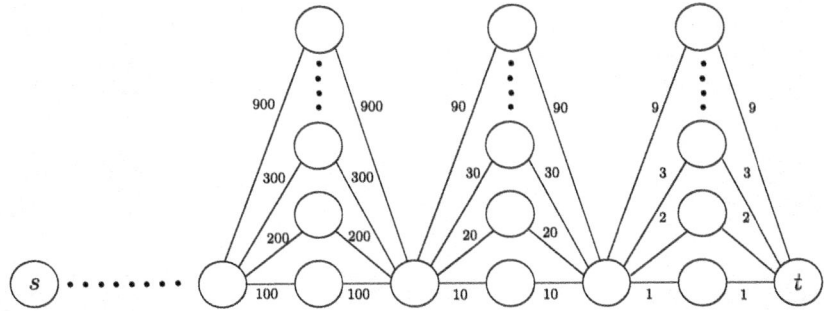

Fig. 2. Example 1 with $r = 9$ and $\bar{q}$ "towers" between s and t.

first $\sum_{i=1}^{q} \prod_{j=0}^{i-1} (\bar{q} - j) + 1 = \Omega((\bar{q} - q + 1)^q)$ near-shortest paths, the maximal edge/node distances between two paths are $2q + 2$ edges and $\min\{2q; \bar{q}\}$ nodes, for any $q \leq \bar{q}$.

Let $G_0 = (V, E, w)$ be an arbitrary weighted graph with a *single shortest path* from s to t. Denote that path by p^* and the second shortest path by p_2^*. We set $\epsilon = (w(p_2^*) - w(p^*))/(\bar{q} + 2) > 0$. Let v be an arbitrary node at p^*, breaking p^* into p_1 and p_2. We split v into two nodes v' and v'', so that p_1 finishes at v', p_2 starts from v'', and all edges originally incident to v, except for that lying on p_1, are now incident to v''. We now add a complete graph, whose nodes are v', v'' and $\bar{q}$ new nodes and whose edges are of weight ϵ each, to the obtained graph. See Fig. 3 for an illustration.

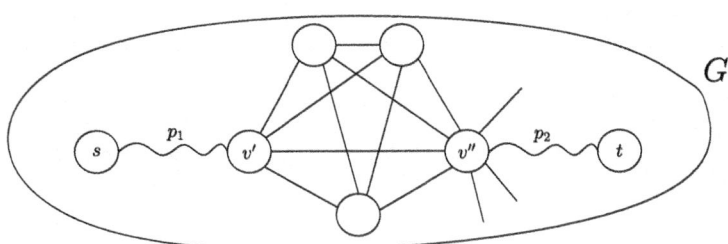

Fig. 3. Example 2 with $\bar{q} = 3$.

Denote the resulting graph by G. The shortest path from s to t in G is $p_1 \circ (v', v'') \circ p_2$ of weight $w(p^*) + \epsilon$. The near-shortest paths in G, in order, are of the form $p_1 \circ p \circ p_2$, where p goes over all paths from v' to v'' of lengths 2, after that 3, and so on up to $\bar{q} + 1$. The weights of all of those paths are strictly less than $w(p_2^*)$. The claimed exponential properties of G are easy to validate.

The main variant of Example 2 arises if we similarly "spoil" G_0 by inserting the graph of Example 1 between nodes v' and v'', with the edge weights proportionally decreased, instead of the complete graph as above. The properties of this variant are also exponential, though a bit weaker than with using the complete graph. Importantly, if the original graph G_0 is planar, then the resulting graph G is also planar, and the lengths of all involved near-shortest paths in G are different.   □

4 k Shortest Paths Trees: A Combinatorial Approach

We consider the problem of finding k shortest paths trees of a graph that was introduced by Sedeño-Noda and González-Martín [22]. They consider spanning trees of a given graph, all rooted at the same given node. The optimization objective considered for such trees is the *total length of the paths* (in the tree) *from the root to the rest of the nodes*. The goal is to find k best trees w.r.t. this objective. Sedeño-Noda and González-Martín proved that such k trees can be found by a greedy algorithm, and provided its polynomial implementation. They proved the correctness of the greedy algorithm using linear programming theory. We present an alternative, combinatorial proof of the correctness of this greedy algorithm.

We begin with definitions and preliminaries. Let $G = (V, E, w)$ be a weighted graph and $s \in V$ be a given node called the root. The graph may be either directed or undirected (where each edge can be used in paths in either direction). Assume, w.l.o.g., that every node of G is reachable from s. Consider any spanning tree $T = (V, E(T))$ of G rooted at s. For definiteness, orient the tree edges *from* s, that is along the paths from s to the rest of the nodes. Also, in the case of an undirected graph, orient the tree edges in the same way. An edge (u, v) is assumed to be oriented from u to v. If $(u, v) \in E(T)$, then node u is the *parent* of node v, and v is the *child* of node u. Also, if u is on the (unique) path from s to v in T, then node u is the *ancestor* of node v, and v is the *descendant* of node u. A node v is considered both ancestor and descendant of itself.

Define the *weight* of tree T, denoted $W(T)$, to be the sum of the weights of the (unique) paths in T from s to all other nodes. We consider $W(T)$ as the objective function to be minimized. Clearly, any shortest paths tree is an optimum ("shortest") one. We are interested in finding k first spanning trees rooted at s in the order of non-increasing weight W (henceforth called k *near-shortest trees*); we break ties arbitrarily.

From now on, we use the term "tree" to refer to a spanning tree in G rooted at s, if this does not cause ambiguity. Let T and T' be two trees. We define the *distance* between them be $d(T, T') = |E(T) \setminus E(T')| = |E(T') \setminus E(T)|$. In case $d(T, T') = 1$, T and T' are called *neighbors*. If $\mathcal{T}$ is a set of trees, then the *distance* $d(T, \mathcal{T})$ is the minimum of $d(T, T')$ over all trees $T' \in \mathcal{T}$.

Recall that in any spanning tree rooted at node s, for any node v other than s, there is *exactly one edge entering* v (from its *parent*). Therefore, if edge (u, v) of T is absent in T', then some other edge (u', v) is present in T'. As a consequence, the entire symmetric difference $E(T) \oplus E(T')$ consists of pairs of edges $((u, v) \in E(T), (u', v) \in E(T'))$, at most one for each node $v \in V$, so that u and u' are not descendants of v in T and T', respectively. Note that the number of these pairs equals $d(T, T')$. If T and T' are neighbors, then their edge sets differ by a single *parent switch* of some node v.

We describe now the greedy algorithm of [22] for solving the problem. We initialize a list $List$ to a single tree T_1 that is a shortest paths tree from s. The general state of $List$ is a sequence $(T_j = (V, E_j))$, $1 \leq j \leq i$, of i first near-shortest paths trees, and at each iteration, we augment it by one more tree. After $k - 1$ iterations, $List$ will contain the desired output.

In list $Next$, we maintain the set of all neighbors of all the current trees in $List$ that are pairwise distinct and distinct from the trees in $List$. At each stage of the algorithm, we move the tree of the minimum weight in $Next$ to the end of $List$ and update $Next$ accordingly. Algorithm 1 contains the pseudo-code of the algorithm, at the coarse level.

Algorithm 1: Finding k near-shortest trees in a graph

    Input: weighted graph $G = (V, E, w)$, $w \geq 0$, node $s \in V$, and an integer k
    Output: the k near-shortest trees in G sorted by their weight W
1 find a shortest paths tree $T_1 = (V, E_1)$ from s in G, by Dijkstra's algorithm
2 $List \leftarrow T_1$; $Next \leftarrow \emptyset$
3 **for** $i = 1$ **to** $k - 1$ **do**
4   **for each edge** $(u', v) \in E \setminus E_i$**, where** u' **is not a descendant of** v **in** T_i **do**
5         $E' \leftarrow E_i \setminus (u, v) \cup (u', v)$
6   **if** (V, E') *not in $Next$ and not in $List$* **then**
7   add tree (V, E') to $Next$;
8   find a tree T in $Next$ with minimum weight
9   move T to the end of $List$
10 return $List$

See Fig. 4 for a running example, where k equals three. Graph G is shown in the first column. The state of $Next$ at iteration i, $i = 1, 2$, is shown in column $i + 1$ under tree T_i. At the first iteration, the first tree in $Next$ is a neighbor of T_1 obtained by replacing the parent of the bottom node; this impacts only the length of the path to the bottom node that increases by 3. The second tree in $Next$ is the neighbor of T_1 obtained by replacing the parent of the central node; the paths lengths to it and to the bottom node increase by 3 each. At the second iteration, the single tree in $Next$ is the neighbor of T_2 obtained by replacing the parent of the central node; unlike the first iteration, in this iteration, this impacts only the length of the path to the central node that increases by 3. The additional element of $Next$ is removed, since it is not needed for generating the last, third, near-shortest tree.

Next, we prove the correctness of the algorithm.

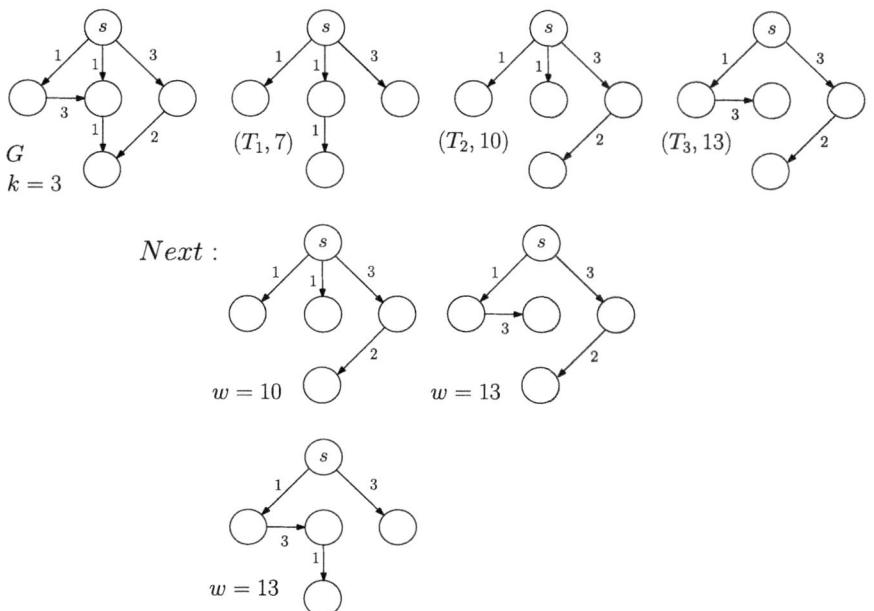

Fig. 4. Running example of finding the first three near-shortest trees in graph G

Lemma 1. *Let G be either a directed or an undirected graph. For any two distinct spanning trees T and T' of G rooted at node s, there exists a pair of edges $(u, v) \in E(T)$ and $(u', v) \in E(T')$ such that the entire (unique) path from s to u' in T' is also in T.*

Proof. Consider any pair of edges $(u, v) \in E(T)$ and $(u', v) \in E(T')$. Denote by P' the unique path from s to u' in T'. The claim holds trivially if path P' is the empty path, i.e., $u' = s$. If the claim does not hold for P', then there are edges in it not belonging to T. Let (x', y) be the edge closest to s among them. Then, there exists edge $(x, y) \in E(T) \setminus E(T')$. By our assumption, the pair $((x, y), (x', y))$ satisfies the condition of the claim. For illustration, see Fig. 5(a). □

Lemma 2. *For any directed graph, at the end of the i-th iteration of the algorithm, List contains the first $i + 1$ trees of a sequence of the near-shortest path trees sorted by their weight.*

Proof. We prove the claim by induction. We show that if at the start of iteration i of the algorithm, *List* contains the first i trees of a sequence of the near-shortest path trees sorted by their weight, then this iteration adds the $(i + 1)$-st tree in the sequence to *List*. The induction basis for $i = 1$ holds since at the start of the first iteration *List* is initialized to contain the shortest paths tree.

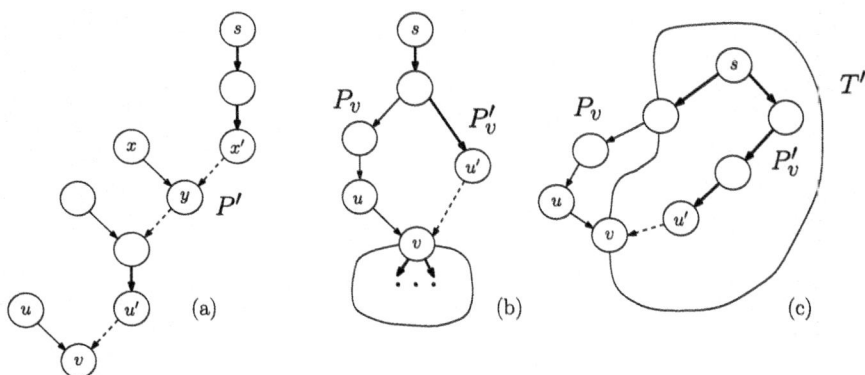

Fig. 5. Illustrations to the correctness proof

Recall that the distance of every tree in *Next* from *List* is 1. We prove the claim for iteration i by contradiction. Assume that after updating *Next* at iteration i, there exists a tree which is neither in *List* nor in *Next* whose weight is strictly smaller than the weight of any tree in *Next*. Among all trees that are not in *List*, let $T = (V, E(T))$ be such a tree with the minimum weight. The tree T is chosen also to minimize its distance to *List* among the trees not in *List* with weight $W(T)$. Let T' be a tree in *List* such that $|E(T) \setminus E(T')| \geq 2$ attains this minimum distance. By our induction hypothesis, $W(T)$ cannot be strictly lower than the weight of any tree in *List*.

Let a pair of edges $(u, v) \in E(T)$ and $(u', v) \in E(T')$ be as in Lemma 1. Denote by P_v and P'_v the unique paths from s to v in T and T', resp. (their last edges are (u, v) and (u', v), resp.) We distinguish two cases depending on the lengths of these paths.

Case 1: $w(P_v) \geq w(P'_v)$ Consider the tree $\tilde{T}$ given by swapping edge (u, v) in T by edge (u', v). The only nodes whose distance from s may differ in T and $\tilde{T}$ are the nodes in the subtree rooted at v in $T \cap \tilde{T}$. By our choice of (u, v) and (u', v), the path from s to v in $\tilde{T}$ is P'_v. Thus, for any node in the subtree rooted at v in T, the change of its distance from s caused by the edge swapping is $w(P'_v) - w(P_v) \leq 0$. We conclude that $W(\tilde{T}) \leq W(T)$. See Fig. 5(b).

If $\tilde{T}$ is not in *List* then by our choice of T, $W(T) = W(\tilde{T})$. By our choice of T', $d(T, List) = d(T, T')$. However, $d(\tilde{T}, T') = d(T, T') - 1$. Hence, $d(\tilde{T}, List) < d(T, List)$, a contradiction to our choice of T. If $\tilde{T}$ is in *List* then since $|E(\tilde{T}) \setminus E(T)| = 1$, the tree T is in *Next*, a contradiction to our assumption.

Case 2: $w(P_v) < w(P'_v)$ Consider graph $\tilde{G} = (V, E(T') \cup P_v)$. Certainly, every node is reachable from s in $\tilde{G}$. Let $\tilde{T}$ be a shortest paths tree from s in $\tilde{G}$. Since the shortest path from s to v in $\tilde{G}$ is strictly shorter than the path from s to v in T', and since for any other node $u \in V$, the shortest path from s to u in $\tilde{G}$ is not longer than the path from s to u in T', we have $W(\tilde{T}) < W(T')$. Since T' is in *List*, tree $\tilde{T}$ is also in *List*.

Next, we compare $|E(T) \setminus E(T')|$ to $|E(T) \setminus E(\tilde{T})|$. Note that $|E(T) \setminus E(T')| = |E(T') \setminus E(T)|$ and $|E(T) \setminus E(\tilde{T})| = |E(\tilde{T}) \setminus E(T)|$. It is easier to compare $|E(T') \setminus E(T)|$ to $|E(\tilde{T}) \setminus E(T)|$. First, consider the contribution of the edge (u', v) to both set differences. Recall that (u', v) belongs to $E(T')$ but not to $E(T)$; thus, $(u', v) \in E(T') \setminus E(T)$. On the other hand, (u', v) cannot be in $E(\tilde{T})$ for the following reason. Recall that by Lemma 1, $P'_v \setminus \{(u', v)\}$ is contained in both $E(T)$ and $E(T')$. Hence, it is the unique path from s to u' in $\tilde{G}$. So, if $(u', v) \in E(\tilde{T})$, then the entire (unique) path from s to v in $\tilde{T}$ is P'_v. However, it is longer than P_v, a contradiction to the definition of $\tilde{T}$. Thus, $(u', v) \notin E(\tilde{T}) \setminus E(T)$. Next, consider any edge in P_v. Since $P_v \subseteq E(T)$, this edge contributes to neither of $|E(T') \setminus E(T)|$ nor $|E(\tilde{T}) \setminus E(T)|$. Finally, consider any other edge $(x, y) \in E(\tilde{T}) \setminus E(T)$. Since (x, y) is not in P_v, we have $(x, y) \in E(T')$. So, (x, y) belongs also to $E(T') \setminus E(T)$. It follows that $|E(T') \setminus E(T)| \geq |E(\tilde{T}) \setminus E(T)| + 1$, and $|E(T) \setminus E(\tilde{T})| < |E(T) \setminus E(T')|$, a contradiction to our choice of T'. For illustration, see Fig. 5(c). $\qquad\square$

Theorem 7. *The algorithm finds k near-shortest trees for both directed and undirected graphs.*

Proof. By Lemma 2, the algorithm is correct for directed graphs. If the input graph $G = (V, E)$ is undirected, we construct a directed graph G' by replacing every edge $[u, v] \in E$ by two directed edges (u, v) and (v, u). Observe that the set of spanning trees rooted at s is the same in G and in G', keeping the weights. Therefore, any algorithm solving the problem on G' solves it also on G. $\qquad\square$

Acknowledgment. This research was supported by the BGU-NJIT Institute for Future Technologies (seed grant), the Israeli Science Foundation (Grant No. 465/22), the Rita Altura trust chair in computer science, and by the Lynne and William Frankel Center for Computer Science.

References

1. Akhmedov, M.: Faster 2-disjoint-shortest-paths algorithm. In: Computer Science—Theory and Applications—15th International Computer Science Symposium in Russia (CSR 2020), pp. 103–116 (2020)
2. Bentert, M., Nichterlein, A., Renken, M., Zschoche, P.: Using a geometric lens to find k disjoint shortest paths. In: 48th International Colloquium on Automata, Languages, and Programming (ICALP 2021), pp. 26:1–26:14 (2021)
3. Bérczi, K., Kobayashi, Y.: The directed disjoint shortest paths problem. In: 25th Annual European Symposium on Algorithms (ESA 2017), pp. 13:1–13:13 (2017)
4. Chondrogiannis, T., Bouros, P., Gamper, J., Leser, U.: Alternative routing: k-shortest paths with limited overlap. In: 23rd SIGSPATIAL International Conference on Advances in Geographic Information Systems, pp. 68:1–68:4 (2015)
5. Chondrogiannis, T., Bouros, P., Gamper, J., Leser, U.: Exact and approximate algorithms for finding k-shortest paths with limited overlap. In: 20th International Conference on Extending Database Technology (EDBT 2017), pp. 414–425 (2017)

6. Deng, Y., Guo, L., Huang, P.: Exact algorithms for finding partial edge-disjoint paths. In: 24th International Computing and Combinatorics Conference (COCOON 2018), pp. 14–25 (2018)

7. Dinitz, Y.: Algorithm for solution of a problem of maximum flow in a network with power estimation. Soviet Math. Dokl. **11**, 1277–1280 (1970)

8. Dinitz, Y., Dolev, S., Kumar, M.: Polynomial time k-shortest multi-criteria prioritized and all-criteria-disjoint paths. In: Cyber Security Cryptography and Machine Learning—5th International Symposium (CSCML 2021). Lecture Notes in Computer Science. Springer (2021)

9. Eilam-Tzoreff, T.: The disjoint shortest paths problem. Discret. Appl. Math. **85**(2), 113–138 (1998)

10. Erdmann, M.A., Lozano-Pérez, T.: On multiple moving objects. Algorithmica **2**, 477–521 (1987)

11. Gajjar, K., Jha, A.V., Kumar, M., Lahiri, A.: Reconfiguring shortest paths in graphs. In: AAAI, pp. 9758–9766. AAAI Press (2022)

12. Guo, L., Deng, Y., Liao, K., He, Q., Sellis, T., Hu, Z.: A fast algorithm for optimally finding partially disjoint shortest paths. In: 27th International Joint Conference on Artificial Intelligence (IJCAI 2018), pp. 1456–1462 (2018)

13. Halperin, D., Latombe, J., Wilson, R.H.: A general framework for assembly planning: the motion space approach. Algorithmica **26**(3–4), 577–601 (2000)

14. Hopcroft, J.E., Wilfong, G.T.: Reducing multiple object motion planning to graph searching. SIAM J. Comput. **15**(3), 768–785 (1986)

15. Jennings, J., Whelan, G., Evans, W.: Cooperative search and rescue with a team of mobile robots. In: 8th International Conference on Advanced Robotics (ICAR'97), pp. 193–200 (1997)

16. Lochet, W.: A polynomial time algorithm for the k-disjoint shortest paths problem. In: ACM-SIAM Symposium on Discrete Algorithms (SODA 2021), pp. 169–178. SIAM (2021)

17. Mataric, M.J., Nilsson, M., Simsarin, K.T.: Cooperative multi-robot box-pushing. In: IEEE/RSJ International Conference on Intelligent Robots and Systems (IROS 1995), pp. 556–561 (1995)

18. Papadimitriou, C.H., Raghavan, P., Sudan, M., Tamaki, H.: Motion planning on a graph (extended abstract). In: 35th Annual Symposium on Foundations of Computer Science (FOCS 1994), pp. 511–520. IEEE Computer Society (1994)

19. Robertson, N., Seymour, P.D.: Graph minors .xiii. the disjoint paths problem. J. Comb. Theory, Ser. B **63**(1), 65–110 (1995)

20. Rodríguez, S., Amato, N.M.: Behavior-based evacuation planning. In: IEEE International Conference on Robotics and Automation (ICRA 2010), pp. 350–355 (2010)

21. Rus, D., Donald, B.R., Jennings, J.: Moving furniture with teams of autonomous robots. In: IEEE/RSJ International Conference on Intelligent Robots and Systems (IROS 1995), pp. 235–242 (1995)

22. Sedeño-Noda, A., González-Martín, C.: On the K shortest path trees problem. Eur. J. Oper. Res. **202**(3), 628–635 (2010)

23. Smith, B.S., Egerstedt, M., Howard, A.M.: Automatic deployment and formation control of decentralized multi-agent networks. In: IEEE International Conference on Robotics and Automation (ICRA 2008), pp. 134–139 (2008)

24. Yen, J.Y.: Finding the k shortest loopless paths in a network. Manage. Sci. **17**(11), 712–716 (1971)

Brief Announcement: Towards Proportionate Fair Assignment

Baruch Schieber$^{(\boxtimes)}$ ⓘ

New Jersey Institute of Technology, Newark, NJ, USA
sbar@njit.edu

Abstract. The well known *assignment problem* finds an optimal assignment of tasks to agents. Optimal assignment is applicable to the placement of virtual machines in cloud systems to optimize resource allocation and ensure efficient operation. It is also applicable to fault-tolerant computation to ensure continued operation in case of component failures. In some instances the assignment needs to satisfy extraneous fairness constraints. For example, in a multi-tenant cloud environment the assignment has to guarantee equitable treatment of customers. We initiate the study of *proportionate fair assignment*. In the multi-tenant setting such an assignment ensures proportionate representation of the customers over all servers. We show that even the simple case of computing an assignment that ensures equal representation of two customers is hard. On the positive side, we present a 1/2-approximation algorithm for computing an assignment that ensures equal representation of two customers.

Keywords: Assignment · Perfect matching · Proportionate fairness · Cloud system · Approximation algorithms

1 Introduction

The classical (balanced) assignment problem that was first practically addressed by Kuhn [10] is defined as follows. Given n tasks and n agents and the effectiveness of each agent for each task, the goal is to assign each agent to one and only one task in such a way that the total measure of effectiveness is optimized (maximized or minimized). For simplicity we will concentrate on the maximization version and assume that the measure of effectiveness of assigning an agent to a task is given by a *weight*. The balanced assignment problem is equivalent to finding a perfect matching in a weighted bipartite graph, in which the sum of weights of the matching edges is maximized.

Optimal assignment is applicable to task allocation in distributed systems [1,6], the placement of virtual machines in cloud systems to optimize resource allocation and ensure efficient operation [5,15], and fault-tolerant computation to ensure continued operation in case of component failures [13].

T. Masuzawa et al. (Eds.): SSS 2024, LNCS 14931, pp. 255–259, 2025.
https://doi.org/10.1007/978-3-031-74498-3_18

In some instances the assignment needs to satisfy extraneous fairness constraints. For example, in a multi-tenant cloud environment the assignment has to guarantee equitable treatment of customers [4,6]. Similar fairness requirements apply also to assignment problems in other fields, such as allocation of online ads [3] and block allocation of gas pipeline operated by a "common carrier" [16].

The fairness criterion that we adapt is formally known as *proportionate fairness* or *p-fairness* [2]. P-fairness was first considered in the classical Chairman Assignment Problem [14] that studies how to rotate a chairperson of a union of states such that at any time the accumulated number of chairpersons from each state is proportional to the state size. In our case we assume that every task t has a *protected attribute* with value a_t. Each possible value a of the protected attribute is associated a proportion $p(a)$, where all proportions sum to 1. In addition, the agents are ordered according to their desirability. The goal is to assign each agent to one and only one task in such a way that the total weight of the assigned pairs is maximized, subject to the p-fairness constraint that requires that for any prefix of $\ell \leq n$ agents (in their sorted order) and for any value a of the protected attribute, the number of agents assigned to tasks with attribute value a is either $\lfloor p(a) \cdot \ell \rfloor$ or $\lceil p(a) \cdot \ell \rceil$. For example, in a multi-tenant cloud system in which applications need to be assigned to servers, the protected attribute of an application may be its Customer Id, and the proportion of the attribute is the fractional share of its associated customer in the site. Assume that the servers are ordered by their computing power. The p-fairness constraint ensures that every customer gets a "fair share" of the powerful servers.

We start by considering the simplest case: bivalued protected attribute where each of the two values has proportion $\frac{1}{2}$. We refer to the two "protected groups" defined by their protected attribute value as *red* tasks and *blue* tasks. Assume n red tasks, n blue tasks, and $2n$ agents. The p-fairness constraint translates in this case to the constraint that the number of red tasks (and thus also blue tasks) assigned to any prefix of ℓ agents is either $\lfloor \frac{1}{2}\ell \rfloor$ or $\lceil \frac{1}{2}\ell \rceil$. It is easy to see that this is equivalent to the constraint that for every $i \in [1, n]$, the agents numbered $2i - 1$ and $2i$ are assigned tasks from different protected groups. Next, we show that finding an assignment even in this simple case is NP-Hard and then show a $\frac{1}{2}$-approximation assignment algorithm.

2   Hardness

We prove that finding an optimal p-fair assignment of 2 equally protected groups is NP-Hard, even in case the weight matrix is $0 - 1$. Note that a $0 - 1$ weight matrix can be viewed as adjacency matrix of a bipartite graph and thus the problem translates to finding p-fair maximum matching.

2.1   P-Fair *Perfect* Matching with 2 Equally Protected Groups

The p-fair *perfect* matching with 2 equally protected groups decision problem is defined as follows.

Input: A bipartite graph $G(X, Y, E)$, where $|X| = |Y| = 2n$. The set of vertices $X = \{x_1, \ldots, x_{2n}\}$ is *ordered*, and the set of vertices Y is partitioned into two subsets Y_1 and Y_2, where $|Y_1| = |Y_2| = n$. Call the vertices in Y_1 *red* vertices and the ones in Y_2 *blue* vertices.

Problem: Is there a prefect matching in G with the property that for every $i \in [1, n]$, the pair of vertices x_{2i-1} and x_{2i} are not matched to monochromatic vertices in Y?

Theorem 1. *The p-fair perfect matching with 2 equally protected groups problem is* NP-Complete.

Proof. Checking whether a given perfect matching satisfies the p-fairness property is trivial and thus the problem is in NP. We show that it is NP-Hardby a reduction from the NP-Hardvariant of SAT called *Cubic-Positive-1-in-3-SAT* [9, 11, 12]. This problem is defined as follows.

Input: A CNF Boolean formula in which each clause contains 3 positive literals, and each Boolean variable b appears 3 times in the clauses.

Problem: Is there a satisfying assignment in which exactly one literal in each clause is true?

Consider an instance S of the Cubic-Positive-1-in-3-SAT problem. Since each clause contains 3 literals and each variable appears 3 times, the number of clauses, denoted by c, is equal to the number of variables. We define a bipartite graph $G(X, Y, E)$, where $|X| = |Y| = 12c$. The graph has one gadget for every variable, shown in the grayed area in Fig. 1. Additionally, the graph has one blue vertex for every clause and $2c$ "surplus" blue vertices.

The gadget associated with variable v: There are 12 vertices of X, 6 red vertices, and 3 blue vertices in the variable gadget. The pairing of the vertices of X that is induced by their order is denoted in the figure. We distinguish between the 3 *top* vertex pairs and 3 *bottom* vertex pairs. Each top pair is associated with a distinct clause that contains the variable v and the right vertex in the pair is connected to the blue vertex corresponding to its associated clause. The right vertex in each bottom pair is connected to every one of the $2c$ surplus blue vertices.

We show that there is a p-fair perfect matching in G, that is a perfect matching in which each one of the defined pairs of vertices of X is matched to one red and one blue vertex, if and only if S has a satisfying assignment in which exactly one literal in each clause is true.

It follows from the structure of the gadget that in any p-fair perfect matching the vertices of X in every gadget have to be matched in one of two ways, denoted (**M1**) and (**M2**). This is determined by the matching edge of the leftmost (red) vertex in the gadget. In (**M1**) this red vertex is matched to its neighbor in the "top" part, and in (**M2**) it is matched to its neighbor in the "bottom" part. The two possible ways are as follows. (**M1**) the *left* vertex in each "top" pair is matched to its adjacent red vertex, the *right* vertex in each "bottom" pair is matched to its adjacent red vertex, the *right* vertex in each "top" pair is matched to the blue vertex corresponding to its associated clause, and the *left* vertex in

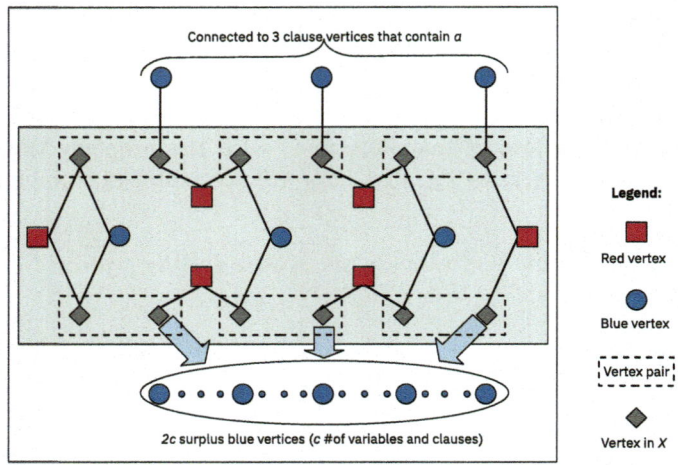

Fig. 1. A variable gadget used in the reduction

each "bottom" pair is matched to its adjacent blue vertex in the gadget. (**M2**) the *left* vertex in each "top" pair is matched to its adjacent blue vertex in the gadget, the *right* vertex in each "bottom" pair is matched to one of the surplus blue vertices, the *right* vertex in each "top" pair and the *left* vertex in each "bottom" pair are matched to their adjacent red vertex.

Suppose that there is a p-fair perfect matching in G. In this case since all the surplus blue vertices have to be matched, exactly $\frac{2}{3}c$ variable gadgets have to be matched in way (**M2**) and $\frac{1}{3}c$ variable gadgets have to be matched in way (**M1**). (Note that c must be divisible by 3.) Moreover, the sets of associated clauses of any pair of variable gadgets matched in way (**M1**) must be disjoint to ensure that all the blue vertices associated with clauses are matched. This implies that assigning "true" to all the variables whose gadgets matched in way (**M1**) and "false" to the others yields a satisfying assignment of S in which exactly one literal in each clause is true. On the reverse direction, given a satisfying assignment of S in which exactly one literal in each clause is true, match all the gadgets of variables with "true" value in way (**M1**) and the rest in way (**M2**) to yield a p-fair perfect matching.

3   An Approximation Algorithm

We give a simple $\frac{1}{2}$-approximation algorithm for p-fair assignment in case of 2 equally protected groups. In this case we are given a set of agents $x_1, \ldots, x_{2n}$, a set of tasks that is partitioned into a *red* task set and a *blue* task set of size n each, and a weight matrix. The goal is to find a p-fair assignment of agents to tasks that maximizes the total weight of the assigned pairs. The p-fairness constraint translates in this case to the constraint that for every $i \in [1, n]$, the agents x_{2i-1} and x_{2i} are not assigned to tasks of the same color.

We solve two instances of the assignment problem. In the first instance we restrict red tasks to be assigned to agents x_{2i-1} and blue tasks to be assigned to agents x_{2i}, for $i \in [1, n]$. This can be done by modifying the weight of "forbidden" pairs to a large negative weight and then using one of the known (weakly or strongly) polynomial time assignment algorithms [7,8]. In the second instance we exchange the roles of the red and blue tasks. It is easy to see that both solutions are p-fair and that the sum of the weights of the two solutions is at least the optimal solution. Thus, one of these two solutions is a $\frac{1}{2}$-approximation.

References

1. Ahmad, I., Dhodhi, M., Ghafoor, A.: Task assignment in distributed computing systems. In: International Phoenix Conference on Computers and Communications, pp. 49–53 (1995)
2. Baruah, S., Cohen, N.K., Plaxton, C.G., Varvel, D.A.: Proportionate progress: a notion of fairness in resource allocation. Algorithmica **15**, 600–625 (1996)
3. Baumann, J., Sapiezynski, P., Heitz, C., Hannak, A.: Fairness in online ad delivery. In: ACM Conference on Fairness, Accountability, and Transparency, pp. 1418–1432 (2024)
4. Chen, Y.T., Lai, K.C.: A fairness-aware load balancing strategy in multi-tenant clouds. Front. Comput. Ind. Appl. **4**, 222–233 (2024)
5. Dai, L., He, A., Sun, G., Pan, Y.: Fair and stable matching virtual machine resource allocation method. Intell. Autom. Soft Comput. **32**(3), 1831–1842 (2022)
6. Doncel, J., De La Pisa, L., Santos, A., Anta, A.F.: A fair and scalable mechanism for resource allocation: the multilevel QPQ approach. IEEE Access **9**, 19439–19456 (2021)
7. Fredman, M.L., Tarjan, R.E.: Fibonacci heaps and their uses in improved network optimization algorithms. J. ACM **34**(3), 596–615 (1987)
8. Gabow, H.N., Tarjan, R.E.: Faster scaling algorithms for network problems. SIAM J. Comput. **18**(5), 1013–1036 (1989)
9. Garey, M.R., Johnson, D.S.: Computers and Intractability: A Guide to the Theory of NP-Completeness (Series of Books in the Mathematical Sciences), 1st edn. W. H. Freeman (1979)
10. Kuhn, H.W.: The Hungarian method for the assignment problem. Naval Res. Logist. Q. **2**(1–2), 83–97 (1955)
11. Moore, C., Robson, J.M.: Hard tiling problems with simple tiles. Discr. Comput. Geom. **26**, 573–590 (2001)
12. Schaefer, T.J.: The complexity of satisfiability problems. In: 10th ACM Symposium on Theory of Computing (STOC), pp. 216–226 (1978)
13. Solouki, M.A., Angizi, S., Violante, M.: Dependability in embedded systems: a survey of fault tolerance methods and software-based mitigation techniques. arXiv:2404.10509 (2024)
14. Tijdeman, R.: The chairman assignment problem. Discret. Math. **32**(3), 323–330 (1980)
15. Wang, J.V., Fok, K.Y., Cheng, C.T., Tse, C.K.: A stable matching-based virtual machine allocation mechanism for cloud data centers. In: 2016 IEEE World Congress on Services (SERVICES), pp. 103–106 (2016)
16. Wen, K., Gao, W., Hui, X., Li, L., Yang, B., Nie, C., Miao, Q., Li, Y., Li, C., Hong, B.: Allocation of transportation capacity for complex natural gas pipeline network under fair opening. Energy **291**, 130330 (2024)

BLINDEXTEE: A Blind Index Approach Towards TEE-Supported End-to-End Encrypted DBMS

Louis Vialar⬤, Jämes Ménétrey⬤, Valerio Schiavoni$^{(\boxtimes)}$⬤, and Pascal Felber⬤

University of Neuchâtel, Neuchâtel, Switzerland
{Louis.Vialar,Valerio.Schiavoni,Pascal.Felber}@unine.ch

Abstract. Using cloud-based applications comes with privacy implications, as the end-user looses control over their data. While encrypting all data on the client is possible, it largely reduces the usefulness of database management systems (DBMS) that are typically built to efficiently query large quantities of data. We present BLINDEXTEE, a new component that sits between the application business-logic and the database. BLINDEXTEE is shielded from malicious users or compromised environments by executing inside an SEV-SNP confidential VM, AMD's trusted execution environment (TEE). BLINDEXTEE is in charge of end-to-end encryption of user data while preserving the ability of the DBMS to efficiently filter data. By decrypting and re-encrypting data, it builds *blind indices*, used later on to efficiently query the DBMS. We demonstrate the practicality of BLINDEXTEE with MySQL in several micro- and macro-benchmarks, achieving overheads between 36.1% and 462% over direct database access depending on the usage scenario.

Keywords: Database security · Privacy-Enhancing Technologies · Trusted Execution Environments · Blind Index

1 Introduction

The high convenience of the modern web-based application, accessible everywhere from any device, comes with important privacy downsides. Once data is off-loaded to third-party service providers, one never knows its future usage. Many solutions exist to protect data stored in remote untrusted database management systems (DBMS) [3, 11, 21, 27, 29]. These solutions protect data stored by the service provider from malicious database administrators, but don't protect user data from the service provider itself.

We present BLINDEXTEE, a novel approach for database encryption. In a nutshell, data is encrypted in such a way that only the data owners (end users of the system) can access it, while preserving the possibility to retrieve the data efficiently from the database system. BLINDEXTEE is a database proxy that transparently sits between the database client and database server. It handles

T. Masuzawa et al. (Eds.): SSS 2024, LNCS 14931, pp. 260–276, 2025.
https://doi.org/10.1007/978-3-031-74498-3_19

Table 1. Comparison of the state-of-the-art protected databases.

	CryptDB [21]	Crypt-SQLite [29]	Enclave DB [22]	Always Encrypted [3]	Gabel et al. [12]	StealthDB [27]	This paper
Can run TPC-C	●	●	●	●	○	●	○
Query public/private data	●	○	○	●	○	●	●
Per-user/app keys	●	●	○	●	○	○	●
Unmodified DBMS	●	○	○	○	●	○	●
DBMS outside TCB	●	○	○	●	●	●	●
Avoid OPE	○	—	—	●	●	●	●
Avoid deterministic enc.	○	—	—	●	●	●	●
End-to-end encryption	○	○	○	○	○	◐	●
Encryption granularity	column	database	table	column	table	column	column
Supported TEEs	—	SGX	SGX	SGX	SGX	SGX	SEV

on-the-fly encryption and decryption of data for confidentiality and makes use of *blind indices* [1] for efficient retrieval of encrypted data in the database. A blind index is a kind of bloom filter [9] made using a fixed-length truncated hash: two identical values always return the same blind index value; however two different values may also give an identical blind index value. By using blind indices of sufficient length, we can meaningfully filter data in a large database table. By keeping this length low enough, we can maintain a sufficiently high number of *collisions* (false positives), that prevents an adversary from inferring equality of values.

BLINDEXTEE needs to decrypt and encrypt data, and it must be protected against its own environment (i.e., no adversary must be able to extract keys from its memory). To enforce such guarantees, we leverage SEV-SNP [2], a trusted execution environment (TEE) offered on modern AMD EPYC server-grade CPUs, and widely available for use in cloud providers.

TEEs are hardware-protected memory areas (often referred to as *enclaves*) that are fully isolated from the host operating system. SEV-SNP is a virtual-machine based TEE, which means it runs VMs with encrypted memory, protected execution state (CPU registers), and strong integrity protection. Hence, malicious hosts/hypervisors cannot read nor write in the memory of a confidential SEV-SNP VM. In addition, TEEs offer multiple ways for external observers to *attest* that a particular piece of software is indeed running in a TEE (and not in an untrusted environment), and that it has not been altered in any way.

Roadmap. In Sect. 2, we survey related work on protected database systems. In Sect. 3, we introduce the terminology of the different components that intervene in a typical modern internet application. Sect. 4 presents the architecture of BLINDEXTEE. Our security analysis of BLINDEXTEE is presented in Sect. 5. We present the experimental evaluation of BLINDEXTEE in Sect. 6, before concluding in Sect. 7.

2   Related Work

We survey state-of-the-art protected database solutions, comparing their cornerstone features in Table 1. Each feature is assessed as either non-applicable (—), missing (○), partially (◑), or fully (●) available. We consider their support for TPC-C, if queries can combine non-encrypted public data and encrypted sensitive data, if the DBMS supports per-user app keys or if it required modifications, and if they use any TEE. We include potential native support for order-preserving encryption (OPE) [14], or if the DBMS avoids deterministic encryption due to known security issues [14]. We distinguish between systems with end-to-end encryption schemes for data stored in the database. Partial availability (◑) signifies that the application backend must be trusted, and full availability (●) indicates that only the end-user device requires trust.

We observe the following. Database systems can be protected by software and hardware-based techniques [11], extensively explored by academia and industry. CryptDB [21] uses a trusted proxy between clients and the database system. This approach offers the benefit of abstraction, allowing the proxy to interface with various database engines seamlessly. BLINDEXTEE follows a similar approach for its trusted proxy. Hardware-assisted TEEs offer strong security guarantees, protecting data confidentiality and integrity even when hosted in untrusted environments, tackling a more powerful threat model than CryptDB. However, we observe how most of them lack abstraction capabilities, requiring tight coupling with specific database engines, or failed to provide robust end-to-end encryption. Authors in [14] showed vulnerabilities of CryptDB's encryption schemes, such as order-preserving and deterministic encryption, to approximate database recovery attacks. Order-preserving encryption maintains the plaintext order in the ciphertext, and deterministic encryption produces the same ciphertext for identical plaintexts when using the same encryption scheme repeatedly. To address these security concerns, BLINDEXTEE combines the benefits of a proxy-based architecture with a stronger threat model. Our approach ensures end-to-end encryption to protect user data and leverages TEEs to provide confidentiality, integrity and trust in stored data.

The execution of database systems within TEEs is challenging, due to the inherent constraints of these secure environments. Consequently, two main approaches have emerged to design such systems. The first approach involves fully encapsulating the database system within the TEE, exposing its services through secure communication channels (i.e., network interfaces). Alternatively, one can partition the DBMS, shielding only critical components within the trusted environment. The choice between these two implementation strategies is a subject of ongoing debate [19,30], as it represents a fundamental tradeoff between minimizing the trusted computing base (TCB) to reduce the attack surface, and the ease of deploying off-the-shelf database systems with modified interfaces. While a smaller TCB enhances security by limiting potential vulnerabilities, the latter approach simplifies adopting existing database solutions in TEE-protected architectures.

CryptSQLite [29] ensures data confidentiality and integrity by fully encapsulating the database system within SGX enclaves using AES-GCM 128-bit encryption for each database page. In contrast, BLINDEXTEE minimizes encryption operations by supporting protected columns, thereby reducing the performance impact.

EnclaveDB [22] can manage both public and sensitive data, storing the latter within an SGX enclave using table-level encryption granularity. It leverages a modified version of Hekaton [10] for secure data management within the TEE and establishes secure communication channels. However, EnclaveDB requires a trusted client machine to compile database queries, aiming to minimize the TCB at the cost of client-side modifications. Our system, on the other hand, supports column-level encryption granularity for sensitive data and enables the processing of both public and sensitive data within a single query, addressing a limitation of EnclaveDB. Furthermore, BLINDEXTEE parses standard SQL queries within the enclave without requiring a database engine inside the TEE, further reducing the TCB size.

Always Encrypted (AE) [3] extends Microsoft SQL Server to store encrypted data with column-level granularity in the regular database engine. It notably uses SGX enclaves to execute queries on encrypted data, decrypting it only within the enclave memory. BLINDEXTEE introduces a proxy that abstracts the underlying database engine, enabling adaptability to various database systems. Moreover, our solution realizes an end-to-end encryption scheme, ensuring that data is never decrypted on the same infrastructure hosting the database engine.

Gabel and Mechler's secure database outsourcing approach [12] shares similarities with BLINDEXTEE, using an SGX enclave to host a proxy that intercepts client-database communication. They protect sensitive data tables by concatenating and encrypting each row's values, storing the resulting ciphertext in a single column while leaving the row identifier unencrypted. In contrast, our work encrypts columns individually, eliminating the need to decrypt entire rows when accessing a subset of columns, thereby improving querying efficiency. Our end-to-end encryption scheme uses per-user keys, ensuring secure communication between clients and the proxy, while [12] does not mention this security aspect when secured within SGX enclaves. In addition, our encryption scheme selectively encrypts sensitive data, avoiding the encryption of the entire dataset, as instead required by [12].

StealthDB [27] relies on proxies inside an Intel SGX enclave, separating responsibilities among multiple enclaves for authentication, query preprocessing, and database operations. It supports column-level encryption granularity and introduces new encrypted data types and functions as extensions to the underlying PostgreSQL database engine. BLINDEXTEE supports a stronger threat model, by introducing blind indexes that obfuscate access patterns and prevent leakage of sensitive data from the index structure, such as record ordering. Our solution offers better end-to-end security guarantees than StealthDB: client data is encrypted locally (e.g., from within a web browser) via per-user keys, similar to privacy-focused products [8,23]. Our approach is DBMS-agnostic and it avoids the need for new data types, unlike StealthDB's engine-specific extensions to PostgreSQL.

3   System Model

We consider the following three roles: the *end-user*, the *service provider* and the *database administrator*. The end-user is the data owner and client of the system, intending to upload data in the system. The service-provider builds, distributes, and sells access to the application. Finally, the database administrator hosts and manages the database. These roles can be shared, e.g., a single entity can act simultaneously as service provider and database provider.

We model a typical internet application using the following three components: the *client*, the *application backend*, and the *database*. The client is the interface used by the end-user to access the application, i.e., a website or software to install on their device. The client communicates with the application backend operated by the service-provider, handling the core business logic. In turn, the application backend stores its data in a database management system (DBMS), hosted and maintained by the database administrator. Note that while we build BLINDEXTEE atop the MySQL DMBS, the architecture is flexible and can be easily ported to alternative SQL systems.

We introduce a fourth component, the *database proxy*, sitting between the application backend and the database. It handles transparent encryption and decryption of data and may rewrite queries. This component is detailed in Sect. 4.

BLINDEXTEE provides *end-to-end encryption*, i.e., some data can only be decrypted by the client and database proxy. The database proxy is a trusted application running in a TEE, and as long as it does not reveal user data, the data is effectively only accessible by the *client*. We provide a detailed security analysis in Sect. 5.

4   The BLINDEXTEE Database Proxy

Design Goals. Our database encryption system has the following main design goals.

(a) *End-to-end data security:* The end user's confidential data should only be accessible to the user and the proxy.
(b) *Ease of implementation:* The approach should require minimal modification to the client and application backend. This is partially achieved by distributing client libraries for the proxy system.
(c) *Efficient use of the DBMS capabilities:* Wherever possible, BLINDEXTEE should rely on the DBMS native capabilities for filtering data.

Overview. Figure 1 shows the architecture of BLINDEXTEE, including the flows of data across the components. The *end-user* interacts with the *client*, which

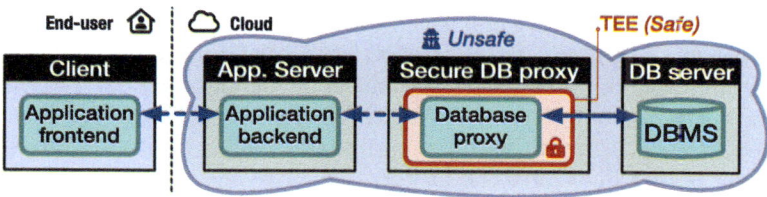

Fig. 1. Architecture of BLINDEXTEE. Dashed arrows denote data encrypted using a session key, the full arrow denotes data encrypted using a long-term key

encrypts and sends queries to the *application backend*, receives and decrypts their results, and present them to the *end-user*. The queries are encoded using an application specific serialization format (e.g.,, JSON or XML), and transmitted using an application specific protocol (i.e.,, HTTP). Only part of a query or response is typically sensitive: the *application backend* may need a cleartext view of some parts of a query or response to correctly operate (e.g.,, to check for correct permissions, to send emails). Hence, only some fields are encrypted in the queries and responses, matching the encrypted columns in the database. The schema of the query is transmitted in clear.

The *application backend* receives partially encrypted queries from the *client*, applies business logic based on the cleartext fields, then transmits SQL queries to the *proxy*. The SQL queries may contain encrypted values extracted from the client query, and the results to the queries contain a mix of encrypted and cleartext values. The backend can also apply business logic based on the cleartext parts of the SQL response, before retransmitting data back to the client. As with the request, encrypted and cleartext values are mixed in the application specific data serialization format. Only minimal modifications are required in the backend (e.g., for login and registration), achieving design goal (b).

The *proxy* receives partially encrypted SQL queries from the *application backend*, communicates with the *database server*, and generates partially encrypted SQL responses. When it receives a simple non encrypted query, the proxy trivially forwards it to the database server, and directly forwards the response to the client, without any further processing. If the query contains encrypted fields, tries to access encrypted data, or uses one of the custom functions of the proxy, then the proxy must handle it (see Sect. 4). The proxy may encrypt or decrypt data, and may submit additional SQL queries to the *database server*. The use of encryption achieves our design goal (a), and the use of blind indices design goal (c). The rationale for encryption is further detailed in Sect. 5.

Note that all communication between the two trusted components (the *client* and the *application proxy*) goes through the *application backend*. Moreover, all encryption happens at the level of individual values, while the communication channel is not protected. For a single client request containing a single encrypted value, the *application backend* may issue multiple SQL queries containing that same value.

Establishing a Trusted Secure Channel. Upon start, the *client* ensures it can trust the *proxy* and establishes an encrypted tunnel with it. These two steps are implemented as a single key exchange, similar to TLS [24]. Since the *application backend* only interacts with the *proxy* via SQL, the key exchange is implemented as a custom SQL function, KEY_EXCHANGE, which is intercepted by the proxy. Figure 2 illustrates the protocol.

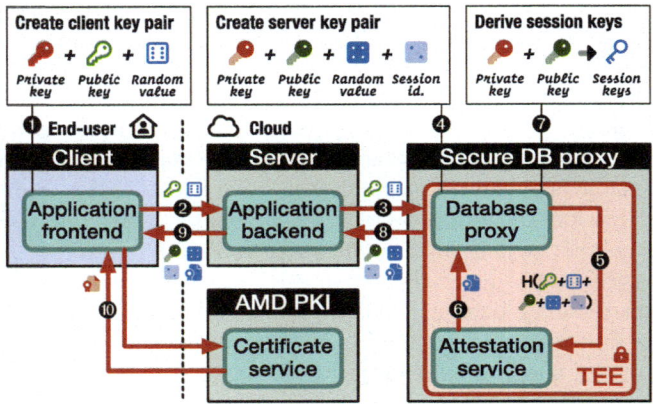

Fig. 2. Exchange of packets during session establishment

To initiate the key exchange, the *client* generates a random value and an ephemeral keypair (❶). It encodes the public key and random value and sends them to the *application backend* (❷), which sends them to the *proxy* using the custom SQL query (❸). The proxy generates its own random value, ephemeral keypair, and a session ID to identify this particular session in further communication (❹). To attest its trustworthiness, the proxy then requests an attestation to the TEE secure processor, passing a hash of both random values, both public keys, and the session ID, as custom data (❺–❻).

To complete the key exchange, the *client* and *proxy* each combine their own ephemeral private key with the other party's ephemeral public key, and derive two final *session keys*, one for each protocol direction (client to proxy and proxy to client), which they store in memory (❼). Finally the response is sent back to the backend (❽) and the frontend (❾). In addition, the *client* needs to assess the trustworthiness of the *proxy*. To do so, it fetches the root certificates and hardware public key from the TEE vendor (❿), and ensures the attestation returned by the *proxy* matches the key exchange, to be signed by the correct public key.

Because the client typically runs in a browser, we choose cryptographic primitives that are available in the Web Cryptography API [15]: we use Elliptic-Curve Diffie-Hellman (ECDH) on the NIST P-256 curve [4] for the key exchange, we derive the key using HKDF [16] instantiated with SHA-256. Finally we encrypt subsequent messages exchanged in the session with AES-GCM.

Obtaining Long-Term Keys. The *session keys* used to encrypt data in transit between them for the duration of a communication session have a short lifetime (i.e., a few hours). For long-term storage in the database, the proxy derives a different set of *long-term* keys, that are preserved across sessions. For each user, the proxy stores in the database an encrypted master long-term key, which can only be unlocked by logging in on an established session. Then, each confidential column in the database is encrypted using a different key, derived from such master long-term key. After successfully initiating a session, the client can use the custom procedures REGISTER and LOGIN, to register a long-term key and unlock a previously stored long-term key, respectively. Both procedures take as arguments a username and a password, the latter encrypted using the session key.

Registration. To register a user given its username and password, the proxy generates the master long-term key randomly, then encrypts it using a key derived from the user password and a salt, and finally stores the encrypted key and salt in the database. This method simplifies user password's changes, as only the master key needs to be re-encrypted. To derive the temporary key from the password, we use the Argon2 [7] PBKDF, with Chacha20-Poly1305 [5,6] AEAD cipher to encrypt the long-term key.

Login. To log a user in, the proxy retrieves its associated record (salt and encrypted master key) from the database, recomputes the intermediate key using the password and salt, and uses that to decrypt the long-term key. On a successful login, the proxy associates the decrypted long-term key with the session ID and keeps this mapping in secure memory during the whole session. It also returns a success response code to the application backend, used to log the user in the application with the same request.

Persistent Storage and Blind Indices. BLINDEXTEE uses two different sets of keys: session keys to encrypt data in transit between the client and proxy, and long-term keys to encrypt data at rest in the database. We detail here how we encrypt the confidential data and how to derive blind indices, thus enabling processing some filter queries directly in the database without leaking sensitive data.

Encrypting confidential columns. Data stored in a confidential column is encrypted using a key specific to the user and the column in which the data is inserted. That key is derived using HKDF [16] with SHA-256, using the user long-term key as the key, and the table and column names as *info*. The chosen encryption scheme guarantees *semantic security*, meaning that multiple encryptions of the same plaintext with the same key give different ciphertexts. In addition, it is authenticated, so any modification to the ciphertext prevents future decryption. The single-use nonce is randomly generated when encrypting data and added as prefix to the encrypted value.

Blind indices. Because the encryption scheme preserves semantic security, it is not feasible to use the encrypted columns directly to filter data, even for simple equality queries. A simple solution to this problem is to process filters over encrypted data entirely in the proxy, by decrypting all rows and only returning those for which the plaintext matches the filters. This approach works well when the number of records to filter is small, e.g., because the plaintext filters in the query reduced that number. However, this solution does not fully leverage the capabilities of the database server to efficiently filter data.

A more efficient solution is to insert additional data alongside encrypted columns, to allow the database server to filter the data partially, called *blind indices*[1]. A blind index is a hash derived from the user key and the plaintext value and is truncated, hence multiple different plaintexts can give the same blind-index value. Given two identical blind-index values, an adversary cannot confirm if the original plaintexts are also identical. When data is inserted or updated in the database, the associated blind-indices are automatically computed by BLINDEXTEE and transparently added to the query. When querying over an encrypted column, BLINDEXTEE computes the associated blind index and replaces it in the query. It then decrypts the values and re-filters the plaintexts, as multiple plaintexts give the same blind index.

The size of the blind index is determined manually on a per-column basis, as a function of the size of the plaintext space of that column. Given a blind index of size n bits with r records, the average number of collisions (that is, identical blind index values for different plaintexts) is $r \times 2^{-n}$, assuming that the size of the input space is itself bigger than 2^n. To achieve an average number of collision $C \geq 2$ over r records or more, we should therefore have $n \geq \log_2 r - \log_2 C$ bits. The average number of collisions should be kept higher than 2 for security, and lower than $\sqrt{r}$ to meaningfully impact filtering performance [1].

Query Processing. When the proxy receives a query, it parses it and accesses its configuration to check if it operates over encrypted columns. If so, the proxy identifies the client from which the query originates, decrypts and re-encrypts its data, sends it to the database, and then decrypts and re-encrypts its results.

Internally, the response parsing is implemented using an iterator. Whenever the client is ready to receive a new row, it is pulled from the iterator, which reads the next row from the database, decrypts it, applies filters, and then re-encrypts it. Using iterators allows to process arbitrary numbers of rows without being bounded by the proxy's memory. It also reduces latency, as the proxy can send the first row without waiting on all subsequent ones.

Retrieving the long-term keys from a query. A single proxy is designed to handle thousands of end-users connected at the same time. It must discern between the different end-users issuing queries, to encrypt and decrypt values for its owner and to recover the correct long-term key. A 64 bits *session ID* is issued to the client during key-establishment, and associated with the long-term key at login. That session ID is additionally used as associated data when authenticating each encrypted value in each query.

To associate a request with a session ID, we offer two options: prefixing the encrypted values, or providing the ID directly. In the first case, the *client*, when it performs a query with confidential data, prefixes each encrypted value with its session ID before transmitting them. The proxy then tries to recover a session ID in each encrypted value it receives. In the second case, the *application backend* appends a special function call to its query, SESSION_ID(*sid*). The proxy detects this function in the WHERE clause of a query and removes it.

We observe that the proxy should only decrypt values for confidential columns. If the proxy decrypts all received encrypted values, it may write a decrypted value to a cleartext column. Similarly, the proxy should only allow filtering over encrypted columns using encrypted values, to prevent an adversary from verifying if a value is present in the encrypted values and to which record it corresponds.

Double-filtering. When the proxy receives a SELECT or UPDATE query that filters over encrypted columns, it needs to filter the data in two steps. First, the filters need to be replaced with their blind indices equivalents (if available) in the request transmitted to the database. Second, the rows returned by the database need to be re-filtered by the proxy. To ensure the practicality of the second step, the proxy transparently adds the columns used in the WHERE clause of the query to its projection, ensuring these columns will be returned by the database server.

We note that this filtering method prevents computing aggregation operations (e.g., SUM, COUNT, AVG) directly in the database server if the filters include encrypted columns, as the database server will include values that should be excluded in its aggregate. For UPDATE queries, we transform them into a transaction: first select the individual identifiers of all the rows matching the filter, and then update rows matching these identifiers.

5   Security Analysis

Threat Model. We assume a powerful adversary with entire control over the software and hardware stack. His goal is to gain information about the confidential data transmitted by the *end user*. Denial of service attacks, or other attacks altering the correct behavior of the application, such as dropping queries, cloning encrypted data or rolling back data, are out of scope. These attacks are generally addressed by orthogonal solutions, e.g., monotonic counters [13]. In the following, we further refine the threat model.

DBMS. The DBMS and the server on which it runs are entirely untrusted. An adversarial database administrator can execute arbitrary read and write queries on stored data, and may also use the application as an end-user to try to access data.

Database proxy. The database proxy runs in a TEE. The adversary has access to the physical machine on which the TEE runs, and can modify the disk image of the virtual machine before launching it. The TEE prevents the adversary from reading or modifying the memory of the virtual machine while it's running. Side-channel attacks targeting TEEs are outside the scope of our threat model, and mitigations exist [17,18,25,26,28,31].

Application backend. The application backend is fully untrusted. An adversarial application developer can inspect any data going through the application backend, and can implement arbitrary modifications to the backend.

Client. The client code can be audited by the end user before execution and is therefore entirely trusted. Attacks targeting the client code are out of scope of this threat-model.

Security Measures. Considering the aforementioned thread model, our approach implements the following security measures.

Data at rest. Confidentiality of data stored in the DBMS is ensured by the use of a semantically secure authenticated encryption scheme, ChaCha20-Poly1305. The security of the encryption keys depends on the strength of the user's password, but the use of a memory-hard key derivation function, Argon2, makes offline brute force attacks inefficient. The use of incorrectly sized blind indices may reveal when two plaintexts are equal, therefore boundaries specified in Sect. 4 must be respected to reduce this risk. The use of different keys for different users prevents confused deputy attacks in which the adversary changes the user associated with a record to gain access to it.

Data in the proxy. The proxy is a critical component as it accesses the session keys and long-term keys of any logged-in user. It is protected from its environment by the use of the AMD SEV-SNP [2] TEE, which prevents compromised OS processes from accessing its memory and, therefore, the keys or confidential data. The risk of accidentally leaking cryptographic materials through programming errors is reduced by using a memory-safe language (Rust) and by relying on standard and audited cryptography libraries. In addition, the proxy only keeps keys in memory for users currently logged in, reducing the impact of a critical TEE vulnerability to only those users.

Key exchange. Its role is both to establish a secure channel between the client and proxy, and for the client to establish trust in the proxy before exchanging sensitive data. We establish trust in the content of the virtual hard drive, including the OS and system files, and the compiled code of the database proxy. The integrity of these components is asserted through remote attestation [20], and the client contains the necessary code measurements to assess the authenticity of the proxy. A hash of the entire key exchange is included in the attestation to prevent man in the middle attacks.

Data in transit. Data in transit between the client and the proxy is encrypted using the AES cipher in GCM mode. To prevent catastrophic nonce reuse, each direction of transit uses a different key and a counter which is incremented for each encrypted value. The client does not exchange any confidential data with the proxy until it has completed the key exchange and appraised the attestation.

6   Evaluation

We present here our extensive experimental evaluation of BLINDEXTEE. We implemented our prototype in 5500 lines of Rust, used for its low level interface and memory safety features, both desirable in a TEE. The proxy uses generic data types which are DBMS agnostic, and a DBMS specific implementation layer that translates these data-types to the underlying DBMS protocol. We demonstrate an integration with the MySQL protocol running in the AMD SEV-SNP [2] TEE for its guest attestation features. Approximately 1700 lines of code are specific to the MySQL implementation, and about 50 are specific to AMD SEV, including the attestation data structure.

Experimental Setup. Two servers were used: an AMD EPYC 9124 (16 cores, 3 GHz) hosting the database and BLINDEXTEE, and an AMD EPYC 7302P (16 cores, 3 GHz) hosting the application backend and benchmarking client. MySQL 8.2 ran in a Docker container; BLINDEXTEE operated in an 8-core AMD SEV-SNP VM with Ubuntu 24.04 LTS and AMD-SEV Linux kernel 6.8.0-35. NodeJS 20.15 was used for the application backend and benchmarking suite.

End-To-End Performance Tests. We simulated a real-world use case with the components in §3 using a custom-built NodeJS server application and benchmarking client for managing patients. We generated test data including confidential names and social security numbers (SSN), encrypted in the tests.

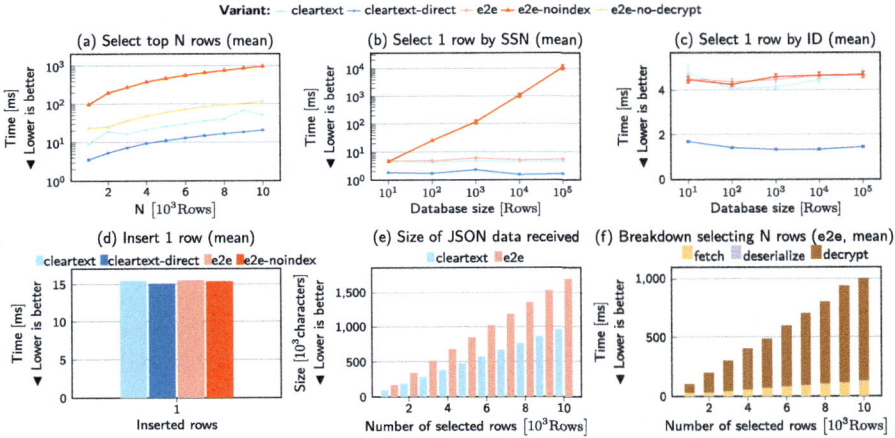

Fig. 3. Results of end-to-end performance test

Figure 3(a)–(e) compares the cleartext app (`cleartext-direct`), its variant through BLINDEXTEE without encryption (`cleartext`), and our encryption proxy with (`e2e`) and without (`e2e-noindex`) SSN blind indexing. Graph (a) includes `e2e-no-decrypt`, a variant of `e2e` where the client does not decrypt the data after receiving it, which isolates proxy performance. We measure the time between request encryption and issuance, and between response reception and decryption.

Graph (a) shows a `SELECT * ... LIMIT N` query with no filter and a random offset. Total time grows linearly with the number of rows returned for all variants. Client-side decryption dominates `e2e` time, confirmed by graph (f), which breaks down the time for selecting N rows with `e2e` into HTTP query `fetch`, JSON deserialization `deserialize`, and client-side decryption `decrypt`. The `e2e-no-decrypt` variant, which removes client decryption time, shows an average overhead of 115% compared to `cleartext`, or 7 µs per row. This overhead is likely dominated by row parsing and encryption, over initial query parsing. Comparing `cleartext` to `cleartext-direct` shows an average overhead of 169% (4 µs per row) for the proxy processing and additional network roundtrips. The complete overhead of the encrypted proxy (`e2e-no-decrypt` compared to `cleartext-direct`) is therefore 462%, or 11 µs per row. Running BLINDEX-TEE outside of SEV, we observed a much smaller overhead of 54% between `cleartext` and `cleartext-direct`, suggesting that the cost of emulation and of SEV in particular is a big part of the overall overhead of our system.

Plots (b) and (c) represent `SELECT *` queries with filters on SSN and ID, respectively, returning a single row. Except for `e2e-noindex`, query time remains constant in both tests, independent of database size. Comparing `e2e` and `cleartext` in (c) shows a 3% average slowdown (0.12 ms), while comparing with direct database access `cleartext-direct` shows a 237% slowdown (3.19 ms), suggesting query processing by BLINDEXTEE dominates the overhead. Graph (b) demonstrates the usefulness of blind indices when filtering over an encrypted column: in `e2e-noindex`, BLINDEXTEE retrieves and decrypts all rows to filter them, and the variant's runtime growing linearly with database size. Meanwhile in `e2e`, only rows that match the blind index are decrypted by the proxy, and the runtime remains constant.

Plot (d) shows an `INSERT` query for a single row. All variants achieve comparable times within the margin of error, suggesting that the base cost of inserting a row in the SQL database dominates.

Finally, in (e), we present the JSON payload size received by the client for cleartext or encrypted queries. The response comprises a JSON array with fields `id`, `doctorOfficeId`, `name`, `SSN`. Encryption consistently increases body size by 75%. This overhead is expected to decrease for larger encrypted values due to the fixed-size prefix nonce. However, base64 encoding inherently imposes a minimum 33% size increase.

The previous tests use small encrypted columns, each < 20 characters, and overhead is dominated by fixed costs such as query parsing and general row processing. Figure 4 illustrates the results of selecting N rows of varying sizes using the previously described variants to show how the overhead evolves with data size.

`e2e-no-decrypt`'s average overhead compared to `cleartext-direct` decreases from 225.4% for datasize 10^2 to just 36.1% for datasize 10^4. The overhead of `cleartext` compared to `cleartext-direct`, that is the pure overhead of our system without any encryption, also decreases with data size, from +60% for size 10^2 to 2% for size 10^4. This indicates that higher data-sizes reduce the impact of BLINDEXTEE's query processing and additional round trips, and the remaining encrypted overhead can be primarily attributed to decryption and encryption of data.

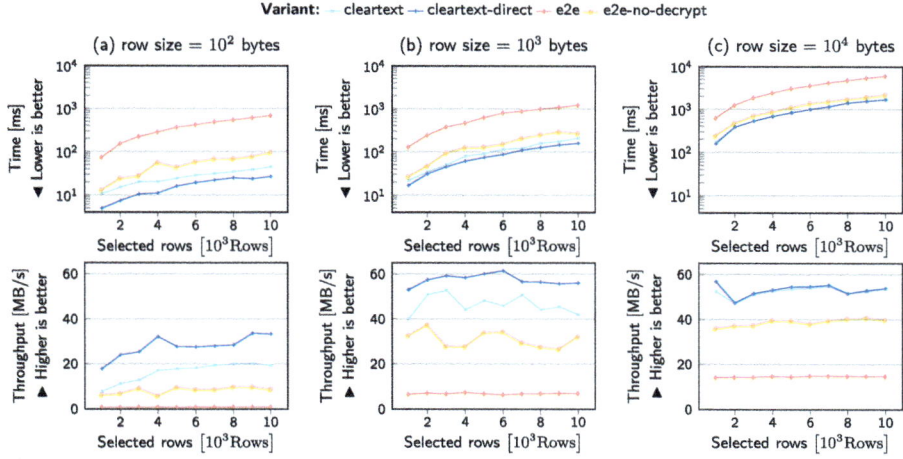

Fig. 4. End-to-end test results in various plaintext data sizes (expressed in bytes)

Micro-benchmarks To better understand how BLINDEXTEE behaves in certain specific conditions, we carried micro-benchmarks on the code handling the selection of rows. We present our results in Fig. 5.

In (a) and (b), we present the time and throughput for a SELECT query that returns N rows of different *row size*, respectively encrypted and in cleartext. To isolate the proxy overhead, we replace the underlying database call with an injected crafted response containing randomized data of the correct size. Because we test in isolation, we do not stream the results to the client, but instead iterate on each of them to consume the stream.

In (c), we test how long it takes our filtering and encryption/decryption iterator to process a single row, after its packet has been parsed but before it is passed to the caller, in different configurations. In the No filter variant, rows are decrypted and re-encrypted, but no filtering or projection is performed. The Transparent filter variant simulates a case of a query with a cleartext filter, so rows are decrypted, passed to a filter that does nothing, then re-encrypted. The 2-step Filter variant presents a query with an encrypted filter, in which rows are decrypted, then compared to a decrypted value, then re-encrypted. The counterpart is the Reject all filter variant, which is similar but in which the comparison value is different from the compared value, and so all rows are rejected. Finally, the No encryption variant quantifies the overhead of our custom iterator when no encryption or decryption operation happens.

We observe in both (a) and (b) that the processing time of the proxy mostly depends on the number of rows. This matches expectations and end-to-end tests, as each protocol packet must be parsed and its content decrypted and re-encrypted (in graph (a)), which leads to a per-row overhead. In addition, the throughput graphs reveal that this per-row cost is the dominant cost, as multi-

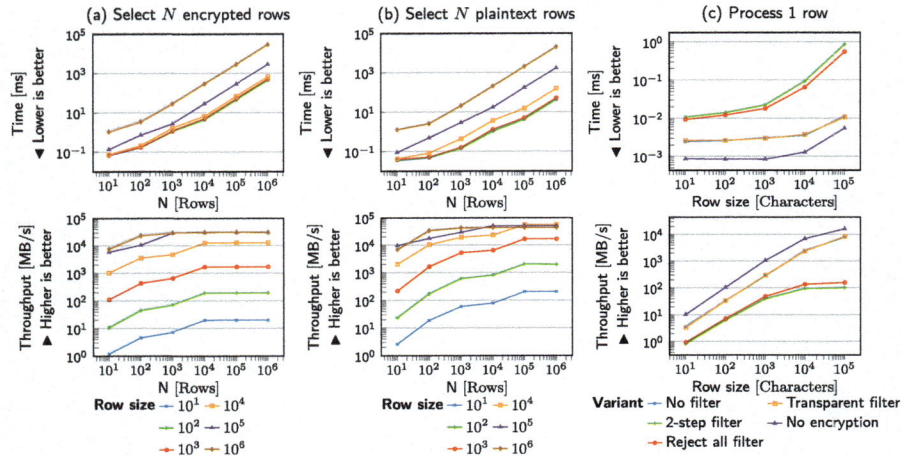

Fig. 5. Results of our micro-benchmarks

plying the row-size by ten roughly multiplies the throughput by the same factor, although this seems to reach a ceiling at 10^5 sized rows.

Graph (c) reveals the breakdown of the different costs involved in encryption and two-step filtering of a single row. As expected, using no encryption nor filtering gives the highest throughput of all variants. Variants with encryption/decryption but no filtering come just after. Finally, variants in which each row must be compared to a value for filtering are the slowest, and their throughput doesn't grow as fast as other variants, as using larger strings causes longer comparison times for each string.

7 Conclusion and Future Work

In this paper, we have presented an experimental system that enables encrypting data in an application using user specific keys that are not accessible to the operators of the application. We have shown how a TEE-enabled proxy can decrypt and re-encrypt this data, without leaking confidential information, in order to compute blind indices, which make further retrieval of data more efficient by leveraging the capabilities of the DBMS. We have demonstrated the practicality of BLINDEXTEE and presented some performance data, but because of its limited scope we could not compare our solution with other database protection solutions, thus limiting the relevance of our analysis.

Future work on this topic includes implementing some complex missing aggregations, such as aggregates or joins. Additionally, we intend to explore the benefits of using a TEE for encryption to implement other features, such as encrypting for groups of users, or computing statistics over encrypted data with differential privacy guarantees.

Acknowledgment. This work was supported by the Swiss National Science Foundation under project P4: Practical Privacy-Preserving Processing (No. 215216).

References

1. CipherSweet: Searchable Encryption Doesn't Have to be Bitter - Paragon Initiative Enterprises Blog (2019), https://paragonie.com/b/HXPUHJZVau577-Zg
2. AMD SEV-SNP: Strengthening VM Isolation with Integrity Protection and More. White paper (2020), http://bit.ly/4bJepse
3. Antonopoulos, P., Arasu, A., Singh, K.D., Eguro, K., et al.: Azure SQL database always encrypted. In: SIGMOD'20 (2020). https://doi.org/10.1145/3318464.3386141
4. Barker, E.: Digital signature standard riptsize (DSS). NIST (2013). https://doi.org/10.6028/NIST.FIPS.186-4
5. Bernstein, D.J.: The Poly1305-AES message-authentication code. In: Fast Software Encryption. Berlin, Heidelberg (2005). https://doi.org/10.1007/11502760_3
6. Bernstein, D.J.: ChaCha, a variant of Salsa20 (2008)
7. Biryukov, A., Dinu, D., Khovratovich, D.: Argon2: the memory-hard function for password hashing and other applications. Tech. rep. (2017)
8. Bitwarden: Bitwarden security whitepaper. White paper (2021), https://bitwarden.com/help/bitwarden-security-white-paper
9. Bloom, B.H.: Space/time trade-offs in hash coding with allowable errors. Commun. ACM **13**(7) (1970)
10. Diaconu, C., Freedman, C., Ismert, E., Larson, P., et al.: Hekaton: SQL server's memory-optimized OLTP engine. In: SIGMOD'13 (2013). https://doi.org/10.1145/2463676.2463710
11. Fuller, B., Varia, M., Yerukhimovich, A., Shen, E., et al.: SoK: cryptographically protected database search. In: SP'17 (2017). https://doi.org/10.1109/SP.2017.10
12. Gabel, M., Mechler, J.: Secure database outsourcing to the cloud: side-channels, counter-measures and trusted execution. In: CBMS'17 (2017). https://doi.org/10.1109/CBMS.2017.141
13. Gregor, F., Ozga, W., Vaucher, S., Pires, R., et al.: Trust management as a service: enabling trusted execution in the face of byzantine stakeholders. In: DSN'20 (2020). https://doi.org/10.1109/DSN48063.2020.00063
14. Grubbs, P., Lacharite, M.S., Minaud, B., Paterson, K.G.: Learning to reconstruct: statistical learning theory and encrypted database attacks. In: SP'19 (2019). https://doi.org/10.1109/SP.2019.00030
15. Huigens, D.: Web Cryptography API. W3C Editor's Draft (2024), https://w3c.github.io/webcrypto/
16. Krawczyk, H.: Cryptographic extraction and key derivation: the HKDF scheme (2010), https://eprint.iacr.org/2010/264, cryptology ePrint Archive, Paper 2010/264
17. Li, M., Wilke, L., Wichelmann, J., Eisenbarth, T., et al.: A systematic look at ciphertext side channels on AMD SEV-SNP. In: SP'22 (2022). https://doi.org/10.1109/SP46214.2022.9833768
18. Li, M., Zhang, Y., Wang, H., Li, K., et al.: CIPHERLEAKS: breaking constant-time cryptography on AMD SEV via the ciphertext side channel. In: USENIX Security 2021 (2021)

19. Lind, J., Priebe, C., Muthukumaran, D., O'Keeffe, D., et al.: Glamdring: automatic application partitioning for Intel SGX. In: ATC'17. USENIX (2017). https://doi.org/10.5555/3154690.3154718

20. Ménétrey, J., Göttel, C., Khurshid, A., Pasin, M., et al.: Attestation mechanisms for trusted execution environments demystified. In: DAIS'22. Lecture Notes in Computer Science, vol. 13272. Springer (2022).https://doi.org/10.1007/978-3-031-16092-9_7

21. Popa, R.A., Redfield, C.M.S., Zeldovich, N., Balakrishnan, H.: CryptDB: protecting confidentiality with encrypted query processing. In: SOSP'11 (2011). https://doi.org/10.1145/2043556.2043566

22. Priebe, C., Vaswani, K., Costa, M.: EnclaveDB: a secure database using SGX. In: SP'18 (2018). https://doi.org/10.1109/SP.2018.00025

23. Proton: What is end-to-end encryption and how does it work? (2023), https://proton.me/blog/what-is-end-to-end-encryption

24. Rescorla, E.: The Transport Layer Security (TLS) Protocol Version 1.3 (2018). https://doi.org/10.17487/RFC8446

25. Schlüter, B., Sridhara, S., Bertschi, A., Shinde, S.: WeSee: using malicious #VC interrupts to break AMD SEV-SNP. In: SP'24 (2024). https://doi.org/10.1109/SP54263.2024.00262

26. Schlüter, B., Sridhara, S., Kuhne, M., Bertschi, A., et al.: HECKLER: breaking confidential VMs with malicious interrupts. In: USENIX Security 2024 (2024). https://doi.org/10.48550/arXiv.2404.03387

27. Vinayagamurthy, D., Gribov, A., Gorbunov, S.: StealthDB: a scalable encrypted database with full SQL query support. Proc. Priv. Enhancing Technol. **2019**(3) (2019). https://doi.org/10.2478/POPETS-2019-0052

28. Wang, W., Li, M., Zhang, Y., Lin, Z.: Pwrleak: exploiting power reporting interface for side-channel attacks on AMD SEV. In: DIMVA'23. Lecture Notes in Computer Science, vol. 13959 (2023). https://doi.org/10.1007/978-3-031-35504-2_3

29. Wang, Y., Liu, L., Su, C., Ma, J., et al.: CryptSQLite: protecting data confidentiality of SQLite with Intel SGX. In: NaNA'17 (2017). https://doi.org/10.1109/NANA.2017.48

30. Yuhala, P., Ménétrey, J., Felber, P., Schiavoni, V., et al.: Montsalvat: Intel SGX shielding for GraalVM native images. In: Middleware '21. ACM (2021). https://doi.org/10.1145/3464298.3493406

31. Zhang, R., Gerlach, L., Weber, D., Hetterich, L., et al.: CacheWarp: software-based fault injection using selective state reset. In: USENIX Security 2024 (2024)

Tight Bounds for Constant-Round Domination on Graphs of High Girth and Low Expansion

Christoph Lenzen$^{(\boxtimes)}$ and Sophie Wenning

CISPA Helmholtz Center for Information Security, Saarbrücken, Germany
{lenzen,sophie.wenning}@cispa.de
lenzen@cispa.de, sophie.wenning@cispa.de

Abstract. A long-standing open question is which graph class is the most general one permitting constant-time constant-factor approximations for dominating sets. The approximation ratio has been bounded by increasingly general parameters such as genus, arboricity, or expansion of the input graph. Amiri and Wiederhake considered k-hop domination in graphs of bounded k-hop expansion and girth at least $4k+3$ [2]; the k-hop expansion $f(k)$ of a graph family denotes the maximum ratio of edges to nodes that can be achieved by contracting disjoint subgraphs of radius k and deleting nodes. In this setting, these authors to obtain a simple $O(k)$-round algorithm achieving approximation ratio $\Theta(kf(k))$. In this work, we study the same setting but derive tight bounds: – A $\Theta(kf(k))$-approximation is possible in k, but not $k-1$ rounds. – In $3k$ rounds an $O(k + f(k)^{k/(k+1)})$-approximation can be achieved. – No constant-round deterministic algorithm can achieve approximation ratio $o(k + f(k)^{k/(k+1)})$. Our upper bounds hold in the port numbering model with small messages, while the lower bounds apply to local algorithms, i.e., with arbitrary message size and unique identifiers. This means that the constant-time approximation ratio can be *sublinear* in the edge density of the graph, in a graph class which does not allow a constant approximation. This begs the question whether this is an artefact of the restriction to high girth or can be extended to all graphs of k-hop expansion $f(k)$.

1 Introduction

Given a graph $G = (V, E)$, a k-hop Minimum Dominating Set (MDS) $M \subseteq V$ minimizes $|M|$ under the constraint that all nodes are within distance at most k of a node in M. The classic case of $k = 1$ has been well-studied in the distributed setting, resulting in constant-time constant-factor approximations for a variety of sparse graph families: (i) outerplanar [3], (ii) planar [14,17,20,21], (iii) bounded genus [1,8], (iv) excluded K_t minor [7], (v) excluded $K_{2,t}$ minor [9],[1] and (vi)

Some proofs and figures are omitted from this version. See [18] for the full article.

[1] This work addresses k-hop domination. Moreover, the authors point out that the result can be generalized to require only exclusion of shallow $K_{2,t}$-minors.

T. Masuzawa et al. (Eds.): SSS 2024, LNCS 14931, pp. 277–291, 2025.
https://doi.org/10.1007/978-3-031-74498-3_20

bounded 1-hop expansion[2] [15]. All of these algorithms share the property that their approximation ratio is at least linear in the edge density of the input graph (family).

The same applies to a recent result by Amiri and Wiederhake [2], who consider the more general setting of k-hop domination, but restrict the input graphs to not only have bounded k-hop expansion, but also girth at least $4k + 3$; the girth of a graph is the length of a shortest cycle. They show how to achieve an approximation ratio of $O(kf(k))$ in $3k$ rounds.

In this work, we provide a number of results for k-hop domination on such graphs that are tight in various ways. Most prominently, for any fixed k, we provide a constant-time algorithm with approximation ratio *sublinear* in $f(k)$; surprisingly, a matching lower bound shows that there is no deterministic constant-time approximation independent of $f(k)$.

Detailed Contribution. All our results are presented in the port numbering model. Concretely, the network is represented by a connected graph $G = (V, E)$ of k-hop expansion at most $f(k)$ and girth at least $4k + 3$. Each node $v \in V$ initially only knows its incident edges, locally labeled by $1, 2, \ldots, \delta(v)$, where $\delta(v)$ is the degree of v. Computation proceeds in *synchronous* rounds, where in each round, $v \in V$ sends messages to neighbors, receives the messages of its neighbors (being aware of the port number of the connecting edge), and performs arbitrary deterministic local computations. The running time of an algorithm is the maximum number of rounds until all nodes terminate and output whether they belong to the selected k-hop dominating set.

Our algorithms use messages of size at most $\lceil \log \Delta \rceil$, where Δ is the maximum degree of a node. Our lower bounds extend to the Local Model, in which nodes have unique identifiers and messages unbounded size, which follows from known results [13]. However, they do not hold when randomization is available.

As a warm-up result, we provide a simpler and faster algorithm for achieving the approximation ratio of $O(kf(k))$ given in [2]. It runs for precisely k rounds and sends only empty messages. In each round, the algorithm logically deletes nodes of degree 1 (checking for the special case that all nodes are deleted); all nodes that remain are in the dominating set. The fact that this procedure results in approximation ratio $O(kf(k))$ is implicit in Wiederhake's thesis [22].

Theorem 1. *Algorithm 2 runs for k rounds, sends at most one empty message in each direction over each edge, and returns a k-hop dominating set that is at most by factor $2kf(k) + 1$ larger than the optimum.*

By considering a selection of trees of depth k, we establish that k rounds are necessary to achieve an approximation ratio that is independent of the number of nodes n and the maximum degree Δ; trees satisfy that $f(k) < 1$ for any k.

---

[2] A graph has r-hop expansion $f(r)$, if deleting nodes and contracting disjoint subgraphs of radius at most r results in ratio of at most $f(r)$ between edges and nodes. Contracting a subgraph means to replace it by a single node sharing an each with each other node that was adjacent to a node of the subgraph.

Theorem 2. *No $(k-1)$-round algorithm can an approximation ratio smaller than Δ, even if the input graph is guaranteed to be a tree.*

We remark that the fact that this lower bound is shown by invoking a family of trees is no coincidence. In fact, we show that the task becomes easier when excluding trees, as this eliminates the special case that the above simple procedure logically deletes all nodes.

Theorem 3. *Algorithm 1 runs for $k-1$ rounds and sends at most one empty message over each edge. If the input graph is further constrained to not be a tree, it returns a k-hop dominating set that is at most by factor $2kf(k)+1$ larger than the optimum.*

However, the improvement in running time is limited to a single round.

Theorem 4. *No $(k-2)$-round algorithm can achieve an approximation ratio smaller than Δ, even when G is guaranteed to not be a tree, have maximum degree $\Delta \in \mathbb{N}$, k-hop expansion $f(k) \leq 1$, and girth at least $g \leq n(\Delta-1)/\Delta^{k+1}$.*

Our main results are matching bounds on the approximation ratio that can be achieved by constant-round algorithms. The upper bound is obtained by an algorithm that is similar to that of Amiri and Wiederhake. We also use their key argument, that the optimum solution induces a Voronoi partition into trees, whose outgoing edges must connect to different cells and hence fully contribute to $f(k)$. However, new ideas are required to obtain the better approximation ratio. We exploit that the number of high-degree nodes that can be selected obeys a smaller bound due to the fact that they contribute (directly or indirectly) many edges leaving the Voronoi cell of their dominator. On the other hand, the number of selected low-degree nodes is limited due to a preference for choosing high-degree dominators. As the distinction between "low" and "high" is only needed in the analysis, our algorithm remains agnostic of $f(k)$.

Theorem 5. *Running Algorithm 3 and then Algorithm 4 requires message size $\lceil \log \Delta \rceil$ and $3k$ rounds, returning an $O(k+f(k)^{k/(k+1)})$-approximate k-hop MDS.*

Despite the similarities between the algorithms, our changes are necessary to achieve the stronger bound: in his thesis [22], Wiederhake constructs a graph on which the approximation ratio of the algorithm from [2] is at least $kf(k)$.

Our matching unconditional lower bound is derived from a graph of large girth g and uniform degree Δ, where Δ is the largest integer satisfying $\Delta(\Delta-1)^k/2 \leq f(k)$. By ensuring that the graph is also bipartite, we obtain a Δ-coloring of the edges. Thus, we can assign port numbers by edge color, resulting in identical views of all nodes, regardless of running time. This forces any port numbering algorithm to select all nodes.

Graphs of uniform degree and large girth are known to exist [12], cf. [5] for an English version we base our proof on. However, in addition we need to make sure that the graph actually has a small k-hop dominating set. To this end, we slightly modify Erdös' proof of existence. Instead of inductively increasing the degree of a uniform graph of high girth, we "plant" a small k-hop dominating set M by starting from a forest of $|M|$ trees of depth k and uniform degree Δ of inner nodes. We then follow Erdös' approach, but limited to the set of leaves of such trees, inductively increasing their degree to Δ while maintaining the original trees.

Theorem 6. *Any algorithm has approximation ratio* $\Omega(f(k)^{k/(k+1)})$.

We note that an $\Omega(k)$ lower bound follows from the inability to break symmetry on rings; see [19] for the classic $\Omega(\log^* n)$ lower bound for the setting with IDs.
Further Related Work. Above, we exclusively discussed constant-time constant-approximation algorithms, i.e., neither running time nor approximation ratio should depend on the number of nodes n or the maximum degree Δ of the graph. Due to the sheer size of the body of work on distributed dominating set approximation, a survey would be required to do it justice. Accordingly, we confine ourselves to a few points putting our work into context.

- An equivalent definition of a k-hop dominating set is to ask for a dominating set of the k-th power of the input graph. Therefore, up to a possible factor of at most k in running time, considering $k > 1$ makes a difference only when message size is bounded.
- Approximating minimum dominating sets better than factor $\ln \Delta$ is NP-hard [10]. More general "sparse" graph classes, e.g., graphs of bounded arboricity α, include all graphs of maximum degree α. Therefore, one should not expect constant approximations by simple—more precisely, computationally efficient—algorithms. However, this might be different for more restrictive graph classes, and this hardness result leaves a lot of room for sublinear approximations in terms of density parameters.
- As little as $O(\log^* n)$ rounds make a substantial difference when unique identifiers are available. For instance, deterministic k-hop domination on rings with approximation ratio independent of k requires $\Theta(\log^* n)$ rounds [4,19]. Moreover, $\Theta(\log^* n)$ rounds are enough to turn several, if not all, of the above constant-factor approximations into approximation schemes [1,6,7,9].
- In general graphs, there is a constant $c > 0$ such that in r rounds, one cannot achieve approximation ratio better than $\Omega(n^{c/r^2}/r)$ or $\Omega(\Delta^{1/(r+1)}/r)$ as function of n and Δ, respectively [16]. These bounds are matched up to small factors by randomized algorithms using large messages [16].
- In graphs of arboricity α within $O(\log \Delta)$ rounds, a deterministic $O(\alpha)$-approximation is feasible [11]. A reduction from the lower bound in [16] shows that this running time is asymptotically optimal. Note that the class of graphs of bounded arboricity contains all the above graph classes for which constant-time constant-factor approximations are known.

2   Time-Optimal Algorithms

In this section, we characterize the minimum round complexity for achieving a constant-factor approximation, i.e., one that depends on k and $f(k)$ only. Due to space constraints, most proofs are deferred to the appendix.

Structural Characterization and Pruning Algorithms. We start by specifying a set of nodes that is "safe" to select in the sense that it is at most factor $2kf(k)$ larger than a minimum k-hop dominating set. In contrast to later material, the results in this subsection are mostly implicit in [22]. Wiederhake proved the structural properties shown here to bound the approximation ratio of the algorithm given in [2], without taking the step of presenting the resulting simpler and faster algorithms we give here. Moreover, we provide what we consider a simplified exposition with minor additions serving our needs in later sections. The easiest way to describe the node set we are interested in is by an extremely simple algorithm: delete nodes of degree 1 for k iterations and keep what remains. Note that this requires only $k - 1$ rounds of communication, cf. Algorithm 1.

Algorithm 1 Pruning algorithm, code at node v.

1: **for** $k - 1$ rounds **do**
2:     **if** exactly one neighbor has not sent "del" to v yet **then**
3:         send "del" to this neighbor
4: **if** sent "del" or exactly one neighbor has not sent "del" to v **then**
5:     **return** false
6: **else**
7:     **return** true

Note that a node outputting *true* deletes himself. For all presented algorithms, the return values denote the output of a node $v \in V$. For convenience, we introduce the following notation.

Definition 1. (*Pruned Graph*) Throughout this work, we denote by $G' = (V', E')$ the input graph and by $G = (V, E)$ the *pruned* (sub)graph induced by the nodes that output true when running Algorithm 1.

It is possible that G is empty, but this happens only if G' is a tree of depth at most k.

Observation 1. *If G' is not a tree, then $V \neq \emptyset$.*

We address the special case $V = \emptyset$ later, in Algorithm 2. Otherwise, the pruning algorithm results in a subgraph containing a k-hop MDS of the input graph G'.

Lemma 1. *If $V \neq \emptyset$, there is a k-hop MDS $M \subseteq V$ of G' with the following property. For $v \in V$, let*

$$r_v := \begin{cases} 0 & \text{if } v \text{ has no neighbor in } V' \setminus V \\ r & \text{if } r \text{ is the latest round in which a neighbor of } v \text{ sent "del"} \\ k & \text{if no neighbor of } v \text{ in } V' \setminus V \neq \emptyset \text{ sent a "del" message,} \end{cases}$$

i.e., r_v is the latest iteration of Algorithm 1 when a neighbor of v is deleted. Then there is $m \in M$ within distance $k - r_v$ of v; note that this distance as well as the shortest path depends on G. Moreover, any set with these properties is a k-hop dominating set of G'.

We now show that, due to the high girth and small k-hop expansion of G' and thus also G, $|V|$ is not much larger than $|M|$. To this end, we partition G with respect to M.

Definition 2. (*MDS Voronoi decomposition*). Let $\{T_m\}_{m \in M}$ be a Voronoi partition of G with respect to the k-hop MDS M given by Lemma 1, i.e., T_m contains all $v \in V$ for which m is the closest dominating node (ties are broken arbitrarily). For $v \in V$, denote by $m_v \in M$ the node such that $v \in T_m$.

Since the girth is at least $4k + 3$, for any two distinct $m, m' \in M$, there can be at most one edge between the trees the partitions T_m and $T_{m'}$ induce. On the other hand, the pruning ensures that each leaf in such a tree has a neighbor in a different tree.

Lemma 2. *For each $m \in M$, T_m induces a tree of depth at most k (when rooted at m). Unless $T_m = \{m\}$ is a singleton, each leaf of T_m has at least one edge to a node outside of T_m. No two edges leaving T_m connect to the same $T_{m'}$ for some $m' \in M \setminus \{m\}$.*

For ease of notation, we will use T_m to refer to both the node set and its induced tree.

Together, the above properties imply that there are at most $2f(k)|M|$ leaves in the trees T_m, $m \in M$, in total. Using that the depth of each T_m is at most k, this bounds $|V|$.

Lemma 3. $|V| \leq (2kf(k) + 1)|M|$.

Proof. (*of Theorem 3*). By Observation 1, $V \neq \emptyset$. By Lemma 1, V thus is a k-hop dominating set of G'. By Lemma 3, $|V| \leq (2kf(k) + 1)|M|$.

As we will discuss shortly, the exclusion of trees is indeed necessary for $k - 1$ rounds to be sufficient to obtain a good approximation ratio. However, unsurprisingly this special case can be addressed with little overhead. Adding one more round of communication, it can be checked whether $V = \emptyset$. If this is the case, this can be locally fixed by re-adding the (at most two) nodes that have been deleted last.

Algorithm 2 Pruning algorithm with fallback for trees, code at v.

1: Execute Algorithm 1.
2: **if** v returned false and did not send "del" **then**
3:     send "del" to the neigbhor from which no "del" was received
4: **if** v returned true or both sent and received a "del" message in the same round **then**
5:     **return** true
6: **else**
7:     **return** false

Lemma 4. *If $V \neq \emptyset$, Algorithm 2 returns V. If $V = \emptyset$, Algorithm 2 returns a k-hop MDS of G' of size at most 2.*

Proof. *(Theorem of 1).* If $V \neq \emptyset$, by Lemma 4 the returned set is V. By Lemma 1, V then is a k-hop dominating set of G' and by Lemma 3 the approximation guarantee is met. If $V = \emptyset$, by Lemma 4 the returned set is a k-hop dominating set of size 2. As the graph contains at least 2 nodes and is connected, $|E'|/|V'| \geq (|V'|-1)/|V'| \geq 1/2$. We conclude that $(2kf(k)+1)|M| \geq (2 \cdot 1 \cdot 1/2 + 1) \cdot 1 = 2$, proving the approximation guarantee.

Lower Bounds Showing Strict Time-Optimality. To show that the above simple pruning algorithms are indeed time-optimal for achieving a constant approximation ratio, we present matching lower bounds. First, we show that k rounds are needed to do better than a Δ-approximation, even in trees of maximum degree Δ.

Proof. *(Theorem of 2).* Fix any port numbering algorithm that outputs a k-hop dominating set on input graphs from a family that contains trees of maximum degree Δ. For $i \in \{1, \ldots, \Delta\}$, denote by T_i the unique properly Δ-edge-colored $(\Delta - 1)$-ary tree of depth $k - 1$ in which the "missing" color among the edges incident to the root is i. Consider the following properly port-numbered trees.

– For $i \in \{1, \ldots, \Delta\}$, take two copies of T_i, connect their roots with an edge of color i, and assign port numbers to match edge colors.
– Take one copy of T_i for each $i \in \{1, \ldots, \Delta\}$ and connect it by an edge of color i to a single root node. Again, we assign port numbers to match edge colors.

Observe that for each node in each T_i, the port-numbered $(k-1)$-hop neighborhood in the first graph for i and the second graph are isomorphic: all non-leaves have Δ neighbors with the port numbers of each traversed edge being the same at its endpoints, and the leaves within distance $k - 1$ are exactly those of T_i.

For each $i \in \{1, \ldots, \Delta\}$, in the first graph the algorithm must select *some* node to ensure k-hop domination. By indistinguishability, for each i the algorithm will select the corresponding node in the copy of T_i in the second graph as well. Hence, at least Δ nodes are selected in the second graph, yet the root node k-hop dominates the entire graph.

In other words, even infinite girth and $f(k) = 1$ is not good enough for a constant approximation within $k - 1$ rounds. In [13], it is shown how to lift this result to the local model.

Corollary 1. (of Theorem 1.3 in [13]). *Theorem 2 applies also when nodes have unique IDs.*

At first glance, it might seem odd that the lower bound in the port numbering model solely relies on trees. However, as demonstrated by Theorem 3, excluding trees enables us to compute a constant-factor approximation within $k-1$ rounds. As it turns out, this improvement is limited to precisely one round.

Proof. *(Theorem of 4).* The special case $\Delta = 2$ is addressed by considering cycles of $n \in (2k+1)\mathbb{N}$ nodes with alternating port numbers, in which a port numbering algorithm selects all nodes, but a k-hop MDS has size $n/(2k+1)$. Hence, suppose that $\Delta \geq 3$ in the following.

Denote by T a complete rooted tree of depth k and uniform degree $\Delta \geq 3$ at all inner nodes. Properly edge-color T with Δ colors, giving rise to a port numbering in which each port number is given by the color of the corresponding edge. Now take a cycle C of length g, replace each of its nodes by a copy of T, and choose an arbitrary leaf in each such copy. For each edge in C, connect the chosen leaves of the trees corresponding to the endpoints of the edge. We make the following observations about the resulting graph H.

- H is a pseudo-forest with one cycle of length g, consisting of the edges corresponding to those of C.
- H has a k-hop dominating set of g nodes, consisting of the roots of the trees.
- H has $f(k) = 1$: pseudo-forests are closed under taking minors and have no more edges than nodes.
- The $(k-2)$-hop neighborhoods of the tree roots and their children are all isomorphic to one another, as any (former) leaf is in distance k from the root.

It follows that any port numbering algorithm running in $k-2$ or fewer rounds must either select all roots and their children or none of these nodes. The former results in approximation ratio at least $\Delta + 1$.

We claim that the latter implies approximation ratio at least Δ. To see this, select in each of the above subtrees rooted at the child of a root one leaf, where for the subtree with the leaf receiving additional cycle edges we maximize the distance to this leaf. Observe that

- any pair of such leafs in the same tree is in distance $2k$, with the only node within distance k of both of them being the root, and
- any pair of such leafs in different trees is in distance at least $2 \cdot 2(k-1)+1 > 2k$; here we used that $k \geq 2$ or trivially no $(k-2)$-round algorithms exist.

As the roots are not selected, this entails that no selected node k-hop dominates more than one of these leaves. The claimed lower bound of Δ also follows in this case. Since H satisfies all constraints imposed by the theorem (girth $g = n/|T| \geq n(\Delta-1)/\Delta^k$, $f(k) \leq 1$, maximum degree Δ, and containing a cycle), this completes the proof.

Corollary 2. (of Theorem 1.4 in [13]). *Theorem 4 applies also when nodes have unique IDs.*

3   Tight Approximation

In this section, we show that the approximation ratio that can be achieved in constant time is $\Theta(k + f(k)^{k/(k+1)})$. In fact, it turns out that in the port numbering model no algorithm can do better. In the graph we construct for the lower bound, all views are identical, regardless of the number of rounds of communication. Hence, unique identifiers or randomization are required to achieve better approximation guarantees. Several proofs are deferred to the appendix.

Upper Bound. To improve over Algorithm 2, we first apply the same pruning procedure. We then select dominators from the candidate set given by the nodes returning true. To this end, each node $v \in V$ with at least one deleted neighbor takes note of the maximum distance r_v of a node that needs to be covered via its deleted incident edges; we set $r_v := 0$ for nodes $v \in V$ without neighbor in $V' \setminus V$. Observe that the components of the subgraph induced by deleted nodes are trees, each of which is connected by a single deleted edge to a non-deleted node. Hence, there is a minimum k-hop dominating set without deleted nodes. Moreover, for any k-hop dominating set that contains no deleted nodes, it is necessary and sufficient that it contains for each $v \in V$ a dominator within distance $k - r_v$. If for $v \in V$ it holds that $r_v \neq 0$, it simply equals the latest round in which a neighbor is deleted, cf. Lemma 1. A simple modification of Algorithm 2 outputs r_v for each $v \in V$, see Algorithm 3.

Algorithm 3 Distance-aware pruning, code at v.

1: $r_v := 0$
2: **for** rounds $r \in \{1, \ldots, k\}$ **do**
3:     **if** exactly one neighbor has not sent "del" to v yet **then**
4:         send "del" to this neighbor
5:     **if** received "del" **then**
6:         $r_v := r$
7: **if** sent "del" and did not receive "del" the same round **then**
8:     **return** false
9: **else**
10:     **return** r_v

Lemma 5. *When executing Algorithm 3, $v \in V$ returns $r_v \in \{0, \ldots, k\}$ if and only if it returns true when Algorithm 2 is run instead. Moreover, if $V \neq \emptyset$, this return value is equal to the value r_v as defined in Lemma 1.*

Corollary 3. *If $V = \emptyset$, the nodes not returning false when executing Algorithm 3 constitute a k-hop dominating set of size 2. If $V \neq \emptyset$, any $D \subseteq V$ containing for each $v \in V$ some node in distance at most $k - r_v$ of v is a k-hop dominating set. Furthermore, there is a minimum k-hop dominating set $M \subseteq V$ of G' with this property.*

Except for the special case that $V = \emptyset$, in its second phase our algorithm selects a dominator for each $v \in V$. We choose a dominator of maximum degree (with respect to the subgraph induced by non-deleted nodes) within distance $k - r_v$, where ties are broken by distance and port numberings. Intuitively, this choice is good, because together with the high girth constraint selecting many high-degree nodes implies a high expansion. On the other hand, low-degree nodes can only be chosen by nodes for which all ancestors in T_{m_v}, the tree of the MDS Voronoi decomposition they participate in, also have low degree. Together, for any fixed k, this results in a bound on the approximation ratio that is sublinear in $f(k)$.

Using suitable pipelining, the implementation of the selection procedure requires only $2k$ rounds and messages of size $\lceil \log \Delta \rceil$. Broadcasting the maximum known degree in the subgraph induced by non-deleted nodes for k rounds lets each node learn the highest degree node within each distance $d \in \{0, \ldots, k\}$. Breaking ties by smallest round number and port number for which the respective value was received in the given round over the respective edge, nodes can correctly route selection messages in a pipelined fashion. The pseudocode for the second step is given in Algorithm 4.

Algorithm 4 Selection procedure, code at v. Only nodes which did not output false after running Algorithm 3 participate. Their output and the information on which ports a "del" message was received serves as input. For notational convenience, nodes may fictively send messages "to themselves" using port number 0.

1: $\Delta_0 := |\{p \in \{1, \ldots, \delta(v)\} \mid v \text{ did not receive "del" on port } p\}|$
2: **for** rounds $r \in \{1, \ldots, k\}$ **do**
3:     send Δ_{r-1} to all neighbors
4:     set Δ_r to the maximum of Δ_{r-1} and all received values and store the smallest respective port number (0 if $\Delta_r = \Delta_{r-1}$).
5: set $\Delta(v) := \Delta_{k-r_v}$ ▷ here r_v computed by Algorithm 3 is used
6: **for** rounds $r \in \{k+1, \ldots, 2k\}$ **do**
7:     **if** $\bot \neq \Delta(v) > \Delta_{2k-r}$ **then**
8:         send $\Delta(v)$ to the port stored for Δ_{2k-r+1} ▷ port is 0 if $\Delta_{2k-r+1} = \Delta_{2k-r}$
9:         set $\Delta(v) := \bot$
10:    **if** any values are received **then**
11:        set $\Delta(v)$ to the minimum received value ▷ includes own value if port was 0
12: **if** $\Delta(v) \neq \bot$ **then**
13:     **return** true
14: **else**
15:     **return** false

We first point out that Algorithm 4 correctly handles the special case that $V = \emptyset$.

Corollary 4. *If $V = \emptyset$, Algorithm 4 selects two nodes, which form a k-hop dominating set.*

Hence, in the following assume that $V \neq \emptyset$ in all statements.

Observation 2. *For each $v \in V$, the local variables Δ_r, $r \in \{0, \ldots, k\}$, store the maximum degree within r hops (with respect to G).*

Lemma 6. *The set of nodes returning true when executing Algorithm 4 after Algorithm 3 is a k-hop dominating set of G'. For each selected node $v \in V$, there is some $w \in V$ within distance $k - r_w$ of v such that $\Delta_{k-r_w}(w) = \Delta_0(v)$.*

It remains to prove the approximation guarantee. Denote by $D \subseteq V$ the set of nodes returning true when executing Algorithm 4 after Algorithm 3. For $d \in D$, arbitrarily fix a node $s_d \in V$ that "selected" it, i.e., that satisfies that d has maximal degree among all nodes within distance $k - r_{s_d}$ of s_d; such a node exists by Observation 2 and Lemma 6. Let $S := \bigcup_{d \in D}\{s_d\}$ and $\Delta \in \mathbb{N}$ be a degree threshold that we will fix later in the analysis. Abbreviate $D_\Delta := \{d \in D \mid \delta(d) \leq \Delta\}$ and $S_\Delta := \{s \in S \mid s = s_d \text{ for some } d \in D_\Delta\}$. Fix a k-hop MDS M such that for each $v \in V$, its dominator m_v is in distance at most $k - r_v$; such an M exists by Corollary 3.

We account for low-degree nodes in D by bounding the available number of nodes that might select them.

Lemma 7.
$$|D_\Delta| \leq \begin{cases} (2k+1)|M| & \text{if } \Delta \leq 2 \\ \frac{\Delta}{\Delta-2} \cdot (\Delta - 1)^k|M| & \text{else.} \end{cases}$$

Proof. Observe that $|D_\Delta| = |S_\Delta|$ by construction, so it is sufficient to bound $|S_\Delta|$. By definition of D_Δ and S_Δ, all nodes within distance $k - r_s$ of $s \in S_\Delta$ have degree at most Δ. In particular, this applies to s and its ancestors in T_{m_s}, i.e., in the tree induced by the Voronoi cell of its dominator, because m_s is within distance $k - r_s$ of s. Thus, each $s \in S_\Delta$ is contained in the connected component of m_s in T_{m_s} induced the nodes of degree at most Δ.

We conclude that $|S_\Delta|$ is bounded by $|M|$ times the maximum size of a tree of depth k whose inner nodes have uniform degree Δ. For $\Delta \leq 2$ this number is at most $2k + 1$, as the tree is then a path. For $\Delta \geq 3$, we get that

$$\frac{|S_\Delta|}{|M|} \leq 1 + \Delta \sum_{i=1}^{k}(\Delta - 1)^{i-1} \leq \Delta(\Delta - 1)^{k-1} \sum_{i=0}^{\infty}(\Delta - 1)^{-i} = \frac{\Delta}{\Delta - 2} \cdot (\Delta - 1)^k.$$

High-degree nodes in D imply a large number of edges leaving their dominator's tree. This contributes to increasing the k-hop expansion of the graph, i.e., we can bound the number of such nodes by means of the k-hop expansion $f(k)$.

Lemma 8. *If $\Delta \geq 2$, then $|D \setminus D_\Delta| \leq 4f(k)|M|/(\Delta - 1)$.*

Proof. Let $D' \subseteq D \setminus D_\Delta$ the set of nodes $d \in D \setminus D_\Delta$ satisfying that there is at most one descendant $d' \in T_{m_d}$ that is also in $D \setminus D_\Delta$. For such a $d \in D'$, let T_d be the subtree rooted at d after removing the subtree rooted at d' (if there is such a descendant d'). Denote by E_d the set of edges that connect a node in T_d to a node outside T_m. We attribute all edges that leave T_m from the subtree rooted at d after removing the subtree rooted at the child of d containing a descendant in $T_m \cap (D \setminus D_\Delta)$ (if there is one). Note that by Lemma 2, each leaf of T_m has at least one incident edge leaving T_m (unless this leaf is m and has no incident edge, which implies that $V = M = \{m\}$). As d has degree at least $\Delta + 1$, this implies that there are at least $\Delta - 1$ such edges; apart from "losing" up to one edge due to d', one connects to the parent of d.

Observe that each edge attributed to d can be attributed to at most one other node, an ancestor of the edge's endpoint in $T_{m'} \in D \setminus D_\Delta$, where $m' \neq m$ is the dominator of the endpoint of the edge. With this in mind, we contract T_m for all $m \in M$. As the depth of each T_m is at most k, the resulting graph has at most $f(k)|M|$ edges. Invoking Lemma 2 once more, none of the edges between trees are removed by the contractions, as each of them connects trees T_m and $T_{m'}$ for a unique pair $m, m' \in M$.

Therefore, after the contraction the sum of degrees is at least

$$\sum_{d \in D'} |E_d| \geq \sum_{d \in D'} \Delta - 1 = (\Delta - 1)|D'|.$$

Applying the upper bound on the number of edges, we conclude that $|D'| \leq 2f(k)|M|/(\Delta - 1)$.

Finally, it remains to bound $|D \setminus (D_\Delta \cup D')|$. To this end, observe that by definition of D', each $d \in D \setminus (D_\Delta \cup D')$ has at least 2 descendants in $D \setminus D_\Delta$. Thus, contracting in each T_m edges with at most one endpoint in $D \setminus D_\Delta$ until this process stops, we end up with a forest with the following properties.

- The node set can be mapped one-on-one to $D \setminus D_\Delta$.
- Nodes corresponding to those of D' have degree at most 2.
- Nodes corresponding to those of $D \setminus (D_\Delta \cup D')$ have degree at least 3.

Further removing degree-2 nodes by contractions thus results in a forest in which inner nodes map one-on-one to those of $D \setminus (D_\Delta \cup D')$ and the number of leaves is bounded from above by $|D'|$. Therefore,

$$|D \setminus D_\Delta| = |D'| + |D \setminus (D_\Delta \cup D')| < 2|D'| \leq \frac{4f(k)|M|}{\Delta - 1}.$$

Choosing Δ suitably, we arrive at Theorem 5.

Proof. *(Theorem of 5).* By Corollary 4 and Lemma 6, running Algorithm 3 followed by Algorithm 4 results in a k-hop dominating set. The running time and message size are immediate from the pseudocode. If $f(k)^{1/k+1} < 2$, choosing $\Delta = 2$ and applying Lemmas 7 and 8 yields approximation ratio $O(k + f(k)) = O(k + f(k)^{k/(k+1)})$. If $f(k)^{1/(k+1)} \geq 2$, choosing $\Delta := \lfloor f(k)^{1/(k+1)} \rfloor$ and applying Lemmas 7 and 8 yields approximation ratio $O(k + f(k)/\Delta) = O(k + f(k)^{k/(k+1)})$.

Lower Bound Our lower bound is based on a graph of uniform degree Δ. We first prove a helper statement relating the k-hop expansion to this degree bound.

Lemma 9. *Any graph has k-hop expansion $f(k) \leq \Delta(\Delta - 1)^k/2$.*

Lemma 10. *Let $\Delta, k \in \mathbb{N}$ with $\Delta \geq 3$. Then there is a graph of uniform degree Δ, girth at least $g := 4k + 3$, k-hop expansion $f(k) \leq \Delta(\Delta - 1)^k/2$, and a k-hop dominating set of size $O(|V|\Delta/f(k))$.*

Proof. Suppose that $m := m(\Delta, k) \in 2\mathbb{N}$ is sufficiently large. Let G_1 be the disjoint union of m rooted trees of depth k, where inner nodes have degree Δ. Clearly, G_1 has a k-hop dominating set of m nodes, and adding edges between leaves of the trees will not change this. Note that each tree has $1 + \sum_{i=0}^{k-1} \Delta(\Delta - 1)^i \in \Theta(\Delta(\Delta - 1)^{k-1}) = \Omega(f(k)/\Delta)$ nodes. Hence, $m \in O(|V|\Delta/f(k))$.

We claim that we can add edges between leaves in G_1 to obtain a graph G_Δ in which these nodes have all degree Δ and the girth is at least g. By Lemma 9, G_Δ then also has k-hop expansion $f(k)$. As adding edges cannot remove the property that a set is a k-hop dominating set, proving the claim will prove the lemma.

We prove the claim inductively, by constructing G_i, $i \in \{2, \ldots, \Delta\}$, where G_i is G_1 with additional edges between leaves of G_1 such that these nodes all have degree i. The base case of $i = 1$ is given by G_1. For the step from $i \in \{1, \ldots, \Delta-1\}$ to G_{i+1}, let G be a graph maximizing the number of edges among all graphs that are G_1 with additional edges between leaves so that (i) the degree in G_i of leaves of G_1 is between i and $i + 1$ and (ii) girth is at least g. By the induction hypothesis, G_i is G_1 with additional edges and meets conditions (i) and (ii), implying that G is well-defined. We will show that the minimum degree of G is $i + 1$, completing the induction step and thereby the proof.

To this end, assume for contradiction that there is a node of degree i in G. Denote by $G' = (V', E')$ the subgraph of G induced by the leaves in G_1. It contains $\ell \geq 1$ nodes of degree $i - 1$ and $|V'| - \ell$ nodes of degree i. As the number of leaves in G_1 is a multiple of $m \in 2\mathbb{N}$, $|V'|$ is even. Using that $2|E'| = (i - 1)\ell + i(|V'| - \ell) = i|V'| - \ell$, we conclude that ℓ is also even. In particular, $\ell \geq 2$, i.e., there are at least two nodes of degree i in G.

Denote by $v, w \in V'$ two nodes of degree i in G. Denote by N the set of nodes with distance at most $g - 2$ in G from both v and w. If there is $x \notin N$ of degree $i - 1$, we can add an edge from x to v or w without decreasing the girth below g, contradicting the maximality of G. Thus, each node in $V' \setminus N$ has degree $i + 1$ in G and i neighbors in V'. On the other hand, the constraints on G imply that nodes in $V' \setminus N$ have at most i neighbors in V'.

Observe that the size of N is bounded by $1 + \sum_{i=1}^{g-2} \Delta(\Delta - 1)^{i-1} = 1 + \sum_{i=1}^{4k} \Delta(\Delta - 1)^{i-1}$. Thus, as $m(\Delta, k)$ is sufficiently large, $2|N| \leq |V'|$. It follows that $i|N \cap V'| < i|N| \leq i|V' \setminus N|$. Therefore, G must contain an edge $\{x, y\}$ with both endpoints $x, y \in V' \setminus N$. However, this also leads to a contradiction to the maximality of G: We can delete $\{x, y\}$ and add $\{v, x\}$ and $\{w, y\}$; there can be no short cycle involving only one of the two new edges, since the distances from v to x and from w to y are at least $g - 1$, and there can be no short cycle involving both $\{v, x\}$ and $\{w, y\}$, since otherwise $G \setminus \{x, y\}$ would contain a short path from x to y, i.e., G would have too small girth.

Considering the bipartite double cover, we can ensure that the constructed graph can be Δ-edge colored. □

Lemma 11. *There is a graph as in Lemma 10 that is also balanced bipartite.*

Corollary 5. *The graph given by Lemma 11 can be properly Δ-edge colored.*

Theorem 6 now readily follows, as choosing port numbers in a graph of uniform degree Δ according to a Δ-edge coloring results in identical views at all nodes.

Proof. *(Theorem of* 6*)*. Consider the graph given by Lemma 11 with port numbers matching the colors of a proper Δ-edge coloring, which is feasible by Corollary 5. In this graph, all nodes have identical views, regardless of the number of rounds of computation. In more detail, initially all nodes have identical state. Thus, on each port $i \in \{1, \ldots, \Delta\}$, each node sends the same message, meaning that each node receives this message on its port i. By induction, this implies that for each round of computation, all nodes end up in the same state. Hence, any port numbering algorithm must select all nodes into a k-hop dominating set, yet only $O(|V|\Delta/f(k))$ many are required. As the graph of Lemma 11 satisfies that $f(k) = \Delta(\Delta - 1)^k/2$, for any choice of $\Delta \geq 2$ the approximation ratio is $\Omega((\Delta - 1)^k) = \Omega(f(k)^{k/(k+1)})$.

Applying [13], this extends to constant-round algorithms in the Local model.

Corollary 6. *Every algorithm with running time depending on k and $f(k)$ has approximation ratio $\Omega(f(k)^{k/(k+1)})$, even if nodes have unique identifiers.*

References

1. Amiri, S.A., Schmid, S., Siebertz, S.: Distributed dominating set approximations beyond planar graphs. ACM Trans. Algorithm. **15**(3), 39:1–39:18 (2019)
2. Amiri, S.A., Wiederhake, B.: Distributed distance-r covering problems on sparse high-girth graphs. Theor. Comput. Sci. **906**, 18–31 (2022)
3. Bonamy, M., Cook, L., Groenland, C., Wesolek, A.: A tight local algorithm for the minimum dominating set problem in outerplanar graphs. In: 35th international symposium on distributed computing, DISC 2021, 4–8 Oct 2021, Freiburg, Germany (Virtual Conference). LIPIcs, vol. 209, pp. 13:1–13:18. Schloss Dagstuhl - Leibniz-Zentrum für Informatik (2021)
4. Cole, R., Vishkin, U.: Deterministic coin tossing with applications to optimal parallel list ranking. Inf. Control **70**(1), 32–53 (1986)
5. Coupette, C., Lenzen, C.: A Breezing Proof of the KMW Bound. CoRR, abs/2002.06005 (2020)
6. Czygrinow, A., Hanckowiak, M., Wawrzyniak, W.: Fast distributed approximations in planar graphs. In: Taubenfeld, G. (ed.) Distributed Computing, 22nd International Symposium, DISC 2008, Arcachon, France, 22–24 Sept 2008. Proceedings. Lecture Notes in Computer Science, pp. 78–92. Springer (2008)
7. Czygrinow, A., Hanckowiak, M., Wawrzyniak, W., Witkowski, M.: Distributed approximation algorithms for the minimum dominating set in K_h-Minor-Free graphs. In: Hsu, W.-L., Lee, D.-T., Liao, C.-S. (eds.) 29th International Symposium on Algorithms and Computation, ISAAC 2018, 16–19 Dec 2018, Jiaoxi, Yilan, Taiwan. LIPIcs, vol. 123, pp. 22:1–22:12. Schloss Dagstuhl - Leibniz-Zentrum für Informatik (2018)
8. Czygrinow, A., Hanckowiak, M., Wawrzyniak, W., Witkowski, M.: Distributed CONGESTBC constant approximation of MDS in bounded genus graphs. Theor. Comput. Sci. **757**, 1–10 (2019)

9. Czygrinow, A., Hanckowiak, M., Witkowski, M.: Distributed distance domination in graphs with no K_2, t-minor. Theor. Comput. Sci. **916**, 22–30 (2022)
10. Dinur, I., Steurer, D.: Analytical approach to parallel repetition. In Shmoys, D.B. (ed.) Symposium on Theory of Computing, STOC 2014, New York, NY, USA, May 31 – June 03, 2014, pp. 624–633. ACM (2014)
11. Dory, M., Ghaffari, M., Ilchi, S.: Near-optimal distributed dominating set in bounded arboricity graphs. In: Milani, A., Woelfel, P. (eds.) PODC '22: ACM Symposium on Principles of Distributed Computing, Salerno, Italy, July 25–29, 2022, pp. 292–300. ACM (2022)
12. Erdös, P., Sachs, H.: Eine Abschätzung für die minimale Knotenzahl regulärer Graphen, welche keinen Kreis der Länge <l enthalten. Wiss. Z. Univ. Halle **XII**(3), 251–258 (1963)
13. Göös, M., Hirvonen, J., Suomela, J.: Lower bounds for local approximation. J. ACM **60**(5), 39:1–39:23 (2013)
14. Hilke, M., Lenzen, C., Suomela, J.: Brief announcement: local approximability of minimum dominating set on planar graphs. In: Halldórsson, M.M., Dolev, S. (eds.) ACM Symposium on Principles of Distributed Computing, PODC '14, Paris, France, 15–18 July 2014, pp. 344–346. ACM (2014)
15. Kublenz, S., Siebertz, S., Vigny, A.: Constant round distributed domination on graph classes with bounded expansion. In: Jurdzinski, T., Schmid, S. (eds.) Structural Information and Communication Complexity—28th International Colloquium, SIROCCO 2021, Wrocław, Poland, June 28–July 1, 2021, Proceedings. Lecture Notes in Computer Science, vol. 12810, pp. 334–351. Springer (2021)
16. Kuhn, F., Moscibroda, T., Wattenhofer, R.: Local computation: lower and upper bounds. J. ACM **63**(2), 17:1–17:44 (2016)
17. Lenzen, C., Pignolet, Y.-A., Wattenhofer, R.: Distributed minimum dominating set approximations in restricted families of graphs. Distrib. Comput. **26**(2), 119–137 (2013)
18. Lenzen, C., Wenning, S.: Tight Bounds for Constant-Round Domination on Graphs of High Girth and Low Expansion (2024). https://arxiv.org/abs/2408.12998
19. Linial, N.: Locality in distributed graph algorithms. SIAM J. Comput. **21**(1), 193–201 (1992)
20. Wawrzyniak, W.: A strengthened analysis of a local algorithm for the minimum dominating set problem in planar graphs. Inf. Process. Lett. **114**(3), 94–98 (2014)
21. Wawrzyniak, W.: A local approximation algorithm for minimum dominating set problem in anonymous planar networks. Distrib. Comput. **28**(5), 321–331 (2015)
22. Wiederhake, B.: Pulse propagation, graph cover, and packet forwarding. PhD thesis, Saarland University, Saarbrücken, Germany (2022)

Adding All Flavors: A Hybrid Random Number Generator for dApps and Web3

Ranjith Chodavarapu[1], Rabimba Karanjai[2], Xinxin Fan[3], Weidong Shi[2], and Lei Xu[1(✉)]

[1] Kent State University, Kent, OH 44224, USA
rchodava@kent.edu, xuleimath@gmail.com
[2] University of Houston, Houston, TX 77204, USA
rkaranja@cougarnet.uh.edu, wshi3@central.uh.edu
[3] IoTeX, Menlo Park, CA 94025, USA
xinxin@iotex.io

Abstract. Random numbers play a vital role in many decentralized applications (dApps), such as gaming and decentralized finance (DeFi) applications. Existing random number provision mechanisms can be roughly divided into two categories, on-chain, and off-chain. On-chain approaches usually rely on the blockchain as the major input and all computations are done by blockchain nodes. The major risk for this type of method is that the input itself is susceptible to the adversary's influence. Off-chain approaches, as the name suggested, complete the generation without the involvement of blockchain nodes and share the result directly with a dApp. These mechanisms usually have a strong security assumption and high complexity. To mitigate these limitations and provide a framework that allows a dApp to balance different factors involved in random number generation, we propose a hybrid random number generation solution that leverages IoT devices equipped with trusted execution environment (TEE) as the randomness sources, and then utilizes a set of cryptographic tools to aggregate the multiple sources and obtain the final random number that can be consumed by the dApp. The new approach only needs one honest random source to guarantee the unbiasedness of the final random number and a user can configure the system to tolerate malicious participants who can refuse to respond to avoid unfavored results. We also provide a concrete construction that can further reduce the on-chain computation complexity to lower the cost of the solution in practice. We evaluate the computation and gas costs to demonstrate the effectiveness of the improvement.

1 Introduction

Random numbers play an important role in a wide range of applications, such as machine learning, simulation, and cryptography. When it comes to blockchain, random numbers are critical for not only the blockchain construction itself (e.g., implementation of proof-of-stake), but also dApps deployed on top of it (e.g.,

T. Masuzawa et al. (Eds.): SSS 2024, LNCS 14931, pp. 292–306, 2025.
https://doi.org/10.1007/978-3-031-74498-3_21

gaming and NFT). An adversary can gain extra advantages if it can manipulate or influence the random numbers involved.

A good random number, or a good random number sequence, should meet two requirements: (i) *Unpredictable*. One should not learn the value before it is released or guess future a random number from historical information; and (ii) *Unbiased*. The value should follow the uniform distribution, i.e., each value is selected with the same probability; These two requirements are not independent. It is easy to see that if the random numbers generated are biased, it is easier for the adversary to predict it. When the random number is generated in a hostile environment, there are extra security related requirements. For instance, an adversary should not be able to influence the value directly/indirectly. In the context of blockchain and dApp, it is also desirable that the random number generation is verifiable by all participants. In this paper, we focus on random number generation used in a decentralized environment, especially dApps.

Random number generation can be classified by various criteria. Based on the original source, random number generators can be divided into *pseudo random number generator (PRNG)* and *true random number generation (TRNG)*. A TRNG utilizes unpredictable physical sources to generate random numbers, and a PRNG uses a mathematical algorithm and a seed value to generate a sequence of numbers. While it is widely believed that TRNG offers random numbers with better quality, most random numbers used in practice are created by PRNGs, because TRNG requires hardware that is not always available. From the perspective of dApps, the random number generation mechanisms are usually divided into two categories, on-chain RNG and off-chain RNG. An on-chain RNG usually uses the blockchain information as a seed to produce random numbers. This approach is convenient for the dApp to verify and consume the generated random number but the seed is suspectable to the influence of an adversary. On the other hand, an off-chain RNG does not rely on the blockchain contents to generate random numbers but the process is usually opaque to the dApp and the dApp participants need to trust that this process is not manipulated.

To better serve the demands of dApps on random numbers, we propose HRNG, a hybrid RNG solution that keeps the advantages of all four types of RNGs while minimizing related limitations. HRNG takes advantage of the emerging DePIN technology (decentralized physical infrastructure network [1]) to solve the problem of introducing physical randomness sources to the blockchain world. The idea of DePIN is to connect a large number of IoT devices to the blockchain and allow them to trade their generated data on the blockchain [2]. When an IoT device is equipped with a hardware-based trusted execution environment (TEE), it can utilize the hardware to work as a TRNG. If all nodes of the blockchain trust the IoT device, the problem of obtaining a random number becomes trivial. For instance, the dApp can request a random number from the IoT device, and the TEE hardware of the IoT device responds with a generated number and a digital signature. As the device is trusted, one only needs to verify whether the provided number and the digital signature are consistent, and does not need to worry that the device colludes with the dApp.

However, as a hardware-aided security enhancement mechanism, TEE is not 100% reliable. In other words, it is possible that an adversary compromises the TEE hardware and manipulates the generation of random numbers [3]. The good thing is that existing attacks against TEE are usually not scalable, i.e., it is hard to compromise a large number of IoT devices equipped with TEE simultaneously. Therefore, HRNG utilizes a group of TEE-enabled IoT devices and aggregates their outputs to provide random numbers to dApps to tolerate the potential risk of compromised hardware. Specifically, HRNG uses multiple TEE devices to create a pool of random numbers, and allows a dApp to combine multiple values from the pool using a preferred method to obtain the final random number it needs as long as it meets the security requirement. Even if some the input random numbers are manipulated by an adversary, the combination mechanism can still guarantee that the final result is not biased. Compared with existing random number generation mechanisms for dApps, HRNG provides the following benefits: (i) *TRNG-based*. HRNG utilizes TRNGs as the source to build the final random numbers for dApps, and it enjoys all the advantages of TRNG. (ii) *Robustness*. HRNG can not only tolerate IoT devices with compromised TEE, but also other malicious participants in the random number generation process under reasonable security assumptions. (iii) *Verifiability*. Except that any third party can verify that the final random number that is consumed by the dApp is obtained following the predefined aggregation method.

In summary, our contributions in the paper include: (i) We provide the details of the design of HRNG that utilizes TRNGs as a building block to generate random numbers for dApps. The design is generic and highly customizable, i.e., one can easily instantiate the system with preferred cryptographic tools based on the demands; and (ii) We propose a concrete construction of HRNG that utilizes the homomorphic feature of the underlying commitment scheme to significantly reduce the on-chain computation complexity, which makes the random numbers more affordable for dApps.

2   Background and Related Works

Existing random number generation mechanisms for dApps are roughly divided into two categories, off-chain random number generation and on-chain random number generation. In this section, we briefly review these mechanisms and compare them with HRNG.

Off-chain random number generation. Using an off-chain source for random number generation is the most convenient approach from an engineering perspective. Randao [4] is a random number generation mechanism used in the beacon chain to provide in-protocol randomness in Ethereum 2.0 [5]. The design of Randao follows the commit-reveal approach to incrementally gather randomness from participants. Motivated by the design of RanShare [6], the Near protocol described a randomness beacon scheme [7] that inherits the randomness properties of Randshare and can tolerate up to 2/3 malicious actors before one can influence the output. Chainlink built a provably fair and verifiable random number generator based on verifiable random functions [8]. Chainlink VRF relies on a decentralized oracle network and can generate one or multiple random numbers together with a cryptographic proof in a single smart contract call. The proof, which can attest to the correctness of the random number generation process, is published and verified on-chain before a dApp consumes the generated random numbers. The off-chain collaborative random number generation schemes like Randao often require multiple participants and communication rounds, thereby incurring significant costs and delays. The VRF-based approach, on the other hand, relies on strong protection of private keys used for computing VRFs.

On-chain random number generation. Creating random numbers on-chain is another option. One straightforward approach for on-chain random number generation is using the block contents (e.g., block headers, block heights, timestamps, smart contract states, etc.) as the random source [9,10]. The ERC721R [11,12] standard describes the process for producing random numbers using the aforementioned on-chain metadata. While it is easy for a smart contract to access various on-chain data, it is extremely difficult, if not impossible, to prevent on-chain random number generators from being exploited by attackers [9,10].

Table 1. Comparison of HRNG with existing random number generation methods.

Method	True Random source	Verifiability	Tolerance of adversaries
On-chain	○	●	◑
Off-chain	○	◑	◑
HRNG	●	●	●

●: Yes   ○: No   ◑: Partially

This work. Compared with existing on-chain and off-chain random number provision methods, HRNG provides a new way to split the on-chain and off-chain workload, and utilizes TEE technology to incorporate TRNG into the process of random number generation, which is not available with existing on-chain/off-chain approaches. Furthermore, HRNG also considers the associated centralization risk in the design. Table 1 summarizes the comparison of these methods, and more detailed analyses are provided in later sections.

3 System Overview, Security Assumptions and Problem Statement

In this section, we first present the overview of the decentralized pseudo-random number generation with TEE technology, and then explain the technique challenges.

Overview of HRNG. Fig. 1 shows the major participants involved in the random number generation process. (i) *IoT device.* An IoT device utilizes equipped TEE to create random numbers to contribute to the provision of final random numbers to dApps. An IoT device only interacts with the gateway it connects with. (ii) *Gateway*. A gateway works as the proxy of a group of IoT devices. The gateway has more computation/communication capacity. It can relay messages generated by connected IoT devices to other destinations and process received IoT messages before sending them out. (iii) *Random number pool*. The random pool is a storage service that holds information submitted by gateways and allows the blockchain nodes/gateways to retrieve information. (iv) *dApp*. A dApp is an application deployed on the blockchain, and its execution requires a random number as input. Different random numbers may lead to completely different execution results. (v) *Blockchain*. The blockchain serves as an immutable ledger to store the necessary information and it is also responsible for the execution of certain computations to support the operation of the whole protocol.

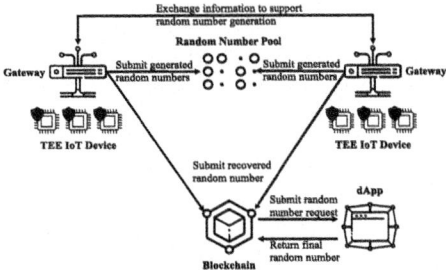

Fig. 1. Overview of HRNG. Each gateway connects with multiple IoT devices equipped with TEE to provide random sources to the system, and the final random number is generated in a verifiable manner on the blockchain. Although the dApp is demonstrated as a different entity, it is actually deployed on the same blockchain and maintained by the same set of blockchain nodes.

The execution of the protocol consists of four major steps: (i) *Publish the random numbers pool*. During this step, each gateway collects random numbers generated by connected IoT devices, processes them, and publishes processing results to the pool; (ii) *Obtain instructions from the dApp*. The dApp creates a request for a random number, which includes specifications such as the number of sources and the aggregation method; (iii) *Respond to the dApp request on a random number*. The gateways obtain the request of the dApp, and work together to generate responses, which are published to the blockchain; and (iv) *Construct the final random number*. The blockchain (through the smart contract deployed on the blockchain) follows the request provided by the dApp to build the final random number, which can be consumed by the dApp. Note that the execution of the protocol follows the sequence strictly, i.e., one step only starts when the previous one finishes.

Security assumptions. The major security concern is that an adversary influences the random number generation process and forces the protocol to generate a biased number based on its preference, which may cause various consequences depending on the nature of the dApp. In the extreme case, the adversary controls all components of the system (i.e., all gateways and connected IoT devices), and there is no way to prevent it from choosing a preferred random number. In this work, we consider a more practical scenario where the adversary can only compromise a fraction of each type of participant. Specifically, (i) *Compromised IoT devices*. We assume hardware-aided TEE is breakable but hard to compromise at a large scale. Although we assume each IoT device is equipped with TEE, we consider the case that the TEE hardware can be compromised and the IoT device provides biased random numbers or refuses to. At the same time, we assume that only a subset of IoT devices in the system are compromised. (ii) *Compromised gateways*. We assume a subset of gateways in the system are malicious but not all of them. A malicious gateway may refuse to respond to the system and/or deviate from the protocol to maximize the benefit of the colluded dApp. (iii) *Trusted random number pool*. We assume the random number pool is a reliable storage system, i.e., it is always available and can safely store received information. (iv) *Trusted blockchain*. Following the common assumption of a blockchain system, we assume that some blockchain nodes can be malicious but the blockchain is trusted as a whole. This also applies to the dApp deployed on the blockchain. Note that we also assume the security statuses of all participants are static during one round execution of the protocol. For ease of description, we summarize notations that are used in the rest of the paper in Table 2.

Table 2. Summary of notations.

Notation	Description
n_g	Number of gateways in the system
n_i	Number of IoT devices that connect to a gateway
n_r	Number of random numbers an IoT device generates in one round of protocol
n_{mg}	Number of malicious gateways in the system, and $n_{mg} < n_g$
n_{mi}	Number of malicious IoT devices connected to the same gateway, and $n_{mi} < n_i$
ℓ	Number of source random numbers selected by the dApp for final random number generation

Problem statement. HRNG aims at guaranteeing generated numbers are uniformly distributed and unpredictable by an adversary under the security assumptions. Specifically, HRNG needs to address the following risks: (i) Compromised gateways. A gateway controlled by an adversary can abandon an unfavored random number generated by a connected IoT device, and refuse to respond according to the protocol based on the preference of the adversary; (ii) Compromised IoT devices. An IoT device (and the associated TEE) controlled by an adversary can generate biased random numbers, or discard a generated random number if the adversary cannot take advantage of it.

4   Detailed Random Number Generation Protocol

We present the detailed design of HRNG in this section.

4.1   Random Number Generation on IoT Devices

We assume each of the IoT devices d_i creates a random number rn_i. To facilitate the verification of the authenticity of rn_i, the device d_i also calculates a digital signature of the random number $\sigma_{d_i}(rn_i)$. We assume that the IoT device is equipped with TEE, both the random number generation and signature generation are done inside the TEE and it is not likely that an attacker can manipulate these operations.

As the IoT device d_i does not connect to the blockchain network directly, the generated message $(rn_i, \sigma_{d_i}(rn_i))$ is sent to the corresponding gateway for further processing.

4.2   Random Number Processing on Gateways

After receiving a message of a random number from a connected IoT device, the gateway cannot put it into the random number pool directly as the dApp may take advantage of it. In other words, the numbers should only be visible to the dApp when it cannot make any choices. To achieve this goal, HRNG utilizes a commitment scheme, which satisfies two features: (i) Hiding. Hiding prevents one from learning information about the original message by observing the commitment. (ii) Binding. Binding guarantees that it is hard/impossible to open a commitment to a different value. These two security features can be formalized in different flavors, such as computation hiding/binding and perfect hiding/binding. The selection of a commitment scheme will not affect HRNG, as long as the commitment scheme's security features are acceptable. The only requirement is that the commitment scheme is non-interactive [13]. Specifically, the gateway utilizes a *commitment scheme* to hide the received random number before sending it to the pool. Each gateway runs the `Setup` when it joins the system. During this stage, the gateway runs the `Commit` algorithm on the generated random numbers one by one and submits all committed values to the pool. The *hiding* feature of the commitment scheme guarantees that no one can extract information from the original random number. At the same time, the *binding* feature of the commitment scheme prevents the gateway from modifying the committed value at a later stage.

The committed value has to be opened in a later stage when the dApp uses this particular random number. The gateway prepares for the opening operation with a threshold secret-sharing scheme. A secure (k, t)-threshold secret sharing scheme guarantees that with t or more splits of the message, one can reconstruct the message. But with less than t splits, it is hard/impossible to rebuild the original message. For each committed random number, the corresponding gateway uses the (k, t)-threshold secret sharing scheme to break the committed value into k pieces, which are distributed to gateways in the system. The values k and t are related to the security assumptions and system configuration. For instance, there must be enough gateways (i.e., $n_g \geq k$) and the number of honest gateways needs to be greater than the threshold ($n_g - n_{mg} \geq t$) to prevent potential denial-of-service attacks. More details will be discussed in Sect. 5.

4.3   Random Number Request by dApp

The dApp is responsible for generating a request for a random number, which is aggregated using multiple inputs from the pool for security purposes. The request must include two major parts that define the aggregation process, i.e., the way to select numbers in the pool and the method to aggregate selected numbers. This request is distributed to all gateways in the system. For security purposes, HRNG verifies the request based on two criteria: (i) The request involves enough number of gateways. The minimum number of gateways needed in the request is determined by the security assumption, i.e., the number of gateways that are compromised and collude with the dApp in the system. (ii) The aggregation algorithm only needs a few unbiased inputs to guarantee the output is unbiased. This criterion essentially means the aggregation algorithm needs to be able to tolerate biased inputs as much as possible. There are different ways to achieve this goal. The simplest one is treating each input as a binary string and using XOR operator to connect all operands, which only requires one unbiased operand to guarantee the output is unbiased. In most cases, a dApp does not have special requirements on the final random number except unbiasedness, and this simple XOR-based aggregation algorithm is enough.

4.4   Response to the Request on Gateways

After a request is created and submitted by the dApp, the gateways collaborate to help the dApp to compute the final random number. The final random number is calculated through two steps:

– Revealing committed random numbers in the pool. For a committed value, the opening information is shared with multiple gateways, and a subset of gateways can collaborate to open the committed random numbers. Because of the threshold feature, as long as
– Aggregating numbers in the pool to produce the final random number. This operation is straightforward when all involved elements in the pool are correctly opened. When the pool is accessible by the public, every blockchain node can easily verify the correctness of the final random number as the aggregation method is usually simple.

4.5   Some Implementation Considerations

We discuss the implementation of some components of HRNG. These design and implementation choices do not affect the architecture of HRNG, though they may affect its performance and reliability. We do not consider parallel protocol execution in this work, and use the term "round" for a complete execution of the protocol.

Construction of the random number pool. The random number pool serves as a temporary information storage system to hold generated random numbers of the current round (in the form of commitments). It can also be used to store splits of commitments open information (each split and its corresponding commitment should be cryptographically protected). There are several options to build this random number pool. For instance, a centralized FTP server can work as a pool, though it will reduce the decentralization level. It is also possible to let each gateway to store everything, or utilize a decentralized storage system such as IPFS or other peer-to-peer storage mechanism [14] as the random number pool. Each option for a random number pool has its pros and cons, and can be chosen based on the actual environment.

The blockchain system. HRNG can work with all types of blockchain systems and dApps deployed atop. As the blockchain is involved in the random number generation process (i.e., verification/opening of commitments and aggregation), its security features affect the reliability of these operations. For instance, if the blockchain is constructed using a BFT style consensus protocol [15], an adversary controls more than 1/3 nodes participating the consensus can disable HRNG or manipulate the final random numbers arbitrarily.

The TEE hardware. HRNG uses TEE as the source of random numbers but does not rely on any specific TEE solution. Compared with other parts of the protocol, the cost associated with TEE is negligible.

5   Security and Performance Analysis

5.1   Security Analysis

There are two major strategies for the dApp and its colluders to manipulate the random number generation to maximize its benefits:

- Active attack. For this type of attack, the dApp and its colluders actively modify the random number generation process to obtain preferred random numbers. Major active attacks include: (i) An IoT device always generates random numbers that the dApp prefers; and (ii) The dApp always selects gateways that collude with it in the request for a random number.
- Passive attack. For this type of attack, the dApp and its colluders do not actively change any information but skip certain numbers or ignore unfavorable requests. Major passive attacks include: (i) An IoT device discards a generated random number that is unfavorable for the dApp. While this attack requires the IoT device to actively skip generated numbers, the device does not need to modify anything. Therefore, we still classify this attack as a passive attack; (ii) A gateway refuses to collaborate to reveal committed numbers that are unfavorable for the dApp to force the protocol to terminate.

HRNG can prevent both types of attacks under the security assumptions discussed earlier.

Prevention of Active Attack. It does not matter what strategy the adversary takes, its ultimate goal is to manipulate one or more steps in the random number generation process to make the final value biased in a preferred manner. HRNG can thwart this type of attacks because it enforces a checking on the dApp's random number request, which guarantees two features: (i) The adversary cannot control all inputs. According to the security assumptions, only a subset of IoT devices/gateways can be compromised. HRNG only needs to ensure that the aggregation algorithm involves more inputs than the maximum number values the adversary can control. (ii) The aggregation output cannot be influenced by a subset of inputs. Given an arbitrary aggregation algorithm, it is a hard problem to determine whether any single input can guarantee unbiasedness. But considering the normal requirements of the dApp on the final random number, HRNG only needs to support XOR operation for aggregation, and this feature is preserved.

Prevention of Passive Attacks. The nature of a passive attack is "denial-of-service", i.e., when the adversary expects that an operation will lead to non-preferred results, it controls compromised entities to stop responding. This type of attack is usually hard to prevent as it is not easy to force an entity to do certain things. Most existing decentralized systems use an incentive mechanism to discourage such behaviors (e.g., offering rewards for taking actions and enforcing penalties for not taking actions). HRNG adopts a different strategy to deal with this type of attack. (i) Prevention of passive attacks in the process of random number generation on the IoT device. From a high level, HRNG divides the random generation into two stages. In the first stage, IoT devices/gateways contribute random numbers to a pool and the adversary does not know which will be included in aggregation. Therefore, the adversary does not have the motivation to control compromised IoT devices/gateways to refuse to participate in the protocol. (ii) Prevention of passive attacks in the process of final random number generation. HRNG prevents passive attacks in the aggregation stage by utilizing the threshold secret-sharing scheme to distribute the response capability to multiple entities. Based on security assumptions, there is always a subset of honest gateways. As long as the number of honest gateways exceeds the threshold, random numbers in the pool can be recovered.

5.2   Performance Analysis

In this section, we consider the performance of HRNG with a generic commitment scheme and threshold secret sharing scheme. A dApp has two performance requirements for HRNG: (i) Low latency. The latency is the period between the submission of the random number request and the receiving of the result. The dApp expects a quick response (i.e., low latency), especially when the dApp is time-sensitive. (ii) Low cost. The on-chain computation is usually expensive and the cost is proportional to the amount of computation. To reduce the execution cost, the dApp expects to minimize the on-chain computation involved in the random number generation protocol.

Computation cost analysis. The generation of the final random number includes both on-chain and off-chain works. While on-chain computation cost is more sensitive, it is better to minimize both on-chain and off-chain computation.

Off-chain computation. HRNG keeps most of the computation off-chain, which includes three components: (i) Random number commitment (`Commit`). Gateways convert all random numbers collected from IoT devices to the form of commitments. The total number of commit operations is $n_g \times n_i \times n_r$. (ii) Secret sharing (`Split`). Each gateway splits the opening information of each commitment and distributes to other $n_g - 1$ gateways. The total number of secret sharing operations is $n_g \times n_i \times n_r$. (iii) Secret recovering (`Reconstruct`). Only ℓ committed random numbers need to be recovered, so it involves ℓ `Reconstruct` operations.

On-chain computation. The on-chain part is mainly related to the aggregation phase, i.e., each blockchain node needs to verify the correctness of the aggregation. This verification involves two parts: (i) Verification of numbers in the pool. This operation is equivalent to the opening of a set of committed values and the cost is determined by the commitment scheme, i.e., ℓ `Open` operations of the selected commitment scheme. (ii) Verification of aggregation process. The simplest way to verify the aggregation result is by repeating the process. If `XOR` is the only operator used for aggregation, the computation cost is $\ell - 1$ `XOR` operations.

Communication cost analysis. There are two communication expensive steps in HRNG, and both are related to the (k, t)-threshold secret sharing. (i) Distribution of shares of committed random numbers. Each committed number is split into k pieces, so the overall communication complexity is $\mathcal{O}(n_g \times n_i \times n_r \times k)$. (ii) Reconstruction of shares of committed random numbers. As only ℓ values need to be reconstructed, the overall communication complexity is $\mathcal{O}(\ell \times t \times k)$. Here we consider a worse situation, where each gateway distributes its splits to all k other gateways. If all gateways are collaborative, this complexity can be reduced to $\mathcal{O}(\ell \times t)$.

6   Improvement of HRNG

The construction described in Sect. 4 is generic and does not consider the special features of the building blocks.

Utilization of commitment scheme features. While any secure commitment scheme can meet the security requirement of HRNG, they are not equivalent from a performance perspective. As analyzed in Sect. 5, the on-chain computation cost of the generic construction includes ℓ commitment open operation, and $\ell - 1$ `XOR` operations which are employed to produce the final random number. If the Pedersen commitment scheme [16] is adopted, each commitment open operation involves two exponents and one multiplication on the given finite field. While this cost is not a big concern for a modern computer, it is not cheap as on-chain computation.

To further reduce the on-chain computation complexity, we utilize the additive homomorphic feature of the Pedersen commitment scheme. Specifically, given two commitments $c_1 = g^{m_1} h^{r_1}$ and $c_2 = g^{m_2} h^{r_2}$, we can build the commitment of $m_1 + m_2$ with one multiplication, i.e., $c_{1+2} = c_1 \cdot c_2 = g^{m_1+m_2} h^{r_1+r_2}$.

With this feature, we can tweak the design of HRNG, and a blockchain node does not need to open each committed value to verify its correctness before aggregating them to obtain the final random number. Specifically, a blockchain node can aggregate the commitments utilizing the homomorphic feature first, and only verify the validity once. With this strategy, the total on-chain computation cost is reduced to one aggregation of ℓ commitments ($\ell - 1$ multiplications) and one open operation (two exponent operations and one multiplication). Depending on the selection of $\mathbb{G}$, the exponent and multiplication can mean different arithmetic operations. If $\mathbb{G}$ is a multiplicative group of a finite field, multiplication is a multiplication of two big integers modulo a prime number. If $\mathbb{G}$ is a group of elliptic curve points, multiplication is a scalar multiplication of elliptic curve points. In both cases, the complexity of a typical efficient exponent operation is equivalent to $\mathcal{O}(\log |\mathbb{G}|)$ multiplication operations. In other words, the exponent operation is much more expensive than multiplication, and the new approach can greatly reduce the on-chain computation complexity. We analyze the security and performance of the improved HRNG in the rest of this section.

Security analysis of improved HRNG. With the improved HRNG, the blockchain can only obtain the addition of ℓ random numbers

$$m_1 + \cdots + m_\ell, \tag{1}$$

not the original bitwise XOR result

$$m_1 \oplus \cdots \oplus m_\ell. \tag{2}$$

Without loss of generality, we assume each m_i has the same bit length. For Eq. (2), as long as one of the input values $m_i, 1 \leq i \leq \ell$ is an unbiased random bit string, the aggregation result is an unbiased random value. The situation of Eq. (1) is more complex. If all m_is are defined as integers within a given range and selected randomly, the sum of these values follows a variant of Irwin-Hall distribution [17,18] other than uniform distribution in that range. Fortunately, m_is are defined in a finite group that is isomorphic to $\mathbb{Z}/p\mathbb{Z}$ where p is a prime number (actually we have $|\mathbb{G}| = p$). Accordingly, the addition given in Eq. (1) is addition modulo p. Addition in this structure offers a similar feature as XOR on bit strings. Specifically, as long as one element m_i is uniformly selected from $\mathbb{Z}/p\mathbb{Z}$, the result of Eq. (1) is uniformly distributed in $\mathbb{Z}/p\mathbb{Z}$.

To prove this feature, we assume m_1 is randomly selected from $\mathbb{Z}/p\mathbb{Z}$, and the adversary manages to set the sum of other elements as a preferred value $a \in \mathbb{Z}/p\mathbb{Z}$. Given an arbitrary value $b \in \mathbb{Z}/p\mathbb{Z}$,

$$\Pr[m_1 + a = b] = \Pr[m_1 = b - a] = \frac{1}{p}. \tag{3}$$

Equation (3) states that it does not matter what strategies the adversary uses to choose the value a, the addition result will be uniformly distributed among all the possible values in $\mathbb{Z}/p\mathbb{Z}$.

Note that there is a tiny difference between a random number in $\mathbb{Z}/p\mathbb{Z}$ and a random bit string in $\{0, 1\}^{\lceil \log(p) \rceil}$. Hardware-based TRNGs usually create random bit strings other than random elements in $\mathbb{Z}/p\mathbb{Z}$. To convert a binary string to an element of $\mathbb{Z}/p\mathbb{Z}$, we can simply treat the bit string as an integer and apply a modulo p operation. When the length of the bit string is longer than $\lceil \log(p) \rceil$, the converted result roughly follows uniform distribution in $\mathbb{Z}/p\mathbb{Z}$. Given a value p, a longer bit string can lead to a more uniform distribution in $\mathbb{Z}/p\mathbb{Z}$. If the dApp needs a random binary string other than an element in $\mathbb{Z}/p\mathbb{Z}$, we can treat the element as a binary string and reduce the leading bits to guarantee it roughly follows the uniform distribution.

On-Chain performance analysis of the improved HRNG on Ethereum. The improved HRNG relies on a smart contract to open/verify a set of Pedersen commitments and aggregate all the random numbers using modular additions over a prime finite field. Here we assume the Pedersen commitment scheme is built on an elliptic curve points group, which is common in blockchain. Table 3 summarizes the gas cost for performing arithmetic operations on the Ethereum Virtual Machine (EVM), where the gas cost for computing point addition and scalar multiplication is based on the precompiled contracts on the alt_bn128 elliptic curve [19].

Table 3. Gas cost for performing arithmetic operations in EVM of Ethereum [20]

Arithmetic Operation	EVM Opcode	Gas Cost
Modular Addition	addmod	8
Modular Multiplication	mulmod	8
EC Point Addition	ecadd	150
EC Scalar Multiplication	ecmul	6000

Table 4 provides a comparison of the gas cost between the non-optimized HRNG construction and the optimized one as described in Sect. 4 and 6, respectively. Thanks to the homomorphic property of the Pedersen commitment scheme, it is not difficult to find that the optimized implementation of HRNG can save the gas cost significantly with the increasing of aggregated random numbers. Note that the gas cost is directly associated with computation complexity, i.e., a method is more expensive in gas if it requires more arithmetic operations.

Table 4. Performance evaluation and gas cost for opening and aggregating ℓ random numbers in EVM of Ethereum.

HRNG	Operation Calls			Gas Cost
	#ecadd	#ecmul	#addmod	
Non-optimized	ℓ	$2 \cdot \ell$	$\ell - 1$	$12,158 \cdot \ell - 8$
Optimized	ℓ	2	$2 \cdot (\ell - 1)$	$166 \cdot \ell + 11,984$

Figure 2 shows the gas costs of the two versions of HRNG. It is easy to see that the gas cost of the non-optimized HRNG grows linearly as more random numbers are used to produce the final result, while the gas cost of the improved HRNG is almost constant. This difference is important from security perspective, as it is harder for an adversary to manipulate the final result when more numbers are involved. When the system uses 12 random numbers to produce the final result, the improved HRNG only consumes about 10% of the gas of HRNG without optimization.

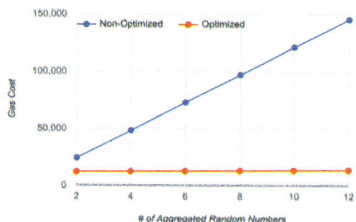

Fig. 2. The gas costs for non-optimized and improved HRNG. Here both versions use the elliptic curve-based Pedersen commitment scheme but the unoptimized HRNG does not utilize the homomorphic feature. The x-axis is the number of values to be aggregated and the y-axis is the amount of gas required to finish the work (commitment openings and aggregation).

7 Conclusion

Random numbers play a critical role in blockchain and dApps, and various random number generation mechanisms have been designed to meet this demand. Compared with previous methods, HRNG works in a hybrid manner, i.e., it utilizes both hardware-based TRNG and on-chain/off-chain algorithms for random number generation. This hybrid approach allows HRNG to enjoy the benefits of different random number generation strategies and can tolerate various malicious participants. As on-chain computation is usually expensive, we also design an efficient on-chain aggregation mechanism that does not sacrifice the quality of generated random numbers and decentralization level. The analysis and evaluation show that this improvement can save significant cost (both computation burden and gas fee if deployed for Ethereum) compared with the non-optimized construction.

Acknowledgements. We thank the anonymous reviewers for the valuable comments.

References

1. Ballandies, M.C., Wang, H., Law, A.C.C., Yang, J.C., Gösken, C., Andrew, M.: A taxonomy for blockchain-based decentralized physical infrastructure networks (DePIN). In: Proceedings of the 1st International Workshop on Decentralized Physical Infrastructure Networks (DePIN 2023). IEEE (2023)

2. Xu, L., Fan, X., Hall, L., Chai, Q.: New gold mine: harvesting IoT data through DeFi in a secure manner. In: International Conference on Blockchain, pp. 43–58. Springer (2021)
3. Mutlu, O., Kim, J.S.: Rowhammer: a retrospective. IEEE Trans. Comput. Aided Des. Integr. Circuits Syst. **39**(8), 1555–1571 (2019)
4. Randao: Verifiable random number generation [Online]. Available at: https://www.randao.org/whitepaper/Randao_v0.85_en.pdf
5. Edgington, B.: Upgrading Ethereum—a technical handbook on Ethereum's move to proof of stake and beyond [Online]. Available at: https://eth2book.info/capella/part2/building_blocks/randomness/
6. Syta, E., Jovanovic, P., Kogias, E.K., Gailly, N., Gasser, L., Khoffi, I., Fischer, M.J., Ford, B.: Scalable bias-resistant distributed randomness. In: 2017 IEEE Symposium on Security and Privacy (SP), pp. 444–460 (2017)
7. Skidanov, A.: Near protocol randomness beacon [Online]. Available at: https://pages.near.org/downloads/NearRandomnessBeacon.pdf
8. Chainlink: Chainlink vrf [Online]. Available at: https://docs.chain.link/vrf
9. Chatterjee, K., Goharshady, A.K., Pourdamghani, A.: Probabilistic smart contracts: secure randomness on the blockchain. In: 2019 IEEE International Conference on Blockchain and Cryptocurrency (ICBC), pp. 403–412 (2019)
10. Blaut, G., Ma, X., Wolter, K.: Exploring randomness in blockchains. In: 2023 IEEE International Conference on Blockchain and Cryptocurrency (ICBC), pp. 1–5 (2023)
11. ERC721R: Bringing greater accountability to NFT creators [Online]. Available at: https://erc721r.super.site/
12. Lehman, T.: Erc721r: a new erc721 contract for random minting so people don't snipe all the rares! [Online]. Available at: https://medium.com/@dumbnamenumbers/erc721r-a-new-erc721-contract-for-random-minting-so-people-dont-snipe-all-the-rares-68dd06611e5
13. Naor, M.: Bit commitment using pseudo-randomness. In: Conference on the Theory and Application of Cryptology, pp. 128–136. Springer (1989)
14. Daniel, E., Tschorsch, F.: IPFS and friends: a qualitative comparison of next generation peer-to-peer data networks. IEEE Commun. Surv. Tutor. **24**(1), 31–52 (2022)
15. Long, J., Wei, R.: Scalable BFT consensus mechanism through aggregated signature gossip. In: 2019 IEEE International Conference on Blockchain and Cryptocurrency (ICBC), pp. 360–367. IEEE (2019)
16. Pedersen, T.P.: Non-interactive and information-theoretic secure verifiable secret sharing. In: Annual International Cryptology Conference, pp. 129–140. Springer (1991)
17. Hall, P.: The distribution of means for samples of size n drawn from a population in which the variate takes values between 0 and 1, all such values being equally probable. Biometrika 240–245 (1927)
18. Irwin, J.: On the frequency distribution of the means of samples from populations of certain of Pearson's types. Metron **8**(4), 51–105 (1930)
19. Salazar Cardozom, A., Williamson, Z.: Eip-1108: reduce alt_bn128 precompile gas costs [Online]. Available at: https://eips.ethereum.org/EIPS/eip-1108
20. EVM codes—an Ethereum virtual machine opcodes interactive reference [Online]. Available at: https://www.evm.codes/

SUPI-Rear: Privacy-Preserving Subscription Permanent Identification Strategy in 5G-AKA

K. Sowjanya[1]([envelope]) [iD], Pabitra Pal[2] [iD], Aman Verma[3] [iD], Bijoy Das[1] [iD], Dhiman Saha[3] [iD], Anand M. Baswade[3] [iD], and Brejesh Lall[1] [iD]

[1] Indian Institute of Technology Delhi, Delhi, India
{sowjanya.kandisa,mantunsec}@gmail.com, brejesh@ee.iitd.ac.in
[2] Maulana Abul Kalam Ajad University of Technology, Kolkata, West Bengal, India
pabipaltra@gmail.com
[3] Indian Institute of Technology Bhilai, Bhilai, India
{amanverma,dhiman,anand}@iitbhilai.ac.in

Abstract. Security and privacy concerns are crucial for the success of any new technology. With the global rollout of 5G networks, new use cases are continually emerging. The 3GPP consortium mentioned the authentication and key agreement protocol for the 5th generation (5G) mobile communication system (i.e., 5G-AKA) in the technical specification (TS) 33.501. It introduces public key encryption to conceal the so-called Subscription Permanent Identifier (SUPI) to enhance mobile users' privacy. However, the user's permanent identity i.e., SUPI is available in cleartext to the Serving Network (SN) after the successful primary authentication. SUPI availability is required for the operational and regulatory perspective of SUPI usage. In 5G-AKA, the SUPI is available in cleartext to the Serving Network (SN). Since the SNs are considered semi-trusted because the long-term secret key and the sequence numbers are not revealed with SNs, only SUPI is provided in cleartext for proper billing. Hence, SUPI availability in cleartext under a zero-trust, multi-tenant-based 5G network compromises the user's privacy. This work provides a way to enhance privacy and security during communication between the Home Network (HN) and the SN without compromising the original SUPI. Furthermore, the proposed solutions (termed collectively as SUPI-Rear) are also applicable to various use cases where SUPI privacy is required, like Public Land Mobile Network (PLMN) hosting Non-Public Network (NPN) scenario. Moreover, it abides by the lawful requirements and 5G AKA authentication procedure.

Keywords: Subscription Permanent Identifier (SUPI) · SUPI Privacy · 5G-AKA

1 Introduction

The ability to provide mutual authentication between users and the network, as well as the generation of cryptographic keys to protect both signaling and user

T. Masuzawa et al. (Eds.): SSS 2024, LNCS 14931, pp. 307–321, 2025.
https://doi.org/10.1007/978-3-031-74498-3_22

plane data, make authentication and key management crucial elements of cellular network security. Every generation of cellular networks has a minimum of one defined authentication mechanism. As an illustration, whereas 4G EPSAKA is described in 4G, 5G provides three authentication techniques: 5G-AKA, EAP-AKA', and EAP-TLS. The 3GPP has specified the Authentication and Key Agreement (AKA) protocol and processes to offer entity authentication, message integrity, and message secrecy, among other security features. The 3GPP 5G-AKA protocol as depicted in Fig. 1 is a challenge-and-response-based authentication mechanism between a subscriber and a home network, and results in a shared symmetric key, K_{SEAF} [1]. Cryptographic keying materials are obtained after mutual authentication between a subscriber and a home network to safeguard future communication between a subscriber and a serving network, including signaling messages and user plane data (e.g., over radio channels).

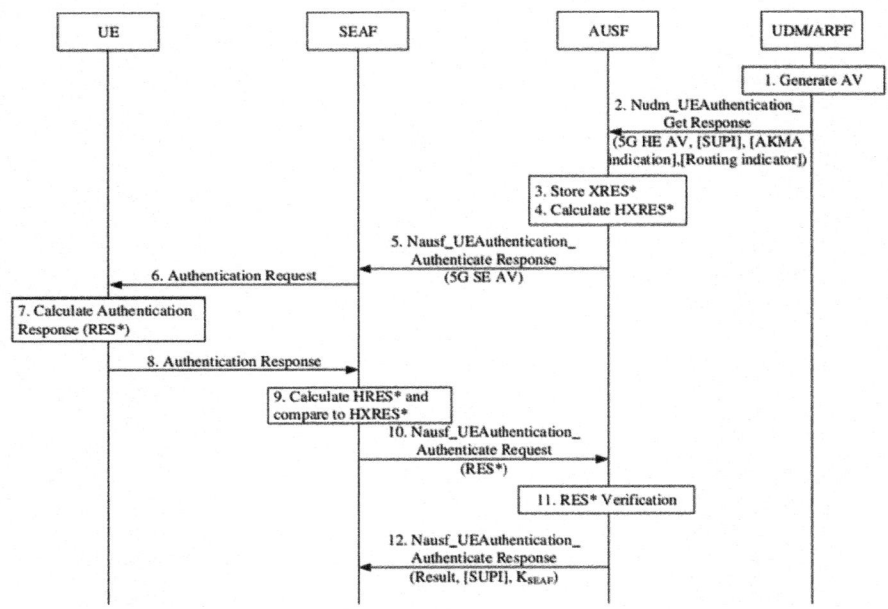

Fig. 1. 5G AKA Procedure [1]

In telecommunications systems, network operators assign each SIM card a unique ID: an IMSI (International Mobile Subscriber Identity) up to the 4G era, and a SUPI (Subscription Permanent Identifier) in the 5G era. User identification is required before authentication can take place since it relies on a shared symmetric key between the user and its network provider. However, users can be recognized, located, and tracked using these permanent identifiers if the IMSI/SUPI values are transmitted in plaintext over the radio access channel. To prevent this privacy breach, the SIM card is assigned temporary identifiers by the visited network: TMSI in 3G systems, and GUTI in 4G and 5G systems. These temporary identifiers, which change frequently, are used for identification purposes over the radio access link. A Subscription Concealed Identifier (SUCI) is a privacy-preserving identifier that contains the concealed SUPI. The UE generates the SUCI using an Elliptic Curve Integrated Encryption Scheme (ECIES)-based protection scheme with the Home Network's public key, which is securely provisioned to the Universal Subscriber Identity Module (USIM) during USIM registration. Although the SUPI is protected over the radio interface using SUCI, but during the primary authentication procedure (5G-AKA), SUPI is available in cleartext to SN along with the anchor key (K_{SEAF}). According to the 3GPP standards, the provisioning of SUPI to SN is required for the purpose of accurate billing and for other lawful interception requirements. Nowadays, the subscriber's private information and/or identity is seemly a captivating target for online advertisements and other connected organizations. Furthermore, SNs are considered semi-trusted (honest but curious), and under a zero-trust and multi-domain architecture of 5G, the compromise of a user's permanent identity may result in privacy leakage.

Another scenario documented by the SA1 working group of 3GPP involves Non-Public Network (NPN) security considerations, as detailed in clause 8.2 of TS 22.261 [2]. In this scenario, the 5G system shall enable a Public Land Mobile Network (PLMN) to host an NPN without compromising the PLMN's security. In customer premises, dedicated network entities of NPN can be deployed that are beyond the control of the PLMN operator. According to the primary authentication and authorization procedure specified in TS 33.501 [1], if the SUPI is available in clear text to network functions on customer premises, it could potentially lead to security threats, UE location tracking, privacy breaches, and targeted attacks. Additionally, with the evolution of roaming architectures (Roaming Hub) and core networks (NPN, Edge computing), as well as distributed CN (multi-site CN), there is no direct trust relationship between the HN and the SN/VPLMN/Edge network (i.e., between different security domains). In this context, the HN must carefully consider the exposure of permanent and/or sensitive identifiers/parameters to network functions in different security domains.

To enhance security in this context, zero-trust security (ZTS) [3] is one of the recommended approaches by NIST [4], and when combined with other security standards, a comprehensive security framework can be constructed for future networks. As data privacy regulations become more strict, a zero-trust architecture can provide a better framework for future 6G networks. In this zero-trust security model [4], if a SUPI is available in cleartext to the entities outside the MNO premises (in other security domains), it can potentially raise numerous security threats.

In this context, the proposed work has the following technical contributions:

- User privacy and identity security are improved in the home network (or PLMN) and SN (or NPN) communication.
- It proves to be a more effective security solutions because it requires minimal modification in the 5G AKA authentication mechanism.

- The law enforcement agencies can still hold a track record of the users by exchanging the identity with the required service.
- The backward compatibility issue is resolved using this solution.

The rest of this paper is structured as follows. Section 2 illustrates the literature review. Section 3 provides an overview of 5G comprising the details of SUPI, SUCI, and 5G AKA protocol. The proposed method is described in Sect. 4 with its possible outcomes. The paper is concluded with final remarks in Sect. 7. List of acronyms used in this work is illustrated in Table 1.

Table 1. List of acronyms

3GPP	Third Generation Partnership Project
5G-AKA	5G Authentication and Key Agreement
EAP	Extensible Authentication Protocol
EAP-TLS	Extensible Authentication Protocol-Transport Layer Security
GUTI	Globally Unique Temporary Identifier
IMSI	International Mobile Subscriber Identity
SEAF	Security Anchor Function
HN	Home Network
NPN	Non-Public Network
PLMN	Public Land Mobile Network
SN	Serving Network
SUCI	Subscription Concealed Identifier
SUPI	Subscription Permanent Identifier
TMSI	Temporary Mobile Subscriber Identity
TS	Technical Specification
TR	Technical Report
USIM	Universal Subscriber Identity Module
VPLMN	Visiting Public Land Mobile Network
ZTS	Zero-trust security

2 Literature Review

Many recent works focus on the privacy of mobile subscribers through various methods. A number of mechanisms [6,7] use pseudonyms to improve the privacy of the subscribers. These schemes preserve the subscriber's privacy by using temporary pseudonyms instead of SUPI. In 2019, Borgaonkar et al. [8] revealed a new privacy attack against all variants of the AKA protocol, including 5G AKA, that breaches subscriber privacy more severely than known location privacy attacks do. They conducted a security analysis and demonstrated the practical feasibility of privacy attacks using low-cost and widely available setups. This attack enables the subscriber's activity monitoring by breaking the confidentiality of the sequence numbers (SQN). In 2020, Ouaissa et al. [9] proposed an improved and efficient authentication and key agreement protocol for 5G mobile networks that overcomes the weaknesses discovered in the existing 5G-AKA and uses lightweight cryptographic methods. They used an Automated Validation of Internet Security Protocols and Applications (AVISPA) tool to demonstrate the proposed authentication scheme. In the same year, Braeken [10] presented a new version of the 5G AKA protocol that is resistant to all known threats while also providing security characteristics like as anonymity, unlinkability, mutual authentication, and secrecy. The suggested protocol uses only symmetric keys and cryptographic primitives that are currently accessible in the universal subscriber identity module's hardware (USIM).

In 2021, Liu et al. [11] proposed a solution for the SUPI guess attack and SUCI replay attack. Their method relies on the UE's public key to defend against the SUCI replay attack and mitigates the SUPI guess attack based on the shared key between the UE and the network. In 2021, Wang et al. [12] proposed a privacy-preserving solution for the AKA protocol of 5G system denoted by 5G-AKA'. It is resistant to linkability attacks performed by active attackers and is compatible with the SIM cards currently deployed Serving Networks.

Khan and Martin [13] surveyed the state of subscription privacy in 5G systems. Their study shows that new and more rigorous privacy protection mechanisms are required to guarantee robust subscription privacy in 5G. Also, they considered the serving networks semi-trusted as the long-term key, and sequence numbers are not shared with them. Hence, the availability of SUPI in cleartext with the serving network compromises the subscriber's privacy. The SUPI is a unique permanent identifier, and while it doesn't contain the content of communications, it serves as a key to correlate and track user activity across sessions, networks, and services. The SUPI itself is less detailed than metadata (like the contents of calls, messages, emails, etc.) but can be used in conjunction with metadata to track or profile a user, especially if the SN uses the SUPI to link different data sources. In this context, there is a need for an alternative method that would preserve the subscriber's privacy on the serving network's side and abide by the lawful interception requirements.

3   Background

This section presents the background knowledge needed to understand the proposed scheme.

3.1   5G Subscription Identifier: Subscription Permanent Identifier (SUPI)

As defined in 3GPP TS 23.501 [5], a SUPI is a globally unique Subscription Permanent Identifier assigned to each subscriber in 5G. Apart from USIM, the

SUPI value is provisioned to UDM/UDR NFs of 5G Core. A Valid SUPI can be one of the following:

- As defined in TS 23.503 [14], IMSI (International Mobile Subscriber Identifier) for 3GPP RAT.
- As defined in TS 23.003 [15], NAI (Network Access Identifier) for non-3GPP RAT as defined in RFC 4282 based on user identification.

A SUPI typically consists of a sequence of 15 decimal digits, as shown in Fig. 2. The first three digits signify the Mobile Country Code (MCC), while the subsequent two or three digits indicate the Mobile Network Code (MNC), identifying the network operator. The remaining digits, either nine or ten in total, are referred to as Mobile Subscriber Identification Numbers (MSIN) and uniquely identify the individual user within that specific operator's network. The SUPI is analogous to the IMSI, both serving to uniquely identify the Mobile Equipment (ME), and each consisting of a 15-digit string.

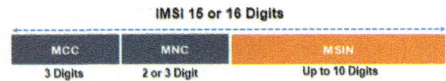

Fig. 2. Subscription Permanent Identifier (SUPI) [5]

Fig. 3. Subscription Concealed Identifier (SUCI) [5]

3.2   5G Identifier: Subscription Concealed Identifier (SUCI)

A SUCI is an identifier designed to preserve privacy, encompassing the concealed SUPI, as depicted in Fig. 3. During USIM registration, the UE generates a SUCI using an ECIES-based protection scheme with the public key of the Home Network, securely provisioned to the USIM.

The protection scheme conceals only the MSIN part of the SUPI, while the home network identifier (MCC/MNC) is transmitted in plain text.

3.3   5G AKA Protocol

In 5G-AKA, the serving network starts the authentication procedure by sending a registration request to the Home Network (HN). The request includes the

SUCI or SUPI and the Serving Network Name (SNN). At the UDM (which is a part of the home network), SUPI is de-concealed from SUCI. An authentication Vector (AV) is generated and sent in a response message along with SUPI to the serving network. The general call flow mechanism is shown in Figure. 1.

As per the Security architecture and procedures for 5G System - 3GPP TS 33.501 V18.0.0 (2022–12), the transmission of SUPI in the clear text is permitted over various scenarios:

- Between 5G core network entities of the same serving network domain.
- Between 5G core network entities of the different serving network domains.
- Between 5G and Evolved Packet System (EPS) core network entities if it has the form of an IMSI.

Our survey on state-of-the-art 5G authentication reveals that most of the work is done on the security enhancement of UE and Serving Network communication. However, some security issues related to the communication between the Serving Network and the Home Network have gone unnoticed. The general perception is to consider the Serving Network and Home Network communication in the semi-trusted setting. Our proposed solution will address those issues and mitigate attacks even in the untrusted scenario.

3.4   PLMN Hosting NPN Scenario

A Public Land Mobile Network (PLMN) has the capability to host Non-Public Networks (NPNs). Customers of NPNs can request dedicated Network Functions (NFs) to be deployed on their premises for reasons related to performance and privacy. PLMN hosting NPN is a concept in the telecommunications industry where a public mobile network operator (MNO) hosts a private network on its infrastructure. This can enable businesses and organizations to deploy private 5G networks with the reliability, coverage, and scale of a public network while maintaining control and customization of their own network slice.

The 5G system must facilitate a PLMN hosting an NPN without compromising the security of either the PLMN or the NPN. This is based on the following security assumptions: 1. There is no mutual trust between the PLMN and the dedicated Network Functions within the NPN, and 2. Attacks may occur from both the NPN to the PLMN and from the PLMN to the NPN.

4   Proposed Scheme

3GPP has introduced a service-based architecture for core networks, featuring novel network entities and services aimed at supporting a unified authentication framework. 3GPP has outlined noval authentication-related services tailored for 5G. This authentication procedure is divided into two phases:

Phase 1: 5G authentication initiation and selection of authentication method, i.e., 5G-AKA or EAP-AKA' [1].

Phase 2: Mutual authentication between the UE and the core network.

During Phase 1, 5G-AKA is selected as an authentication and key agreement mechanism. Further, for Phase 2, three mechanisms are proposed in this work to preserve the privacy of the SUPI and these solutions are collectively termed as *SUPI-Rear.*

4.1   Scheme 1: Home Network Controlled Mechanism

In this scheme, an HN-controlled (or on-demand) privacy-preserving scheme is proposed. When the registration request is received at the Home Network (UDM: Unified Data Management), SUPI is de-concealed from SUCI. Here, UDM generates a Virtual SUPI corresponding to the received SUPI, and stores the mapping. Here the Virtual-SUPI is generated using the SHA3 hash function of the cryptographic approach along with one random value for each Virtual SUPI corresponding to the original SUPI. $VirtualSUPI = SHA3(OriginalSUPI\|Randomvalue)$, here random value is generated using any standardized pseudo-random number generators (PRNGs) (like NIST, ANSI-based standard PRNGs).

And finally, after the successful primary authentication, the Authentication Server Function (AuSF) will share the Virtual SUPI in the Authentication Response and the Authentication Vector (AV) will be sent as a response message to the serving network as depicted in Fig. 4.

After receiving the response from the home network, the serving network generates a GUTI and keeps a mapping of Virtual SUPI and GUTI. Furthermore, the Virtual SUPI is shared with UE to generate the key K_{AMF} for mutual authentication with the network.

Serving Network requirement (or NPN): Proper and accurate billing issue of the SN (or NPN) will be resolved by using the virtual SUPI.

Lawful Interception (LI) requirement: If LI needs the original SUPI for any necessity, then through an on-demand query with the virtual SUPI, LI can request the HN to provide the original SUPI corresponding to the specified virtual SUPI. After receiving the request, HN will send the original SUPI using the mapping table.

4.2   Scheme 2: LI Controlled Mechanism

In some use cases, where the LI does not want any dependency on HN, this second proposed SUPI privacy-preserving mechanism can be useful. After successful primary authentication, AuSF sends the virtual SUPI along with the mapping table (SUPI-Virtual SUPI). Here, the mapping table is encrypted using the public key of the LI as illustrated in Fig. 5. Hence, when LI needs the original SUPI corresponding to the virtual SUPI, it can decrypt the mapping table using its own private key. As a result, the privacy of the SUPI is preserved at the SN and the requirement of the LI is also fulfilled. The selection of PKE and key generation/distribution will be governed by the policy. The proposed mechanism is agnostic and can accommodate any standardized PKE and key distribution mechanism.

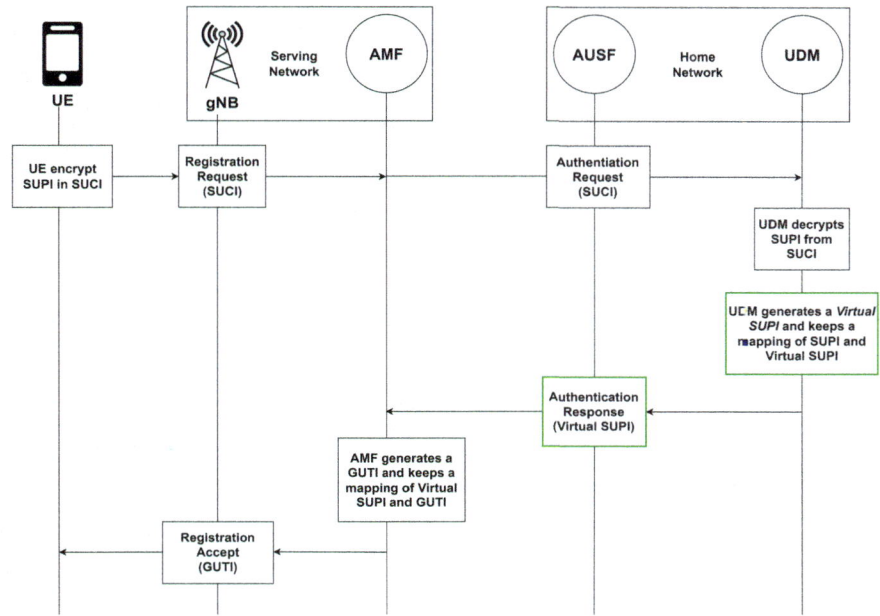

Fig. 4. Solution 1—HN controlled Virtual SUPI

4.3 Scheme 3: No UE Impact: Home Network Controlled Mechanism

In Scheme 1 (Sect. 4.1), Virtual SUPI is shared with UE in order to compute the key K_{AMF}, this key is further used to generate the subsequent keys for the network. K_{AMF} is generated using the values: SUPI, anchor key K_{SEAF} and other specific parameters. If Virtual SUPI is not shared with UE then the key K_{AMF} computed by the SN's (NPN's) AMF (Using Virtual SUPI) and UE (Using Original SUPI) will be inconsistent. Hence, Scheme 1 uses Virtual SUPI to compute the key K_{AMF}, for this, the Virtual SUPI is transferred to UE. Hence, the solution has an impact on UE. In order to mitigate the UE impact, Scheme 3 provides SUPI privacy using temporary/virtual SUPI without impacting the UE.

After the successful primary authentication procedure as mentioned in clause 6.1.3.2 [1] (also in Fig. 1), here in this solution 3, the Home Network (HN)/PLMN sends the key K_{AMF} (instead of K_{SEAF}) along with Virtual SUPI instead of the original SUPI. K_{AMF} is sent towards customer premises instead of K_{SEAF} in order to overcome the impact on UE. At first, AUSF derives $K_{SEAF} = KDF(K_{AUSF}, SNN)$ and further derives

$$K_{AMF} = KDF(K_{SEAF}, SUPI, SNN, ABBAparameter)$$

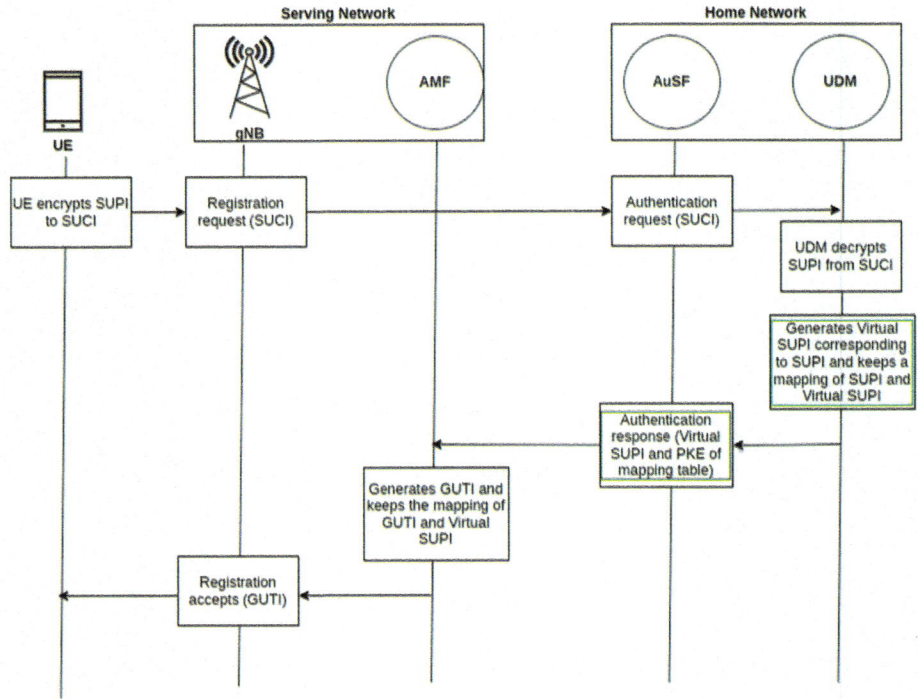

Fig. 5. Solution 2—LI controlled Virtual SUPI

Finally, AUSF sends Virtual SUPI along with K_{AMF} to SN (NPN). Here, KDF refers to the Key Derivation Function (specified in 3GPP standards), SNN refers to the Serving Network Name, ABBA refers to the Anti-Bidding down Between Architectures (ABBA) parameter.

4.4 SUPI-Rear Outcomes

Virtual SUPI solution is helpful in the following way:

- In the Roaming scenario, when a subscriber travels outside their home network coverage area, other service providers with roaming agreements in place with the subscriber's home operator step in to provide SN services. These services ensure uninterrupted connectivity for the subscriber while they are away from their primary network. In the traditional setup, it's essential for these roaming partners to access certain subscriber information, particularly the SUPI, to enable the delivery of the services. However, there is a need to balance the provision of these services with the protection of subscriber privacy. This challenge is especially relevant when SUPI is available as plaintext between different network operators, potentially exposing sensitive information.

To address this privacy concern and maintain the security of subscriber data, Virtual SUPI acts as a privacy-enhancing mechanism that allows operators to fulfill the conditions required for SN services without disclosing excessive personal information to the roaming service provider.

- In the Migration scenario, where a subscriber transitions from a 5G network to a 4G network, potential security vulnerabilities like downgrade attacks can be effectively mitigated through the use of Virtual SUPI. This technology plays a crucial role in ensuring that the SUPI remains secure and cannot be compromised during the migration process.

  Downgrade attacks are a security risk when a subscriber switches from a more advanced network, such as 5G, to an older generation network like 4G. Attackers may attempt to force the connection to occur at the less secure 4G level, where they could exploit vulnerabilities to intercept sensitive information.

  Virtual SUPI acts as a protective layer during this migration, adding an extra level of security. When a subscriber moves from 5G to 4G, instead of transmitting the real SUPI, a virtual identifier is used for the transition. This virtual identifier is unique to the migration process and does not disclose the actual SUPI or any sensitive subscriber details.

- Lawful Interception requirement is also fulfilled using Virtual SUPI, as this solution does not intervene in the work of law enforcement agencies.

5   Security Analysis

In this section, we illustrate the attack model and consequently provide the potential threats and their corresponding informal security analysis.

5.1   Attack Model

The attack model assumes a scenario where the SN has some level of control over the communication between the user and the network, making it a powerful and potentially dangerous adversary if not adequately mitigated by strong encryption and authentication protocols.

- **Adversary:** A malicious or compromised SN with the ability to intercept or access the SUPI and potentially other user data.
- **Target:** The user's SUPI and any services, accounts, or data linked to it.
- **Methods:** Interception of the SUPI, session hijacking, correlation of SUPI with other data, impersonation, and long-term tracking.
- **Impact:** Unauthorized access to services and accounts, identity theft, loss of privacy, and potential financial loss.

5.2   Potential Threats and Informal Analysis

When the SUPI is available in cleartext to the SN in a 5G environment, it can potentially introduce several security threats. Here are some of the key threats to consider.

Privacy Breach. The SUPI stands for Subscriber Permanent Identifier, which is a unique identifier associated with a subscriber's personal information. For instance, if the Serving Network (SN) possesses unrestricted access to the SUPI, it has the potential to result in a violation of privacy. Unauthorized access to or exposure of the SUPI could grant the ability to compromise the confidentiality of the subscriber. This further leads to

- Linkability (the ability to link multiple activities of the same identity).
- Traceability (the process of identifying various actions of the same identity).

Attacks might involve the tracking of the user's activities and locations, as well as engaging in identity theft or profiling endeavors.

Hence, the concept of virtual SUPI of the proposed work overcomes the mentioned privacy breach. Furthermore, linkability and traceability attacks are not possible with virtual SUPI because in each and every primary authentication, the virtual SUPI is different for the same SUPI. In other words, virtual SUPI is used in replace of the actual SUPI. It provides a layer of anonymity for the user. Hence, it enhances privacy because it does not explicitly reveal the user's identity.

Location Tracking. The SN can able to track the subscriber's real-time location. This location information can be misused for surveillance, stalking, and other malicious purposes.

With virtual SUPI, the permanent identity of the subscriber is separated from the location of the device. Therefore, the location of the user is not revealed directly with the proposed technique.

Impersonation Attacks.   The presence of SUPI accessible to the SN can potentially lead to impersonation attacks. In such scenarios, these attacks might exploit to masquerade as legitimate subscribers, thereby illicitly obtaining access to services, financial accounts, or confidential data. Some systems or services might rely on the SUPI as a key part of their authentication process. If the SUPI is intercepted or illicitly obtained, the malicious SN might use it to access those services fraudulently. The SUPI might be linked to financial accounts or services (like mobile payment systems) in certain cases. With the SUPI, a malicious SN could try to access these accounts, especially if combined with other data obtained from the network or subscriber.

Virtual SUPIs are dynamically allocated to users, changing over time or under specific conditions. This makes it more challenging for malicious actors to track users based on their identifiers.

Insider Threats. The serving network may faces a potential risk from insider threats. Individuals such as employees or insiders who possess SUPI access could misuse their granted privileges, compromise the privacy of subscribers or engage in unauthorized activities, and may use it for advertisements.

Subscribers have more control over their privacy with virtual SUPI. They can choose when and how their virtual SUPI is used, giving them greater control over their personal information.

Network Attacks.   Gaining access to SUPI increases the vulnerability to network-based attacks. Attacks like communication interception, man-in-the-middle attacks, or unauthorized access to sensitive data may be launched by malicious actors using vulnerabilities in the SN infrastructure.

The use of virtual SUPIs is a crucial step in protecting user privacy and preventing tracking. Frequently changing identifiers makes it difficult for malicious entities to monitor and trace a user's activities.

Service Disruption.   The services offered to the subscriber could perhaps be disrupted if the SN has access to SUPI. This may entail network configuration manipulation or unauthorized service deactivation.

Since virtual SUPI is randomly assigned to the permanent SUPI with each authentication, the SN faces difficulty in launching the service disruption for the specified user.

Lawful Interception (LI) Requirement.   Virtual SUPI successfully meets the need for accurate billing and other lawful requirements. Furthermore, if LI of the SN needs the SUPI then it may query for the same using the virtual SUPI with the HN (according to the proposed scheme 1) or it can decrypt the mapping table using its own private key (according to the proposed scheme 2) and obtains the original SUPI.

6   Performance Analysis

The performance of the proposed solution is evaluated using the following hardware setup: Intel(R) Core(TM) i7-8565U CPU @ 1.80GHz*8 with 16 GB memory. Operating system: Ubuntu 22.04 with type 64-bit. Here, we have analyzed the proposed mechanism using the computation overhead w.r.t. the generation of the mapping table using C++ language because the only extra feature in the 5G-AKA is the proposed scheme's mapping table. Here, it can be observed that the computation overhead for generating the virtual SUPI corresponding to the permanent SUPI is negligible with respect to the computation power of the UDM.

Furthermore, the communication overhead is a bit high, since instead of the original SUPI (15 or 16 digits), the virtual SUPI (same size) and/or PKE of the mapping table is sent to SN (or NPN) after the successful primary authentication. Moreover, the storage overhead is circumvented, since the mapping table is stored at UDM/LI side, which is not constrained w.r.t. storage, computation, and energy.

The performance comparison between Scheme 1, Scheme 2, and Scheme 3 (see Sect. 4) is depicted in Table 2. Scheme 1 needs less computation and communication overhead than Scheme 2 because no PKE is required and only virtual SUPI is sent to SN. But Scheme 1 has an impact on UE and Scheme 3 overcomes this impact by generating K_{KMF} at SN (or PLMN) side. But schemes 1 and 3

Table 2. Comparison of the proposed schemes

Features	Scheme 1	Scheme 2	Scheme 3
UE Impact	Yes	Yes	No
On-demand activity	Yes	No	Yes
Extra Communication Overhead	Yes	Yes	No
LI controlled	No	Yes	No

are prone to information leakage attacks because of the on-demand activity. For scheme 2, the mapping table is sent in encrypted form. Therefore, the following benefits are gained from Scheme 2:

- LI can successfully track the user by decrypting the mapping table.
- Privacy and secrecy of the SUPI are maintained since no third party (except LI) can decrypt the table to get the SUPI.

7 Conclusion

To improve the subscribers' privacy in SNs (or NPNs), the Virtual SUPI solution proposed in this work provides an efficient way to protect the users' permanent identity. This solution also abides by the LI requirements as well as mitigates the downgrade attacks. Hence, this solution is compatible with 5G as well as the 4G communication mechanism. Moreover, the proposed solution is compatible with the existing 5G-AKA procedure, the only change required is to maintain one table at UDM that maps the original SUPI with virtual SUPIs. Therefore, the proposed solution can be efficiently implemented in order to preserve the privacy of the subscriber's permanent identity with negligible computational overhead. Scheme 1 and 3 are HN/PLMN controlled privacy-preserving mechanisms, whereas scheme 2 is LI-controlled. Further, in scheme 2, we achieve security with anonymity under the zero-trust security model. The proposed solutions (SUPI-Rear) not only help the LI to rack the subscriber but also maintain the privacy and secrecy of the SUPI.

In the future, we are planning to implement the proposed solution using the Open Air Interface (OAI) 5G core setup and consequently evaluate the performance (w.r.t. computational cost, latency, long-term cost, etc.) of the proposed schemes. Furthermore, a quantitative analysis of the proposed scheme will be provided.

References

1. 3GPP TS 33.501: Security architecture and procedures for 5G system, V18.0.0 (2022)
2. 3GPP TS 22.261: Service requirements for the 5G system, Release 18

3. 3GPP TR 33.894: Study on applicability of the Zero Trust Security principles in mobile networks, Release 18
4. NIST Special Publication 800-207: Zero Trust Architecture, Zero Trust Architecture (nist.gov)
5. 3GPP TS 23.501: System architecture for the 5G System (5GS), v18.1.0 (2023)
6. Khan, M., Ginzboorg, P., Järvinen, K., Niemi, V.: Defeating the downgrade attack on identity privacy in 5G. In: Fourth International Conference on Research in Security Standardisation, pp. 95–119. Springer (2018)
7. Broek, F.V.D., Verdult, R., Ruiter J.: Defeating IMSI catchers. In: Proceedings of the 22nd ACM SIGSAC Conference on Computer and Communications Security, pp. 340–351 (2015)
8. Borgaonkar, R., Hirschi, L., Park, S., Shaik, A.: New privacy threat on 3G, 4G, and upcoming 5G AKA protocols. In: Proceedings on Privacy Enhancing Technologies, IACR Cryptology ePrint Archive, p. 1175 (2018)
9. Ouaissa, M., Ouaissa, M.: An improved privacy authentication protocol for 5G mobile networks. In: IEEE International Conference on Advances in Computing, Communication & Materials (ICACCM), pp. 136–143 (2020)
10. Braeken, A.: Symmetric key based 5G AKA authentication protocol satisfying anonymity and unlinkability, Comput. Netw. **181**, 107424 (2020)
11. Liu, F., Su, L., Yang, B., Du, H., Qi, M., He, S.: Security enhancements to subscriber privacy protection scheme in 5G systems. In: IEEE International Wireless Communications and Mobile Computing (IWCMC), pp. 451–456 (2021)
12. Wang, Y., Zhang, Z., Xie, Y.: Privacy-preserving and standard-compatible AKA protocol for 5G. In: 30th USENIX Security Symposium (USENIX Security 21), pp. 3595–3612 (2021)
13. Khan, H., Martin, K.M.: A survey of subscription privacy on the 5G radio interface—the past, present and future. J. Inf. Secur. Appl. **53**, 102537 (2020)
14. 3GPP TS 23.503: Policy and charging control framework for the 5G System (5GS), V18.4.0, Release 18
15. 3GPP TS 23.003: Numbering, addressing and identification, V18.4.0, Release 18

Anomaly Detection Within Mission-Critical Call Processing

Sean Doris[1], Iosif Salem[2], and Stefan Schmid[2]($^{(\boxtimes)}$)

[1] Motorola Solutions, Inc., 2600 Glostrup, Denmark
sean.doris@motorolasolutions.com
[2] TU Berlin, 10587 Berlin, Germany
iosif.salem@inet.tu-berlin.de, stefan.schmid@tu-berlin.de

Abstract. With increasingly larger and more complex telecommunication networks, there is a need for improved monitoring and reliability. Requirements increase further when working with mission-critical systems requiring stable operations to meet precise design and client requirements while maintaining high availability. This paper proposes a novel methodology for developing a machine learning model that can assist in maintaining availability (through anomaly detection) for client-server communications in mission-critical systems. To that end, we validate our methodology for training models based on data classified according to client performance. The proposed methodology evaluates the use of machine learning to perform anomaly detection of a single virtualized server loaded with simulated network traffic (using SIPp) with media calls. The collected data for the models are classified based on the round trip time performance experienced on the client side to determine if the trained models can detect anomalous client side performance only using key performance indicators available on the server. We compared the performance of seven different machine learning models by testing different trained and untrained test stressor scenarios. In the comparison, five models achieved an F1-score above 0.99 for the trained test scenarios. Random Forest was the only model able to attain an F1-score above 0.9 for all untrained test scenarios with the lowest being 0.980. The results suggest that it is possible to generate accurate anomaly detection to evaluate degraded client-side performance.

Keywords: anomaly detection · call processing system ·
mission-critical systems · virtualized environments · machine learning ·
random forest

1 Introduction

Anomaly detection has become a key focus for many industries as a method of producing accurate monitoring and analysis. Their applications range from

Supported by German Research Foundation project ReNO, SPP 2378, 2023–2027.

assisting medical diagnosing and establishing a prognosis of cancer [29] to monitoring the system health of servers and routers in various telecommunication applications [14]. While the applications differ, many anomaly detection applications are interested in being used in mission-critical applications where safety and reliability are essential. The use of machine learning (ML) in these scenarios allows for improved data analysis by detecting complex patterns that could have otherwise gone unnoticed and could be used as predictors for future failures.

Mission-critical telecommunication systems have become increasingly advanced to handle the growing demand for higher data throughput and scalability. New features and bug fixes are regularly being deployed through releases of updates putting systems in a state of flux. Scalability is often implemented using several multi-card chassis, each having numerous virtual machines or containers to handle call volume [20]. All the while, the systems must be thoroughly tested to ensure that telecommunication systems have a high degree of reliability to ensure regular availability. Various potential failures can occur on these systems, such as reduced link quality, resource contention, orphan processes, hardware failures, and thermal limiting. Even minor hard-to-detect failures could cause performance degradation beyond safe operating conditions. Furthermore, when one virtualized instance fails, it can affect the performance of other connected virtualized instances by increasing response times and delays [19]. This degradation can be difficult to diagnose in a large, complex, modularized network. Modularizing systems using virtual machines and docker containers allows for improved maintainability and load balancing [4].

Mission-critical applications' uptime requirements can require a 99.999% uptime with no more than 6 min of downtime per year [3]. These mission-critical systems are often used in emergency response to support emergency workers to safely and quickly reach and provide the required assistance to those in need. As such, mission-critical telecommunications require effective methods to help ensure the system's stability can meet the performance required by its applications. For real-time applications, downtime also includes periods when performance is degraded beyond the levels of its service requirements. Due to the limited downtime or anomalous instances, it would be more efficient to develop a method generating simulated anomalous data than trying to collect anomalous instances from active systems. Additionally, by providing autonomous monitoring for each virtualized system, it can reduce the complexity of applying self-healing methods since fixes can be applied without affecting the larger system [17].

The use of ML to detect faults such as intrusion, DDOS attacks, memory or hardware faults is an area of focus in many research efforts today [15,26]. Another application of ML models is that they can assist in maintaining the uptime in mission-critical systems by training a model that can detect if a system's performance has degraded beyond acceptable specifications. These specifications change from application to application; however, within call processing, this can be the server's processing time required to set up the communication. This paper aims to create a ML model that accurately detects abnormal performance degradation that exceeds the normal operating parameters of a virtualized system.

Scope. This paper describes a procedure for simulating call processing under different scenarios with varying loads. The call processing setup is two virtual machines loaded with SIPp call traffic with media (SIPp simulates the SIP call and messaging protocol [10]). One virtual machine will be the server, and the other will be the client. The two virtual machines will exist on the same host system, with all communications taking place over the local building network. A definition of degraded performance and how it is classified will be determined by analyzing the normal operating behaviour of the test setup. By defining the specifications of what is expected for the system, a level can be set for what resulting performance on the client side is classified as anomalous or non-anomalous.

Various stressors will be applied to the system using stress-ng (open source software for stressing systems [16]). Each type of stressor shall have two different load levels applied, a low level that will not result in degraded performance on the client side and one that will. The data is classified as anomalous if the round trip time (RTT) reaches anomalous levels due to performance degradation caused by stressors active on the server side. The RTT consists of the client-server-client communication delay and the server processing time and we will assume reliable client-server communication. Thus, increased RTT indicates increased server processing time due to anomalous server behaviour. While other client or design parameters could be used for classification, RTT is selected as a classifier for validation as it is a metric that could be applied to many mission-critical approaches in communication systems. In theory, the methodology of using measurable design parameters for classifying training data could be adapted to other mission-critical systems by using a different set of more relevant parameters.

The ML models are tested and trained with data from measuring key performance indicators (KPI) on the server side during different scenarios, we will build a training and test data set for the ML models. The scenarios being tested are (i) unstressed scenarios when no stressor is active, (ii) non-anomalous stressed scenarios when a stressor is active, but the RTT does not exceed anomalous levels and (iii) anomalous scenarios where the stressors will cause the RTT to exceed anomalous levels. The models will be tested with untrained scenarios where the data from the stressors are exclusively used for testing. The untrained scenarios will validate that the ML models are accurate for other anomalous scenarios and are not overtrained for the stressors that they were trained with.

From the results, seven well-known ML models were compared over various scenarios using the stressors tested. The diverse types of ML models that were selected include support vector machines, tree-based, cluster-based, and probability-based models. In addition, both classification and regression, as well as both supervised and unsupervised methods, are tested to determine if there are any distinctions in their ability to be used for anomaly detection under this methodology and the generality of the approach.

Contributions. Our main contributions are the following:
1) We leverage client-server RTT in the call-setup protocol as the main classifier for defining anomalies and (non-)anomalous scenarios. The novelty of this approach is that it more accurately classifies anomalies compared to the standard classification of data collected from when the system is stressed or unstressed, since low-stress scenarios may still be tolerated by the call-processing system.
2) We generate accurate anomaly detection that can evaluate degraded client-side performance. All tested supervised ML models achieved F1-scores above 0.99 for trained test scenarios excluding Gaussian Naive Bayes. However, many of the supervised models had reduced performance when comparing the machine-learning results with the stressor test results. Random Forest performed the best across all untrained test scenarios, with the lowest result being 0.980.
Outline. Sect. 2 will provide an overview of the measurement and anomaly generation applications. Section 3 will provide the methodology for the paper. Section 4 will display and discuss the results of the performance for ML models on the tested scenarios. Finally, in Sect. 5, we present related work.

2   Background

Training ML models require applications to measure and store the KPIs being measured on the system. Generating the data for training the model also requires a system that is under load to monitor and another application to create an anomalous state to generate data for training the supervised ML models.
Measurement Tools & Applications. This section will denote what tools will be used to measure the KPI required for anomaly detection. The selected tools are vmstat, iostat, and netstat. vmstat is an application can be used to measure various hardware performance metrics on the virtual machine [27]. It provides access to active/inactive memory, context switching, CPU interrupts, and buffer memory usage. iostat command can be used on Linux systems to monitor the IO usage of connected devices [11]. netstat is a command that can be used to display network related metrics for Linux based systems. netstat can display a large amount of detailed information regarding the system [28].
Anomaly Generation Application. We used stress-ng to generate anomalous data on the systems to be tested. stress-ng is open-source software that can stress a system in numerous ways through its selection of over 280 selectable and customizable stress tests [16]. The tests can be selected to stress targeted pieces of a system. This can be ideal for generating anomalous data for select KPIs to ensure all of the features we are measuring can have anomalous training data that can stand out from a healthy system measurement. Several authors have researched ML using anomalous data generated using stress-ng to simulate various instances such as CPU, memory, and disk faults [6,12,13,21,22].
SIPp. SIPp is a software that can simulate and allow performance testing of SIP protocol [10]. Using SIPp on two different platforms, one can have user agent client (UAC) and user agent server (UAS) communication using different packet types. SIPp can perform operations during a call, such as executing commands

or streaming multimedia such as audio. During communication, it can provide detailed statistics such as RTT, call rate, and other call-related statistics. These statistics allow a user to determine the quality of a given communication. We will use SIPp as the method of producing system and network load on our virtual machine and measuring the RTT for all communication during a period.

3   Data Collection Methodology

This section of the paper will outline the steps and methods used for the data collection. The data collection will include the normal and anomalous system data for training the ML models used in this project. The platform used is a VM running using two cores from an Intel Xeon E5-2670, 3GB of memory, and a virtual operating system running Red Hat Enterprise Linux 8.4 (Fig. 3).

3.1   SIPp

The call scenario that was used is a customized version of the default user agent client packet capture (UAC PCAP) call scenario offered by SIPp. The PCAP scenario simulates media communication between the client and server end. This media communication generates significantly more load than the standard UAC scenario, which has a small exchange of packets. Additionally, the load can be a better simulation of an actual scenario, as in the UAC PCAP scenario, data is streamed between the two. For the call scenario, a call rate of 70 call/s provided a sufficiently high load on the server side but not high enough to cause the server to be overloaded and lead to timeouts due to excess load.

Settings that were adjusted within the default UAC PCAP scenario were the timeout settings that needed to be adjusted to prevent calls from staying open and limiting the startup of new calls. Otherwise, the call rate for the scenario could be unstable. The retransmission time for the invite message was reduced from 500ms to 50ms due to the network's very short latency. The default 500ms caused large spikes in the average due to the long retransmission delay; reducing the retransmission time to 50ms reduces these spikes but still makes them impactful. The RTT is measured from the initiator's or client's end, starting the RTT timer on the invite message and stopping it upon receiving all set-up messages just before the initiator sends the ACK (Fig. 1).

RTT can measure the quality of communication traffic, especially in mission-critical systems where if RTT spikes are left unaccounted for, they could lead to disruptions. Under the assumption that the network connecting two systems is stable and transmission time (t_{tx}) remains unchanged, changes in RTT can be used to quantify the changes in the time to process (t_p) a message provided to the server before replying (Fig. 2). Suppose the connection of a system is stable and the RTT is increased. In that case, the increased RTT is due to increased processing time. This increase in processing time can be due to the server being in an anomalous state (t_{p_a}), increasing the server's time to respond. Monitoring the server's KPIs while measuring the RTT, one can classify the RTT from the

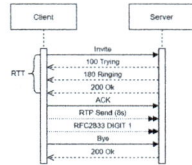 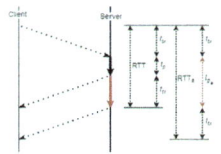 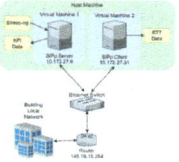

Fig. 1. UAC PCAP call scenario

Fig. 2. Round trip time (RTT) diagram

Fig. 3. Diagram of the test setup

client's end as anomalous or non-anomalous and apply that same classification to the KPIs logged on the server end. With this application, one can build enough data to train and test ML models to monitor the KPIs of the server. This claim of a stable network while monitoring a system under a load of SIPp voice calls is similar to other studies that used a similar methodology for assessing the viability of containers [20] or studying anomaly detection [22]. Under stable network conditions data can be collected to train ML models that could be applied to similar active systems where these assumptions aren't present and RTT cannot be used alone as a suitable metric for determining system stability.

3.2   Anomalous State

A definition for what is classified as anomalous must be made before determining what is anomalous and the acceptable RTT. An anomalous state should have an RTT that exceeds the levels expected for transmission. Unexpectedly high RTT could cause degraded communication quality experienced by the client. With such a definition, ML models can monitor and ensure a system's reliability.

To determine what value of RTT is anomalous, the RTT of an unstressed system running the UAC PCAP scenario is measured for 20 h. The RTT is measured from the client's end to evaluate the quality of the communication from the client's perspective. Meanwhile, the KPIs are measured on the server end. The RTT is averaged over the time frame of the data point in the KPI log so that the averaged RTT can be used to classify the data point. Mathematically, the classification for what is outside normal distribution is any RTT that exceeds the third standard deviation for the averaged RTT data points over the full 20-hour measurement time period.

To measure the effectiveness of the ML models, the metrics that are used are accuracy and F1 score [9, Appendix 6.1]. To perform these calculations, the classification matrices need to be defined for the provided test scenarios that are performed. For any tests where the expected RTT level is below the anomalous RTT level, a positive result is the model correctly predicting a non-anomalous result. For any tests where the expected RTT is above or equal to the anomalous RTT level, a positive result is the model correctly predicting the system's state as anomalous. The detailed classification matrices are in [9, Appendix 6.1].

3.3   Stress Test Selection

The following tests were selected for creating data that will stress the measured KPI of the system and create both anomalous and non-anomalous data for different segments of the system. The collected data from these tests are used for training and testing of the models:

- **CPU** to cause CPU load, stress-ng iterates through a list of loads such as floating point calculations, square root and other computational loads such as the eight queens problem used in computational load tests
- **icache** to simulate a load causing interference in the instruction cache leading to further instruction stalls and delays to the pipeline
- **aio** to cause stress on the io by triggering several small writes and reads to the disk
- **UDP** to cause stress on the local network sockets for UDP
- **rawsock** to create additional stress on the TCP/IP using the raw sockets on the local host

In addition to these stress methods, the four additional stressors are selected. These stressors will affect similar KPIs but will strictly be used to test the ML models and not to train them. With this data, the anomaly detection accuracy can be tested for other similar stressing scenarios. These will test for overtraining and ensure that models can classify any similar stimuli that may cause the system to enter an anomalous state. The chosen untrained stressors are:

- **matrix** this iterates through several matrix operations. By enabling the yx option, the stressor can cause both CPU and cache interference
- **revio** has workers writing to temporary files on the hard disk. The writes are performed in reverse order. This is similar stress usage to aio; however, revio only performs writes
- **rawudp** similar to rawsock, but instead it will stress using UDP
- **rawpkt** sends and receives packets over the localhosts ethernet port

The provided tests are completed using different parameters to allow them to create enough stress to achieve anomalous RTT, as well as create a low-stress version that will have a non-anomalous RTT when performing non-anomalous data generation. The universal parameters for each stressor are the number of workers, and runtime and deadline parameters can be adjusted to cause different load levels. There are a few other stressor-specific parameters that are changed, such as enabling the yx option in the matrix stressor to cause cache interference.

The resulting data from testing shall have at least 97% of the generated data points as anomalous or non-anomalous depending on whether it is an anomalous test or non-anomalous test. The amount of variation in the average RTT for each test is different due to the different stressors. However, for non-anomalous stress loads, the RTT shall also be below the anomalous RTT threshold.

Other stress test options from stress-ng were tested. However, among the tests selected, they were also selected for being reproducible by producing RTT above the anomalous threshold consistently.

4   Evaluation

We present the results from data collection and methodology described in Sect. 3. We present the test results from determining normal RTT with the applied system, the application of the ML models, analysis of the applied stress tests and how the ML models applied to the system react to the stressors.

4.1   Determining Normal Conditions

The calculation of the anomalous threshold was completed by collecting over 20 h of system KPI and RTT data while SIPp was active. A call rate of 70/s was selected since it produced a relatively stable connection while generating enough load on the system. The 20 h of test data were collected in two 10-hour intervals, with the systems being fully reset between the test intervals.

The chosen frame time for the logging was 6 s, allowing enough time to execute the 2 s monitoring intervals required for both `vmstat` and `iostat` as well as having minimal impact on the system KPI. The received RTT data was then parsed and averaged over the 6 s logging period so each logging frame would have a correlated averaged RTT. A shorter logging interval could be selected if a mission-critical application required a quicker reaction time to an anomalous state, however, this would lead to a larger impact on system performance which could be an issue for systems with limited hardware.

From the measured results, the majority of the RTT ended up being less than 25ms, with averages over the frame being between 4–5ms. Some of the spikes in RTT could be explained by network congestion, as the tests are not performed on a closed network. Due to the bursty nature of RTT there were several measurements exceeding 50ms, however, averaging the RTT for the duration of the frame significantly lowered the variance of RTT (Table 1). The RTT around 50ms are caused by retransmissions where the sender didn't receive the ACK packet within the 50ms timeout and would then resend the packet.

By using the third standard deviation for the RTT of a frame from the client side and the KPIs that were taken during that same 6-second frame interval on the server side, a given KPI for that frame can be classified as normal or anomalous to provide training data for the ML models.

4.2   Applying ML Models

We describe the process of tuning of the parameters of the ML models we used. All collected data was pre-processed using linear scalar without using the mean to center the data before the scaling. The libSVM package offered through Python was used for the support vector machine methods [7]. The polynomial kernel was used for both Support Vector Classifier (SVC) [2] and ν-Support Vector Regression (ν-SVR) [23], while One Class Support Vector Machines (OC-SVM) [25] performed best using the radial basis function (RBF) kernel. Due to the unequal number of non-anomalous cases in the training data, an additional non-anomalous test from the unstressed test case weight calculation was enabled to

Table 1. RTT metrics from unstressed system runs

Unstressed RTT Metrics			
Metric	1st run	2nd run	Overall
Average over full duration	4.689 ms	4.662 ms	4.675 ms
Standard deviation using rtt	1.991 ms	1.892 ms	1.942 ms
3rd standard deviation using rtt	10.663 ms	10.339 ms	10.503 ms
Standard deviation using rtt for the frame	1.339 ms	1.372 ms	1.355 ms
3rd standard deviation using rtt for the frame	8.707 ms	8.776 ms	8.741 ms

correct this offset. OC-SVM was trained using only non-anomalous data and was tuned by adjusting the ν parameter to adjust the upper and lower bounds. A balanced approach was chosen to accurately detect anomalous instances while limiting the number of false positives occurring during non-anomalous data.

The Gaussian Naive Bayes (GNB) [30], k-Nearest Neighbors (kNN) [5], Decision Tree (DT) [24], and Random Forest (RF) [2] models were all applied using crates (compilation units) available in Rust programming language. GNB and DT are available through the Linfa-Trees and Linfa-Bayes crates(v.0.6.0), while kNN and RF are available in the Smartcore crate (v.0.2.1) [1]. The GNB model was tuned by adjusting the variance smoothing, the variance smoothing was increased from the default to add a portion of the feature with the largest variance to the stability equation. When training the kNN model, the k parameter for selecting the number of clusters was tested with k values between 3 and 25. The number of clusters was tested using Euclidian, Hamming, Manhattan, and Minkowski distance metrics and found that Manhattan performed the best. The cover tree search algorithm was selected due to its increased speed and efficiency, it also improved accuracy compared to the brute force approach of Linear Search.

Tuning the tree-based models required adjusting the parameters for the split decision-making criteria. Entropy and Gini splitting criteria are available with these rust crates. Gini impurity measures the probability of a data point being misclassified if inserted into a randomly selected point. Entropy measures the disorder or variable composition of a provided node. A split is decided based on which split will decrease the entropy the most and thus increase the level of information the split provides. Both were tested on DT and RF, DT benefited more from the Entropy criterion while RF performance increased with Gini splitting. Both RF and DT had decision making parameters tuned to adjust the weight required to make a split. The default weights were low given the size of our dataset and would result in specialized nodes and trees with larger depths that proved impractical, leading to misclassifying some of the test data.

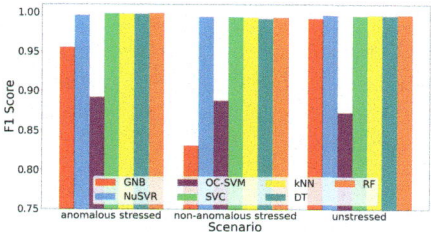

Fig. 4. F1 score of models for different test scenarios

4.3   Model Results

Overall, all models produced an F1 score above 0.8 for unstressed, anomalous, and non-anomalous stressed scenarios. With ν-SVR, SVC, kNN, DT, and RF producing F1-scores scores above 0.99. The RF model has the best F1-score for anomalous scenarios resulting in an overall F1 Score of 0.99898. ν-SVR, kNN, and RF are tied for best overall F1 score for unstressed scenarios with a score of 0.99663 (Fig. 4). In comparison, ν-SVR performs the best in non-anomalous with an F1 score of 0.99663. SVC achieved the highest accuracy overall trained cases despite not having the highest F1 score in the three test scenarios types.

As OC-SVM is unsupervised learning, it was unable to reach the same level of accuracy compared to supervised training methods. It obtained F1 scores of between 0.87287 and 0.89189 for anomalous, non-anomalous and unstressed scenarios (Fig. 4). While having generally lower scores than the other models, OC-SVM did outperform the GNB model during several low-stress test cases.

For a model to be accurate in classifying the state of a system, it is essential to balance classifying both anomalous and non-anomalous scenarios. An example of an improper balance can be seen in the GNB and the OC-SVM models when looking at the icache stressor scenario, both have an accuracy of below 60%, but their accuracy in low_icache is above their overall accuracy (Table 2). In such cases, the boundary between anomalous and non-anomalous states is more within the non-anomalous space resulting in numerous misclassifications.

Test cases with unstable bursty behaviour regarding RTT such as aio and rawsock had lower accuracy due to creating more anomalous points within the non-anomalous stress tests. Models had difficulty classifying these RTT bursts often resulting in misclassifications during the bursts. There are several possible reasons for these misclassification. One possible cause is that there is not enough variance in classifying RTT values in the training data. Depending on the test, there could be a lack of data points closer to the RTT threshold resulting in some misclassifications of near threshold values. An improvement that could be tested is to include data sets that vary around the threshold between anomalous and non-anomalous. This could improve the accuracy of these intermediate results, but it could also lead to more misclassifications of non-anomalous data and more false alarms. The second is that some stressors caused a bursty nature in the RTT, which may be challenging to detect, provided the current KPI. A shorter logging interval may allow for detecting these short bursts at the expenditure of higher computational cost. However, there is an argument that, depending on the system, short bursts of high RTT should not impact the overall system's stability. Furthermore, detecting anomalies that sustainably impact the systems should be the higher priority.

Table 2. Prediction accuracy results for individual trained tests

Test	GNB		ν-SVR		OC-SVM		SVC		kNN		DT		RF	
overall	83.25		99.14		79.93		99.31		99.31		99.03		99.27	
unstressed	98.49		99.33		77.48		99.16		99.33		99.16		99.33	
RTT Level	High	Low	High	Low	High	Low	High	Low	High	Low	High	Low	High	Low
aio	98.95	96.84	98.95	98.74	98.95	68.2	98.95	98.74	98.95	98.74	98.95	98.11	98.95	98.74
cpu	100.0	58.11	100.0	99.58	77.41	86.95	99.81	99.58	100.0	99.58	99.81	98.32	100.0	98.95
icache	56.84	93.41	97.47	99.34	25.89	96.70	100.0	99.34	99.79	99.12	99.37	98.68	100.0	99.34
rawsock	100.0	90.30	100.0	97.23	99.59	90.89	100.0	97.03	100.0	97.03	100.0	97.03	100.0	96.83
udp	100.0	19.2	100.0	99.8	100.0	58.8	100.0	99.8	100.0	99.8	100.0	99.8	100.0	99.8

4.4 Untrained Results

The untrained results were obtained with the same test setup and method with 1000 test data points for each anomalous and non-anomalous scenario. The results showed many ML models could not classify either the anomalous or non-anomalous case for a given stressor. Both SVC and ν-SVR had reduced F1-scores for all untrained stressors. The other models attained performance similar to their trained counterparts for matrix and revio test cases. The networking stressors were more challenging with DT, RF, OC-SVM, and GNB achieving good performance in rawpkt. However, RF was the only model to perform well in the rawudp scenario. Across all the untrained tests, RF achieved an F1-Score of above 0.9 (Fig. 5) with better accuracy regarding CPU, cache, and io interference and lower performance for the network stressors (Table 3).

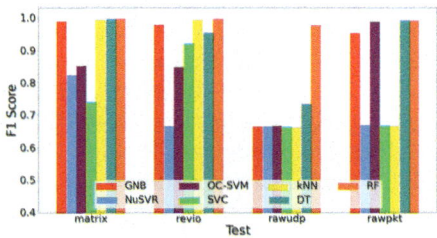

Fig. 5. F1 score of models against untrained test scenarios

The decrease in accuracy for both SVC and ν-SVR could be caused by over-training. For both SVC and ν-SVR, a potential cause of over-training is a high C

value creating very thin margins [2], out of the two models only SVC has a high
C value. However, when testing with lower C values the accuracy was reduced
in accuracy overall trained scenarios and the matrix and revio test cases. Revio
eventually regains accuracy once reducing the C value significantly. However,
the matrix scenario then reduces in accuracy below 1%. Another possible cause
of overtraining is that higher degree polynomial kernels can cause overfitting as
the shape of the support vector will mirror changes that could be caused by
noise rather than fitting the overall trend of the training data [18]. The applied
ν-SVR does have an exponential of 7 on the polynomial kernel. However, upon
testing lowering the exponential resulted in decreased accuracy for both trained
and untrained CPU and io stressors with no effect on the network stressors. The
kNN model also had poor accuracy for the tested untrained networking stressors.
When performing testing the different methods, the use of Hamming distance
metric greatly improves the accuracy of rawpkt and rawudp. However, it also
lowers the accuracy of all other trained and untrained test scenarios with an overall accuracy of 91.316% with higher accuracy towards non-anomalous scenarios.
Switching the search algorithm to linear search had a similar impact but at an
increased computation cost. Moreover, GNB and OC-SVM generally achieved
comparatively higher scores despite having lower accuracy in the trained results
because these models offer more flexibility at the cost of trained accuracy.

A possible cause for the lower performance of SVC, ν-SVR, and kNN is that
the training network stressor scenarios may have different patterns in the KPI
values that cause the untrained scenarios to land on one side of the support
vector. For kNN, this would instead be caused by the majority of nearest data
points within the training data set being classified as anomalous. One possible
way to improve these models would be to perform additional training cases
covering an improved spectrum of anomalous and non-anomalous scenarios for
network stressors. RF achieved high accuracy in trained and untrained results
without any overtraining regarding any of the test cases. In comparison, other
models needed to give up accuracy as a trade-off for better generalization.

Table 3. Prediction accuracy results for individual unknown tests

Test	GNB		ν-SVR		OC-SVM		SVC		kNN		DT		RF	
RTT Level	High	Low	High	Low	High	Low	High	Low	High	Low	High	Low	High	Low
matrix	99.3	97.4	100.0	39.0	55.4	94.8	99.9	15.7	100.0	100.0	99.2	100.0	99.9	100.0
revio	94.9	97.7	100.0	0.2	61.8	86.1	100.0	71.8	100.0	99.9	83.4	99.7	99.8	92.0
rawudp	0.0	100.0	0.0	100.0	100.0	0.6	100.0	0.0	100.0	0.0	99.8	16.2	100.0	92.0
rawpkt	91.6	91.7	99.9	2.0	100.0	95.8	99.9	2.0	100.0	2.0	100.0	97.4	100.0	97.6

4.5 Discussion

The use of RTT as means of classifying training and test data is a different
approach than what is presented in other ML papers that use simulated data

[6,12–14,21,26]. Often, the metric for anomalous is whether or not the stressor is active which for practical applications could require more verification on whether the stressor is causing the system to enter an anomalous state by exceeding the specification of the system. It could be that the system can still handle the additional load of the stressor while maintaining availability. The use of averaging and the third standard deviation of the RTT allows users to define anomalous under the perspective of meeting the system specifications and maintaining availability. From the trained results, this proved to be a viable method as all the models could achieve F1 scores above 0.8 for each scenario.

Analyzing the provided results from the trained scenarios shows that many of the ML models achieve a stable F1 score across all scenarios, excluding GNB, which has the lowest F1 score for non-anomalous despite achieving an F1 score greater than 0.95 for both anomalous and unstressed (Fig. 4). Among OC-SVM and GNB, there is a concern for causing false alarms for non-anomalous scenarios with higher UDP stress or detecting anomalous stress within the instruction cache as these tests had lower accuracy (Table 2). For the other support vector machine, kNN and tree-based models did not display significant levels of lower accuracy for a particular test case in the trained results. They achieved F1-scores above 0.99 for all trained scenarios (Fig. 4).

The untrained results presented a different perspective. They tested how well the ML models could generalize the trained results and allow the accurate detection of anomalous RTT for untrained scenarios. Out of the tested ML models, only RF achieved F1 scores above 0.97 for all test cases (Fig. 5). GNB, DT, and OC-SVM also have notable performances, as each showed lower performance only in the rawudp test case. ν-SVR and SVC showed reduced performance across all test cases, while kNN only showed reduced accuracy during the networking test cases. From the untrained results, kNN, ν-SVR, and SVC showed a greater performance decrease despite their high accuracy across the trained test cases, they proved insufficient in generalizing the approach to detecting anomalous RTT.

The resulting applied methods show that RF will achieve the best results unless the ML is trained using anomalous training data that includes all possible anomalous cases. Using other supervised ML with highly trained accuracy could be worth creating an ensemble method with other ML methods such as DT, GNB, and OC-SVM. These models showed better performance for generalizing the trained results to create an accurate model that could detect when the systems state and provide improved assistance in maintaining availability. However, a more generalized and easy to train model could still be advantageous for self-learning applications where the model needs to completely retrain itself, which can be time-consuming and computationally intensive for large datasets.

The results show that when training ML models using this methodology it is best to use a tree-based model or pair more generalized models with other ML when not able to simulate all possible loads when accuracy is of greater importance. If applying these to a telecommunication system with limited computational resources to adjust the training parameters of models such as reducing the possible depth for tree-based models or increasing the polling interval from 6 s if accuracy is more important than time response.

5   Related Work

There is a growing interest in developing automated monitoring and anomaly detection for various computer systems. Many focus on different networking levels, such as core routers [14] and network function virtualization [22]. Other papers focus on cellular networks evaluated ML techniques using data collected from active systems with anomalous data that human personnel has evaluated [6,8]. These applications do not apply to mission-critical systems and have anomalous events lasting hours and enough data to contribute to the training process.

Regarding virtualized environments, several papers evaluate the application of anomaly detection within docker containers, and virtual machines [12,22]. Their application included a real-time running application that was also evaluated regarding response time. Other papers focus on modularized systems such as cluster or cloud-based services [13,21,26]. These applications offer metric measuring from individual virtual machines that offer modular monitoring. However, the application differs by using the data to determine whether the whole cluster or cloud is anomalous rather than a single virtualized instance. One of the papers has a similar method of training using a low and high-stress application, increasing their models' accuracy. Although it differs in that both stress levels were classified as anomalous [26].

The approach to anomaly classification differs depending on the ML application. Several papers seek to simulate hardware failures and classify anomalous as the stressor being applied[6,12,21]. In this paper, the approach to anomalous classification is unique in that it is instead classified by the resulting impact on the system and whether it degrades the performance beyond the specified RTT.

6   Conclusion

This paper proposed a method of training ML models that can detect when given server-client communication exceeds the expected bounds by measuring only the KPI of the server side. Although different models proved to have different performance levels at detecting whether the system was in an anomalous state when provided with KPI metrics from a system loaded with a SIPp UAC with media simulations. With the provided training and test data, out of the seven models, random forests proved to be the best method to measure both the tested trained and similar untrained scenarios accurately. The evaluated methodology can provide the potential for generating anomaly detection, ensuring a more robust and reliable system for many applications and providing additional assurance for mission-critical applications. For future work, a suggested action would be to explore the accuracy of non-binary classification in order to provide improved feedback from anomaly detection. Further testing could be done to discern the most impactful KPI for training and to validate other possible classification parameters than RTT.

References

1. Rust Package Registry. https://crates.io/ (2022)
2. Aurelien, G.: Hands-On Machine Learning with Scikit–Learn and TensorFlow. O'Reilly Media (2017)
3. Avdotin, E., Bankov, D., Khorov, E., Lyakhov, A.: OFDMA resource allocation for real-time applications in IEEE 802.11ax networks. In: 2019 IEEE International Black Sea Conference on Communications and Networking (BlackSeaCom), pp. 1–3 (2019). https://doi.org/10.1109/BlackSeaCom.2019.8812774
4. Azizi, S., Zandsalimi, M., Li, D.: An energy-efficient algorithm for virtual machine placement optimization in cloud data centers. Clust. Comput. **23**(4), 3421–3434 (2020). https://doi.org/10.1007/s10586-020-03096-0
5. Barber, D.: Bayesian Reasoning and Machine Learning, 1 edn. Cambridge University Press (2012). https://doi.org/10.1017/CBO9780511804779
6. Casas, P., Fiadino, P., D'Alconzo, A.: Machine-Learning Based Approaches for Anomaly Detection and Classification in Cellular Networks. TMA (Apr 2016)
7. Chang, C.C., Lin, C.J.: LIBSVM: a library for support vector machines. ACM Trans. Intell. Syst. Technol. **2**, 27:1–27:27 (2011)
8. Ciocarlie, G.F., Lindqvist, U., Nováczki, S., Sanneck, H.: Detecting anomalies in cellular networks using an ensemble method. In: Proceedings of the 9th International Conference on Network and Service Management (CNSM 2013), pp. 171–174 (2013). https://doi.org/10.1109/CNSM.2013.6727831
9. Doris, S., Salem, I., Schmid, S.: Anomaly detection within mission-critical call processing (2024). http://arxiv.org/abs/2408.14599
10. Gayraud, R., Jacques, O., Robert, D., Charles, W.: SIPp (2014). http://sipp.sourceforge.net/doc/reference.html
11. Godard, S.: iostat(1)—Linux man
12. Gulenko, A., Schmidt, F., Acker, A., Wallschläger, M., Kao, O., Liu, F.: Detecting anomalous behavior of black-box services modeled with distance-based online clustering. In: 2018 IEEE 11th International Conference on Cloud Computing (CLOUD), pp. 912–915 (2018). https://doi.org/10.1109/CLOUD.2018.00134
13. Gulenko, A., Wallschläger, M., Schmidt, F., Kao, O., Liu, F.: Evaluating machine learning algorithms for anomaly detection in clouds. In: 2016 IEEE International Conference on Big Data (Big Data), pp. 2716–2721 (2016). https://doi.org/10.1109/BigData.2016.7840917
14. Jin, S., Zhang, Z., Chakrabarty, K., Gu, X.: Accurate anomaly detection using correlation-based time-series analysis in a core router system. In: 2016 IEEE International Test Conference (ITC), pp. 1–10 (2016). https://doi.org/10.1109/TEST.2016.7805836
15. Khalil, K., Eldash, O., Kumar, A., Bayoumi, M.: Machine learning-based approach for hardware faults prediction. IEEE Trans. Circ. Syst. I **67**(11), 3880–3892 (2020). https://doi.org/10.1109/TCSI.2020.3010743
16. King, C.I.: stress-ng (stress next generation) (2022). https://github.com/ColinIanKing/stress-ng

17. Liu, Y., Zhang, J., Jiang, M., Raymer, D., Strassner, J.: A case study: a model-based approach to retrofit a network fault management system with self-healing functionality. In: 15th Annual IEEE International Conference and Workshop on the Engineering of Computer Based Systems (ECBS 2008), pp. 9–18 (2008). https://doi.org/10.1109/ECBS.2008.30

18. Mostafa, Y.A., Magdon-Ismail, M., Lin, H.T.: Learning From Data, 1st edn. AML-Book (2017)

19. Program, C.N.A. (ed.): Connecting Networks. Companion Guide. 1st edn. Cisco Press, Indianapolis, Indiana (2014)

20. Romanov, O., Nesterenko, M., Fesokha, N., Mankivskyi, V.: Evaluation of productivity virtualization technologies of switching equipment telecommunications networks. Inf. Telecommun. Sci. 1, 53–58 (2020)

21. Samir, A., Pahl, C.: Detecting and predicting anomalies for edge cluster environments using hidden Markov models. In: 2019 Fourth International Conference on Fog and Mobile Edge Computing (FMEC), pp. 21–28 (2019). https://doi.org/10.1109/FMEC.2019.8795337

22. Sauvanaud, C., Lazri, K., Kaâniche, M., Kanoun, K.: Towards black-box anomaly detection in virtual network functions. In: 2016 46th Annual IEEE/IFIP International Conference on Dependable Systems and Networks Workshop (DSN-W), pp. 254–257 (2016). https://doi.org/10.1109/DSN-W.2016.17

23. Schölkopf, B., Smola, A.J., Williamson, R.C., Bartlett, P.L.: New support vector algorithms. Neural Comput. 12(5), 1207–1245 (2000). https://doi.org/10.1162/089976600300015565

24. Segaran, T.: Programming Collective Intelligence. O'Reily Media (2007)

25. Tao, C., Li, T., Huang, J.: Kernel choice in one-class support vector machines for novelty and outlier detection. In: 2020 2nd International Conference on Machine Learning, Big Data and Business Intelligence (MLBDBI), pp. 116–120 (2020). https://doi.org/10.1109/MLBDBI51377.2020.00026

26. Tuncer, O., Ates, E., Zhang, Y., Turk, A., Brandt, J., Leung, V.J., Egele, M., Coskun, A.K.: Online diagnosis of performance variation in HPC systems using machine learning. IEEE Trans. Parallel Distrib. Syst. 30(4), 883–896 (2019). https://doi.org/10.1109/TPDS.2018.2870403

27. Ware, H., Frederick, F.: VMSTAT(8): Report virtual memory statistics-Linux Man

28. Welsh, M., Cox, A., Hoang, T., Eckenfels, B.: NETSTAT(8)-Linux man

29. Wong, D., Yip, S.: Machine learning classifies cancer. Nature 555(7697), 446–447 (2018). https://doi.org/10.1038/d41586-018-02881-7

30. Zhang, H.: The Optimality of Naive Bayes. 2 (2004)

Brief Announcement: Make Master Private-Keys Secure by Keeping It Public

Shlomi Dolev[1], Komal Kumari[2], Sharad Mehrotra[3], Baruch Schieber[2], and Shantanu Sharma[2(✉)]

[1] Ben-Gurion University of the Negev, Be'er Sheva, Israel
`dolev@cs.bgu.ac.il`
[2] New Jersey Institute of Technology, Newark, NJ, USA
`sbar@njit.edu` , `shantanu.sharma@njit.edu`
[3] University of California at Irvine, Irvine, CA, USA
`sharad@ics.uci.edu`

Abstract. The private key associated with a blockchain is the sole means of linking a cryptocurrency asset to its owner, and any loss or compromise of this key could result in significant consequences. Typically, crypto-wallets generate a private key from a string of words, which users are advised to store in a private record, such as a piece of paper. This method poses several security risks, as the private record holding the secret words can leak the private key. Additionally, private records are vulnerable to being lost or destroyed, leading to the potential loss of assets. Moreover, clients have limited control over the generation of their private key, as the wallet generates it. Our approach empowers clients to securely generate and manage their own private keys, minimizing the risk of key loss. We have developed an open-source technique that allows clients to use memorized secrets to store and retrieve their private keys. Our method employs Bloom filters with hash functions, such as SHA-256, to store and retrieve the private key from the Bloom filter securely.

1 Introduction

Private keys of cryptocurrency systems, such as Bitcoin and Ethereum, are the only means of associating the ownership of a client/user with their digital assets. Loss or compromise of the private key can lead to severe consequences, including the permanent loss of funds.[1] To mitigate the risk of losing the private key, clients use cryptographic wallets to store their keys. A crypto wallet can be either a "cold/offline wallet," such as a piece of paper or a flash drive, or a "hot/online wallet," as offered by services like Coinbase and MetaMask. While

---

[1] https://tinyurl.com/yvxkpk95.

This work was supported by the BGU-NJIT Institute for Future Technologies (seed grant), the Israeli Science Foundation (Grant No. 465/22), the Rita Altura trust chair in computer science, and by the Lynne and William Frankel Center for Computer Science. The work of S. Mehrotra is supported by NSF grants 2420846, 2245372, 2133391, 2008993, and 1952247. The work of S. Sharma is supported by NSF grant 2245374.

being offline, adversaries cannot access cold wallets; however, they have the risk of being lost/destroyed, preventing the client from accessing their digital assets.

In contrast, existing online wallets provide secure and easy access to digital assets. For every client, these wallets create a private key (a deterministic sequence of 256 bits) derived from a string of secret words, known as the secret recovery phrase. The online wallet generates the secret recovery phrase, consisting of 12, 18, or 24 words, selected from a list of 2048 words [3]. This secret recovery phrase serves as an alternative to memorize the 256-bit private key and must be securely stored by the client. However, in this process, the client neither generates the private key nor selects the secret recovery phrase. To access the wallet, clients need to present the secret recovery phrase to authenticate themselves. The use of these online wallets presents two significant problems:

- *No control to the client.* This major issue arises because the wallet generates both the private key and the secret recovery phrase. For instance, in Coinbase Wallet, clients are provided with an automatically generated 12-word secret recovery phrase, which represents the private key used to access the wallet and perform transactions. Consequently, clients lack complete control over the generation of their private key and secret recovery phrase.
- *Need to remember the secret recovery phrase.* The scheme's security relies on the client's ability to remember the secret recovery phrase presented by the wallet. If clients fail to remember this phrase, they lose complete access to their crypto assets. [5] shows that humans struggle to remember such combinations of words effectively. Humans often store these phrases on a personal computing device, a piece of paper, or in the cloud [4]. However, these options are prone to being misplaced, damaged, or compromised, leading to asset loss. The brain wallet [1], where a user sets a memorable phrase serving as a key, results in choices that can be easily guessed. [6] discovered 884 brain wallets containing 1,806 bitcoins that were compromised due to predictable phrases. Also, brain wallets suffer from the limitations of human memory, which can result in the loss of bitcoins.

This paper tries to address the above-mentioned security concerns of the private key of crypto-wallets in terms of the ***creation and maintenance of the private key*** and asks the following question:

Is it possible to develop a mechanism that empowers the clients to create their own totally random and never-revealed private keys and store them securely without the risk of being lost?

Our contribution. We develop a technique, entitled R2R (Reminisces to Rescue), that addresses our question. ***The key advantage is that, unlike crypto wallets, our technique leverages the client to create and manage their own private key without the risk of losing it.*** R2R uses memorized (possibly very long) secrets, which are different from the private key, and Bloom filters to store/retrieve the client's private key. Humans showcase a great ability to recall ***memorized facts/reminisces/secrets that are unique and***

known only to the individual, in contrast to remembering random strings of keywords, such as those generated by existing crypto wallets, as explained above. Examples of memorized and/or owner-retrievable (typically long) secrets could be the first stanza of your favorite song, the fourth paragraph of the third chapter of your favorite book, or a dialogue from your favorite movie or TV show. Unlike the traditional keyword-based secrets, *e.g.*, the first name of your favorite teacher, the name of your first pet, or the last four digits of SSNs, these long-memorized secrets are insusceptible to dictionary attacks. R2R uses Bloom Filter to associate "any" true random private key to private memorized secrets.

R2R enables clients to create their own random bits of private keys, use memorized secrets to store, and (later) extract their private keys. R2R appends a private key after memorized secrets and stores bit-by-bit in a Bloom filter. This results in a pseudorandom sequence of zeros and ones. Such random bits hold no value, unless private memorized secrets of the client are known to adversaries. The client can publish/store replicas of the Bloom filter publicly in newspapers/clouds/local files, avoiding the risk of losing the private key.

Full version, pseudocode, code in Python, and demo video of R2R technique: are given in https://tinyurl.com/R2R-Code.

2   R2R Technique

Client and adversarial view. The client generates its private key, knows security questions provided by R2R, and knows memorized secret answers to the questions. The client executes Insert Algorithm on its private key, resulting in a Bloom filter, which is placed in the public domain, and executes Retrieve Algorithm over the Bloom filter to retrieve the private key. Note that remembering these security questions by the client implies the risk of the client forgetting relevant questions; thus, questions and their order used by the client are also public (as long as the answers are private). An adversary knows the security questions used by client, their order, and the Bloom filter. We call this as *adversarial view*. Based on the adversarial view, an adversary wishes to learn the private key of client. As will become clear soon, the Bloom filter merely appears as a pseudorandom sequence of zeros and ones to adversary, with no meaningful information, unless all private memorized secrets are known to adversary.

Assumptions: R2R technique assumes that: (*i*) a client always remembers the memorized secrets, (*ii*) the Bloom filter resides in a public domain, mitigating the risk of it getting lost, (*iii*) the security questions and their order of occurrence are publicly available, and (*iv*) a private key authentication mechanism exists within the online wallet to authenticate client's private key.

2.1   Storing Client's Private Key—Insert Algorithm

Our idea is to use public storage and still benefit from the state-of-the-art pseudorandomness implied by the cryptographic hash function, e.g., Secure Hash

Algorithm (SHA). Insert Algorithm encodes the client's private key using memorized secrets and stores the private key in a Bloom filter, which can be published in a public domain. Let $\mathbb{K}$ be a private key of a client. Let q be the number of security questions. Let a_i be the memorized secret answers to the ith question. Let $\mathbb{B}$ be a Bloom filter using hash function $\mathbb{H}$. The client first concatenates all q memorized secret answers: $answer \leftarrow a_1||a_2||\ldots||a_q$.[2]

Then, each bit of $\mathbb{K}$ is appended at the end of, one by one, and the resultant sequences are inserted into $\mathbb{B}$ using hash function $\mathbb{H}$ (similar to Bloom filter-based lookup table BFLUT [2]). Finally, $\mathbb{B}$ is placed in the public domain.

Example of Insert Algorithm. Suppose, Lisa is a client who wants to store private key 110, selects two questions, and the memorized secrets answers: (i) the first stanza of your favorite song, e.g., "You are somebody..." from the song "You Need to Calm Down" by Taylor Swift, and (ii) the fourth paragraph of the third chapter of the favorite book, e.g., "Thorndike tracked the behavior..." from the book "Atomic Habits". For simplicity, we are providing a few words from each memorized secret; however, in practice, these memorized secrets will comprise a complete stanza or paragraph. Lisa concatenates the two memorized secret answers as "You...Thorndike...". Then, Lisa creates a Bloom filter $\mathbb{B}$, as: $\mathbb{H}($"You...Thorndike...1"$)$, $\mathbb{H}($"You...Thorndike...11"$)$, $\mathbb{H}($"You...Thorndike...110"$)$ by setting one at the corresponding indices. Finally, $\mathbb{B}$ is placed in a public domain. Note that the adversary knows only the two security questions used by Lisa, but not the corresponding secret answers.

2.2   Retrieving Client's Private Key—Retrieve Algorithm

The client uses Retrieve Algorithm to retrieve their private key by downloading $\mathbb{B}$ from public domain for performing lookup operations over $\mathbb{B}$. Similar to Insert Algorithm, client first concatenates q memorized secret answers as: $a_1||a_2||\ldots||a_q$, resulting in $answer$. To $answer$, the client appends bit zero and then bit one and performs a lookup in $\mathbb{B}$ for the appended sequence—note that bits zero and one can be appended and checked in any order. For each successful lookup ($\mathbb{B}$ outputs as one), the client further appends bit zero, and then bit one and performs lookup for the updated sequence. In case of an unsuccessful lookup ($\mathbb{B}$ outputs zero) the process is terminated for the corresponding sequence. The process continues until the number of appended bits equals $|\mathbb{K}|$, to produce $\mathbb{K}$.

Example of Retrieve Algorithm.   Consider that Lisa wants to retrieve the private key 110 from $\mathbb{B}$, using the same memorized secret answers, mentioned in the example of Insert Algorithm. Lisa performs lookups: $\mathbb{H}($"You...Thorndike...0"$)$, $\mathbb{H}($"You...Thorndike...1"$)$. Suppose,

---

[2] An alternative is just to use the output of hash digest, say SHA($answer$), as the private key; however, the result may not be a valid private key for public/private key systems, which is used in the current crypto-wallets. Another alternative to Bloom filter-based solution could use SHA($answer$) to generate a key for AES512, which in turn is used to encrypt and decrypt the signing private key—a private key that is coupled with a paired public key. The Bloom filter solution is more memory efficient when several private keys (say, one for each of the cryptocurrencies) have to be supported. Moreover, the access pattern for retrieving a particular key is less tractable when compared to the access of an entry in a table of encrypted keys.

$\mathbb{H}($ "You...Thorndike...1" $)$ results in one. Then, Lisa appends zero and one to perform lookup for $\mathbb{H}($ "You...Thorndike...10" $)$, $\mathbb{H}($ "You...Thorndike...11" $)$. Note, since the lookup of $\mathbb{H}($ "You...Thorndike...0" $)$ outputs zero, Lisa discontinues the append/lookup process for these sequences. Suppose, $\mathbb{H}($ "You...Thorndike...11" $)$ results in one. The append/lookup continues until finally Lisa gets the output of $\mathbb{H}($ "You...Thorndike...110" $)$ as one. Thus, Lisa retrieves $\mathbb{K}$ as 110.

2.3   Making False Positives to Zero

Bloom filter lookup comes with false positives with some probability. To avoid false positives, we can append a long sequence of ones or concatenate memorized secret answers at the end of private key. Besides performing insertions as per Insert Algorithm, we append either a sequence of ones or *answer* at the end of *answer*$||\mathbb{K}$ that results in either *answer*$||\mathbb{K}||111...$ or *answer*$||\mathbb{K}||$*answer*. The sequence of ones is inserted bit-by-bit into $\mathbb{B}$, while *answer* is inserted word-by-word. Note that insertion will increase size of $\mathbb{B}$. During retrieval, once we extract all candidate private keys, say *candidate*, using Retrieve Algorithm. We check all *candidate* appended with ones or *answer* bit by bit, and discard *candidate* whose lookup operation results zero. The process continues until the client is left with a single *candidate* that is the private key of the client.

2.4   Security Analysis

The adversarial view constitutes the security questions and their order used by client; however, not the memorized secret answers. Thus, the adversary needs to try **all possible** combinations for stanzas of all the songs and the fourth paragraph from the third chapter of every book. Note that using only one question, such as the first stanza of your favorite song, may make the technique less secure, since an adversary can focus to find this information and learn private key. In contrast, using more than one question, enhances the security of the technique, as adversary needs to learn all the correct answers to learn the private key. In particular, there are over 100M songs on Spotify and over 5M English novels. The adversary needs to try all possible $100M \times 5M \approx 2^{25}$ combinations to retrieve client's private key. As the client uses q security questions, such that each question has at least a domain of size of 1M, the complexity to learn the memorized secret answers will be at least $(1M)^q$ or 2^{20q}.[3] Further, the client might consider not disclosing the questions and obfuscating these questions using reminisces in the questions too; e.g., using nicknames or polysemous words. For instance a question "what is best in Israel," could have multiple answers such as city (Haifa), food (Shakshuka), beach (Beit Yanai), actress (Gal Gadot), or TV series (Fauda). The client remembers only one thing that they really like.

---

[3] Selecting memorized secrets from a large domain is *not* a restriction of R2R. A client can also select memorized secrets from a smaller-sized domain, say 50. In this case, the client needs to select multiple questions. Recall that since questions and their order are available in public, it does not pose a risk of forgetting them. For example, for memorized secrets, each with a domain of size 50, a client may select 20 questions. Here, the adversary needs to try 2^{215} combinations, which is computationally infeasible, to learn the memorized secret and then the key.

Suppose, for the client, the best in Israel is Haifa. Based on the keyword "Haifa," the client selects the publicly-known questions $\langle 8, 1, 9, 6, 1 \rangle$ provided by the R2R technique. This method enhances the concealing of the public questions used by clients, hence yielding a practically impossible search for the right answers.

3   Conclusion

R2R technique empowers the clients to create their own totally random and never-revealed private keys and store them securely without the risk of being lost. R2R offers security against alphabetically exhaustive searches, preventing an adversary from learning the key. Further, clients do not need to remember the questions used, as they will become public, as well as, long answers (e.g., a book chapter)—the only need is to remember which chapter and then the client can find the chapter. Making questions and Bloom filters public avoids the possibility of losing the private storage, keeping the key secure.

References

1. Brainwallet. Available at https://en.bitcoin.it/wiki/Brainwallet
2. Dolev, S., et al.: BFLUT bloom filter for private look up tables. In: CSCML (2022)
3. Everything you need to know. https://tinyurl.com/4v682pu
4. What's in the Cloud? Available at https://tinyurl.com/yux3y3yd
5. Ur, B., et al.: "i added '!' at the end to make it secure": Observing password creation in the lab. In: SOUPS, pp. 123–140 (2015)
6. Vasek, M., et al.: The bitcoin brain drain: examining the use and abuse of bitcoin brain wallets. In: FC, pp. 609–618 (2016)

Selection Guidelines for Geographical SMR Protocols: A Communication Pattern-Based Latency Modeling Approach

Kohya Shiozaki[1] and Junya Nakamura[2]

[1] Department of Computer Science and Engineering, Toyohashi University of Technology, Toyohashi, Japan
koya@dsl.cs.tut.ac.jp
[2] Information and Media Center, Toyohashi University of Technology, Toyohashi, Japan
junya@imc.tut.ac.jp

Abstract. State machine replication (SMR) is a replication technique that ensures fault tolerance by duplicating a service. Geographical SMR can enhance its robustness against disasters by distributing replicas in separate geographical locations. Several geographical SMR protocols have been proposed in the literature, each of which is tailored to specific requirements; for example, protocols designed to meet the requirement of latency reduction by either sacrificing a part of their fault tolerance or limiting the content of responses to clients. However, this diversity complicates the decision-making process for selecting the best protocol for a particular service. In this study, we introduce a latency estimation model for these SMR protocols based on the communication patterns of the protocols and perform simulations for various cases. Based on the simulation results and an experimental evaluation, we present five selection guidelines for geographical SMR protocols based on their log management policy, distances between replicas, number of replicas, frequency of slow paths, and client distribution. These selection guidelines enable determining the best geographical SMR protocol for each situation.

Keywords: Fault tolerance · State-machine replication · Geographical SMR · Performance modeling · Performance evaluation · Cloud computing

This work was supported by JSPS KAKENHI Grant Numbers JP20KK0232 and JP22K11971.

1   Introduction

State machine replication (SMR) [14] is a technique that equips a service with fault tolerance by replicating the service across multiple servers called *replicas*. The replicas can reach a consensus on the processing order of the requests using the SMR protocol, thereby ensuring that the replicated service maintains a consistent state among all replicas. Consequently, the service continues to operate even when several replicas fail.

Unlike conventional SMR, which places replicas within a single data center, geographical SMR [1,6,7,12,18] is a specialized form of SMR that distributes replicas across different geographical locations to enhance resilience against disasters. Thus, the service can remain operational even if several replicas become inoperative because of major disasters such as earthquakes or tsunamis.

However, in geographical SMR, the distributed placement of replicas significantly decreases responsiveness, which can be attributed to the increased and unbalanced communication delay resulting from a more significant physical distance between replicas [5,13,15]. In addition, the consensus of the SMR protocol among geographically distant replicas introduces an additional overhead.

Researchers have conducted numerous studies to address the responsiveness challenges of geographical SMR and have proposed several geographical SMR protocols [1,6,7,12,18]. However, each protocol is optimized for different applications, which makes selecting the appropriate SMR protocol from among these developed protocols challenging. The responsiveness of the protocols depends on the environment and requirements of the service. Further, the environment may change during service operations [5,13], thereby affecting the responsiveness.

While a comparative evaluation of SMR protocols under diverse conditions is essential, evaluating every possible combination of SMR protocols and replica placement is unrealistic. In addition, to the best of our knowledge, no study provided a comprehensive overview of SMR protocol characteristics or selection guidelines for the protocols, and therefore, selecting an SMR protocol that aligns with the service environment and requirements becomes highly challenging.

To address these issues, this study presents simulation-based selection guidelines for SMR protocols. These guidelines help service designers select an SMR protocol suited to their environment and requirements efficiently. These guidelines are based on the communication patterns of five representative SMR protocols: MultiPaxos [16], Mencius [11], FastPaxos [9], Domino [18], and EPaxos [12].

For clarifying the guidelines, we constructed a simulation model for each protocol to estimate the latency considering the communication time between clients and replicas. Using the model, we conducted simulations across various scenarios to reveal the characteristics of each protocol and their effects on responsiveness. Further, we deployed an SMR on a cloud service to assess metrics that could not be assessed through simulation. These wide-ranging measurements allowed us to evaluate the responsiveness of the protocols comprehensively and propose guidelines for selecting the most suitable protocol for each scenario.

The key contributions of this study include establishing selection guidelines for geographical SMR protocols, proposing models that can accurately estimate the latency of five representative SMR protocols, and unveiling the detailed characteristics of these protocols. Using the constructed models, we can accurately predict the latency of a service incorporating geographical SMR. Furthermore, clarifying the characteristics of each protocol can significantly contribute to the development of new protocols and related research.

2   Related Work

SMR [14] equips a service with fault tolerance by replicating across n replicas to withstand up to f replica failures. The replicas use an SMR protocol to reach a consensus on the processing order of client requests. Once the replicas agree on the order, they *commit* the commands requested by the clients to their replicated commit logs; each replica then independently executes the commands from the commit log in the same order to ensure consistency across all replicas. This replication technique ensures that the replicas maintain a consistent state, thereby enabling the service to persist even if several replicas fail.

Replicas may experience *crash* or *Byzantine* faults during replication. In the former, a replica stops operating when it fails, and in the latter, a replica behaves arbitrarily when it fails. SMR is classified as a crash fault-tolerant SMR (CFT-SMR) or Byzantine fault-tolerant SMR (BFT-SMR) based on the assumed failure type. The CFT-SMR can tolerate up to f crash faults with $n \geq 2f + 1$ replicas, whereas the BFT-SMR can tolerate up to f Byzantine faults with $n \geq 3f + 1$ replicas. Many SMR [3,11,12,16,18,19] and geographical SMR protocols [6,11,12,18] have been proposed in the literature.

A service designer must have sufficient knowledge of the characteristics of several protocols to select an appropriate protocol for geographical SMR. However, to the best of our knowledge, a comprehensive survey that covers the various aspects of these protocols to help select an SMR protocol suited to the requirements of the designer remains lacking. Although many studies that proposed SMR protocols compared their performances with those of existing protocols, their evaluations focused on demonstrating the superiority of the proposed protocols. These studies did not provide this information, and therefore, guidelines for selecting geographical SMR protocols are required.

Evaluating the performances of several SMR protocols in various scenarios is necessary to design such guidelines. However, this evaluation is time-consuming, expensive, and impractical, and therefore, several attempts have been made to capture the performance without building a geographical replication. Castro presented a latency estimation model for the PBFT protocol considering the communication time between replicas and the processing delay within a replica [4]. Numakura et al. modeled the latency of BFT-SMaRt [3] and proposed an efficient method to determine the optimal replica placement in the geographical SMR [13]. Inspired by Numakura et al.'s approach, we model and evaluate five representative CFT-SMR protocols in this paper. Lorünser et al. modeled PBFT focusing on network layer parameters such as the packet loss rate of communication channels [10]. Their attention layers differ from those in our approach. Another approach involves emulating a wide-area network to measure the responsiveness of the SMR protocols [2,17]. This approach provides more detail than the modeling approach; however, it requires more time and computational resources.

3   SMR Protocol Latency Model

This section constructs the latency estimation models for the five representative SMR protocols, which we will use in the simulations in Sect. 4, and verifies the estimation accuracy of the models.

3.1   Model Construction

We construct latency estimation models for the five representative SMR protocols (MultiPaxos [16], Mencius [11], FastPaxos [9], Domino [18], and EPaxos [12]). These protocols assume an asynchronous network with FIFO channels between processes (i.e., replicas and clients) and use TCP as their transport protocol. The modeling follows the approach proposed by Numakura et al. [13]. The latency is estimated by imitating the communication patterns during the consensus process of the SMR protocols. For the estimation, we use the round-trip time (RTT) collected before estimation as the basic unit in the models.

We can categorize services employing SMR based on their response requirements into *full-response* and *status-response* services. Each type exhibits different responsiveness characteristics. In the full-response service, after committing a requested command, each replica executes the command to update its state and returns the execution result to the client. In contrast, in the status-response service, each replica returns the success or failure of the command immediately after its commit, and then, it executes the command to update its state.

We aim to construct latency estimation models that can be used to compare responsiveness in various scenarios for formulating selection guidelines for geographical SMR protocols, and therefore, hereafter, we construct latency models for the status-response services. This is because, the command execution time, which is considered only in the full-response service, heavily varies depending on each service and its execution environment. Therefore, we focus on the status-response service to simplify the models here and discuss the characteristics of the full-response service later in Sect. 4.2.

The latency estimation function $EL_{avg}(R, C)$ for a replica placement R consisting of n replicas is described first. C represents a set of client locations, indicating the client distribution. This function is the basis of the estimation and invokes the estimation model for each protocol. $EL_{avg}(R, C)$ calculates the average latency of all clients by calculating the estimated latency of each client $c \in C$ as

$$EL_{avg}(R, C) = \sum_{c \in C} EL(R, c)/|C| \tag{1}$$

In Eq. 1, function EL estimates the latency for each client based on the type of request as

$$EL(R, c) = Protocol(R, \ell, c, p_{slow}) \times p_w + Read(R, c) \times (1 - p_w)$$

where p_w is the probability that a write request is sent; thus, the first and second terms of the right-hand side of the equation estimate the latency of write and read operations, respectively.

In the first term, the function *Protocol* takes four arguments R, ℓ, c, and p_{slow}, where ℓ and p_{slow} represent the leader replica and the slow path occurrence ratio due to the concurrent client requests for the same data, respectively. *Protocol* calculates the latency of the write operation for each protocol based on its communication pattern, and its formulation is shown below. Mencius and EPaxos omit the ℓ because they do not require a leader replica. Similarly, Multi-Paxos and Mencius omit p_{slow} because no slow path occurs in these protocols. In the models below, $OWD(a,b)$ represents the communication time between two replicas or clients a and b, which can be calculated from RTT between a and b.

In the second term, the function $Read(R,c)$ estimates the latency of the read operation for each client c. A client issues a read operation to retrieve (possibly a part of) the state of a replicated service. The communication pattern of the read operation varies depending on the consistency guarantees. In cases where the service guarantees linearizability, similar to write operations, reading from the leader replica or a quorum is necessary. Conversely, reading from any replica is feasible if the operation prioritizes responsiveness and ensures eventual consistency. Therefore, the communication pattern for read operations depends on the consistency guarantees provided, and it presents no variance across protocols. Thus, we did not consider read operations when establishing guidelines for protocol selection and instead focused on discussing write operations.

MultiPaxos. MultiPaxos [16] extends the consensus protocol Paxos [8] to realize SMR. MultiPaxos invokes individual Paxos instances for each request to agree on the execution order of its command. Figure 1 shows the communication pattern during a write operation in MultiPaxos. This communication pattern is divided into the request, propose, accept, commit, and response phases. We calculate the timing for each replica when sending and receiving messages in each phase based on RTT, and we determine the latency of the write operation. The elapsed times in each phase are denoted by S_{req}, S_{pro}, S_{acc}, S_{cmt}, and S_{res}. We add the superscript S_{pro}^i to distinguish these times for a replica r_i.

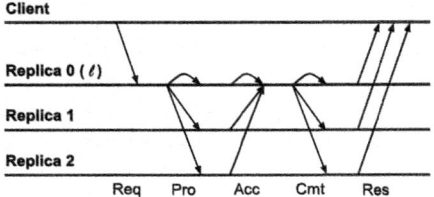
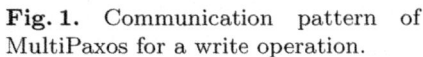

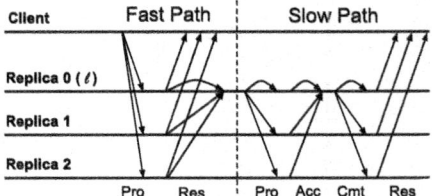

Fig. 1. Communication pattern of MultiPaxos for a write operation.

Fig. 2. Communication pattern of FastPaxos for a write operation.

Initially, the leader replica ℓ received a request from client c at $S_{req} = OWD(c, \ell)$. Upon reception, the leader replica ℓ transmits this request to all replicas as a propose message, and each replica r_i receives the message at $S_{pro}^i = S_{req} + OWD(\ell, r_i)$. When r_i receives the propose message, it replies with an accept message to the leader replica, and the leader receives the majority of the accept messages at $S_{acc} = find\,(T_{acc}, \lceil (n+1)/2 \rceil)$, where $T_{acc} = \{t \mid S_{pro}^i + OWD(r_i, \ell), 0 \le i < n\}$ and the function $find(S, k)$ returns the k-th smallest element from set S. Once the leader replica receives a sufficient number of accept messages, it sends the commit message to all replicas. Consequently, the timing S_{cmt}^i at which r_i receives the commit message is $S_{cmt}^i = S_{acc} + OWD(\ell, r_i)$. Lastly, once the commit message is received, the replica r_i notifies the client of the successful commit via a response message. The client accepts the result upon receiving the first response message. Thus, the model $MultiPaxos(R, \ell, c)$ that estimates the latency of the write operation of MultiPaxos is given by

$$MultiPaxos(R, \ell, c) = S_{res} = \min(T_{res}),$$

where $T_{res} = \{t \mid S_{cmt}^i + OWD(r_i, c), 0 \le i < n\}$.

Mencius. Mencius [11] was developed as an extension of MultiPaxos. Although it maintained the core principles of MultiPaxos, it was tailored for SMR in wide area networks (WANs). In MultiPaxos, a single fixed leader replica is responsible for managing all consensus instances, which causes the clients to send requests to the leader replica, thereby leading to poor responsiveness for clients geographically distant from the leader. To address this issue, Mencius partitions the commit log among all replicas in advance, thereby enabling them to manage consensus in parallel. This enables clients to send requests to replicas that are geographically closer and have a lower RTT. Therefore, Mencius's latency estimation model $Mencius(R, c)$ can be expressed using MultiPaxos model as

$$Mencius(R, c) = MultiPaxos(R, OptLeader(R, c), c),$$

where $OptLeader(R, c) = \arg\min_{r \in R} OWD(c, r)$ is a function that returns replica $r \in R$ with the shortest communication time from client c.

FastPaxos. Fig. 2 shows the communication pattern for the write operation in FastPaxos [9], which has two consensus timings: the *fast* and *slow* paths. In the fast path, client and replicas can quickly reach a consensus on the execution order of a command by proposing a request to all replicas directly without going through the leader. However, if a request from another client for the same data arrives in the fast path (*conflict*), the conflict is resolved in the slow path. Thus, the latency of FastPaxos is estimated by summing the latencies of both the fast and slow paths while considering the occurrence rate p_{slow} of the slow path.

First, replica r_i receives a propose message from a client c in the fast path at $S_{fpro}^i = OWD(c, r_i)$. Upon receiving the propose message, replica r_i sends a response message to the client and the leader replica. The client accepts the results when it receives $3n/4$ response messages. Thus, the latency estimation model of the fast path $FastPath(R, c)$ is given by

$$FastPath(R, c) = find\left(T_{fres}, \lceil 3n/4 \rceil\right),$$

where $T_{fres} = \{t \mid S_{fpro}^i + OWD(r_i, c), 0 \le i < n\}$.

There are two reasons for the occurrence of slow paths in FastPaxos: (1) The leader replica receives multiple requests in response messages from $3n/4$ replicas. (2) The leader replica cannot receive response messages from $3n/4$ replicas within a certain time frame. We assume that the differences between these two timings do not have a significant effect on the latency and that they do not distinguish between them. When a slow path is triggered, the leader replica ℓ sends a propose message to each replica. Replica r_i receives this propose message at $S_{spro}^i = OWD(\ell, r_i)$. After acceptance, the leader replica sends a commit message to each replica, and replica r_i receives this commit message at $S_{acc} = find\left(T_{acc}, \lceil (n+1)/2 \rceil\right)$, where $T_{acc} = \{t \mid S_{spro}^i + OWD(r_i, \ell), 0 \le i < n\}$. After acceptance, the leader replica sends a commit message to each replica. Thus, replica r_i receives this commit message at $S_{cmt}^i = S_{acc} + OWD(\ell, r_i)$. On receiving the commit message, the replica sends a response message to client c, which accepts the result when it receives the first message. Thus, the model for estimating the latency of the slow path $SlowPath(R, \ell, c)$ is:

$$SlowPath(R, \ell, c) = S_{sres} = \min(T_{sres}),$$

where $T_{sres} = \{t \mid S_{cmt}^i + OWD(r_i, c), 0 \le i < n\}$. Therefore, the latency estimation model for the write operation of FastPaxos is

$$FastPaxos(R, \ell, c, p_{slow}) = FastPath(R, c) + SlowPath(R, \ell, c) \times p_{slow}$$

Domino. Domino [18] integrates both Mencius and FastPaxos as core protocols. In Domino, each client measures the network speed between replicas periodically and selects an appropriate protocol based on these measurements. Domino estimates the latencies for the fast paths of FastPaxos and MultiPaxos when each replica becomes a leader. The system selects Mencius when the latency for MultiPaxos is the shortest; otherwise, it selects FastPaxos. This mechanism provides Domino with the capability to dynamically adjust to different network scenarios, thereby enabling a fast consensus. Consequently, the latency estimation model of Domino for the write operation, $Domino(R, \ell, c, p_{slow})$, is

$$Domino(R, \ell, c, p_{slow}) = \begin{cases} Mencius(R, c) & \text{if } M \le FastPaxos(R, \ell, c, 0) \\ FastPaxos(R, \ell, c, p_{slow}) & \text{otherwise} \end{cases},$$

where $M = \min_{r \in R} MultiPaxos(R, r, c)$ is the minimal latency via MultiPaxos across all replicas.

EPaxos. EPaxos [12] is an SMR protocol optimized for WANs. Clients in EPaxos send requests to geographically closer replicas that have shorter communication times, similar to that in Mencius. However, unlike Mencius, EPaxos uses different commit log management methods for replicas. In Mencius, the commit log is partitioned equally among all the replicas in advance, and therefore, each replica is responsible for managing its assigned slot. In contrast, EPaxos allows each replica to manage its commit log asynchronously. Upon receiving a request, the replica takes on the role of a commit log manager and coordinates the consensus with other replicas using the same communication pattern as that of Mencius. This approach may cause conflicts in the command execution order if multiple replicas process requests for the same data simultaneously. EPaxos addresses such conflicts with a slow path. Therefore, the latency of EPaxos is estimated by summing the latencies of both the Mencius and slow paths, while considering the occurrence rate of the slow path p_{slow}. The latency estimation model of EPaxos for a write operation, i.e., $EPaxos(R, c, p_{slow})$, is expressed as

$$EPaxos(R, c, p_{slow}) = Mencius(R, c) + SlowPath(R, OptLeader(R, c), c) \times p_{slow}$$

3.2 Model Validation

We evaluate the validity of the proposed latency estimation models to confirm whether they are sufficiently accurate for discussing the selection guidelines presented in Sect. 4. For the evaluation, we compare the latency estimated by the models with the experimentally measured latency. We investigate the accuracy under two distinct environments for both the normal path (including the fast path) and the slow path. The former environment involves all protocols without any request conflicts, whereas the latter targets the slow paths of FastPaxos and EPaxos under the condition where two or more requests conflict[1].

For validation, we measure the latencies of these protocols using Microsoft Azure[2]. We considered 13 regions[3] as potential locations for placing replicas and clients. We adopted Domino's [18] implementation[4] to construct the SMR. We derived the OWD between the regions from the average RTTs measured using the `ping` command taken 60 s prior to the experiment.

Estimation Accuracy in a Non-Conflict Environment. We investigate the estimation accuracy of the five proposed latency models in an environment without request conflicts (slow paths). We deploy a client in Tokyo and the replicas in three to seven locations, i.e., $n = 3, 4, 5, 6, 7$, randomly selected from the 13 regions. For each n, we randomly selected 100 replica placement patterns, resulting in 500 placements, and measured the actual and estimated latencies.

Generally, configuring an even number of replicas is considered rare because even replicas do not increase the number of tolerable faulty replicas. The number of tolerable faulty replicas in the fast paths of FastPaxos and Domino can be improved by using replicas; for example, the fast path tolerates one faulty replica when $n = 4$. Moreover, Loruenser et al. clarified that additional replicas can suppress leader elections due to packet loss [10]. Thus, we consider even replicas.

---

[1] We omit the verification of Domino's slow path because the slow path is derived from FastPaxos.

[2] https://azure.microsoft.com.

[3] Paris, Iowa, Toronto, Seoul, London, California, Victoria, Gävle, São Paulo, Tokyo, Singapore, Virginia, Chennai.

[4] https://github.com/xnyan/domino.git.

Table 1. Correlation coefficients between the estimated and measured latencies

Protocol	Correlation Coefficient
MultiPaxos	0.988
Mencius	0.987
FastPaxos	0.985
Domino	0.985
EPaxos	0.986

(a) Non-conflicting environment

Protocol	Correlation Coefficient
FastPaxos	0.976
EPaxos	0.982

(b) Conflicting environment

For the actual latency, the client sends write operations (i.e., $p_w = 100\%$) to the replicas repeatedly in a closed loop, where each request is sent only after receiving the response to the previous one, for 10 s. All other parameters of the SMR protocols adopt their default values. We assume a status-response service, and therefore, replicas return responses to the client as soon as they have completed the commit phase. Under these settings, we measure the time taken from sending a request to the reception of a response from the client and average all measured times. The estimated latency for the placement and protocol is calculated using $EL_{avg}(R, C)$, where R represents the placement of three randomly chosen replicas, C represents Tokyo, p_w is 100%, p_{slow} is 0%, and *Protocol* represents the protocol being measured.

Table 1(a) summarizes the correlation coefficients. The correlation coefficients of all protocols are 0.985 or higher. Therefore, the latency estimations are highly accurate in environments where request conflicts do not occur. The actual latency measurement for 1,000 placements required 67,421 s. By contrast, the estimation for the same number of placements required only 85 s when using a Python program running on a PC equipped with an Intel Core i5-12400 CPU.

Based on these results, we can conclude that the estimation model allows us to rapidly and precisely estimate the latency of a placement.

Estimation Accuracy in a Conflict Environment. We investigate the estimation accuracy of the proposed latency models for FastPaxos and EPaxos in an environment in which request conflicts (slow paths) occur. The slow path is triggered when multiple clients simultaneously access the same data. Thus, we place three clients in the same location as the replicas and set α, which represents the Zipfian distribution parameter, to 1000, such that conflicts occur frequently. All other settings were consistent with those used in the non-conflict environment experiment. Under these settings, we measured the latency of the requests in the slow path. The estimated latency for the replica placement and protocol was calculated using the value of $EL_{avg}(R, C)$, where R and C represent the placement of three randomly chosen replicas and same placement as the replicas, respectively; Further, p_w is 100%, p_{slow} is 100%, and *Protocol* represents the protocol being measured. Similar to the results in the non-conflict environment experiment, the correlation coefficients in Table 1(b) for both protocols were 0.976 or higher. The estimations were highly accurate also in conflicting environments.

4   Selection Guidelines for Geographical SMR Protocols

The guidelines for selecting geographical SMR protocols are formulated as follows. In Sect. 4.1, we investigated the characteristics of the protocols in the status-response service under various conditions using the estimation models. In Sect. 4.2, we analyzed the latency in the full-response service, considering the command execution time, through an experimental evaluation. Finally, we present selection guidelines based on these results in Sect. 4.3. Note that we present evaluation results in Sects. 4.1 and 4.2 and discuss them in Sect. 4.3.

4.1   Status-Response Service

We simulated the responsiveness of SMR protocols under various conditions by utilizing the latency models. The simulation results aid in formulating the selection guidelines for the SMR protocols.

We used two client distributions, the *global setting* and the *US setting*, to clarify the effect of the geographical distribution tendencies of clients on responsiveness. The global setting dispersed clients to Paris, São Paulo, Toronto, Victoria, and Tokyo, while the US setting concentrated clients in North America (Virginia, California, and Iowa). We used the RTTs between regions published by Microsoft[5]. We considered half of these values for OWD in our latency model.

We conducted each simulation for protocol P, client distribution C, and slow path occurrence rate p_{slow} as follows:

1. Generate all $_{13}C_n$ replica placements by selecting n from the 13 regions[3].
2. Exclude from the generated placements those where the minimum distance between two regions is less than d km.
3. For each placement R, calculate the latency of protocol P under the condition where $p_w = 100\%$ with given p_{slow} and client distribution C.
4. Choose the minimum latency amongst all placements as the latency of P.

Hereafter, we analyze how much each factor affects the latency of protocols.
Effect of Geographical Distance Between Replicas.   The geographical SMR distributes replicas to enhance resilience against large-scale disasters. Defining the disaster scale that the service must withstand is crucial when designing a service with a geographical SMR; for example, by specifying the minimum geographical distance between the replicas. However, assuming a large-scale disaster can increase the distance, thereby degrading the responsiveness.

Given this background, we investigated the effect of the geographic distance between replicas on responsiveness. In the simulation, we varied the minimum distance d between the replicas from 0 to 5000 km in increments of 1000 km. We set the other parameters as follows: the number of replicas $n = 3$, slow-path occurrence rate $p_{slow} = 0.2$, and client distribution is the global setting. The simulation results are shown in Fig. 3. In this figure and the ones that follow, we show the results of Mencius and Domino with dotted lines to distinguish them from other protocols. The reason for this will be discussed in Sect. 4.2.

---

[5] https://learn.microsoft.com/azure/networking/azure-network-latency.

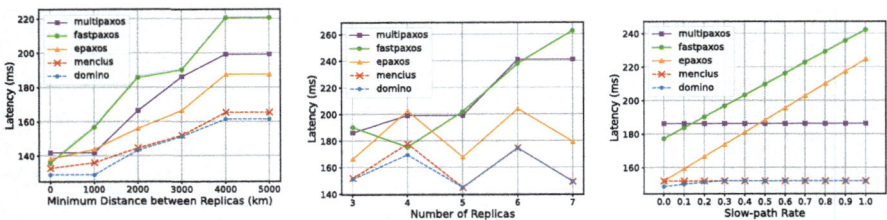

Fig. 3. Latency v.s. geographical distances of the replicas

Fig. 4. Latency v.s. the number of replicas

Fig. 5. Latency v.s. the slow-path occurrence rate

Effect of the Number of Replicas. Key parameters for designing an SMR service include the number of replicas, n, and the number of tolerable faulty replicas, f. More replicas increase fault tolerance but involve more replicas in the consensus process, potentially decreasing responsiveness. We investigated the effect of the number of replicas on responsiveness. We incrementally changed n from three to seven in increments of one with the client distribution of the global setting, $d = 3000$ km, and $p_{slow} = 0.2$. The simulation results for all protocols are shown in Fig. 4. It indicates that the response time decreases with an increase in the number of replicas. This trend differs from the expected behavior of typical SMR systems, where increasing the number of replicas leads to a larger quorum size, which increases the response time. However, in the geographical SMR, the replica placement can reduce response times despite the increase in the number of replicas. When n increases from three to four, the quorum size increases from two to three, thereby resulting in longer response times for Mencius. However, when n increases from four to five, the quorum size remains unchanged and the addition of an extra replica allows clients to form geographically closer quorums more flexibly, thereby resulting in shortened response times.

Effect of the Slow-Path Occurrence Rate. FastPaxos, EPaxos, and Domino trigger a slow path when multiple requests that manipulate the same data simultaneously. Additional communication is required when a slow path is triggered. Consequently, in scenarios where a slow path occurs frequently, responsiveness may decline. We investigated the effect of the slow-path occurrence rate on responsiveness. We incrementally changed the slow-path occurrence rate p_{slow} from zero to one in increments of 0.1 with the client distribution of the global setting, $n = 3$, and $d = 3000$ km. The simulation results are shown in Fig. 5.

Mapping of Optimal Protocols. Building on previous simulations, we explore how the optimal protocol for responsiveness varies by adjusting both the number of replicas n and slow-path occurrence rate p_{slow} under a distinct client distribution. Given $d = 3000$ km, we varied n from three to seven and p_{slow} from zero to one in increments of 0.1. Figure 6 presents the optimal protocol for each combination of n and p_{slow}. In the figure, we fill a cell with two or more colors if multiple protocols are optimal for the cell.

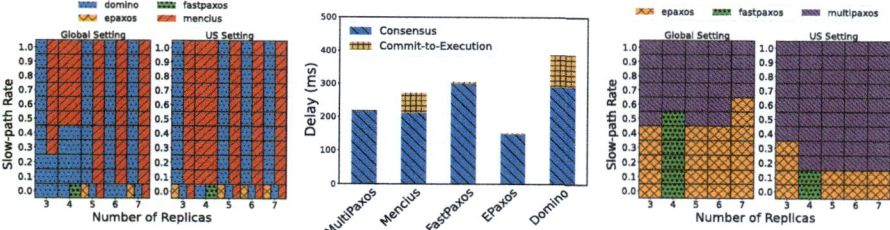

Fig. 6. Optimal protocol for each case for the status-response service

Fig. 7. Each protocol's commit-to-execution delay

Fig. 8. Optimal protocol for each case of the full-response service

In Figs. 3, 4, and 5, Domino demonstrates lower latency compared to Fast-Paxos and Mencius. This is because Domino dynamically selects the appropriate protocol for each individual client.

4.2 Full-Response Service

In Sect. 3, we modeled the latency of SMR protocols for the status-response service assuming that the execution time of commands remains unaffected by the protocol and evaluated them. However, the evaluation does not consider the time from the commit of a command to the start of command execution, which we call the *commit-to-execution delay*. Although several protocols exhibit high performance in simulations of the status-response service, they may experience a performance drop when applied to the full-response service.

For example, in Mencius and Domino, the commit log management role is distributed equally among the replicas. Replicas periodically share the state of the commit log they manage with others. Consequently, each replica obtains a full commit log and executes commands in the log from the oldest to update the state of the replicated service. In this process, a replica cannot execute a command until all the commit logs preceding it are filled with a command or the special command, *no-op*, thereby indicating the absence of a command. Thus, even if a command can be committed quickly, its execution may be delayed.

Effect of the Commit-to-Execution Delay. We conducted an experiment using an actual SMR implementation built on Microsoft Azure to measure the commit-to-execution delay. For this experiment, we deployed three replicas in Tokyo, California, and London and three clients in Seoul, Toronto, and Paris. All other settings were aligned with the non-conflict environment experiments described in Sect. 3.2. Given these settings, we measured the time from when a command was committed until its execution began at the replica. The average duration across all requests, measured in the three replicas, is considered the commit-to-execution delay for each protocol with the client distribution.

The experimental results are presented in Fig. 7. As indicated in the chart, the protocols were divided into two groups: The first group comprises Multi-Paxos, FastPaxos, and EPaxos. Their commit-to-execution delay is small (0.82–

3.11 ms), and therefore, these protocols are suitable for full-response services. The delay of the second group (Mencius and Domino) is large (59.0–96.2 ms) and about 100 times longer than that of the first group. This significant delay arises because the replicas in the second group must gather information from the commit log managed by other replicas to execute a command.

Mapping of Optimal Protocols. These results are used to re-evaluate the optimal protocol for the full-response service similar to that in Sect. 4.1. Figure 8 shows the corresponding results. Although Mencius and Domino performed well in Sect. 4.1, this result indicates that these protocols are not suitable for full-response services due to their long commit-to-execution delay. These aspects must be considered to clarify the selection guidelines for SMR protocols, which is why we distinguish Mencius and Domino from the others in Figs. 3, 4, and 5.

4.3 Selection Guidelines

Finally, we discuss the results from both the simulations and experiments and formulate the selection guidelines for SMR protocols based on these findings.

We present the following guidelines for a full-response service based on the results shown in Fig. 7:

Guideline 1 *For a full-response service, selecting SMR protocols that do not have pre-partitioned commit logs (i.e., MultiPaxos, FastPaxos, EPaxos) can enhance responsiveness.*

For the distance between replicas, Fig. 3 shows that the latency increases with an increase in the distance between replicas. In addition, a larger inter-replica distance resulted in a more significant difference in the latency among protocols, which implies that protocol selection plays an important role, particularly when considering major disasters. Among all protocols, Domino outperformed the others at most distances, with Mencius slightly surpassing Domino at 3000 km. Among the protocols suitable for the full-response service (i.e., MultiPaxos, FastPaxos, and EPaxos), MultiPaxos performs the best with a maximum inter-replica distance of up to 1000 km, whereas EPaxos performs better for greater distances. Based on these findings, we derived the following guideline:

Guideline 2 *For the status-response service, Domino can improve responsiveness regardless of distance constraints. For a full-response service, MultiPaxos and EPaxos should be selected for shorter and larger inter-replica distances, respectively.*

For the effect of the number of replicas on responsiveness shown in Fig. 4, the greater the number of replicas, the longer is the latency. However, EPaxos, Mencius, and Domino can effectively reduce degradation because each client can send a request to the replica near it, not the leader. Conversely, MultiPaxos and FastPaxos exhibited significant performance deterioration with an increasing number of replicas. From these observations, we propose the following guideline:

Guideline 3 *For a service requiring a higher fault tolerance (i.e., more replicas), using protocols that prioritize geographically closer replicas can minimize performance degradation. Mencius and Domino are suitable for the status-response service, and EPaxos for the full-response service.*

In scenarios where massive requests are sent to a small portion of the data, the slow path occurs frequently. Figure 5 shows that the latencies of EPaxos, Fast-Paxos, and Domino increase with an increase in the slow-path occurrence rate. EPaxos and FastPaxos experience significant degradation, whereas the increase in latency for Domino is moderate. This can be attributed to its hybrid approach, where clients can select a suitable protocol between FastPaxos and Mencius. Based on these findings, we propose the following guidelines.

Guideline 4 *In a service where clients share a large amount of data and request frequencies are high (leading to frequent slow paths), it would be advantageous to use protocols that avoid slow paths, such as MultiPaxos or Mencius. This selection can help mitigate performance degradation caused by slow paths.*

Client distribution affects replication responsiveness. The simulation involving all protocols (Fig. 6) indicates that protocols that partition commit logs in advance, such as Domino and Mencius, are superior in most cases, regardless of the client distribution trends. In the global setting (Fig. 6), Mencius is optimal for smaller values of n, whereas Domino is best for larger values of n because Domino's clients choose FastPaxos and Mencius flexibly. In the comparison of the protocols with a small commit-to-execution delay (Fig. 8), MultiPaxos is selected as the optimal protocol in many cases. This trend becomes stronger if the client distribution is dense (as in the US setting). Based on these considerations, we derived the following guideline:

Guideline 5 *If we can predict client distribution in advance, we can select an optimal protocol. For status-response services, Mencius and Domino can improve responsiveness, regardless of client distribution. For a full-response service, EPaxos is beneficial for a dispersed client distribution, whereas MultiPaxos is recommended when the clients are densely located.*

5   Conclusion

In this study, we proposed five selection guidelines for geographical SMR protocols. We constructed latency estimation models for the consensus process of five representative CFT-SMR protocols and simulated their performance under various conditions to formulate these guidelines. Correlation coefficients between the latency estimated by the models and the actual latency were 0.976 or higher, thereby demonstrating its high accuracy. In addition, we implemented SMR on a cloud service to measure metrics that could not be evaluated through simulations. A service designer can easily select the best SMR protocol suited to their service requirements using these guidelines. In addition, our models allowed us to predict the responsiveness of a geographical SMR service with high precision. The classification characteristics of these protocols are expected to contribute to the development of new geographical SMR protocols.

References

1. Berger, C., Reiser, H.P., Sousa, J., Bessani, A.: Aware: adaptive wide-area replication for fast and resilient byzantine consensus. IEEE Trans. Dependable Secure Comput. **19**(3), 1605–1620 (2022). https://doi.org/10.1109/TDSC.2020.3030605
2. Berger, C., Toumia, S.B., Reiser, H.P.: Does my BFT protocol implementation scale? In: Proceedings of DICG'22, pp. 19–24 (2022). https://doi.org/10.1145/3565383.3566109
3. Bessani, A., Sousa, J., Alchieri, E.E.: State machine replication for the masses with BFT-SMART. In: Proceedings of DSN'14, pp. 355–362 (2014). https://doi.org/10.1109/DSN.2014.43
4. Castro, M.: Practical Byzantine Fault Tolerance. Ph.D. thesis, Massachusetts Institute of Technology (2001)
5. Chiba, T., Ohmura, R., Nakamura, J.: Network bandwidth variation-adapted state transfer for geo-replicated state machines and its application to dynamic replica replacement. Concurr. Comput. Pract. Exp. **35**(19), e7408 (2023). https://doi.org/10.1002/cpe.7408
6. Coelho, P., Pedone, F.: Geographic state machine replication. In: Proceedings of SRDS'18, pp. 221–230 (2018). https://doi.org/10.1109/SRDS.2018.00034
7. Enes, V., Baquero, C., Rezende, T.F., Gotsman, A., Perrin, M., Sutra, P.: State-machine replication for planet-scale systems. In: Proceedings of EuroSys'20 (2020). https://doi.org/10.1145/3342195.3387543
8. Lamport, L.: The part-time parliament. ACM Trans. Comput. Syst. **16**(2), 133–169 (1998). https://doi.org/10.1145/279227.279229
9. Lamport, L.: Fast Paxos. Distrib. Comput. **19**(2), 79–103 (2006). https://doi.org/10.1007/s00446-006-0005-x
10. Loruenser, T., Rainer, B., Wohner, F.: Towards a performance model for byzantine fault tolerant services. In: Proceedings of CLOSER'22, pp. 178–189 (2022). https://doi.org/10.5220/0011041600003200
11. Mao, Y., Junqueira, F.P., Marzullo, K.: Mencius: building efficient replicated state machines for WANs. In: Proceedings of OSDI'08, pp. 369–384 (2008)
12. Moraru, I., Andersen, D.G., Kaminsky, M.: There is more consensus in Egalitarian parliaments. In: Proceedings of SOSP'13, pp. 358–372 (2013). https://doi.org/10.1145/2517349.2517350
13. Numakura, S., Nakamura, J., Ohmura, R.: Evaluation and ranking of replica deployments in geographic state machine replication. In: Proceedings of SRDSW'19, pp. 37–42 (2019). https://doi.org/10.1109/SRDSW49218.2019.00014
14. Schneider, F.B.: Implementing fault-tolerant services using the state machine approach: a tutorial. ACM Comput. Surv. **22**(4), 299–319 (1990). https://doi.org/10.1145/98163.98167
15. Sousa, J., Bessani, A.: Separating the WHEAT from the chaff: an empirical design for geo-replicated state machines. In: Proceedings of SRDS'15, pp. 146–155 (2015). https://doi.org/10.1109/SRDS.2015.40
16. Van Renesse, R., Altinbuken, D.: Paxos made moderately complex. ACM Comput. Surv. **47**(3), 42:1–42:36 (2015). https://doi.org/10.1145/2673577

17. Wang, P.L., Chao, T.W., Wu, C.C., Hsiao, H.C.: Tool: an efficient and flexible simulator for byzantine fault-tolerant protocols. In: Proceedings of DSN'22, pp. 287–294 (2022). https://doi.org/10.1109/DSN53405.2022.00038
18. Yan, X., Yang, L., Wong, B.: Domino: using network measurements to reduce state machine replication latency in WANs. In: Proceedings of CoNEXT'20, pp. 351–363 (2020). https://doi.org/10.1145/3386367.3431291
19. Zhao, W.: Fast Paxos made easy: theory and implementation. Int. J. Distrib. Syst. Technol. **6**(1), 15–33 (2015). https://doi.org/10.4018/ijdst.2015010102

Byzantine Reliable Broadcast with One Trusted Monotonic Counter

Yackolley Amoussou Guenou[1], Lionel Beltrando[2], Maurice Herlihy[3],
and Maria Potop-Butucaru[2(✉)]

[1] Université Paris-Panthéon-Assas, CRED, Paris, France
yackolley.amoussou-guenou@u-paris2.fr
[2] Sorbonne Université, LIP6, Paris, France
lbeltrando@finaxys.com , maria.potop-butucaru@lip6.fr
[3] Brown University, Providence, RI, USA

Abstract. Byzantine Reliable Broadcast is one of the most popular communication primitives in distributed systems. Byzantine reliable broadcast ensures that processes agree to deliver a message from an initiator, even if some processes (possibly including the initiator) are Byzantine. In asynchronous settings, it is known since the prominent work of Bracha [4] that Byzantine reliable broadcast can be implemented deterministically if the total number of processes, denoted by n, satisfies $n \geq 3t + 1$ where t is an upper bound on the number of Byzantine processes. Here, we study Byzantine Reliable Broadcast when processes are equipped with *trusted components*, special software or hardware designed to prevent equivocation. Our contribution is twofold. First, we show that, despite common belief, when each process is equipped with a trusted component, Bracha's algorithm still needs $n \geq 3t+1$. Second, we present a novel algorithm that uses a single trusted component (at the initiator) that implements Byzantine Reliable Asynchronous Broadcast with $n \geq 2t + 1$.

1 Introduction

Byzantine reliable broadcast is a fundamental problem in fault-tolerant distributed systems. It consists of ensuring that a correct initiator process broadcasts its value to all correct processes, even in the presence of malicious Byzantine processes. For decades, Byzantine Reliable Broadcast has been at the core of various consensus protocols, and more recently, at the core of certain blockchains.

Byzantine Reliable Broadcast has been addressed in various settings: with fixed and mobile Byzantine nodes, dynamicity or in conjunction with transient faults. Byzantine Reliable Broadcast solutions (e.g. [2, 3, 12, 14, 15]) achieve resilience of at least $n \geq 3t + 1$ processes, where t is the maximum number of Byzantine processes. However, these solutions require strong network assumptions, such as *synchrony* (processes execute in lock-step) or *non-equivocation* (the initiator must send the same message to all processes). More recently, *trusted execution environments* and more generally *trusted components* have emerged

T. Masuzawa et al. (Eds.): SSS 2024, LNCS 14931, pp. 360–374, 2025.
https://doi.org/10.1007/978-3-031-74498-3_26

as a promising protection against Byzantine failures by providing cryptographic primitives that protect participants against equivocation.

Trusted components are especially promising for consensus protocols such as PBFT, and therefore for many recent blockchain algorithms. For example, [7, 8] recently introduced the *TTCB wormhole*, which supports a PBFT protocol that tolerates up to half of the processes to be Byzantine, well beyond the tolerance of classical systems [11]. Nevertheless, the trusted part of these systems makes practical implementations difficult.

A2M (Attested Append-only Memory) [5] provides a small and easy-to-implement abstraction of a trusted append-only log. Each log has a unique identifier and offers methods to append and read values. A value, once added, cannot be rewritten. A2M increases the resilience of PBFT by appending each message to the log and sending that attestation along with the message, which increases resiliency to one-half.

We are not the first to suggest that trusted environments similar to A2M can increase resilience for blockchains: see *HotStuff* [19], *Damysus* [10], and TenderTee [1]. However, none of these works focuses on the Byzantine reliable broadcast primitive. The authors of [1] conjecture that plugging A2M hardware into Bracha's protocol might increase its resilience. In this paper, we refute their conjecture by showing that A2M-Bracha has the same resilience as the original in asynchronous settings.

An alternative to A2M is the use of a *monotonic counter* implemented in a tamper-proof module. TrInc [13] is a trusted component that deals with equivocation in large systems by providing a set of monotonic counters, supported by a trusted hardware unit called a *trinket*.

More recently, [17] proposed *USIG* (Unique Sequential Identifier Generator), a service available to each process (and implemented in a tamper-proof module) that assigns each message a unique counter value, and signs that message. The service offers two functions: one that returns a certificate, and one that validates certificates. These certificates are based on a secure counter: the counter value is never duplicated, and successive counter values are successive integers. To the best of our knowledge, this kind of trusted component has never been used to increase the resilience of reliable broadcast. Here, we prove that using a trusted component can implement Byzantine Reliable Broadcast in asynchronous environments with optimal resilience.

Similarly to our work, [9] proposes a reliable broadcast algorithm tolerant with $n \geq 2f + 1$ processes where f is the number of Byzantine faults. Contrarily to us, they use failure detectors. [18] proposes an algorithm similar to ours but which only tolerates $f < n/3$ Byzantine processes, whereas our algorithm tolerates up to $f < n/2$ Byzantine faults.

Our contribution. This paper presents a study of Byzantine Reliable Broadcast using trusted components. First, we show that, despite popular belief, trusted components cannot improve the resilience of Bracha's algorithm with no modification. Instead, we propose a novel algorithm that uses a single (optimal number) trusted component to implement asynchronous Byzantine reliable broadcast with n processes, $n \geq 2t + 1$ where t is an upper bound on the number of Byzantine processes. Interestingly, this algorithm uses only one simple trusted component that provides a trusted monotonic counter. We abstract the trusted component via a distributed object called *Trusted Monotonic Counter Object*.

Paper organisation. The paper is organized as follows. Section 2 defines the execution model and presents the specification of the Byzantine Reliable Broadcast problem. Section 3 introduces the key component of our Byzantine Reliable Broadcast implementation, the Trusted Monotonic Counter Object. Section 4 discusses the impossibility of improving Bracha's Byzantine Reliable Broadcast resilience even when each process is equipped with a trusted component. Section 5 presents our algorithm for Byzantine Reliable Broadcast using a single Trusted Monotonic Counter Object at the initiator.

2   System Model and Problems Definition

We consider a set of n asynchronous sequential processes. Up to t processes can be *Byzantine*, meaning they can deviate from the given protocol. The rest are *correct* processes.

Processes communicate by exchanging messages through an asynchronous network. We make the usual assumption that there is a public key infrastructure (PKI) where public keys are distributed, each process has a (universally known) public key and a matching private key, moreover, each message is signed by its creator. Messages are not lost or spuriously generated. Each process can send messages directly to any other process, and each process can identify the sender of every message it receives.

We assume processes have access to a broadcast primitive, broadcast(m), which ensures that the message m is received by every correct process in a finite (but unknown) time. When a process initiates a broadcast instance, we call that process the *initiator*.

Following Bracha, [4], we define *Byzantine Reliable Broadcast* as follows:

Definition 1. (Byzantine Reliable Broadcast problem)

- brb-CorrectInit: If the initiator is correct, all correct processes deliver the initiator's value.
- brb-ByzantineInit: If the initiator is Byzantine, then either no correct process delivers any value, or all correct processes deliver the same value.

3   Trusted Monotonic Counter Object

TEEs in general (e.g. A2M [5], TrInc [13], USIG [17]) are reputed to be powerful tools for avoiding equivocation. Although the trusted component abstraction makes protocols immune to equivocation (where the initiator sends different messages to different processes), [6] shows that non-equivocation is not enough to provide $n \geq 2f+1$ resilience nor to support the equivalent of digital signatures.

We now define the Trusted Monotonic Counter Oracle abstraction TMC-Object the core of our novel Byzantine Reliable Broadcast protocol that supports t Byzantine failures among n processes, where $n \geq 2t + 1$, an improvement on the classical $n \geq 3t + 1$ algorithms.

In short, TMC-Object provides a non-falsifiable, verifiable, unique, monotonic, and sequential counter. In particular, TMC-Object provides each process with a read-only local variable, called trustedCounter. Whenever the TMC-Object is invoked, it returns a value for trustedCounter that is strictly greater than any previous value it returned. The difference between two successive counter values is exactly 1, so when a process p receives two messages stamped with counter values, it can detect whether there have been intermediate messages.

The TMC-Object supports the operation **get_certificate()** A process p invokes **get_certificate(m)** with a message m. The object returns a *certificate* and a *unique identifier*. The certificate certifies that the returned identifier was created by the tamper-proof TMC-Object object for the message m. The unique identifier is essentially a reading of the monotonic counter trustedCounter, incremented whenever **get_certificate(m)** is called.

The TMC-Object object guarantees the following properties:

- **Uniqueness:** TMC-Object will never assign the same identifier to two different messages.
- **Sequentiality:** TMC-Object will always assign an identifier that is the successor of the previous one.

4   Bracha's Byzantine Reliable Broadcast with Trusted Components

In this section, we prove that modifying Bracha's reliable broadcast algorithm [4] by (only) equipping each process with a trusted component (here. TMC-Object) does not change the tolerance threshold of Byzantine processes, which remains $1/3$, we do so by using the modular formalism introduced by [16].

To send a message u certified with TMC-Object, a process p first invokes TMC-Object, which creates a certificate $\mathcal{C}_p$ corresponding to the value of the trustedCounter c_p, then the process sends the tuple $(u, \mathcal{C}_p, c_p)$, which can be verified by any other process receiving the message. Each invocation to TMC-Object increments the value of the trustedCounter c_p of process p, and the initial value of the counter is 0. In the following, that sequence of operation is simply called **TMC-Object-Send** u.

In the following, we describe Algorithm 1 which is Bracha's reliable broadcast where each process uses the TMC-Object-Send operation instead of a Send operation. In more detail, the protocol works in sequential steps. In the broadcast primitive described in Algorithm 1, there are three types of messages used in the protocol: *initial*, *echo*, and *ready*. All these messages are sent using the TMC-Object-Send operation. In the initial step (Step 0) of the protocol, when a process p wants to broadcast a value u, it TMC-Object-Sends an initial message for u ($<$ *initial*, u $>$) to all other processes, therefore incrementing its trusted counter. Recall that the process initiating the broadcast is called the *initiator*.

In Step 1, upon receiving a valid[1] initial message with value v from the initiator, a process A2M-Sends an echo message for v ($<$ *echo*, v $>$). An echo message is also sent if instead of receiving the initial message, the process receives enough (here, α) echo messages for the same value from different processes, implying that many processes saw the initiator message. After, and only after the A2M-Send operation, the process moves to Step 2.

In Step 2, each process waits to receive echo messages for the same value, say v, from at least α different processes sent from Step 1. When that is the case, the process TMC-Object-Sends a ready message for the value v ($<$ *ready*, v $>$). After the send operation, the process moves to Step 3.

Step 3 is similar to Step 2. The process waits for β ready messages, when the β ready messages are received for the same value, say v, the process delivers value v and finishes the instance of broadcast.

The broadcast is successful if all correct processes rb-Deliver the same value. Thus, rb-Broadcast and rb-Deliver provide us with a pair of communication primitives.

Algorithm 1 Reliable Broadcast with a Trusted Environment

1: **procedure** RB-BROADCAST(u)
2:     **Step 0**
3:        if p is the initiator **then**
4:            TMC-Object-Send $<$ *initial*, u $>$ to all
5:     **Step 1**
6:        Wait until receipt of
7:            1 valid TMC-Object-message $<$ *initial*, v $>$ message, or
8:            α $<$ *echo*, v $>$ messages
9:            for some v
10:         TMC-Object-Send $<$ *echo*, v $>$ to all
11:     **Step 2**
12:        Wait until receipt of
13:            α $<$ *echo*, v $>$ messages
14:               (including messages received in Step 1)
15:            for some v
16:         TMC-Object-Send $<$ *ready*, v $>$ to all
17:     **Step 3**
18:        Wait until receipt of
19:            β $<$ *ready*, v $>$ messages
20:               (including messages received in Steps 1 and 2)
21:            for some v
22:         rb-Deliver v

---

[1] Here, valid TMC-Object-message. If the value of the counter is strictly greater than 2, the initiator may have equivocated (having already A2M-sent another message).

Lemma 1. *Consider Algorithm 1 with parameter $n \geq \alpha > t$ and $\alpha \geq n/2 + 1$ where t is the number of Byzantine processes. If two correct processes TMC-Object-Send $< echo, v >$ and $< echo, u >$ messages, respectively, then $u = v$.*

Proof. The proof will be conducted by contradiction. Assume there exist two correct processes that TMC-Object-Send $< echo, v >$ and $< echo, u >$ messages, respectively, with $u \neq v$. Let q be the first correct process that TMC-Object-Sends an $< echo, v >$ message, and let r be the first correct process that TMC-Object-Sends an $< echo, u >$ message.

- Case 1: Process q receives an initial message $< initial, v >$ and process r receives an initial message $< initial, u >$. If the initiator is correct, then this situation is impossible since a correct initiator sends only one initial value. If the initiator is Byzantine, then either q or r rejects the initial value since the TMC-Object nominal sequence is invalid (the counter associated with one of these values is strictly greater than 1, hence the message is not valid).
- Case 2: Process q must have received $\alpha < echo, v >$ messages, and process r must have received $\alpha < echo, u >$ messages. Notice that a correct process can send only one echo. Since $\alpha > t$, where t is the number of Byzantine processes in the system, among the α messages some come from correct processes. Since $\alpha \geq n/2 + 1$ then there is at least one correct process that TMC-Object-Sent $< echo, v >$ and $< echo, u >$ messages which is impossible since the process is correct.

$\square$

Lemma 2. *Consider Algorithm 1 with parameter $n \geq \alpha > t$ and $\alpha \geq n/2 + 1$ where t is the number of Byzantine processes. If two correct processes TMC-Object-Send $< ready, v >$ and $< ready, u >$ messages, respectively, then $u = v$.*

Proof. Proof by contradiction. Assume there exist two correct processes which TMC-Object-Send $< ready, v >$ and $< ready, u >$ messages, with $u \neq v$. Let q be the first process that TMC-Object-Sends a $< ready, v >$ message, and let r be the first process that TMC-Object-Sends a $< ready, u >$ message. Process q must have received more than $\alpha < echo, v >$ messages and process r must have received more than $\alpha < echo, u >$ messages. Since $\alpha > t$ and $\alpha \geq n/2 + 1$ it follows that at least one correct process must have TMC-Object-Sent $< echo, v >$ and at least one correct process must have TMC-Object-Sent $< echo, u >$ messages. Following Lemma 1, we then have $u = v$. $\square$

Lemma 3. *Consider Algorithm 1 with parameter $n \geq \alpha > t$ and $\alpha \geq n/2 + 1$ where t is the number of Byzantine processes. If two correct processes, q and r, deliver the values v and u, respectively, then $u = v$.*

Proof. If q delivers the value v then it must have received $\alpha < ready, v >$ messages, and therefore a $< ready, v >$ message is from at least 1 correct process. Similarly, r must have received a $< ready, u >$ message from at least 1 correct process. By Lemma 2, $u = v$. $\square$

We now show that Algorithm 1 satisfies the property brb-CorrectInit of the Byzantine reliable broadcast.

Theorem 1. *Consider Algorithm 1 with parameters α and β. If the initiator is a correct process Algorithm 1 satisfies the* **brb-CorrectInit** *property if $\alpha = \beta$ and $n/2 + 1 \leq \alpha \leq n$ and $n \geq 2t + 1$ where t is the number of Byzantine processes and n is the total number of processes.*

Proof. The proof follows directly from Lemmas 1, 2 and 3. Let p be the initiator. Since the initiator is correct, the value broadcast by p, u, will eventually be received by all other correct processes (at least $n - t = t + 1$ correct processes). These processes will echo that value u (TMC-Object-Send an echo message for u). Since all correct processes echo the same value (Lemma 1), and their number is sufficient to make the protocol advance (there are at least $n/2 + 1$ correct processes), each correct process will receive enough echoes to send a ready message, and the same one (Lemma 2). By the same argument and applying Lemma 3, all correct processes will receive enough ready messages for the initiator value u, and then will rb-Deliver the initiator message.   □

Unfortunately, the following result shows that in the presence of a Byzantine initiator, Algorithm 1 could produce undesirable behaviour, and hence does not implement the Byzantine reliable broadcast.

Lemma 4. *Let n be the number of processes, and t be an upper bound on the Byzantine processes with $n \geq 2t + 1$. Consider Algorithm 1 with parameters α and β with $\alpha = t + 1$ and $t + 1 \leq \beta < 2t + 1$ Algorithm 1 does not satisfy the* **brb-ByzantineInit** *property when the initiator is Byzantine.*

Proof. If the initiator is a Byzantine process, Byzantine processes could force a subset of correct processes to deliver a value, and another subset of correct processes to never deliver any value. Note that even though all processes use a TMC-Object abstraction such that Byzantine processes cannot equivocate, Byzantine processes still can send a message to some processes but not to others.

Let p be the Byzantine initiator. p TMC-Object-sends a value u to $1 \leq x \leq t$ correct processes $q_1, q_2 \ldots q_x$ but not to the other $n - t - x$ processes. Denote by Q this set of x correct processes receiving the initiator's initial message. Since processes in Q receive the message from the Byzantine initiator, they TMC-Object-Send an echo message for u. Assume now that all the Byzantine processes TMC-Object-Send an echo message for value u only to processes in Q but not to the others. It follows that all correct processes but those in Q have no message from the initiator and only the $< echo, u >$ from processes in Q. Those processes cannot advance past Line 6 of Algorithm 1 since they need at least $\alpha = t + 1 > x$ echo messages.

All correct processes but those in Q have only x echo messages that come from the processes in Q. Processes in Q on the other hand would have the echo messages from all Byzantine processes in addition to their own echo messages, which sums to $t + x$ echo messages for the value u. Therefore, processes in Q will advance and TMC-Object-Send a ready message for u.

In the same spirit, Byzantine processes can TMC-Object-Send a ready message for value u to processes in Q only. The other correct processes are still blocked at 6 of Algorithm 1. In addition to the ready messages from the Byzantine processes, processes in Q also get their own ready messages for value u, so each process $q \in Q$ has a total of $t + x$ ready messages and hence delivers the value u (rb-Delivery). The other correct processes only receive the values TMC-Object-Sent by processes in Q, meaning x echo message and x ready message, both for u, hence, they can never reach an acceptance decision in Algorithm 1. It follows that Algorithm 1 does not satisfy the brb-ByzantineInit property when the initiator is Byzantine. □

When $\alpha = \beta = t + 1$, Lemma 4 violates the brb-ByzantineInit property of the Byzantine reliable broadcast (Definition 1), and, hence, does not satisfy the Byzantine reliable broadcast (Definition 1) as stated by the following Corollary.

Corollary 1. *Let n be the number of processes, and t an upper bound of the Byzantine processes. If $n \geq 2t + 1$, Algorithm 1 does not implement the Byzantine reliable broadcast.*

However, for Algorithm 1 to implement the Byzantine reliable broadcast, we show in Theorem 2 that we must have $\beta = 2t + 1$.

Theorem 2. *Necessary conditions for 1 with parameters α and β to implement the Byzantine reliable broadcast are $\alpha = t + 1$, $\beta = 2t + 1$, and $n - t \geq \beta$, where t is the upper bound of the number of Byzantine processes and n is the total number of processes.*

Proof. If the initiator is correct, all correct processes decide to deliver the initiator message, by Theorem 1.

It remains to show that when the initiator is Byzantine, either no correct process delivers any value or all correct processes deliver the same value.

- By Lemma 3, if two correct processes deliver a value, they must deliver the same one.
- Now, let us turn to the case where only one correct process reaches a decision.

Assume that process q reaches a decision and delivers a value u. It means that q received at least β ready messages, from which at least $\beta - t$ are from correct processes.

At least $\beta - t$ correct processes sent a ready message. However, to TMC-Object-Send a ready message, a correct process must have reached Step 2, and must have completed Step 1 of Algorithm 1. In fact, if a correct process does not TMC-Object-Send an echo message, it cannot enter Step 2. Therefore, we know that at least $\beta - t$ correct processes have TMC-Object-Sent an echo message. By Lemma 1, all correct processes that TMC-Object-Sent an echo message have TMC-Object-Sent it for the same value, hence it means that they all TMC-Object-Sent an echo message for the value u.

We would like to have that, with at least $\beta - t$ correct processes TMC-Object-Sending an echo message for the same value, say a value u, all correct processes must have received at least α echo messages for value u. Hence, we have that $\beta - t \geq \alpha \implies \beta \geq \alpha + t \geq 2t + 1$. For lower bounds, now assume that $\alpha = t + 1$ and $\beta = 2t + 1$. The rest of the proof shows that it is sufficient for Algorithm 1 to implement the Byzantine reliable broadcast.

Hence they will all TMC-Object-Send an echo message for u. In that case, all correct processes (at least $2t + 1$ processes) TMC-Object-Sent an echo for u, all correct processes (at least $2t + 1$) will then TMC-Object-Send a ready message for u, which leads to all correct processes eventually delivering value u. Hence, if one correct process delivers a value u, all other correct processes eventually deliver the same value u.

$\square$

5 Byzantine Reliable Broadcast with Optimal TMC-Object

Algorithm 2 Byzantine Reliable Broadcast with a unique Trusted Environment for the initiator

---

1: **procedure** BRB-BROADCAST(u)
2:     **Step 0**
3:         if p is the initiator **then**
4:             TMC-Object-Send $< initial, u, id\_initiator >$ to all
5:     **Step 1**
6:         Wait until receipt of
7:             1 valid TMC-Objectmessage $< initial, v, id\_initiator >$ message
8:             for some v
9:         Send $< echo\_in, v, (< initial, v, id\_initiator >, \mathcal{C}_{initiator}, c_{initiator}) >$ to all //*the process broadcasts back the initiator's message with the associated certificate and trusted counter.*
10:     **Step 2**
11:         Wait until receipt of
12:             $t + 1 < echo\_in, v >$ messages   //*Projection of the echoes received keeping only the value of the message.*
13:             (including messages received in Step 1)
14:             for some v
15:         Send $< ready, v >$ to all
16:     **Step 3**
17:         Wait until receipt of
18:             $t + 1 < ready, v >$ messages
19:             (including messages received in Steps 1 and 2)
20:             for some v
21:         brb-Deliver v

In this section, we present Algorithm 2, which contains a small modification of Bracha's algorithm [4]. Algorithm 2 solves the reliable broadcast problem tolerating $t < n/2$ Byzantine processes. Algorithm 2 uses the trusted component setups to increase the security threshold of Byzantine reliable broadcast from 1/3 of Byzantine processes to 1/2.

Moreover, to reduce the use of the trusted component, which can be resource-intensive, only the initiator is required to send certified messages. The other processes simply check the validity of the message and its certification but do not require the equipment to send certified messages. In this section, we consider the use of TMC-Object defined in Sect. 3.

In Algorithm 2, only the initiator sends a message using the light TMC-Object abstraction for its send operation. We say that the initiator TMC-Object-Sends a message. The other processes send their message classically. The other difference with Algorithm 1 is that during Step 1, when a process receives the initiator's message, say for value u, it sends an echo message for u coupled with the initiator message (meaning the message, and the certificate and counter sent by the initiator). In such a way, it ensures that all other processes will eventually receive the initiator's certified message. The rest of the algorithm proceeds as in Algorithm 1 where the TMC-Object-Sends are replaced by classical Send operations).

The broadcast is successful if all the correct processes brb-Deliver the same value, say u. Thus, brb-Broadcast and brb-Deliver provide a pair of communication primitives resilient to $t < n/2$ Byzantine processes.

We can now prove the correctness of Algorithm 2 against the Byzantine reliable broadcast abstraction.

Recall that we say that a process *accepts* a message if it receives and adds the "valid" messages in terms of the validity of the TMC-Object, meaning that there was no message before in that same category, hence the value of the trusted counter is the lowest. Notice that if the message received is not expected to be a TMC-Object message, and is indeed not part of a TMC-Object operation (e.g., a send which is not TMC-Object-Send), such a message is valid by default and, therefore, is accepted.

Finally, for any value v, the message $< echo\_in, v, (< initial, v, id\_initiator >$, $C_{initiator}, c_{initiator}) >$ should be understood as two messages bundled together, i.e., the echo message $< echo, v >$ sent after the reception of the initiator message and sending back the initiator's message $< initial, v, id\_initiator >$, along with the certificates and trusted counter, i.e., $C_{initiator}, c_{initiator}$. The message is exactly the initiator's message, the TMC-Object message will be correctly validated.

Lemma 5. *In any execution of Algorithm 2, with $n \geq 2t + 1$ where t is the number of Byzantine processes, a correct process sends each type of message (initial, echo_in, ready) at most once.*

Proof. In Steps 1 and 2 of Algorithm 2, the protocol requires sending exactly 1 message, and then moving to the subsequent phase. A correct process cannot send more messages.

In Step 0, a correct process TMC-Object-Sends a message if and only if it is the initiator, and after that moves to Step 1. If the process is not the initiator, it does not (TMC-Object-)Send anything, but moves directly to Step 1. Hence, in Step 0, at most 1 message is sent. $\qquad \square$

Lemma 6. *Consider Algorithm 2 with $n \geq 2t + 1$ where t is the number of Byzantine processes. If two correct processes p and q receive and accept respectively $< initial, u, id\_initiator >$ and $< initial, v, id\_initiator >$, then $u = v$.*

Proof. This holds thanks to the properties of the TMC-Object. Since equivocation is not possible at the initiator level, thanks to the use of the counter in TMC-Object if two correct processes p and q accept an initiator message, it means they received the same message, and they then echo the accepted initial message (Line 9 of Algorithm 2). Therefore, they receive and accept the same message. $\qquad \square$

By Lemma 6, we know that if two correct processes accept an initiator message, then they accept the same message. The only thing that could happen is for one correct process to receive the initiator message, while another process does not receive such a message.

Lemma 7. *Consider Algorithm 2 with $n \geq 2t + 1$ where t is the number of Byzantine processes. If one correct process sends $< echo\_in, v >$ for some v, then all correct processes will eventually send $< echo\_in, v >$.*

Proof. Let p and q be processes. Without loss of generality, assume that p is the first correct process to do an echo. If a correct process echoes a message, it means it accepts the initiator message, and will send the echo along with the TMC-Object-Send of the initiator (Line 9 of Algorithm 2). Two cases can arise. Either the initiator sent the initial message to both p and q or the initiator did not send a message to q. Notice that it is not possible for the initiator not to have TMC-Object-Sent a message to p, since p echoed the initiator message.

First, consider the case where p received the initiator message, but not q. The process p sent an echo message containing the initiator's initial message respecting the TMC-Object format. By assumption, a message sent by a correct process will eventually be received by all the other correct processes. Therefore, eventually, q will receive p's echo message, containing the initial message. q will be able to assess the validity of the initial message (according to the initiator signatures), will accept it, and will send an echo for the message too. Because before receiving p's echo message, q is still waiting in Step 1.

Finally, if the initiator sends a message to both p and q, therefore, either it sends the same message to p and q, and so q will echo that same message (by Lemmas 5 and 6, it is not possible for q to echo something else), or the message sent to q is invalid and not accepted. That last case is equivalent to the above situation, since an invalid message is not registered nor considered, and is equivalent to not having received a message. $\qquad \square$

Thanks to Lemmas 5 and 7, we know that whenever a correct process sends an *echo_in* message, all other correct processes will also echo a message (and more accurately, the same message).

Lemma 8. *If two correct processes p and q send respectively $< ready, v >$ and $< ready, u >$, then $u = v$.*

Proof. By contradiction. Assume $u \neq v$. Without loss of generality, let p be the process that sends a $< ready, v >$ message, and let q be a process that sends a $< ready, u >$ message. To send a ready message for value x, a correct process must have received from at least $t + 1$ different processes the message $< echo\_in, x >$ (Line 15 of Algorithm 2). Therefore, p must have received the message $< echo\_in, v >$ from at least $t+1$ different processes, and process q must have received the message $< echo\_in, u >$ from at least $t + 1$ different processes. Since there are at most t Byzantine processes, at least one correct process must have sent an echo message for both u and v, which is impossible by Lemma 5. Therefore, it is impossible to have $u \neq v$.   ☐

Lemma 9. *Consider Algorithm 2 with $n \geq 2t + 1$ where t is the number of Byzantine processes. If two correct processes, p and q, brb-deliver the values v and u, respectively, then $u = v$.*

Proof. This proof is similar to the proof for 8. We proceed by contradiction. Assume two messages containing respectively values u and value v such that $u \neq v$. Without loss of generality, let p be the process that brb-delivers the message containing v, and let q be a process that brb-delivers u. To brb-deliver a value x, a correct process must have received from at least $t + 1$ different processes the message $< ready, x >$ (Line 18 of Algorithm 2). Therefore, p must have received the message $< ready, v >$ from at least $t + 1$ different processes, and process q must have received the message $< ready, u >$ from at least $t + 1$ different processes. Since there are at most t Byzantine processes, it means that at least one correct process sent ready messages for both values u and v, which is impossible by Lemma 5. Therefore, it is impossible to have $u \neq v$.   ☐

Lemma 10. *Consider Algorithm 2 with $n \geq 2t + 1$ where t is the number of Byzantine processes. If a correct process p delivers the value v then every other correct process will eventually deliver v.*

Proof. If p brb-Deliver v then p received the message $< ready, v >$ from at least $t + 1$ different processes. Since there are at most t Byzantine processes, it means that at least one correct process sent a message $< ready, v >$. Since one correct process sent a ready message for v, it means that it received an $< echo, v >$ message from at least $t + 1$ different processes; hence, (since there are at most t Byzantine processes) it means that at least one correct process sent a message $< echo, v >$. Therefore, by Lemma 7, all other correct processes will (eventually) send an echo message for value v. Those will be received by all the correct processes. This will lead to having at least $t + 1$ different processes sending it. All correct processes will, therefore, eventually send a ready message for value v (by Lemma 8, since we already know that one correct sent a ready for v). Hence, at least $t + 1$ ready messages will be received by all correct processes, which will lead them to brb-Deliver v.   ☐

Lemma 11. *Consider Algorithm 2 with $n \geq 2t + 1$ where t is the number of Byzantine processes. If a correct process p broadcasts v then all correct processes brb-deliver v.*

Proof. The proof of the lemma is straightforward. If a correct process broadcasts an initial message, it does so to all processes. All processes in Step 1 will echo_in the initiator message containing value v, thanks to Lemma 7. Since correct processes are the majority, and the network is eventually synchronous, they will all eventually receive at least $t + 1$ echo_in message for value v and send each a ready message for value v. Thanks to Lemma 8, since one correct process sends a ready for v, v is the only value correct process will send a ready for. That value will, therefore, be present in at least $t + 1$ ready messages, hence in Step 3, a correct process will brb-deliver v. By Lemma 10, all correct processes will eventually brb-deliver v.   □

We can now prove that the algorithm implements the Byzantine reliable broadcast.

Theorem 3. *Let n be the number of processes, and t an upper bound of the Byzantine processes. If $n \geq 2t + 1$, Algorithm 2 implements Byzantine reliable broadcast.*

Proof. By Lemma 11, when a correct initiator broadcasts a value, all correct processes brb-deliver that value.

By Lemma 7, if a correct process brb-delivers an initiator message (even if the initiator is Byzantine), all correct processes will eventually brb-deliver the same initiator message. In that case, the rest of the proof follows thanks to Lemma 11.

Otherwise, no correct process brb-delivers any value. In more detail, if a Byzantine initiator does not send an initial message to any correct process, no correct process will deliver anything. That is because no correct process will send an echo message (then none will send ready messages). Since all advances require $t + 1$ messages from different processes, and Byzantine processes are at most t, the correct processes will be stuck in Step 1 and will make no decision.   □

Theorem 4. *Let n be the number of processes, and t be an upper bound of the Byzantine processes. If $n \geq 2t + 1$, in Algorithm 2, the number of TMC-Object used is optimal, in the sense that if we remove the only TMC-Object (initiator), Algorithm 2 does not implement the Byzantine reliable broadcast.*

Proof. In Algorithm 2, only 1 TMC-Object is used, the one at the initiator. If the TMC-Object is removed (instead of doing a TMC-Object-Send, the initiator does only a Send operation), then the algorithm resembles the Bracha's Byzantine reliable broadcast protocol [4] where instead of a $2t + 1$ bound to advance, we have only a $t + 1$ bound. Since Bracha's Byzantine reliable broadcast protocol is optimal in the number of faults [4, 16], Algorithm 2 with no TMC-Object cannot implement Byzantine reliable broadcast.   □

6   Conclusion

We focus on Byzantine Reliable Broadcast in trusted components. First, we show that adding a TMC-Object to all processes to prevent equivocation does not improve the security threshold or security guarantees of Bracha's Byzantine Reliable Broadcast, thanks to the formalism introduced in [16]. Second, we propose an optimal trusted component-based algorithm that implements Byzantine Reliable Broadcast in asynchronous settings with resilience $n \geq 2t + 1$. Our algorithm employs a very simple trusted component that provides a trusted monotonic counter. Moreover, our solution needs only one trusted monotonic counter at the initiator.

References

1. Beltrando, L., Potop-Butucaru, M., Alfaro, J.: Tendertee: increasing the resilience of tendermint by using trusted environments. In: 24th International Conference on Distributed Computing and Networking, ICDCN 2023, Kharagpur, India, January 4–7, 2023, pp. 90–99. ACM (2023). https://doi.org/10.1145/3571306.3571394, https://doi.org/10.1145/3571306.3571394
2. Bonomi, S., Decouchant, J., Farina, G., Rahli, V., Tixeuil, S.: Practical byzantine reliable broadcast on partially connected networks. In: 2021 IEEE 41st International Conference on Distributed Computing Systems (ICDCS), pp. 506–516. IEEE (2021)
3. Bonomi, S., Farina, G., Tixeuil, S.: Reliable broadcast despite mobile byzantine faults. In: Bessani, A., Défago, X., Nakamura, J., Wada, K., Yamauchi, Y. (eds.) 27th International Conference on Principles of Distributed Systems, OPODIS 2023, December 6–8, 2023, Tokyo, Japan. LIPIcs, vol. 286, pp. 18:1–18:23. Schloss Dagstuhl - Leibniz-Zentrum für Informatik (2023). https://doi.org/10.4230/LIPICS.OPODIS.2023.18, https://doi.org/10.4230/LIPIcs.OPODIS.2023.18
4. Bracha, G.: Asynchronous byzantine agreement protocols. Inf. Comput. **75**(2), 130–143 (1987). https://doi.org/10.1016/0890-5401(87)90054-X
5. Chun, B., Maniatis, P., Shenker, S., Kubiatowicz, J.: Attested append-only memory: making adversaries stick to their word. In: Proceedings of the 21st ACM Symposium on Operating Systems Principles 2007, SOSP 2007, Stevenson, Washington, USA, October 14–17, 2007, pp. 189–204. ACM (2007). https://doi.org/10.1145/1294261.1294280
6. Clement, A., Junqueira, F., Kate, A., Rodrigues, R.: On the (limited) power of non-equivocation. In: ACM Symposium on Principles of Distributed Computing, PODC '12, Funchal, Madeira, Portugal, July 16–18, 2012, pp. 301–308. ACM (2012). https://doi.org/10.1145/2332432.2332490
7. Correia, M., Neves, N.F., Veríssimo, P.: How to tolerate half less one byzantine nodes in practical distributed systems. In: 23rd International Symposium on Reliable Distributed Systems (SRDS 2004), 18–20 October 2004, Florianpolis, Brazil, pp. 174–183. IEEE Computer Society (2004). https://doi.org/10.1109/RELDIS.2004.1353018
8. Correia, M., Neves, N.F., Veríssimo, P.: BFT-TO: intrusion tolerance with less replicas. Comput. J. **56**(6), 693–715 (2013). https://doi.org/10.1093/comjnl/bxs148

9. Correia, M., Veronese, G.S., Lung, L.C.: Asynchronous byzantine consensus with 2f+1 processes. In: Shin, S.Y., Ossowski, S., Schumacher, M., Palakal, M.J., Hung, C. (eds.) Proceedings of the 2010 ACM Symposium on Applied Computing (SAC), Sierre, Switzerland, March 22–26, 2010, pp. 475–480. ACM (2010). https://doi.org/10.1145/1774088.1774187, https://doi.org/10.1145/1774088.1774187
10. Decouchant, J., Kozhaya, D., Rahli, V., Yu, J.: DAMYSUS: streamlined BFT consensus leveraging trusted components. In: EuroSys'22: Seventeenth European Conference on Computer Systems, Rennes, France, April 5–8, 2022, pp. 1–16. ACM (2022). https://doi.org/10.1145/3492321.3519568
11. Fischer, M.J., Lynch, N.A., Paterson, M.: Impossibility of distributed consensus with one faulty process. In: Proceedings of the Second ACM SIGACT-SIGMOD Symposium on Principles of Database Systems, March 21–23, 1983, Colony Square Hotel, Atlanta, Georgia, USA, pp. 1–7. ACM (1983). https://doi.org/10.1145/588058.588060
12. Guerraoui, R., Komatovic, J., Kuznetsov, P., Pignolet, Y.A., Seredinschi, D.A., Tonkikh, A.: Dynamic byzantine reliable broadcast [technical report]. arXiv preprint arXiv:2001.06271 (2020)
13. Levin, D., Douceur, J.R., Lorch, J.R., Moscibroda, T.: Trinc: Small trusted hardware for large distributed systems. In: Proceedings of the 6th USENIX Symposium on Networked Systems Design and Implementation, NSDI 2009, April 22–24, 2009, Boston, MA, USA, pp. 1–14. USENIX Association (2009), http://www.usenix.org/events/nsdi09/tech/full_papers/levin/levin.pdf
14. Maurer, A., Tixeuil, S.: Self-stabilizing byzantine broadcast. In: 2014 IEEE 33rd International Symposium on Reliable Distributed Systems, pp. 152–160. IEEE (2014)
15. Raynal, M.: Fault-Tolerant Message-Passing Distributed Systems: An Algorithmic Approach. Springer (2018)
16. Raynal, M.: On the versatility of Bracha's byzantine reliable broadcast algorithm. Parallel Process. Lett. 31(3), 2150006:1–2150006:9 (2021). https://doi.org/10.1142/S0129626421500067
17. Veronese, G.S., Correia, M., Bessani, A.N., Lung, L.C., Veríssimo, P.: Efficient byzantine fault-tolerance. IEEE Trans. Comput. 62(1), 16–30 (2013). https://doi.org/10.1109/TC.2011.221
18. Wattenhofer, R.: Distributed systems. Lecture notes. https://disco.ethz.ch/courses/hs21/distsys/lnotes/DistSys_Script.pdf (2022)
19. Yandamuri, S., Abraham, I., Nayak, K., Reiter, M.K.: Communication-efficient BFT protocols using small trusted hardware to tolerate minority corruption. IACR Cryptol. ePrint Arch., 184 (2021). https://eprint.iacr.org/2021/184

Brief Announcement: On the Feasibility of Local Failover Routing on Directed Graphs

Erik van den Akker$^{(\boxtimes)}$ [iD] and Klaus-Tycho Foerster [iD]

TU Dortmund, Dortmund, Germany
erik.vandenakker@tu-dortmund.de

Abstract. Local failover mechanisms are used to achieve fast recovery from link failures in communication networks. These mechanisms are typically implemented using static routing tables at the nodes of a network, only relying on failures of outgoing links, as well as the label of the source and target node of a packet (called $s - t$-routing). Static failover $s - t$-routing on undirected graphs has been shown to be able to tolerate at most 2 failures, denoted 2-resilient, with 3-resiliency being impossible without additional rewritable bits in the packet header. In this work, we investigate local failover routing on directed graphs with n nodes and show lower and upper bounds on the number of bits required. Even 1-resilience cannot be achieved on all topologies without additional bits and we prove that 1-resilience can be obtained with $\lceil \log(n) \rceil$ bits. For $k > 1$ failures, we show that at least $\lceil \log(k + 1) \rceil$ bits are necessary, but that $k(\lceil \log(|E|) \rceil)$ bits are sufficient to obtain k-resiliency.

Keywords: routing · directed graphs · networks · failover · connectivity

1 Introduction and Background

Maintaining routing connectivity is crucial in communication networks [2] and hence modern networked systems implement some form of fast failover routing, to quickly react to equipment failures [8] until global routing tables are (relatively slowly [7]) re-computed and re-distributed. However, past research has shown [2] that there is a trade-off w.r.t. coverage (how many failures can be tolerated) and the number of communication rounds needed. Ideally this fast failover routing is implemented without any communication post-failures in the form of static local forwarding rules, to guarantee immediate reaction.

In this setting on undirected graphs $G = (V, E)$, Feigenbaum et al. [5] proved that without source-information (denoted as t-routing) or rewritable bits in the packet header, at most 1 edge failure can be tolerated (denoted 1-resilience). Here Chiesa et al. [3] showed that resilience to 2 edge failures is impossible in

Table 1. Overview of the lower and upper bounds for directed failover routing

Number of Failures	Lower Bound	Upper Bound
1	1 (Theorem 1)	$\lceil \log(n) \rceil$ (Theorem 3)
$k > 1$	$\lceil \log(k+1) \rceil$ (Theorem 2)	$k(\lceil \log(|E|) \rceil)$ (Theorem 4)

general, though in practice a good amount of topologies can still be 2-resilient [1]. When matching on the source as well (denoted as $s - t$-routing), Dai et al. [4] proved that 2-resilience can always be obtained, but that 3-resilience is impossible in general without rewritable bits in the packet header. Moreover, Lakshminarayanan et al. [6] gave a scheme that can survive any k number of arc failures, assuming the network stays connected, by carrying information about all encountered failed links in the packet header, requiring $k \cdot \lceil \log |E| \rceil$ rewritable bits to succeed.

Wada et al. [9] considered a similar problem of fault-tolerant fixed routings on directed graphs, where they studied the size of surviving route graphs. Their model allowed only up to k arc failures for $k + 1$-connected graphs, while our model allows an arbitrary amount of arc failures, as long as no deadends are constructed. Also, instead of local forwarding rules for the next hop at any node, their model uses a set of paths between each pair of nodes in the graph to construct the routing.

Contribution. In this work we investigate fast failover $s - t$-routing with static routing tables on *directed graphs*. To the best of our knowledge, there are no known upper or lower bounds in this setting, except for the above mentioned results by Lakshminarayanan et al. [6], which can be applied to directed graphs as well. A summary of our results is given in Table 1.

2   Model

We consider a network as a directed, simple graph $G = (V, E)$, in which nodes represent routers and edges represent links, and nodes are labeled with unique ids. For each node we define a set of *local forwarding rules*, having the form $in \times f \times s \times t \times b \rightarrow out, b'$, where on the left side in is the incoming arc the packet came from (or $\perp$ if the packet originates from the current node), f is the set of locally failed outgoing arcs (routers are assumed to keep track of local failures of outgoing links), s is the id of the source, t the id of the target of the packet and b is the bitstring in the packet header. On the right side, out is the outgoing arc chosen under these conditions and b' is the new bitstring written to the packet header. Whenever a packet arrives at a node (router) that is not the target t, it uses the information available locally and in the packet header, finds the corresponding forwarding rule, rewrites the bitstring and sends the packet to the next node. Directed Local Failover Routing can be described as a game between a player 1 and a player 2 (the adversary):

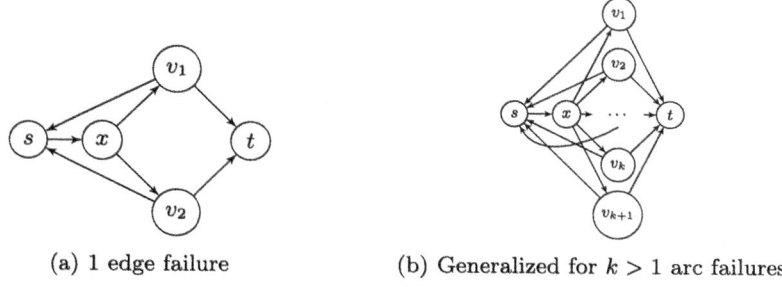

(a) 1 edge failure   (b) Generalized for $k > 1$ arc failures

Fig. 1. Lower bound constructions for 1 and k arc failures

1. Two players are given a network as a directed, simple graph $G = (V, E)$ with $n = |V|$ nodes and a value k.
2. For each node, player 1 defines the set of local forwarding rules, which must be able to route from s to t, as long as these nodes are connected in G.
3. An adversary (player 2) can now remove a subset $F \subseteq E$ in G, containing up to k arcs.
4. We now have to check: can our local forwarding rules still route any packet from s to t, as long as any node that can be reached from s still has a connection to t[1]
5. If routing still works: player 1 wins, else the adversary (player 2) wins.

The question is: is there some set of local forwarding rules for G, so that an adversary cannot find a set F of size at most k, that can break the forwarding pattern, or: does player 1 always have a winning strategy for the input (G, k)? If this is the case, we call the forwarding-pattern k-resilient.

3 Lower Bounds

In this section, we cover the lower bounds for the number local failover routing on directed graphs, first in Theorem 1 looking at the lower bound with exactly one failed arc, showing that we already need a rewritable bit in the case of a single failure, while in the undirected case for $s - t$-routing, two failures can generally be tolerated. We then show in Theorem 2, that for $k \geq 1$ failures, we generally need at least $\lceil \log(k + 1) \rceil$ rewritable bits in the packet header.

Theorem 1. Lower Bound for a single failure
$s - t$-routing strategies with local forwarding rules on directed graphs with one failed arc need at least one rewritable bit in the packet header.

---

[1] Without this assumption, the adversary could easily send the player 1 into deadends.

Proof. Consider the graph shown in Fig. 1a: From s any routing strategy can only send the packet to x, where the routing strategies can only differ in sending the packet to either v_1 or v_2. W.l.o.g we assume v_1 is chosen. The adversary can now remove the arc (v_1, t), which leads the packet to be sent via (v_1, s). From s the only option is to send the packet to x. If there is no rewritable information in the packet header, the packet will now repeatedly follow the loop s, x, v_1, s.

With one rewritable bit in the packet header, the bit could be flipped when traversing the arc (v_1, s), influencing the decision at x, where then the arc (x, v_2) could be chosen. Since the arc (v_2, t) must be intact for t to be reachable from s, we can reach t using one rewritable bit in the packet header in this case.

Theorem 2. Lower Bound for multiple failures
$s-t$-routing strategies on directed graphs with local forwarding rules with k failed arcs need at least $\lceil \log(k + 1) \rceil$ rewritable bits in the packet header.

Proof. Consider a graph as in Fig. 1b, with $k + 4$ nodes $s, x, t, v_1, ..., v_k, v_{k+1}$. If an adversary is allowed to delete up to k arcs and our strategy has less than $\lceil \log(k + 1) \rceil$ rewritable bits in the packet header available, then we can only encode up to k different configurations $c_1, ..., c_k$ in the bitstring of the packet header. The adversary can now remove the arcs (v_j, t) for each arc (x, v_j) that is chosen in any of the k configurations, which will lead to the packets from s not reaching t.

Note that with $\lceil \log(k + 1) \rceil$ rewritable bits, the bitstring has $k + 1$ possible configurations, which can then be used to try $k + 1$ different arcs (x, v_i), until a node with an arc to t is available (as there are at most k failures).

4   Upper Bounds

In this section, we look at the maximum number of rewritable bits in packet headers needed, first showing in Theorem 3, that for a single failure, we need at most $\lceil \log(n) \rceil$ rewritable bits, where n is the number of nodes, which are used to encode the node where the default path failed, implicitly marking the failed arc. Afterwards, we show in Theorem 4 that for k failures at most $k(\lceil \log(|E|) \rceil)$ bits are needed, encoding each of the k failed arcs.

Theorem 3. Upper Bound for a single failure
There exists a routing strategy for $s - t$ routing with local forwarding rules on directed graphs with n nodes and one failed arc, that needs at most $\lceil \log(n) \rceil$ rewritable bits in the packet header.

Proof. A forwarding pattern can define a default path from s to t, using local forwarding rules. In an attempt to break the routing pattern, the adversary must remove an arc f from the default path. However, the forwarding pattern can write the id of the node where the failed arc was encountered to the bitstring b with $\lceil \log(n) \rceil$ bits. The bitstring is initially set to the id of s, meaning no failure has occured yet, and will be set to the id of t, if the failure already occurs at s. Since every node that is reachable from s still must have a path to t in $G - \{f\}$, there must still be some path from any reachable node to t, that does not use the arc f. We can then encode this path with local forwarding rules, which are used when b is set to the id of the node from the default path. Since only one arc can fail in this case, this implicitly means, that the arc f failed.

Theorem 4. Upper Bound for multiple failures
There exists a routing strategy for $s - t$ routing with local forwarding rules on directed graphs with n nodes and k failed arcs, that needs at most $k(\lceil \log(|E|) \rceil)$ rewritable bits in the packet header.

Our proof of Theorem 4 is based on the idea of Lakshminarayanan et al. [6], who also encoded all so far encountered arc failures as packet header bits.

Proof. We can encode fallback paths using information about all failed arcs we have already encountered, by encoding fallback paths for each node under each possible failure set up to size k. We can use the bitstring of size $k(\lceil \log(|E|) \rceil)$, to encode all k failed arcs. Whenever we encounter an error, we add the newly found failure to the bitstring, which can be initially set to all zeroes and try the fallback path defined for the currently found set of failures. If we have not reached t yet when finding the last failed arc, we can reach t using the fallback path for the complete failure set.

5   Conclusion

We investigated upper and lower bounds on the number of bits required for local failover routing on directed graphs, showing that even for $s - t$-routing, resilience to one (k) failure(s) is not possible with 0 ($\lceil \log(k) \rceil$) bits, while in the undirected case two failures can always be tolerated without the use of any additional bits. We also showed that one (k) failure(s) can always be tolerated with logarithmic multiplicative overhead. Besides closing the gaps between upper and lower bounds in general, it would also be interesting to investigate the resilience and number of bits required for restricted graph classes, like planar graphs, since even for $k = 2$ failures, the graph with 6 nodes used in our lower bound construction (Fig. 1b) is not planar.

References

1. van den Akker, E., Foerster, K.: Short paper: towards 2-resilient local failover in destination-based routing. In: ALGOCLOUD (2024)
2. Chiesa, M., Kamisinski, A., Rak, J., Rétvári, G., Schmid, S.: A survey of fast-recovery mechanisms in packet-switched networks. IEEE Commun. Surv. Tutorials **23**(2), 1253–1301 (2021)
3. Chiesa, M., Nikolaevskiy, I., Mitrovic, S., Gurtov, A.V., Madry, A., Schapira, M., Shenker, S.: On the resiliency of static forwarding tables. IEEE/ACM Trans. Netw. **25**(2), 1133–1146 (2017)

4. Dai, W., Foerster, K., Schmid, S.: A tight characterization of fast failover routing: resiliency to two link failures is possible. In: SPAA, pp. 153–163. ACM (2023)
5. Feigenbaum, J., Godfrey, B., Panda, A., Schapira, M., Shenker, S., Singla, A.: Ba: On the resilience of routing tables. In: PODC, pp. 237–238. ACM (2012)
6. Lakshminarayanan, K., Caesar, M., Rangan, M., Anderson, T., Shenker, S., Stoica, I.: Achieving convergence-free routing using failure-carrying packets. In: SIGCOMM, pp. 241–252. ACM (2007)
7. Liu, J., Panda, A., Singla, A., Godfrey, B., Schapira, M., Shenker, S.: Ensuring connectivity via data plane mechanisms. In: NSDI, pp. 113–126. USENIX (2013)
8. Rak, J., Hutchison, D. (eds.): Guide to Disaster-Resilient Communication Networks. Springer, Computer Communications and Networks (2020)
9. Wada, K., Kawaguchi, K.: Efficient fault-tolerant fixed routings on $(k + 1)$-connected digraphs. Discret. Appl. Math. **37**(38), 539–552 (1992)

TRAIL: Cross-Shard Validation for Byzantine Shard Protection

Joseph Oglio, Mikhail Nesterenko$^{(\boxtimes)}$, and Gokarna Sharma

Department of Computer Science, Kent State University, Kent, OH 44242, USA
{joglio,mikhail,sharma}@kent.edu

Abstract. We present *TRAIL*: an algorithm that uses a novel consensus procedure to tolerate failed or malicious shards within a blockchain. Our algorithm takes a new approach of selecting validator shards for each transaction from those that previously held the asset being transferred. This approach ensures the algorithm's robustness and efficiency. *TRAIL* is presented using *PBFT* for internal shard transaction processing and a modified version of *PBFT* for external cross-shard validation. We describe *TRAIL*, prove it correct, analyze its message complexity, and evaluate its performance. We propose various *TRAIL* optimizations: we describe how it can be adapted to other Byzantine-tolerant consensus algorithms, how a complete system may be built on the basis of it, and how *TRAIL* can be applied to existing and future sharded blockchains.

1 Introduction

In this paper, we present *TRAIL*—an algorithm for robust blockchain design. A blockchain is a shared, immutable, append-only distributed ledger, typically maintained by a peer-to-peer network [2,3]. This design eliminates centralized control over transaction processing and makes the system potentially more scalable, flexible, and efficient.

Blockchains are usually designed to tolerate Byzantine faults [4]. A Byzantine peer may deviate from the algorithm and behave arbitrarily. Therefore, such faults encompass a variety of failures and security threats. Despite the faults, correct peers need to be able to arrive at consensus on proposed transactions.

Popular blockchains use proof-of-work based consensus algorithms [2] in which peers compete for the right to publish records on the blockchain by searching for solutions to cryptographic challenges. Such algorithms tend to be conceptually simple and robust. However, they are resource intensive and environmentally harmful [5]. Therefore, modern blockchain designs often focus on cooperative consensus algorithms.

In these cooperative consensus algorithms, rather than compete, peers exchange messages to arrive at a joint decision. Such algorithms may tolerate some number f of faulty processes. This number is called tolerance threshold. It

---

A technical report [1] contains a more extensive version of the paper.

T. Masuzawa et al. (Eds.): SSS 2024, LNCS 14931, pp. 381–397, 2025.
https://doi.org/10.1007/978-3-031-74498-3_28

is usually a fraction of the network size n. One of the most widely used algorithms in this category is *PBFT* [6].

Scaling up a Byzantine-robust algorithm is challenging as it usually involves system-wide broadcasts. Such broadcasts are expensive in large systems. A prominent approach of improving scalability in blockchains is sharding. In sharding, the network peers are divided into committees or shards. Each shard is made responsible for a subset of the processing done or the data stored by the network. Every shard internally runs a consensus algorithm, such as *PBFT*, and coordinates with other shards to achieve global consistency. Thus, the overall workload is distributed and the processing of records is potentially accelerated.

However, such sharding is at cross-purposes with fault tolerance: the network is only as reliable as any of its shards. For example, given a fixed number of peers, decreasing the shard size increases the number of available shards. This results in greater parallelism in transaction processing. Yet, a small shard is more vulnerable to failure since it has lower tolerance threshold f of its internal consensus algorithm. The sharded blockchains presented in the literature usually assume that no shard tolerance threshold is breached. This places a limit on the efficiency of the sharding approach to performance improvement since shards need to be made large enough to ensure that they never fail.

In this paper, we address the handling of complete shard failures which potentially allows aggressively small shards and removes the shard size scalability obstacle. A naive approach would be to group shards into static meta-shards. Such a meta-shard would treat individual shards as peers and run a meta-consensus algorithm among them to validate transactions across shards to withstand individual shard failures. However, concurrent transactions that are assigned to different shards would be verified by the same static meta-shard, regardless of the transactions' nature or history. This may create a performance bottleneck.

Paper contribution. We propose *TRAIL*: a novel approach to cross-shard validation. With this technique, a trail of shards dynamically tracks each coin according to its transaction history. The source shard runs an internal shard consensus algorithm to validate and linearize transactions. The trail of shards runs a cross-shard consensus algorithm to confirm the transaction and fortify it against shard failure. We present *TRAIL* using *PBFT* for both internal shard transaction processing and external cross-shard validation. We utilize *PBFT* since it is well-known and widely used. Our solution may use various *PBFT* efficiency enhancements such as parallel transaction processing and transaction pipelining. Moreover, *TRAIL* is independent of the specifics of sharding operation and may be adapted to enhance the robustness of consensus algorithms other than *PBFT*.

We evaluate the performance of *TRAIL* using an abstract simulator and study its transaction confirmation rate, scalability and robustness against peer and shard failure. Our experiments indicate that *TRAIL* adds shard failure protection with relatively modest resource expenditure.

2   Network Model, Problem Statement, *PBFT*

System model. We assume a peer-to-peer network. Each peer has a unique identifier. A peer may send a message to any other peer so long as it has the receiver's identifier. Peers communicate through authenticated channels: the receiver of the message may always identify the sender. The communication channels are FIFO and reliable.

Network peers are grouped into *shards*. Every shard has a unique identifier. For simplicity, we assume that all shards are the same size s. Each shard maintains a portion of the blockchain's data. Each shard peer stores a copy of its shard's data. Any peer may determine the shard identifier of any other peer in the network. Peers are either correct or faulty. Faults are *Byzantine* [4]: a faulty peer may behave arbitrarily. A *peer tolerance threshold* f is the maximum number of faulty peers that a shard can tolerate. A shard is correct if it has at most f faulty peers. The shard is faulty otherwise.

Data model and the problem. A *coin* is a unit of ownership whose movements are recorded by the network. Each coin has a unique identifier, which can track fungible assets like currency (e.g., UTXO) or non-fungible assets like NFTs or smart-contracts, without affecting the fungibility of the currency. A *wallet* is a collection of coins. Each shard is responsible for storing and updating a disjoint subset of the network's wallets. A *client* is an entity that owns a wallet. We assume a client is external to the peer-to-peer network but may communicate with any peer. Clients may submit transactions to the network requesting a coin to be moved from a *source wallet* to a *target wallet*. The peers are able to authenticate the wallet owner; the peers accept transaction requests for the wallet owned by the client. The approval of the target wallet owner is not required.

A blockchain algorithm constructs a sequential ledger of transactions reflecting coin movements. Two transactions $t1$ and $t2$ are *consequent* in this ledger if they operate on the same coin and there is no transaction $t3$ also operating on this coin such that $t3$ comes after $t1$ and before $t2$.

An algorithm *state* is an assignment of values to variables in all processes. Algorithm code contains a sequence of actions guarded by boolean guard predicates. An action whose guard evaluates to **true** is *enabled*. An algorithm *computation* is a sequence of steps such that for each state s_i, the next state s_{i+1} is obtained by executing an action enabled in s_i.

To make the *TRAIL* correctness argument more rigorous, we formally state the problem that it solves.

Definition 1. An algorithm solves *the Coin Transmission Problem* if it constructs a transaction ledger satisfying the following two properties:
ownership continuity—for any pair of consequent transactions $t1$ and $t2$, the target of $t1$ is the source of $t2$;
request satisfaction—if the owner requests a coin movement from its wallet, this request is eventually satisfied.

Ownership continuity is a safety property that requires that a coin can only be moved out of a wallet once for each time it is moved into it. This is crucial because it prevents the same coin from being spent more than once-thereby precluding double-spend attacks and disallowing spending money that the client does not have. The request satisfaction property guarantees liveness: the client request is eventually fulfilled.

PBFT. PBFT is a Byzantine-robust consensus algorithm. Its tolerance threshold is $f = \lfloor (n-1)/3 \rfloor$, where n is the total number of peers in the system.

Peers communicate directly with each other via message broadcast. One of the peers is a *leader*. The leader linearizes client requests. A period of single leader continuous operation is a *view*. PBFT is in *normal operation* if the leader is correct. Normal operation has three phases: pre-prepare, prepare, and commit.

Once the leader receives a client transaction request, it assigns it a unique sequence number and starts the *pre-prepare* phase by broadcasting a pre-prepare message containing the transaction and the sequence number to all peers. In the *prepare* phase, each peer receives the pre-prepare message and broadcasts a prepare message containing the information that it received from the leader. If a peer receives $n - f - 1$ prepare messages (excluding itself) that match the initial pre-prepare, this peer is certain that correct peers agree on the same transaction. In this case, the peer starts the *commit* phase by broadcasting a commit message. Once a peer receives $n - f$ commit messages, normal *PBFT* operation concludes and the peer informs the client that the transaction is committed. The transaction is confirmed once the client receives $f + 1$ commits.

If the leader is faulty, the client requests may not be carried out. In this case, the client or the peers initiate a *view change* to replace the leader. The view change process is designed to maintain transaction consistency through the transition: if a transaction is committed by a correct peer in the old view, the new leader submits it with the same sequence number so that the rest of the peers commit it in the new view.

If the number of faulty peers does not exceed the peer tolerance threshold f, *PBFT* guarantees the following three properties: *agreement*—if a correct peer confirms a transaction, then every correct peer confirms this transaction; *total order*—if a correct peer confirms transaction $t1$ before transaction $t2$, then every correct peer confirms $t1$ before $t2$; and *liveness*— if a transaction is submitted to a sufficient number of correct peers, then it is confirmed by a correct peer. We assume these properties also apply to correct clients.

The first two properties are satisfied regardless of the network synchrony. The liveness property is guaranteed only if the network is partially synchronous, meaning that the message transmission delay does not indefinitely grow without a bound.

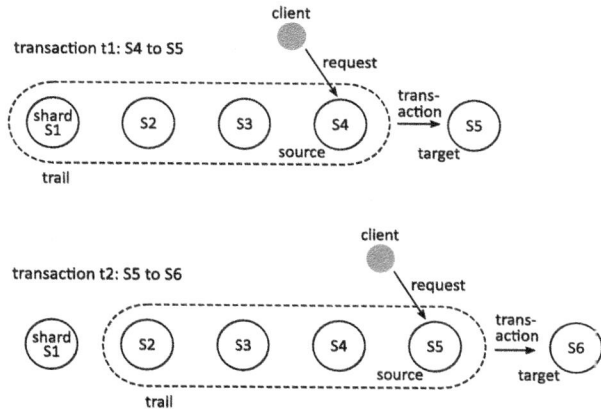

Fig. 1. Trail membership modification under consequent transactions for $t = 4$. The first transaction moves a coin from a wallet in shard $S4$ to a wallet in shard $S5$. The second moves the same coin from $S5$ to $S6$.

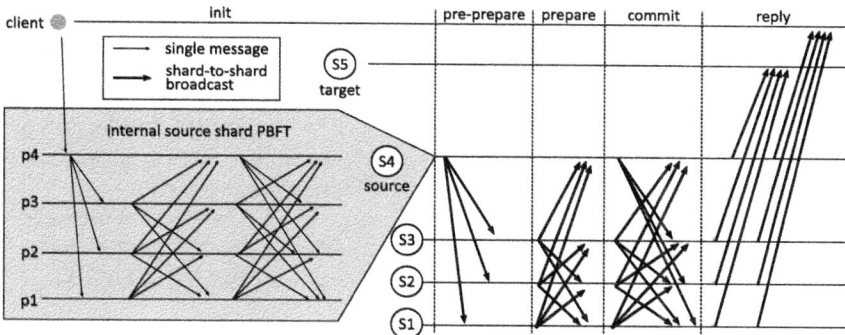

Fig. 2. Message transmission in $TRAIL$'s normal operation. The coin trail contains shards: $S1 - S4$. The coin is located in a wallet stored by shard $S4$. A client sends a transaction moving the coin from $S4$ to a wallet of shard $S5$. First, the $S4$ runs internal $PBFT$; then, the phases of external shard $PBFT$. After committing, the shards notify the client and the target shard.

3   *TRAIL* Description

Algorithm outline. The objective of the algorithm is to ensure the validity of coin transitions between wallets despite faulty peers and shards. To counter malicious behavior of faulty shards, $TRAIL$ requires a collection of shards to agree on coin movement. This collection is called a *trail*. A trail is composed of the t unique shards whose wallets the coin visited most recently. Refer to Fig. 1 for an illustration. Notice that this trail is specific to a coin and changes as the coin moves from wallet to wallet. At any point in the computation, each coin may have its own separate trail of shards. We assume that each client knows the identities of the coins in its wallet as well as their trails. *Shard tolerance threshold* F is the maximum number of faulty shards that $TRAIL$ may tolerate. The length of the trail, $t \geq 3F + 1$. Note that F and the peer tolerance threshold f are not related.

$TRAIL$ consists of two parts: (1) *internal* source shard $PBFT$ and (2) *external* trail $PBFT$. To initiate the movement of a coin, the client that owns the source wallet sends a transaction request to the shard that holds it. To linearize received transaction requests and ensure that each individual transaction has an agreed-upon sequence number, the source shard peers execute internal $PBFT$, see Fig. 2 for illustration, in which shard $S4$ is the source shard.

Once the source shard peers agree on this transaction, they initiate a modified external trail $PBFT$. For that, each source shard peer broadcasts a *pre-prepare* message to every peer of every trail shard. Once a trail peer receives $s - f$ such *pre-prepares* from the source shard, it initiates the next $PBFT$ phase by sending *prepare* messages to every peer in every trail shard. In this way, each shard-to-shard broadcast emulates an individual message transmission in classic $PBFT$. This continues until the external $PBFT$ instance commits. After that, each trail shard peer records the transaction in its ledger and notifies the target shard and the client.

Once the target shard and the client are notified by $t - F$ trail shards, they record the transaction in their ledgers. The target shard, which is shard $S5$ in Fig. 1, becomes the source shard for the next transaction. If the leader of the source shard, $S4$, is faulty, the other peers of the source shard execute a view change, switch to a new leader, and continue with internal $PBFT$.

If the source shard as whole is not faulty, i.e. the number of faulty peers in the source shard is below the tolerance threshold f, then the faulty peers may not influence the trail shard. Indeed, for each external $PBFT$ message, each peer of the trail shard expects at least $s - f$ individual messages.

If the number of peers in the source shard exceeds the tolerance threshold f then the whole shard is faulty. In this case, the individual messages of the faulty source shard peers are equivalent to the faulty messages of the source shard. The external $PBFT$ guarantees that, despite the faulty shard, no spurious transactions will be recorded by the trail shards and that eventually the faulty leader shard is replaced.

Specifically, the trail shards execute a view change, switch to a new shard as a leader, and continue with the consensus process, including a new internal $PBFT$ instance being performed within the new leader shard. Note that in the latter case, the record of the transaction may be placed in the trail shards but not in the faulty source shard that is nominally responsible for maintaining the source wallet record. This is an essential feature of our algorithm: the faulty source shard that stores the client wallet may be bypassed.

Let us now describe the algorithm in detail.

TRAIL constants, variables, and functions. These constructs are shown in Algorithm 1. Each peer with id p knows the following constants: f—peer tolerance threshold (the maximum number of faulty peers in a correct shard); F—shard tolerance threshold (the maximum number of faulty shards); s—shard size; and t—trail size.

Several variables are common across transactions. We list them in a single place for convenience. Each transaction uses a coin identifier *coin*; a source wallet id *sWallet*; a target wallet is *tWallet*; a transaction sequence number *seq* assigned by the source shard; and the sequence of shard ids *trail* that indicates the trail shards for this coin at its present location. Each peer maintains a *ledger*, which is a sequence of transaction records that the peer confirmed in a trail or received as a target.

TRAIL functions are shown in Algorithm 2. They are grouped by their purpose. Ledger maintenance functions are in Lines 16–27. *TRAIL* has two such functions.

Function RECORD appends the transaction record to the ledger. Function ISPRESENT(*coin*, *wallet*) returns **true** if the ledger's most recent transaction record about *coin* moved it to *wallet*, i.e. there is a transaction where *wallet* is the target wallet and this record is not followed by a transaction moving *coin* from *wallet* to a different target wallet.

TRAIL uses several functions for wallet lookup and communications. They are shown in Lines 28–38. Function GETSHARD(*wallet*) returns the id of the shard that stores *wallet*. We assume that every peer is able to identify which shard maintains each *wallet*.

Functions SEND and RECEIVE are single-message transmissions to the specified sender and receiver with straightforward functionality. In function SENDTOSHARD(*shard*, *message*), the sender peer broadcasts a *message* to all peers in *shard*. Function RECEIVEFROMSHARD(*shard*, *message*) returns **true** once the peer receives *message* from $s - f$ unique peers of *shard*.

The internal source shard *PBFT* is represented by two functions in *TRAIL*. They are shown in Lines 39–46. Function STARTSHARDPBFT initiates the *PBFT* operation. The last function COMPLETESHARDPBFT signifies that the internal *PBFT* is completed and the peers assigned sequence number *seq* to the transaction.

Algorithm 1: *TRAIL*: Normal Operation, Variables

---

1 **Constants**
2 p ▷ process id
3 f ▷ peer tolerance threshold
4 F ▷ shard tolerance threshold
5 s ▷ shard size; $s \geq 3f + 1$
6 t ▷ trail size; $t \geq 3F + 1$

7 **Transaction variables**
8 *coin*   ▷ coin to be transmitted
9 *sWallet*   ▷ coin owner wallet
10 *tWallet*   ▷ coin recipient wallet
11 *seq*   ▷ transaction sequence number
12 *trail*   ▷ sequence of confirming shard

13 **Process variables**
14 *ledger*   ▷ sequence of records of committed
15   transactions
   ▷ record format: $\langle coin,\ sWallet, tWallet, seq, trail \rangle$

TRAIL phases. The actions for the algorithm are presented in Algorithms 2 and 3. We only show normal operation code for *TRAIL*. View change code is added accordingly. Client and target code is not shown. See Fig. 2 for the illustration of algorithm operation.

TRAIL phases execute the internal source *PBFT* and the external trail *PBFT*. **Phase 0: Init** (see Lines 48–53) starts when a peer receives a transaction request from a client. If the peer contains the source wallet, i.e. it is the source shard for the transaction, the peer initiates internal shard *PBFT*. After the source shard runs classic *PBFT*, if the shard is not faulty, all the source shard peers agree on the transaction and its sequence number. The completion of internal *PBFT* starts **Phase 1: Pre-prepare** (Lines 54 through 58). Each source shard peer sends a *pre-prepare* message to a peer of every trail shard.

The receipt of $s-f$ messages from the source shard starts **Phase 2: Prepare** (Lines 59–64) in all the trail shards. Once a peer of the trail shard ascertains that the coin is present in the source wallet, i.e. the transaction is valid, the peer sends a *prepare* message to all of the trail shards.

In **Phase 3: Commit** (Lines 65–74), each trail peer assembles the *prepare* messages. Variable *prepShards.coin.seq* collects the identifiers of the shards from which this peer has received $s - f$ *prepare* messages. If the number of these identifiers is $t - F$, the peer sends *commit* message to all trail shards, signifying that it is ready to commit.

Phase 4: Reply (Lines 75–90) is similar to the **Commit** phase. Once enough *commit* messages from trail shards arrive, the peer records the committed transaction to its ledger and notifies the peers of the target shard and the client.

Algorithm 2: *TRAIL*: Normal Operation, Functions

---

16 **Ledger functions**
17 **function** RECORD(*coin, sWallet, tWallet, seq, trail*)
18   ▷ add transaction record to ledger
19 **function** ISPRESENT(*coin, wallet*):
20 | **if** $\exists r1 \equiv \langle coin, x, wallet, \cdots \rangle \in ledger$ *and*
21 | $\forall r2 \equiv \langle coin, wallet, y, \cdots \rangle \in ledger \Rightarrow$
22 | *r2 precedes r1 in ledger* **then**
23 | | **return true**
24 | **else**
25 | | **return false**

26 **function** GETTRAIL(*coin*):
27 | $r \equiv \langle coin, sWallet, tWallet, seq, trail \rangle$ such that r is the last record
     for *coin* in *ledger* **return** *trail*

28 **Communication and wallet functions**
29 **function** GETSHARD(*wallet*)
30   ▷ returns the id of the shard that stores *wallet*
31 **function** SEND(*peer, message*)
32   ▷ send *message* to *peer*
33 **function** RECEIVE(*peer, message*)
34   ▷ receive *message* from specific *peer*
35 **function** SENDTOSHARD(*shard, message*)
36   ▷ send *message* to all peers of *shard*
37 **function** RECEIVEFROMSHARD(*shard, message*)
38   ▷ receive *message* from $s - f$ unique peers of *shard*

39 **Internal PBFT functions**
40 **function** STARTSHARDPBFT(*coin, sWallet, tWallet*)
41   ▷ initiate internal shard PBFT
42   ▷ by sending request to shard leader
43 **function** COMPLETESHARDPBFT()
44   ▷ finish internal shard PBFT
45   ▷ return $\langle coin, sWallet, tWallet, seq \rangle$
46   ▷ with unique sequence number *seq*

47 **Basic PBFT operation**
48 **Phase 0: Init**   ▷ done by source shard leader
49 **upon** RECEIVE (*client,*
50 | *request*$\langle coin, sWallet, tWallet \rangle$):
51 | **if** $p \in$ GETSHARD(*sWallet*) *and*
52 | ISPRESENT(*coin, sWallet*) **then**
53 | | STARTSHARDPBFT(*coin, sWallet, tWallet*)

Algorithm 3: *TRAIL*: Normal Operation, Cross-Shard Actions

---

54 **Phase 1: Pre-prepare**   ▷ done by source shard
55 **upon** ⟨*coin, sWallet, tWallet, seq*⟩ ←
56 │   COMPLETESHARDPBFT() :
57 │   **forall** *shard* ∈ GETTRAIL(*coin*) **do**
58 │   └   SENDTOSHARD(*shard, prePrepare*⟨*coin, sWallet, tWallet, seq*⟩)

59 **Phase 2: Prepare**   ▷ done by non-source shards
60 **upon** RECEIVEFROMSHARD(*senderID*,
61 │   *prePrepare*⟨*coin, sWallet, tWallet, seq*⟩):
62 │   **if** ISPRESENT(*coin, sWallet*) **then**
63 │   │   **forall** *shard* ∈ *GetTrail*(*coin*) **do**
64 │   └   └   SENDTOSHARD(*shard, prepare*⟨*coin, sWallet, tWallet, seq*⟩)

65 **Phase 3: Commit**   ▷ done by all trail shards
66 *prepShards.coin.seq* ← ∅
67 **upon** RECEIVEDFROMSHARD (
68 │   *senderID* ∈ GETTRAIL(*coin*),
69 │   *prepare*⟨*coin, sWallet, tWallet, seq*⟩):
70 │   *prepShards.coin.seq* ←
71 │   *prepShards.coin.seq* ∪ {*senderID*}
72 │   **if** |*prepShards.coin.seq*| = *t* − *F* − 1 **then**
73 │   │   **forall** *shard* ∈ GETTRAIL(*coin*) **do**
74 │   └   └   SENDTOSHARD(*shard, commit*⟨*coin, sWallet, tWallet, seq*⟩)

75 **Phase 4: Reply**   ▷ done by all trail shards
76 *cmtdShards.coin.seq* ← ∅
77 **upon** RECEIVEDFROMSHARD (
78 │   *senderID* ∈ GETTRAIL(*coin*),
79 │   *commit*⟨*coin, sWallet, tWallet, seq*⟩):
80 │   *cmtdShards.coin.seq* ←
81 │   *cmtdShards.coin.seq* ∪ {*senderID*}
82 │   **if** |*cmtdShards.coin.seq*| = *t* − *F* **then**
83 │   │   **if** GETSHARD(*tWallet*) ∉ GETTRAIL(*coin*) **then**
84 │   │   │   *newTrail* ← GETSHARD(*tWallet*) ‖
85 │   │   │   all of GETTRAIL*(coin)* except last
86 │   │   **else**
87 │   │   └   *newTrail* ← GETTRAIL(*coin*)
88 │   │   RECORD(*coin, sWallet, tWallet, seq, newTrail*)
89 │   │   SENDTOSHARD(GETSHARD(*tWallet*),
   │   │   *reply*⟨*coin, sWallet, tWallet, newTrail*⟩)
90 │   └   SEND(*client, reply*⟨*coin, sWallet, tWallet, newTrail*⟩)

---

4 *TRAIL* Correctness and Efficiency

Theorem 1. *Algorithm* TRAIL *solves the Coin Transmission Problem with at most F Byzantine shards and at most f individual Byzantine faults in each correct shard.*

See [1] for the proof of the theorem.

4.1 *TRAIL* Algorithmic Extensions and Implementation Considerations

Parallelizing transactions, splitting, merging and mining coins. The same source shard may run multiple external or internal transactions so long as they concern different coins.

Multiple coins may be merged and a coin may be split to accommodate more complex transactions: this is analogous to creating 1 dollar out of 100 cents or vice versa. Both operations may be convenient to simplify transactions or make them more efficient. For example, a client wants to send 100 people each 1 cent of their 1 dollar, with coin splitting this can be done in a single transaction. If a coin is split, all its portions inherit the old coins' trail. Coin merging is a bit more involved since the merging coins, even if they are located in the same wallet, may have different trails. To merge, the two coins are marked as merging and their movement transactions are executed jointly. The coins are finally merged once they travel together for the length of the trail. At this point, they share all shards of the trail. This allows clients to combine coins and transfer them to another wallet in a single transaction.

To create, or mine, a new coin, it needs to acquire a trail of length t. This may be accomplished by forming a committee of arbitrary t shards and running a *PBFT* on this committee to agree on the new coin's trail. These processes guarantee that all coins have a trail of the required length regardless of the origin of the coin.

Optimizing internal transaction validation. To decrease message overhead, transactions are divided into internal and external. In an *internal transaction* the source and target wallet are maintained by the same shard. To confirm this transaction, the source shard does not consult the trail shards; it runs internal *PBFT*, and thus relies on the internal shard fault tolerance to maintain wallet integrity. External transactions are processed as usual: with external *PBFT*.

The trade-off for this optimization is decreased shard fault tolerance: the trail shards are not aware of the source shard internal transactions. However, shard failure may be determined by a failure detector [7–12]. Such a detector establishes a shard failure and notifies other shards. In the event of a detected shard failure, the trail shards perform a *failed shard recovery procedure* to restore the integrity of the system: the wallets maintained by the failed shard are moved to other shards and their contents are restored to the last known external transaction. The clients have to re-submit internal transactions.

Wallet location, client data recovery, shard maintenance. While coins move between shards, the wallets are assumed to be stationary. For quick shard lookup by the client, the wallet id might contain the shard number. Alternatively, the wallet-to-shard mapping may be recorded in the same or in a separate ledger. For efficiency, wallets frequently participating in joint transactions may be moved to the same shard.

If a client loses its local information about the coin contents of its wallet, it may be able to recover it by conducting a network-wide query. Note that asking the shard that keeps the wallet information alone is not sufficient: the shard may be faulty. Instead, the complete network broadcast is required. The trail shards that confirmed moving the coin should answer to the recovering client. Again, since some shards may be faulty, the client considers the coin present in its wallet if the trail confirms its location.

In *TRAIL* description, we assumed that the shard sizes are uniform. However, this does not have to be the case. Instead, the shards may grow and shrink as peers join or leave system. Shard sizes may also be adjusted in response to transaction load requirements. Shard membership may be maintained in the shard ledger or, alternatively, in a separate membership ledger.

Algorithm parameter selection. *TRAIL* operates correctly regardless of the concrete values of shard size s and trail size t. These parameters, however, affect the algorithm performance. Larger s makes it less likely that the complete shard fails. Yet, larger s makes the internal consensus algorithm less efficient. The smaller s necessitates larger trail size t to protect against shard failure.

5   Performance Evaluation

Simulation setup. We evaluate the performance of *TRAIL* in an abstract algorithm simulator QUANTAS [13]. QUANTAS simulates multi-process computation, message transmission and has extensive experimental setup capabilities. The simulator is implemented in C++. It is optimized for multi-threaded large scale simulations [14]. The code for our *TRAIL* implementation in QUANTAS as well as our performance evaluation data is available online [15,16].

The simulated network consists of individual peers. Each pair of peers is connected by a message-passing channel. Channels are FIFO and reliable. A computation is modeled as a sequence of rounds. In each round, a peer receives messages that were sent to it, performs local computation, and sends messages to other peers.

Peers are divided into shards. Shard leaders propose transactions; clients are not explicitly simulated. A transaction has a 25% probability of having source and target wallets in separate shards. Internal transactions are not externally verified by the trail of shards. If a shard is Byzantine, it generates invalid cross-shard transactions only. Specifically, the shard creates transactions moving coins that it has already spent.

Experiment description. Figures 3, 4, 5, and 6 show the dynamics of transaction processing during a computation. In these simulations, a computation runs for 500 rounds. The internal faulty peer tolerance threshold f is 7. The shard size is $s = 3 \cdot f + 1 = 22$. The faulty shard tolerance threshold F is 2. This makes the trail size $3 \cdot F + 1 = 7$. In the experiments, the number of actual faulty shards is equal to the shard fault tolerance threshold F; that is, we run the experiments with maximum tolerance. The faulty shards behave correctly at the start of the simulation and fail at round 100. The total number of shards in the system is $S = 50$. Therefore, the network size is $S \cdot s = 1100$. We run 15 experiments per data point and show the average of the results.

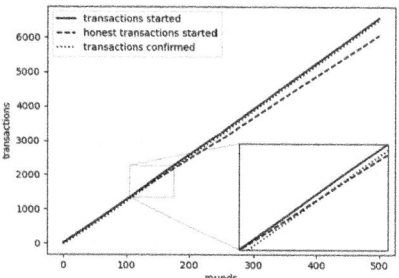

Fig. 3. Transactions approved over time without *TRAIL* shard validation. The network approves both honest and malicious transactions.

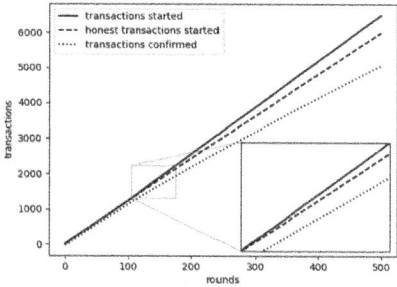

Fig. 4. Transactions approved over time with *TRAIL* shard validation. The network approves honest transactions only.

Figures 3, 4, and 5 show the accumulated counts of started and confirmed transactions. We distinguish between the *honest* transactions generated and the total number of transactions, which includes *malicious* transactions generated by faulty shards. The number of honest confirmed transaction is lower.

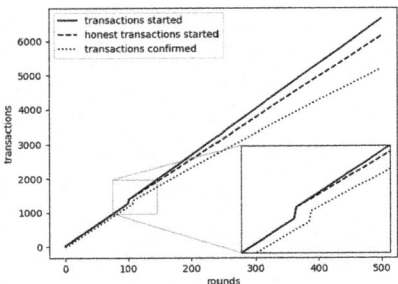

Fig. 5. Transactions approved over time in *TRAIL* with shard validation and wallet recovery from the failed shards. Correct shards detect the failure and submit additional transactions moving coins from the failed shards.

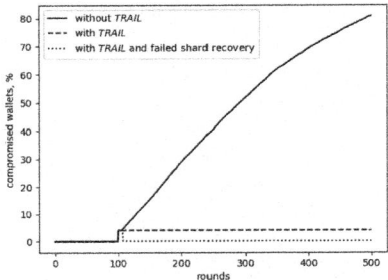

Fig. 6. Percentage of wallets compromised by malicious transactions.

In Fig. 3, no cross-shard validation is performed. In this figure, the number of confirmed transactions matches the total number of transactions; that is, transactions are confirmed whether they are malicious or not. The graph indicates a certain delay before transaction starting and confirmation due to the operation of external and internal *PBFT*.

In Fig. 4, *TRAIL* validates the external transactions. Malicious transactions are not confirmed, and the total number of confirmed transactions only accounts for the honest transactions.

In graph shown in Fig. 5, *TRAIL* uses the failed shard recovery procedure. Specifically, at round 100, when F shards fail, *TRAIL* detects the faults and generates transactions to move the coins from the faulty shard wallets to the correct ones. This explains the increase in the transaction generation and confirmation rates near round 100 in the figure.

Figure 6 shows the effect of malicious transactions on the overall system integrity. A wallet is *compromised* if it is in a faulty shard or if it receives a coin from a compromised wallet that is not possessed by that compromised sender wallet. A wallet is *safe* otherwise. The safety of a compromised wallet can be restored by the failed shard recovery procedure. The solid line in Fig. 6 shows the wallet compromise trend if no cross-shard validation is used. In this case, the failed shards continuously generate malicious transactions, compromising progressively larger number of wallets in the correct shards of the network. In the case in which cross-shard validation is used, the dashed line in Fig. 6, the number of compromised wallets does not exceed the number of wallets in the failed shard. In the case with the failed shared recovery procedure, all wallets eventually become safe again: the correct shards generate coin wallet recovery transactions. The delay in wallet recovery shown in the graph is due to the validation of these transactions.

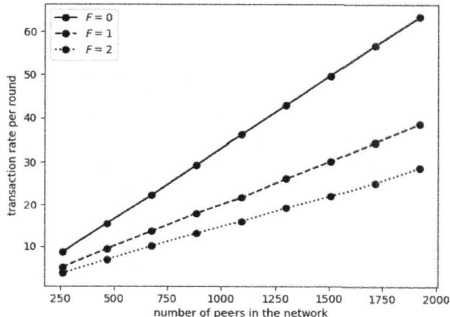

Fig. 7. Throughput with respect to the number of peers in the network for different fault tolerance levels.

Figure 7 shows the performance of *TRAIL* at scale. We run these experiments with a maximum of 148 shards made up of 13 peers per shard. Each data point represents the average throughput from 5 simulations of 200 rounds each. We plot *TRAIL*'s performance with three shard tolerance thresholds F: 0, 1 and 2. There is no cross-shard validation in case of $F = 0$. The figure indicates that the performance of *TRAIL* scales well with network size increase. Larger fault tolerance thresholds incur more overhead. Therefore, the transaction rate is lower for higher values of F.

6   Related Work and Its Application to *TRAIL*

Sharding blockchains. A number of sharding blockchains are presented in the literature. See Le *et al.* [17] for an extensive recent survey. We, however, have not seen an approach where sharding is done on the basis of the coin trail. We are not aware of any blockchain that is robust to shard failure. We believe that most of the published blockchains, even if they do not use *PBFT*, can employ *TRAIL* to fortify themselves against shard failures. For this, consensus on transactions has to be deferred until the transaction's trail confirms it.

PBFT optimizations and replacements. There are numerous proposals to optimize *PBFT* performance. See Wang *et al.* for a survey [18]. Several propose using multiple leaders concurrently [19–21]. *Mir-BFT* [19] and *RCC* [21] suggest accelerating *PBFT* by processing non-conflicting requests concurrently. The algorithms have multiple leaders that process these requests simultaneously. *Big-BFT* [20] further enhances parallelism by pipelining subsequent requests. The above *PBFT* optimizations can be applied in *TRAIL* to the internal shard consensus protocol in a straightforward manner. Most of these optimizations can also be applied to the external *TRAIL* algorithms as well.

7   Future Work

The *TRAIL* algorithm presented in this paper is the first to systematically address Byzantine shard failure protection in blockchains for cryptocurrencies. We foresee that it might be developed into a fully-fledged system. Alternatively, *TRAIL* may be used as an add-on component to fortify existing blockchains against shard failure. As a third alternative, *TRAIL* may be enhanced to handle more challenging conditions, such as network partitioning [22] or dynamic networks [23]. Any and all of these alternative directions will increase the robustness of future blockchains.

Acknowledgment. We would like to thank Mitch Jacovetty of Kent State University for his contributions to the paper.

References

1. Jacovetty, M., Oglio, J., Nesterenko, M., Sharma, G.: Trail: Cross-shard validation for cryptocurrency byzantine shard protection (2024). https://arxiv.org/abs/2405.07146
2. Nakamoto, S.: Bitcoin: a peer-to-peer electronic cash system (2008)
3. Wood, G.: Ethereum: a secure decentralized generalized transaction ledger. Ethereum Project Yellow Paper **151**, 1–32 (2014)
4. Lamport, L., Shostak, R., Pease, M.: The byzantine generals problem. ACM Trans. Progr. Lang. Syst. **4**(3), 382–401 (1982)
5. Wendl, M., Doan, M.H., Sassen, R.: The environmental impact of cryptocurrencies using proof of work and proof of stake consensus algorithms: a systematic review. J. Env. Manage. **326**, 116530 (2023)
6. Castro, M., Liskov, B.: Practical byzantine fault tolerance and proactive recovery. ACM Trans. Comput. Syst. **20**(4), 398–461 (2002)
7. Chandra, T.D., Toueg, S.: Unreliable failure detectors for reliable distributed systems. JACM **43**(2), 225–267 (1996)
8. Chandra, T.D., Hadzilacos, V., Toueg, S.: The weakest failure detector for solving consensus. JACM **43**(4), 685–722 (1996)
9. Bramas, Q., Foreback, D., Nesterenko, M., Tixeuil, S.: Packet efficient implementation of the omega failure detector. Theor. CS **63**, 237–260 (2019)
10. Doudou, A., Schiper, A.: Muteness detectors for consensus with byzantine processes. In: PODC, p. 315 (1998)
11. Kihlstrom, K.P., Moser, L.E., Melliar-Smith, P.M.: Byzantine fault detectors for solving consensus. Comput. J. **46**(1), 16–35 (2003)

12. Baldoni, R., Hélary, J.-M., Raynal, M., Tangui, L.: Consensus in byzantine asynchronous systems. J. Discr. Algorithms **1**(2), 185–210 (2003)
13. Oglio, J., Hood, K., Nesterenko, M., Tixeuil, S.: Quantas: quantitative user-friendly adaptable networked things abstract simulator. In: Applied, pp. 40–46 (2022)
14. Shoshany, B.: A C++17 Thread Pool for High-Performance Scientific Computing, May 2021. arXiv e-prints: arXiv:2105.00613, https://doi.org/10.5281/zenodo.4742687
15. Trail implementation in quantas, Aug 2023. https://github.com/QuantasSupport/Quantas/tree/54ca5de9d556338d1281f85317fb555afd2171fb/quantas/TrailPeer
16. Trail performance evaluation data, Aug 2023. http://www.cs.kent.edu/~mikhail/Research/trail.output.tar.gz
17. Li, Y., Wang, J., Zhang, H.: A survey of state-of-the-art sharding blockchains: models, components, and attack surfaces. J. Netw. Comput. Appl.. 103686 (2023)
18. Wang, X., Duan, S., Clavin, J., Zhang, H.: BFT in blockchains: from protocols to use cases. ACM Comput. Surv. (CSUR) **54**(10s), 1–37 (2022)
19. Stathakopoulou, C., Tudor, D., Pavlovic, M., Vukolić, M.: MIR-BFT: scalable and robust BFT for decentralized networks. J. Syst. Res. **2**(1) (2022)
20. Alqahtani, S., Demirbas, M.: BIGBFT: a multileader byzantine fault tolerance protocol for high throughput. In: IPCCC, pp. 1–10. IEEE (2021)
21. Gupta, S., Hellings, J., Sadoghi, M.: RCC: resilient concurrent consensus for high-throughput secure transaction processing. In: ICDE, pp. 1392–1403. IEEE (2021)
22. Hood, K., Oglio, J., Nesterenko, M., Sharma, G.: Partitionable asynchronous cryptocurrency blockchain. In: ICBC, pp. 1–9. IEEE (2021)
23. Bricker, R., Nesterenko, M., Sharma, G.: Blockchain in dynamic networks. In: SSS, pp. 114–129 (2022)

Softening the Impact of Collisions in Contention Resolution

Umesh Biswas[1], Trisha Chakraborty[2], and Maxwell Young[1]($\boxtimes$)

[1] Department of Computer Science and Engineering, Mississippi State University,
Mississippi State, MS 39762, USA
ucb5@msstate.edu, myoung@cse.msstate.edu
[2] Amazon Web Services, Minneapolis, MN 55401, USA
trichakr@amazon.com

Abstract. Contention resolution addresses the problem of coordinating access to a shared communication channel. Time is discretized into synchronized slots, and a packet can be sent in any slot. If no packet is sent, then the slot is empty; if a single packet is sent, then it is successful; and when multiple packets are sent at the same time, a collision occurs, resulting in the failure of the corresponding transmissions. In each slot, every packet receives ternary channel feedback indicating whether the current slot is empty, successful, or a collision. Much of the prior work on contention resolution has focused on optimizing the makespan, which is the number of slots required for all packets to succeed. However, in many modern systems, collisions are also costly in terms of the time they incur, which existing contention-resolution algorithms do not address. In this paper, we design and analyze a randomized algorithm, COLLISION-AVERSION BACKOFF (CAB), that optimizes both the makespan and the collision cost. We consider the static case where an unknown $n \geq 2$ packets are initially present in the system, and each collision has a known cost $\mathcal{C}$, where $1 \leq \mathcal{C} \leq n^\kappa$ for a known constant $\kappa \geq 0$. With error probability polynomially small in n, CAB guarantees that all packets succeed with makespan and a total expected collision cost of $\tilde{O}(n\sqrt{\mathcal{C}})$. We give a lower bound for the class of fair algorithms: where, in each slot, every packet executing the fair algorithm sends with the same probability (and the probability may change from slot to slot). Our lower bound is asymptotically tight up to a poly($\log n$)-factor for sufficiently large $\mathcal{C}$.

Keywords: Distributed computing · contention resolution · collision cost

1 Introduction

Contention resolution addresses the fundamental challenge of coordinating access by devices to a shared resource, which is typically modeled as a multiple access channel. Introduced with the development of ALOHA in the early 1970s [1], the problem of contention resolution remains relevant in WiFi networks [16], cellular networks [32], and shared-memory systems [8,25].

© The Author(s), under exclusive license to Springer Nature Switzerland AG 2025
T. Masuzawa et al. (Eds.): SSS 2024, LNCS 14931, pp. 398–416, 2025.
https://doi.org/10.1007/978-3-031-74498-3_29

Our problem is described as follows. There are $n \geq 2$ devices present at the start of the system, each with a packet to send on the shared channel, and n is *a priori* unknown. For ease of presentation, we adopt a slight abuse of language by referring to packets sending themselves (rather than referring to devices that send packets). Time proceeds in discrete, synchronized *slots*, and each slot has size that can accommodate the transmission of a packet. For any fixed slot, the channel provides ternary feedback, allowing each packet to learn whether 0 packets were sent, 1 packet was sent (a *success*), or 2+ packets were sent (a *collision*). Sending on the channel is performed in a distributed fashion; that is, *a priori* there is no central scheduler or coordinator. Traditionally, the contention-resolution problem focuses on minimizing the number of slots until all n packets succeed, which is referred to as the *makespan*.

However, in many modern systems, collisions also have a significant impact on performance. In the *standard cost model*, each collision can be viewed as a wasted slot that increases the makespan by 1, since no packet can succeed. Yet, in many settings, a collision wastes *more* than a single slot, and we denote this cost by $\mathcal{C}$. This cost can vary widely across different systems. In intermittently-connected mobile wireless networks [36], failed communication due to a collision may result in the sender having to wait until the intended receiver is again within transmission range. For WiFi networks, a collision may consume time proportional to the packet size [6]. In shared memory systems, concurrent access to the same memory location can result in delay proportional to the number of contending processes [8]. Thus, makespan gives only part of the performance picture, since collisions may add to the time until all packets succeed.

In this work, we account for the cost of collisions, in addition to the makespan. As we discuss later (see Sect. 3), existing contention-resolution algorithms do not perform well in this setting. For instance, randomized binary exponential backoff (BEB) [30], arguably the most popular of all contention-resolution algorithms, incurs a collision cost of $\Omega(n\mathcal{C})$, where $\mathcal{C}$ is the cost of a single collision. New ideas are needed to achieve better performance in this model.

1.1   Model and Notation

Our work addresses the case where an $n \geq 2$ number of packets are active at the start of the system. There is no known upper bound on n *a priori*, and packets do not have identifiers. A packet is *active* if it is executing a protocol; otherwise, the packet has terminated. Each packet must be successfully sent in a distributed fashion (i.e., there is no scheduler or central authority) on a multiple access channel, which we now describe.

Multiple Access Channel. Time is divided into disjoint *slots*, where each slot is sized to accommodate a single packet. In any slot, a packet may send itself or it may listen. For any fixed slot, if no packet is sent, then the slot is *empty*; if a single packet is sent, then it is *successful*; and if two or more packets are sent, then there it is a *collision* and none of these packets is successful. A packet that transmits in a slot immediately learns of success or failure; if it succeeds, the packet terminates immediately. Any packet that is listening to a slot, learns which of these three cases occurred; this is the standard **ternary feedback model**. Likewise, for any slot, a packet that is sending can also determine the channel feedback; either the packet succeeded (which it learns in the slot) or it failed (from which the packet infers a collision occurred). Packets cannot send messages to each other, and no other feedback is provided by the channel.

Metrics. A well-known measure of performance in prior work is the **makespan**, which is the number of slots required for all n packets to succeed. An equivalent metric in the static case is, **throughput**, which is defined as the time for n successes divided by the makespan.

In our model, each collision incurs a known cost of $\mathcal{C}$, where $1 \leq \mathcal{C} \leq n^\kappa$, for a known constant $\kappa \geq 0$. Given a contention-resolution algorithm, the pertinent costs are: (i) the **collision cost**, which is the number of collisions multiplied by $\mathcal{C}$, and (ii) the makespan. Often we refer to "cost", by which we mean the maximum of (i) and (ii), unless specified otherwise.

Notation. Throughout this manuscript, $\lg(\cdot)$ refers to logarithm base 2, and $\ln(\cdot)$ refers to the natural logarithm. We use $\log^y(x)$ to denote $(\log(x))^y$, and we use $\texttt{poly}(x)$ to denote x^y for some constant $y \geq 1$. The notation $\tilde{O}$ denotes the omission of a $\texttt{poly}(\log n)$ factor. We say an event occurs **with high probability** (w.h.p.) **in n** (or simply "with high probability") if it occurs with probability at least $1 - O(1/n^c)$, for any tunable constant $c \geq 1$.

For a random variable X, we say that w.h.p. the expected value $E[X] \leq x$ if, when tallying the events in the expected value calculation, the total probability of those events where $X > x$ is $O(1/n^\kappa)$. In our analysis, the expected collision cost holds so long as the subroutine DIAGNOSIS (described later) gives correct feedback to the packets, and this occurs with high probability.

1.2 Our Results

We design and analyze a new algorithm, **Collision-Aversion Backoff (CAB)**, that solves the static contention resolution problem with high-probability bounds on makespan and the expected collision cost. The following theorems provide our formal result.

Theorem 1. *With high probability in n,* COLLISION-AVERSION BACKOFF *guarantees that all n packets succeed with a makespan of $\tilde{O}(n\sqrt{\mathcal{C}})$, and an expected collision cost of $\tilde{O}(n\sqrt{\mathcal{C}})$.*

How does this compare with prior results? For starters, consider SAWTOOTH-BACKOFF (STB), which w.h.p. has an asymptotically-optimal makespan of $\Theta(n)$, but incurs $\Omega(n)$ collisions. The expected cost for CAB is superior to STB when $\mathcal{C}$ is at least polylogarithmic in n. In Sect. 1.3, we elaborate on how our result fits with previous work on contention resolution.

In a well-known lower bound in the standard cost model, Willard [35] defines and argues about **fair** algorithms. These are algorithms where, in a fixed slot, every active packet sends with the same probability (and the probability may change from slot to slot). Our lower bound applies to fair algorithms as follows.

Theorem 2. *Any fair algorithm that w.h.p. has a makespan of $\tilde{O}(n\sqrt{\mathcal{C}})$ has w.h.p. an expected collision cost of $\tilde{\Omega}(n\sqrt{\mathcal{C}})$ for $\mathcal{C} = \Omega(n^2)$.*

Since CAB is fair and guarantees w.h.p. a makespan of $\tilde{O}(n\sqrt{\mathcal{C}})$, our upper bound is asymptotically tight up to a $\texttt{poly}(\log n)$-factor for sufficiently large $\mathcal{C}$. A more general form of our lower bound is given in Sect. 4, along with additional discussion.

1.3 Why Care About Collision Costs?

Here, we discuss the potential value of our result and exploring beyond the standard cost model for contention resolution.

A Simple Answer. Prior contention-resolution algorithms that optimize for makespan suffer from many collisions. For example, arguably the most famous backoff algorithm is BEB (despite its sub-optimal makespan [9]), which has $\Omega(n)$ collisions. Similarly, STB has asymptotically-optimal makespan, but also suffers $\Omega(n)$ collisions [6]. Thus, these algorithms have cost $\tilde{\Theta}(n + n\mathcal{C})$. In contrast, the expected completion time for CAB is $\tilde{O}(n\sqrt{\mathcal{C}})$, which is superior whenever $\mathcal{C} = \texttt{poly}(\log n)$.

In the extreme, even a hypothetical algorithm that suffers (say, w.h.p.) a single collision would do poorly by comparison if collisions are sufficiently costly. Specifically, such an algorithm pays $\mathcal{C}$, which is asymptotically worse than the expected completion time of CAB for $\mathcal{C} = \tilde{\omega}(n^2)$.

Perhaps the above is reasonable motivation (we think it is), but why not also consider the cost of successes?[1] Below, we argue that reducing collision cost remains important, and that adding a non-unit cost for successes does not seem interesting from a theory perspective.

Connecting to the Standard Cost Model. Let us *temporarily* consider the implications of a cost model where a success and a collision each have cost $P \geq 1$. In the standard cost model $P = 1$. In WiFi networks, both costs are also roughly equal, where $P \gg 1$ [6].

In this context, reconsider STB's performance. Since, n packets must succeed, a cost of at least nP is unavoidable, and the additional $\Theta(nP)$ cost from collisions becomes (asymptotically) unimportant. (Indeed, any algorithm must pay at least nP for successes; given this unavoidable cost, it does not seem interesting from a theory perspective to consider a model where a success costs far more than a collision, since the cost from successes would dominate.) Likewise, CAB must also pay at least nP, which is (asymptotically) no better than STB.

---

[1] Empty slots do not seem to warrant such consideration since nothing is happening in such slots, and so a per-cost of 1 makes sense.

Does reducing collision costs matter here? Yes, and the value of our result is best viewed via throughput (recall Sect. 1.1). STB has cn collisions, for some constant $c > 0$, and (ignoring empty slots) its throughput is less than $nP/(nP + (cn)P) < 1/(1 + c) < 1 - 1/c$; that is, the throughput is bounded away from 1 by a constant amount. Doing a similar calculation for CAB, there is nP cost for n packets to succeed, plus $O(n\sqrt{P})$, which accounts for collisions and empty slots. Thus, the expected throughput is $E[nP/T]$ where T is the completion time. Since, $E[1/T] \geq 1/E[T]$, we have $E[nP/T] \geq nP/(nP + \tilde{O}(n\sqrt{P})) > 1 - \tilde{O}(1/\sqrt{P})$.

Thus, for the standard cost model, with $P = 1$, our result is (unsurprisingly) not an improvement. However, as P grows, our result provides better throughput in expectation, approaching 1. There are settings where we expect P to be large, such as: wireless networks where P can be commensurate with packet size, routing in mobile networks where communication failure increases latency [36], and shared memory systems where concurrent access to the same memory location by multiple processes results in delay [8].

2   Related Work

There is a large body of work on the static case for contention resolution, where n packets arrive together and initiate the contention resolution algorithm; this is often referred to as the *static* setting. Bender et al. [9] analyze the makespan for BEB, as well as other backoff algorithms; surprisingly, they show that BEB has sub-optimal makespan.

An optimal backoff algorithm is STB [27,28]. Since we borrow from STB to create one of our subroutines (discussed further in Sect. 3.1), so we describe it here. Informally, STB works by executing over a doubly-nested loop, where the outer loop sets the current window size w to be double the one used in the preceding outer loop. Additionally, for each such window, the inner loop executes over a *run* of $\lg w$ windows of decreasing size: $w, w/2, w/4, ..., 1$. For each window, every packet chooses a slot uniformly at random (u.a.r.) to send in.

The static setting has drawn attention from other angles. Bender et al. [10] examines a model where packets have different sizes under the binary feedback model. Anderton et al. [6] provide experimental results for several algorithms and argue that packet size should be incorporated into the definition of makespan, since collisions tend to cost time proportional to packet size.

While we address the static setting, we note the dynamic setting has been addressed (under the standard cost model), where packet arrival times are governed by a stochastic process (see the survey by Chlebus [21]). Another direction researchers have explored is the application of adversarial queueing theory [18]; for examples, see [2,3,22,23]. Arrival times are dictated by an adversary, but typically subject to constraints on the rate of packet injection and how closely in time (how bursty) these arrivals may be scheduled. Even more challenging, there is a growing literature addressing packet arrival times that are set by an unconstrained adversary; see [14,24,26]. Recent work in this area addresses additional challenges facing modern networks, such as energy efficiency [15,29], malicious disruption of the shared channel [4,5,7,12,20,31], and the ability to have a limited number of concurrent transmission succeed [13].

3   Technical Overview for Upper Bound

Due to space constraints, our proofs for the upper bound are provided in the full version of our paper [17]. However, we do reference some well-known inequalities based on the Taylor series that are omitted here, but can be found in Section A.1.

Here, we present an overview of our analysis, with the aim of imparting some intuition the design choices behind CAB, as well as highlighting the novelty of our approach. To this end, we first consider two natural—but ultimately flawed—ideas for solving our problem.

Straw Man 1. An immediate question is: *Why can we not use a prior contention-resolution algorithm to solve our problem?* For example, in the static setting, a well-known backoff algorithm, such BEB guarantees w.h.p. that all packets succeed with makespan $\Theta(n \log n)$ [9]. Under BEB, the packets execute over a sequence of disjoint windows, where window $i \geq 0$ consists of 2^i contiguous slots. Every active packet sends in a slot chosen u.a.r. from the current window. Unfortunately, a constant fraction of slots in each window $i \leq \lg(n) + O(1)$ will be collisions. Each collision imposes a cost of $\mathcal{C}$ leading to a collision cost of $\Omega(n\mathcal{C})$. □

Despite yielding a poor result, this straw man provides three useful insights. First, we cannot have packets start with a "small" window, since this leads to many collisions. Many backoff algorithms start with a small window, and will not yield good performance for this same reason.

Second, we should seek to better (asymptotically) balance the costs of makespan and collisions. Under BEB, these costs are highly unbalanced, being $\Theta(n \log n)$ and $\Omega(n\mathcal{C})$, respectively. We may trade off between makespan and collision cost; that is, we can make our windows larger, which increases our makespan, in order to dilute the probability of collisions.

Third, a window of size $\Theta(n\sqrt{\mathcal{C}})$ seems to align with these first two insights; that is, it appears to asymptotically balance makspan and collision cost. To understand the latter claim, suppose that in each slot, every packet sends with probability $\Theta(1/(n\sqrt{\mathcal{C}}))$. We can argue (informally) that, for any fixed slot, the probability of any two packets colliding is at least $\binom{n}{2}\Theta(1/n\sqrt{\mathcal{C}})^2 = O(1/\mathcal{C})$. Thus, over $n\sqrt{\mathcal{C}}$ slots in the window, the expected number of collisions is $O(n\sqrt{\mathcal{C}}/\mathcal{C}) = O(n/\sqrt{\mathcal{C}})$, and each collision has cost $\mathcal{C}$, so the expected collision cost is $O(n\sqrt{\mathcal{C}})$. Of course, this informal analysis falls short, since collisions may involve more than two packets, and we have not shown that all n packets can succeed over this single window (they cannot). Yet, this insight offers us hope that we can outperform prior backoff algorithms by achieving costs that are $o(n\mathcal{C})$.

How should we find a window of $\Theta(n\sqrt{\mathcal{C}})$? Since n is unknown *a priori*, we cannot simply instruct packets to start with that window size. It is also clear from the above discussion that we cannot grow the window to this size using prior backoff algorithms. This obstacle leads us to our second straw man.

Straw Man 2. Perhaps we can estimate n and then start directly with a window of size $\Theta(n\sqrt{\mathcal{C}})$. A well-known "folklore" algorithm for estimating n is the following. In each slot $i \geq 0$, each packet sends with probability $1/2^i$; otherwise, the packet listens. This algorithm, along with improvements to the quality of the estimation, is explored by Jurdzinski et al. [29], but we sketch why it works here.

Intuitively, when i is small, say a constant, then the probability of an empty slot is very small: $(1 - 1/2^{\Theta(1)})^n \leq e^{-\Theta(n)}$ by Fact 1(a). However, once $i = \lg(n)$, then the probability of an empty slot is a constant: $(1 - 1/2^{\lg(n)})^n = (1 - 1/n)^n \geq e^{\Theta(1)}$ by Fact 1(c). In other words, once a packet witnesses an empty slot, it can infer that its sending probability is $\Theta(1/n)$, and thus the reciprocal yields an estimate of n to within a constant factor.

At first glance, it may seem that we can estimate n, and then proceed to send packets in a window of size $\Theta(n\sqrt{\mathcal{C}})$. Unfortunately, for each slot $i \leq \lg(n)$, this algorithm (and others like it) will likely incur a collision, and thus the expected collision cost is $\Omega(\mathcal{C} \log n)$. To see why this is a problem, suppose that $\mathcal{C} = n^4$. This implies that the expected collision cost is $\tilde{\Omega}(\mathcal{C}) = \tilde{\Omega}(n^4)$, while the makespan is at best $O(n\sqrt{\mathcal{C}}) = O(n\sqrt{n^4}) = O(n^3)$. That is, there is an n-factor discrepancy between the expected collision cost and makespan, which grows worse for larger values of $\mathcal{C}$; this approach is not trading off well between these two metrics. $\square$

Even though the above algorithm racks up a large collision cost, it highlights how channel feedback can provide useful hints. If packets are less aggressive with their sending, we can reduce collisions while still receiving feedback that lets us reach our desired window size of $\Theta(n\sqrt{\mathcal{C}})$.

3.1   Our Approach

Motivating Sample Size. Hoping to avoid the problems illustrated in our discussion above, all packets begin with an initial current window size that can be "large"; the exact size we motivate momentarily. Recall that we aim for packets to tune their sending probability to be $\Theta(1/(n\sqrt{\mathcal{C}}))$, so we are aiming for a window size of $\Theta(n\sqrt{\mathcal{C}})$. However, this must be achieved with low expected collision cost. To do this, the packets execute over a ***sample*** of ***s*** slots. For each slot in the sample, every packet sends with probability 1 divided by the current window size, and monitors the channel feedback.

What does this channel feedback from the sample tell us? Suppose we have reached our desired window size of $\Theta(n\sqrt{\mathcal{C}})$. Then, the window size may be large relative to n, and so we should expect empty slots to occur frequently, while successes *do* occur, but less often; the probability of success is approximately

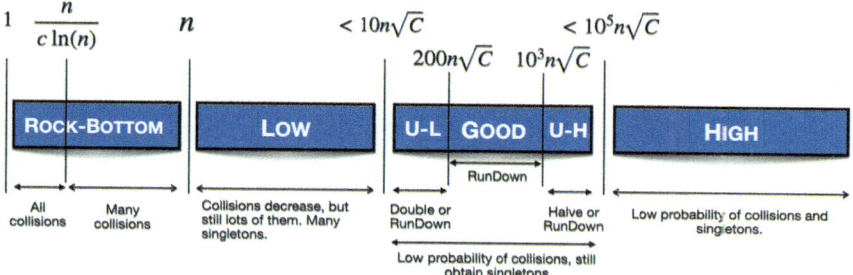

Fig. 1. Illustration of the ranges discussed in Sect. 3.2

$\binom{n}{1}(1/(n\sqrt{C})) \approx 1/\sqrt{C}$. To correctly diagnose w.h.p. (using a Chernoff bound) that $w = \Theta(n\sqrt{C})$, we rely on receiving $\Theta(\log n)$ successes. This dictates our sample size to be $s = \Theta(\sqrt{C}\log(n\sqrt{C})) = \Theta(\sqrt{C}\log(n))$, since $C \leq n^{\kappa}$.

The details behind this intuition are given in Section A.2 of our full paper [17], where we show that samples of size $\Theta(\sqrt{C}\log(n))$ are sufficient to make correct decisions that tune the window size to be $\Theta(n\sqrt{C})$. Furthermore, in Section A.3 of our full paper [17], we show that the corresponding cost from this tuning is $\tilde{O}(n\sqrt{C})$.

Motivating Initial Window Size. Suppose we are not at our desired window size. Then, the feedback from sampling tells us what to do. If our current window size is too large (i.e., our sending probability is too low), then the number of successes will be "small", since most slots are empty, and we should decrease our window size. Else, if the current window size is too small, (i.e., our sending probability is too high) then the number of successes is again "small", since most slots are collisions, and we should increase the window size. But wait, in the latter case, can we tolerate the cost of the resulting collisions? In the worse case, the entire sample might consist of collisions, resulting in a potentially large cost of $sC = \Theta(C^{3/2}\log n)$.

We remedy this problem by setting our initial window size to be C. Then, the probability of a collision in a fixed slot is approximately $\binom{n}{2}(1/C^2)$ and so the expected cost is roughly $T = \binom{n}{2}(1/C^2)C = \Theta(n^2/C)$. Reasonaing informally, if $C > n$, then $T = O(n)$, and over the sample the expected cost is $O(n\sqrt{C}\log(n))$, as desired. Else, if $C \leq n$, then even if our sample does consist entirely of collisions, the resulting cost is $\Theta(C\sqrt{C}\log n) = O(n\sqrt{C}\log n)$. This reasoning is formalized in Lemma 13 of our full paper [17].

In light of the above, we emphasize that CAB does not avoid all collisions; in fact, many collisions may occur when the collision cost is "small' (i.e., $C \leq n$). However, as the collision cost grows "larger" (i.e., $C > n$), CAB expresses an increasing aversion to collisions by having packets tune their respective sending probabilities such that we expect $o(1)$ collisions per sample. The cost analysis for sampling is given in Sect. A.3 of our full paper [17].

Borrowing from Sawtooth Backoff. By using sampling, ultimately a window size of $\Theta(n\sqrt{\mathcal{C}})$ is reached. At this point, every active packet executes the analog of the final run of STB (recall Sect. 2). Specifically, each packet sends with probability $p = \Theta(1/(n\sqrt{\mathcal{C}}))$, which we show is sufficient to have at least half of the active packets succeed (see Lemma 20). Then, the window is halved, and the process repeats where the remaining packets send with probability $p/2$. This halving process continues until all packets succeed. By the sum of a geometric series, the number of slots in this process is $O(n\sqrt{\mathcal{C}})$. Informally, the expected collision cost per slot is roughly $\binom{n}{2}(1/(n\sqrt{\mathcal{C}})^2)\mathcal{C} = O(1)$, and thus $O(n\sqrt{\mathcal{C}})$ over all slots in the window. The analysis of cost is provided in Sect. A.4 of our full paper [17].

3.2   Our Algorithm and Overview of Analysis

This section describes and gives intuition for CAB, whose pseudocode is given in Fig. 2. As stated in our model (Sect. 1.1), when a packet succeeds it terminates immediate; for ease of presentation, we omit this from our pseudocode.

From a high-level view, each packet keeps track of a current window size, w_{cur}; this size is critical, as it dictates the per-slot sending probability of each packet, which is $\Theta(1/w_{\mathrm{cur}})$. Each packet keeps track of its own notion of a current window. However, in each slot, since every packet is either listening or sending (and, thus, learning whether the slot contained a success or a collision), all packets receive the same channel feedback and adjust their respective current window identically (stated formally in Lemma 8). Therefore, for ease of presentation, we refer only to a single current window.

Defining Ranges. In order to describe how CAB works, we define the six size *ranges* that the current window (or just "window" for short), w_{cur}, can belong to during an execution.

- ROCK-BOTTOM: $[1, n)$.
- LOW: $[n, 10n\sqrt{\mathcal{C}})$.
- UNCERTAIN-LOW: $[10n\sqrt{\mathcal{C}}, 200n\sqrt{\mathcal{C}})$.
- GOOD: $[200n\sqrt{\mathcal{C}}, 10^3 n\sqrt{\mathcal{C}})$.
- UNCERTAIN-HIGH: $[10^3 n\sqrt{\mathcal{C}}, 10^5 n\sqrt{\mathcal{C}})$.
- HIGH: $[10^5 n\sqrt{\mathcal{C}}, \infty)$.

These ranges are depicted in Fig. 1. We note that the particular constants in these ranges are not special; they are chosen for ease of analysis. To gain intuition, we now describe the events we expect to witness in the ROCK-BOTTOM, LOW, GOOD, and HIGH ranges when CAB executes; we defer an in-depth discussion of UNCERTAIN-LOW and UNCERTAIN-HIGH until the end of the next subsection.

The ROCK-BOTTOM range captures "tiny" window sizes, starting from 1 up to $n - 1$. In this range, the probability of sending exceeds $1/n$, and we expect that most slots will be collisions. The next range is LOW, and it includes window sizes from n to just below $10n\sqrt{\mathcal{C}}$. This range represents a moderate increase in window size, allowing for more successes than ROCK-BOTTOM, although there can still be many collisions. As discussed in Sect. 3.1, in ROCK-BOTTOM and in the bottom portion of LOW, we can afford to have an single sample consist entirely of collisions, since $\mathcal{C} = O(n)$ (see Lemma 16). However, lingering in these ranges would ultimately lead to sub-optimal expected collision cost.

The GOOD range spans from $200n\sqrt{\mathcal{C}}$ to just below $10^3 n\sqrt{\mathcal{C}}$. In this range, the window sizes are sufficiently large that we expect $o(1)$ collisions, along with handful of successes; notably, less successes than LOW. This turns out to be a "good" operating range for the algorithm, where the balance between collision costs and makespan is achieved, as discussed in Sect. 3 (i.e., our third insight after Straw Man 1).

The HIGH range covers window sizes from $10^5 n\sqrt{\mathcal{C}}$ and above. In this range, the window sizes are very large, leading to a very low probability of collisions, which is good. However, there are also far-fewer successes compared to GOOD, which means that lingering in this range would lead to a sub-optimal number of slots until all packets succeed.

The UNCERTAIN-LOW and UNCERTAIN-HIGH ranges span $[10n\sqrt{\mathcal{C}}, 200n\sqrt{\mathcal{C}})$ and $[10^3 n \sqrt{\mathcal{C}}, 10^5 n\sqrt{\mathcal{C}})$, respectively. Why do we have these ranges? They capture values of w_{cur} where we cannot know w.h.p. exactly what will happen, although we *do* prove that the algorithm is making "progress". We will elaborate on this further after describing our methods for sampling and interpreting channel feedback, which we address next.

Sampling and Diagnosing Feedback. CAB navigates these ranges by using two subroutines: **Collect-Sample** and **Diagnosis**. As motivated earlier in Sect. 3.1, a sample is a contiguous set of $d\sqrt{\mathcal{C}}\ln(w_{\mathrm{cur}})$ slots that are used by each packet to collected channel feedback. Specifically, under COLLECT-SAMPLE, in each slot of the sample, every packet sends independently with probability $1/w_{\mathrm{cur}}$ and records the result if it is a success or a collision.

Based on the number of successes and the number of collisions, DIAGNOSIS attempts to determine the range to which w_{cur} belongs. We discuss these thresholds in order to give some intuition for how DIAGNOSIS is making this determination. To simplify the presentation, we omit discussion of UNCERTAIN-LOW and UNCERTAIN-HIGH until the end of this section. The process by which CAB homes in on the range to which w_{cur} belongs is depicted in Fig. 3.

We begin with the first if-statement on Line 17. This line checks if the number of successes is at least a logarithmic amount (in w_{cur}); if so, this indicates that w_{cur} cannot be in the HIGH range, which would yield far fewer successes (see Lemma 13). Therefore, if meet the conditional on Line 17, it must be the case that w_{cur} belongs to ROCK-BOTTOM, LOW, or GOOD.

Line 18 checks whether the number of successes falls below a "large" logarithmic amount; if not, then we are seeing a "large" number of successes that indicates w_{cur} cannot be in ROCK-BOTTOM or in GOOD, and so must be in LOW (see Lemma 10). Otherwise, the number of successes falls below this logarithmic amount, indicating that w_{cur} is in the ROCK-BOTTOM or in GOOD ranges.

COLLISION-AVERSION BACKOFF

---

1 Initial window size $w_{cur} \leftarrow \mathcal{C}$

2 **repeat**

3 $\quad$ #successes, #collisions $\leftarrow 0$

4 $\quad$ COLLECT-SAMPLE($\mathcal{C}, w_{cur}$)

5 $\quad$ DIAGNOSIS(#successes, #collisions, w_{cur})

6 **until** *true*;

7 **Function** COLLECT-SAMPLE($\mathcal{C}, w_{cur}$):

8 $\quad$ #successes $\leftarrow 0$

9 $\quad$ #collision $\leftarrow 0$

10 $\quad$ **for** *slot* $i = 1$ **to** $d\sqrt{\mathcal{C}}\ln(w_{cur})$ **do**

11 $\quad\quad$ Send with probability $1/w_{cur}$; otherwise, listen

12 $\quad\quad$ **if** *slot is a success* **then**

13 $\quad\quad\quad$ #successes++

14 $\quad\quad$ **else if** *slot is a collision* **then**

15 $\quad\quad\quad$ #collisions++

16 **Function** DIAGNOSIS(*#successes, #collisions, w_{cur}*):

17 $\quad$ **if** *#successes* $> \frac{2d\ln(w_{cur})}{10^5}$ **then**

18 $\quad\quad$ **if** *#successes* $\leq \frac{d\ln(w_{cur})}{20e}$ **then**

19 $\quad\quad\quad$ **if** *#collisions* $\geq \frac{d\sqrt{\mathcal{C}}\ln(w_{cur})}{8e^2}$ **then**

20 $\quad\quad\quad\quad$ $w_{cur} \leftarrow 2w_{cur}$

21 $\quad\quad\quad$ **else**

22 $\quad\quad\quad\quad$ Execute RUNDOWN (w_{cur})

23 $\quad\quad$ **else**

24 $\quad\quad\quad$ $w_{cur} \leftarrow 2w_{cur}$

25 $\quad$ **else**

26 $\quad\quad$ **if** *#collisions* $\geq \frac{d\sqrt{\mathcal{C}}\ln(w_{cur})}{8e^2}$ **then**

27 $\quad\quad\quad$ $w_{cur} \leftarrow 2w_{cur}$

28 $\quad\quad$ **else**

29 $\quad\quad\quad$ $w_{cur} \leftarrow w_{cur}/2$

30 **Function** RUNDOWN(w_{cur}):

31 $\quad$ $w_0 \leftarrow w_{cur}$

32 $\quad$ **while** $w_{cur} \geq 8\sqrt{\mathcal{C}}\lg(w_0)$ **do**

33 $\quad\quad$ **for** *each slot* $j = 1$ **to** w_{cur} **do**

34 $\quad\quad\quad$ Send packet with probability $2/w_{cur}$

35 $\quad\quad$ $w_{cur} \leftarrow w_{cur}/2$

36 $\quad$ **for** $i = 1$ **to** $c\ln(w_0)$ **do**

37 $\quad\quad$ **for** *each slot* $j = 1$ **to** w_0 **do**

38 $\quad\quad\quad$ Send packet with probability $2/w_0$

Fig. 2. Pseudocode for COLLISION-AVERSION BACKOFF

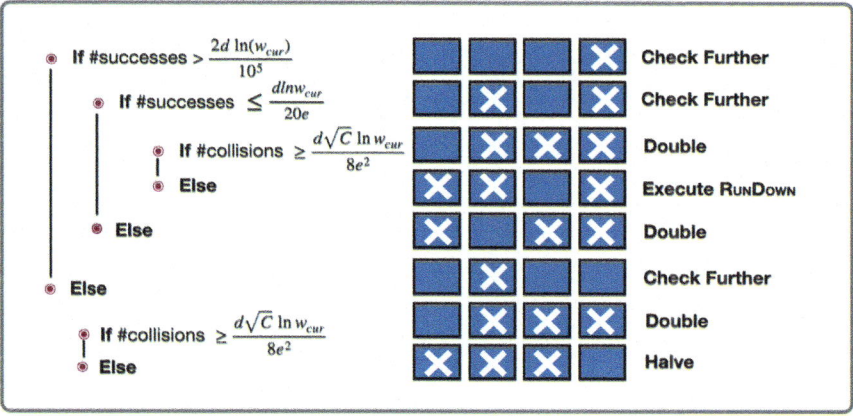

Fig. 3. The elimination of ranges as discussed in Sect. 3.2

To discern between these two cases, Line 19 checks if the number of collisions exceeds $\frac{d\sqrt{\mathcal{C}}\ln(w_{\mathrm{cur}})}{8e^2}$, which indicates $w_{\mathrm{cur}} \in$ ROCK-BOTTOM (see Lemmas 6, 7, and 9). Otherwise, w_{cur} falls within the GOOD and the job of DIAGNOSIS is complete—at this point, RUNDOWN will be executed.

How can an algorithm with low expected collision cost make a decision based on whether $\Theta(\sqrt{\mathcal{C}}\log(w_{\mathrm{cur}}))$ collisions occur? Recall that in this case, we are deciding between $w_{\mathrm{cur}} \in$ ROCK-BOTTOM and $w_{\mathrm{cur}} \in$ GOOD. We will only witness this number of collisions when w_{cur} is in the ROCK-BOTTOM range, where $\mathcal{C} \leq n$ and so we can tolerate this cost. Otherwise, as described above, $w_{\mathrm{cur}} \in$ GOOD and the expected number of collisions is only $\tilde{O}(1/\sqrt{\mathcal{C}})$ (see the discussion preceding Lemma 11 in our full paper [17]).

The remaining portion of DIAGNOSIS starts with the else-statement on Line 25. This line is executed only if the conditional on Line 17 is not met; that is, we have very few successes. This can occur only if $w_{\mathrm{cur}} \in$ ROCK-BOTTOM or $w_{\mathrm{cur}} \in$ HIGH. The former case is (again) diagnosed be checking whether there are many collisions (Line 26) and, if so, the window is doubled; otherwise, we are in the latter case and the window is halved. Again, we only have many collisions if w_{cur} is in the ROCK-BOTTOM or lower portion of the LOW ranges, where $\mathcal{C} \leq n$ and so we can tolerate this cost.

What about the uncertain ranges? In UNCERTAIN-LOW, a sample will contain enough successes to satisfy Line 17. However, it is unclear whether the number of successes will be "large" (if w_{cur} is in the lower portion of UNCERTAIN-LOW) or "moderate" (if w_{cur} is in the upper portion of UNCERTAIN-LOW). In the former case, we fail Line 18, while in the latter case, we satisfy Line 18. Despite this uncertainty, observe that we are nonetheless guaranteed to either double the window or execute RUNDOWN. Either of these outcomes count as progress, since either w_{cur} moves closer to GOOD by doubling, or RUNDOWN is executed on a window of size $\Theta(n\sqrt{\mathcal{C}})$ (in contrast, halving w_{cur} would be counterproductive).

The intuition behind UNCERTAIN-HIGH is similar. If $w_{\mathrm{cur}} \in$ HIGH, then we can show that the number of successes fails Line 17 (see Lemma 13) and that, ultimately, we halve w_{cur}. However, between GOOD and HIGH, this cannot be shown w.h.p.; it may hold if w_{cur} is close to the lower end of HIGH, but not hold if w_{cur} is close to the upper end of GOOD. So, instead, we argue that we either halve w_{cur} or execute RUNDOWN. Either of these actions counts as progress, since either w_{cur} moves closer to GOOD, or RUNDOWN is executed on a window of size $\Theta(n\sqrt{\mathcal{C}})$.

Ultimately, we can show the following key lemma (in Sect. A.3 of our full paper [17]):

Lemma 19. *The executions of* COLLECT-SAMPLE *and* DIAGNOSIS *guarantee that* w.h.p.: *(i)* RUNDOWN *is executed within* $O(\sqrt{\mathcal{C}}\log^2(n))$ *slots, and (ii) the total expected collision cost until that point is* $O(n\sqrt{\mathcal{C}}\log^2(n))$.

All Remaining Active Packets Succeed. The final subroutine, **RunDown**, is executed once $w_{\mathrm{cur}} = \Theta(n\sqrt{\mathcal{C}})$, and it allows all active packets to succeed. In each slot of w_{cur}, every packet sends with probability $2/w_{\mathrm{cur}}$. Otherwise, the current window size is halved and the remaining active packets repeat this process. This continues until the window size reaches $\Theta(\sqrt{\mathcal{C}}\log(w_0)) = \Theta(\sqrt{\mathcal{C}}\log(n))$, where the asymptotic equality holds by recalling that $\mathcal{C} = \mathbf{poly}(n)$. Once this smallest window in the run is reached, $O(\log n)$ active packets remain. To finish these packets, CAB performs an additional $\Theta(\ln(n\sqrt{\mathcal{C}})) = \Theta(\ln(n))$ windows of size $\Theta(n\sqrt{\mathcal{C}})$, where any remaining active packet sends in each slot with probability $\Theta(1/n\sqrt{\mathcal{C}})$.

We prove the following lemmas (in Sect. A.4 of our full paper [17]):

Lemma 21. *When* RUNDOWN *is executed,* w.h.p. *all packets succeed within* $O(n\sqrt{\mathcal{C}}\ln(n))$ *slots.*

Lemma 22. *W.h.p. the expected collision cost for executing* RUNDOWN *is* $O(n\sqrt{\mathcal{C}}\ln(n))$.

Finally, our upper bound in Theorem 1 follows directly from Lemmas 19, 20, and 21.

4   Technical Overview for Lower Bound

In this section, we provide an overview of our argument, which focuses on placing a lower bound on the expected collision cost. Here, we highlight the key lemmas in our argument, and our full proofs are provided in Section B of our full paper [17].

We consider only the set of slots, $\boldsymbol{\mathcal{S}}$, in the execution of any algorithm where at least two active packets remain, since we cannot have a collision with a single active packet. While we do not always make this explicit, but going forward, any slot t is assumed to implicitly belong to $\mathcal{S}$.

Let $\boldsymbol{p_i(t)}$ denote the probability that packet i sends in slot t. Note that, if a packet has terminated, its sending probability can be viewed as 0. For any fixed slot t, the **contention** in slot t is $\text{Con}(t) = \sum_{i=1}^{n} p_i(t)$; that is, the sum of the sending probabilities in that slot.

When Contention is High. We start by showing that any algorithm that has even a single slot t with $\text{Con}(t) > 2$ must have $\Omega(\mathcal{C})$ expected collision cost, which is one portion of our lower bound. This is done by deriving an expression for the probability of a collision in any fixed slot t as a function of $\text{Con}(t)$, which is useful when $\text{Con}(t)$ is "high" (see Lemmas 23, 24, and 25 in Section B of our full paper [17]).

When Contention is Low. What about an algorithm where all slots of the execution have "low" contention (i.e., $\text{Con}(t) \leq 2$)? Our previous expression is hard to work with in this case. So, we derive a different expression for the probability of a collision in a fixed slot t, which can be useful for small values of $\text{Con}(t)$:

Lemma 26. *Fix any slot t and let $\text{Con}(t) \leq 2$. The probability of a collision in t is at least $\left(\frac{1}{110}\right)\left(\text{Con}(t)^2 - \sum_i p_i(t)^2\right)$.*

However, this expression requires some additional work to be deployed in our argument, as we now describe.

For any fixed slot $t \in \mathcal{S}$, let $\boldsymbol{p}_{\max}(t)$ be the maximum sending probability of any packet in slot t. Similarly, let $\boldsymbol{p}_{\sec}(t) \leq p_{\max}(t)$ be the next-largest sending probability of any packet in slot t; note that $p_{\sec}(t) = p_{\max}(t)$, if more than one packet sends with probability $p_{\max}(t)$.

Define $\boldsymbol{\Delta}(t) = p_{\sec}(t)/p_{\max}(t)$. Our analysis ignores any slot t where $\text{Con}(t) = 0$; that is, any slot where every packet has a sending probability of 0. We give such slots to any algorithm for "free"; that is, we do not include them in the cost of the execution. Thus, $\Delta(t)$ is always well-defined, since $p_{\max}(t) > 0$.

Why do we need $\Delta(t)$? It captures a sense of "balance". For our purposes, the situation is most "unbalanced" when exactly one packet has non-zero sending probability, while all other packets have zero sending probability; that is, when the total value of $\text{Con}(t) > 0$ is due to a single packet. Clearly, in such slots, there can be no collision and, correspondingly, $\Delta(t) = 0$. In our argument, $\Delta(t)$ plays a key role in establishing the following:

Lemma 27. *For any fixed slot t, $\text{Con}(t)^2 - \sum_i p_i(t)^2 \geq \Delta(t)\,\text{Con}(t)^2/2$.*

A natural extension of $\Delta(t)$ is $\boldsymbol{\Delta}_{\min}$, which is the minimum $\Delta(t)$ over all slots $t \in \mathcal{S}$. As we show momentarily, our lower bound is parameterized by $\Delta_{\min}$. The last component of our argument addresses the sum of the contention over all slots in $\mathcal{S}$:

Lemma 30. *Any algorithm that guarantees w.h.p. that n packets succeed must w.h.p. have*
$$\sum_{t \in \mathcal{S}} \text{Con}(t) = \Omega(n).$$

We can now establish a lower bound for this low-contention case:

Lemma 31. *Consider any algorithm $\mathcal{A}$ whose contention in any slot is at most 2 and guarantees w.h.p. a makespan of $\tilde{O}(n\sqrt{\mathcal{C}})$. W.h.p. the expected collision cost for $\mathcal{A}$ is $\tilde{\Omega}(\Delta_{\min}n\sqrt{\mathcal{C}})$.*

Proof. Let X be a random variable that is $|\mathcal{S}|$ under the execution of $\mathcal{A}$, and note that w.h.p. $X = \tilde{O}(n\sqrt{\mathcal{C}})$. By Lemma 30, we have:

$$\sum_{t=1}^{X} \text{Con}(t) \geq cn. \tag{1}$$

for some constant $c > 0$. Let $Y_t = 1$ if slot $t \in \mathcal{S}$ has a collision; otherwise, $Y_t = 0$. By Lemmas 26 and 27:

$$Pr(Y_t = 1) \geq \Delta(t) \cdot \text{Con}(t)^2/220$$
$$\geq \Delta_{\min} \cdot \text{Con}(t)^2/220$$

where the second line follows from the definition of $\Delta_{\min}$. The expected collision cost is:

$$\sum_{t=1}^{X} P(Y_t = 1) \cdot \mathcal{C} \geq \sum_{t=1}^{X} \frac{\Delta_{\min}\text{Con}(t)^2 \cdot \mathcal{C}}{220}$$
$$= \frac{\Delta_{\min} \cdot \mathcal{C}}{220} \sum_{t=1}^{X} \text{Con}(t)^2. \tag{2}$$

By Jensen's inequality for convex functions, we have:

$$\frac{\sum_{t=1}^{X} \text{Con}(t)^2}{X} \geq \left(\frac{\sum_{t=1}^{X} \text{Con}(t)}{X}\right)^2 \tag{3}$$

Finally, the expected cost is at least:

$$\frac{\Delta_{\min} \cdot \mathcal{C}}{220} \sum_{t=1}^{X} \text{Con}(t)^2 \geq \frac{\Delta_{\min} \cdot \mathcal{C}}{220} \frac{\left(\sum_{t=1}^{X} \text{Con}(t)\right)^2}{X} \quad \text{by Eqs. 2 and 3}$$
$$\geq \frac{\Delta_{\min} \cdot \mathcal{C}}{220} \left(\frac{c^2 n^2}{X}\right) \quad \text{by Eq. 1}$$
$$= \tilde{\Omega}\left(\Delta_{\min} n\sqrt{\mathcal{C}}\right)$$

where the second line follows by Eq. 1, which was defined with regard to $\mathcal{S}$, and so can be compared to Eq. 2. The last line follows since w.h.p. $X = \tilde{O}(n\sqrt{\mathcal{C}})$. □

Given our analysis of the high- and low-contention cases, we have the following lower bound:

Theorem 4. *Consider any algorithm that w.h.p. guarantees $\tilde{O}(n\sqrt{\mathcal{C}})$ makespan. Then, w.h.p., the expected collision cost for $\mathcal{A}$ is $\tilde{\Omega}(\min\{\mathcal{C}, \Delta_{\min}n\sqrt{\mathcal{C}}\})$.*

What algorithms does our lower bound say something interesting about? We can start with our statement in Theorem 2. For a well-known lower bound under the standard cost model, Willard [35] argues about *fair* algorithms. These are algorithms where, in a fixed slot, every active packet sends with the same probability (and the probability may change from slot to slot). Recall that our analysis ignores any slot t where all packets have sending probability 0 (i.e., $\mathsf{Con}(t) = 0$), giving such slots to the algorithm for free. Thus, here $\Delta_{\min} = 1$, and for a sufficiently large collision cost—specifically, $\mathcal{C} = \Omega(n^2)$—the lower bound becomes $\tilde{\Omega}(n\sqrt{\mathcal{C}})$. Given that CAB is fair and w.h.p. guarantees $\tilde{O}(n\sqrt{\mathcal{C}})$ makespan, we thus have a lower bound that is asymptotically tight to a $\mathsf{poly}(\log n)$-factor with our upper bound.

Our lower bound also applies in a similar way to a generalized notion of fairness, where the sending probabilities of any two packets are within some factor $\delta > 0$. For example, if $\delta = \Theta(1)$, then $\Delta_{\min} = \Theta(1)$. Indeed, we only need such δ-fairness between the two packets with the largest probabilities for our lower bound to apply, although our bound weakens as δ grows larger.

Another class of algorithms that our lower bound applies to is *multiplicative weights update algorithms*, where in each slot every packet updates its sending probability by a multiplicative factor based on channel feedback in the slot. Many of these algorithms are fair (such as [7,11,13,19,31,33,34]), but not all are (such as [11]). Our lower bound applies in a non-trivial way to all such algorithms; that is, $\Delta_{\min} > 0$ given the update rules.

5   Conclusion and Future Work

We considered a model for the problem contention resolution where each collision has cost $\mathcal{C}$. Our algorithm, COLLISION-AVERSION BACKOFF, addresses the static case and guarantees w.h.p. that all packets succeed with makespan and expected collision cost that is $\tilde{O}(n\sqrt{\mathcal{C}})$.

There are several directions for future work. First, we would like to extend this cost model to the dynamic setting (where packets may arrive over time) and design solutions. Here, many existing approaches use collisions to separate the packets into (approximately) disjoint batches, each of which finishes before the next begins. However, the collision cost of such approaches seems prohibitive.

Second, for the lower bound, we would like to derive a more general lower bound. A significant challenge appears to be addressing algorithms that set up slots where exactly one packet has non-zero sending probability, while all others have zero sending probability. For example, an elected leader might "schedule" each packet their own exclusive slot in which to send (with probability 1). Since there can be no collisions after such a schedule is implemented, it seems a lower bound argument must demonstrate that establishing a schedule is costly.

Third, what if collisions are not fully dictated by the actions of packets? Some wireless settings are inherently "noisy", due to weather conditions, co-located networks, or faulty devices. Is there a sensible model for such settings and, if so, can we reduce the cost from collisions?

Acknowledgment. We are grateful to the anonymous reviewers for their feedback on our manuscript. This work is supported by NSF award CCF-2144410.

References

1. Abramson, N.: The ALOHA system: another alternative for computer communications. In: Proceedings of the November 17–19, 1970, Fall Joint Computer Conference, pp. 281–285 (1970)
2. Aldawsari, B.A., Chlebus, B.S., Kowalski, D.R.: Broadcasting on adversarial multiple access channels. In: Proceedings of the IEEE 18th International Symposium on Network Computing and Applications (NCA), pp. 1–4 (2019)
3. Anantharamu, L., Chlebus, B.S., Kowalski, D.R., Rokicki, M.A.: Deterministic broadcast on multiple access channels. In: Proceedings IEEE INFOCOM, pp. 1–5. IEEE (2010)
4. Anantharamu, L., Chlebus, B.S., Kowalski, D.R., Rokicki, M.A.: Medium access control for adversarial channels with jamming. In: Proceedings of the International Colloquium on Structural Information and Communication Complexity, pp. 89–100. Springer (2011)
5. Anantharamu, L., Chlebus, B.S., Kowalski, D.R., Rokicki, M.A.: Packet latency of deterministic broadcasting in adversarial multiple access channels. J. Comput. Syst. Sci. **99**, 27–52 (2019)
6. Anderton, W.C., Chakraborty, T., Young, M.: Windowed backoff algorithms for WiFi: Theory and performance under batched arrivals. Distrib. Comput. **34**, 367–393 (2021)
7. Awerbuch, B., Richa, A., Scheideler, C.: A jamming-resistant MAC protocol for single-hop wireless networks. In: Proceedings of the 27th ACM Symposium on Principles of Distributed Computing (PODC), pp. 45–54 (2008)
8. Ben-David, N., Blelloch, G.E.: Analyzing contention and backoff in asynchronous shared memory. In: Proceedings of the ACM Symposium on Principles of Distributed Computing, pp. 53–62 (2017)
9. Bender, M.A., Farach-Colton, M., He, S., Kuszmaul, B.C., Leiserson, C.E.: Adversarial contention resolution for simple channels. In: Proceedings of the 17th Annual ACM Symposium on Parallelism in Algorithms and Architectures (SPAA), pp. 325–332 (2005)
10. Bender, M.A., Fineman, J.T., Gilbert, S.: Contention resolution with heterogeneous job sizes. In: Proceedings of the 14th Conference on Annual European Symposium (ESA), pp. 112–123 (2006)
11. Bender, M.A., Fineman, J.T., Gilbert, S., Kuszmaul, J., Young, M.: Fully energy-efficient randomized backoff: slow feedback loops yield fast contention resolution. In: Proceedings of the 43rd ACM Symposium on Principles of Distributed Computing (PODC), pp. 231–242 (2024). https://doi.org/10.1145/3662158.3662807
12. Bender, M.A., Fineman, J.T., Gilbert, S., Young, M.: How to scale exponential backoff: constant throughput, polylog access attempts, and robustness. In: Proceedings of the Twenty-Seventh Annual ACM-SIAM Symposium on Discrete Algorithms, pp. 636–654. SODA'16 (2016)
13. Bender, M.A., Gilbert, S., Kuhn, F., Kuszmaul, J., Médard, M.: Contention resolution for coded radio networks. In: Proceedings of the 34th ACM Symposium on Parallelism in Algorithms and Architectures (SPAA), pp. 119–130. ACM (2022)

14. Bender, M.A., Kopelowitz, T., Kuszmaul, W., Pettie, S.: Contention resolution without collision detection. In: Proceedings of the 52nd Annual ACM Symposium on Theory of Computing (STOC), pp. 105–118 (2020)
15. Bender, M.A., Kopelowitz, T., Pettie, S., Young, M.: Contention resolution with log-logstar channel accesses. In: Proceedings of the 48th Annual ACM SIGACT Symposium on Theory of Computing, pp. 499–508. STOC 2016 (2016)
16. Bianchi, G.: Performance analysis of the IEEE 802.11 distributed coordination function. IEEE J. Sel. Areas Commun. **18**(3), 535–547 (2000)
17. Biswas, U., Chakraborty, T., Young, M.: Softening the impact of collisions in contention resolution (2024). https://arxiv.org/abs/2408.11275
18. Borodin, A., Kleinberg, J., Raghavan, P., Sudan, M., Williamson, D.P.: Adversarial queuing theory. J. ACM (JACM) **48**(1), 13–38 (2001)
19. Chang, Y., Jin, W., Pettie, S.: Simple contention resolution via multiplicative weight updates. In: Proceedings of the Second Symposium on Simplicity in Algorithms (SOSA), pp. 16:1–16:16 (2019)
20. Chen, H., Jiang, Y., Zheng, C.: Tight trade-off in contention resolution without collision detection. In: Proceedings of the ACM Symposium on Principles of Distributed Computing, pp. 139–149 (2021)
21. Chlebus, B.S.: Randomized communication in radio networks. Handbook Randomized Comput **1**, 401–456 (2001)
22. Chlebus, B.S., Kowalski, D.R., Rokicki, M.A.: Stability of the multiple-access channel under maximum broadcast loads. In: Proceedings of the Symposium on Self-Stabilizing Systems (SSS), pp. 124–138. Springer (2007)
23. Chlebus, B.S., Kowalski, D.R., Rokicki, M.A.: Maximum throughput of multiple access channels in adversarial environments. Distrib. Comput. **22**(2), 93–116 (2009)
24. De Marco, G., Stachowiak, G.: Asynchronous shared channel. In: Proceedings of the ACM Symposium on Principles of Distributed Computing, pp. 391–400. PODC'17 (2017)
25. Dwork, C., Herlihy, M., Waarts, O.: Contention in shared memory algorithms. J. ACM (JACM) **44**(6), 779–805 (1997)
26. Fineman, J.T., Newport, C., Wang, T.: Contention resolution on multiple channels with collision detection. In: Proceedings of the 2016 ACM Symposium on Principles of Distributed Computing, pp. 175–184. PODC'16 (2016)
27. Geréb-Graus, M., Tsantilas, T.: Efficient optical communication in parallel computers. In: Proceedings of the 4th Annual ACM Symposium on Parallel algorithms and Architectures, pp. 41–48 (1992)
28. Greenberg, R.I., Leiserson, C.E.: Randomized routing on fat-trees. In: Proceedings of the 26th Annual Symposium on the Foundations of Computer Science (FOCS), pp. 241–249 (1985)
29. Jurdziński, T., Kutyłowski, M., Zatopiański, J.: Energy-efficient size approximation of radio networks with no collision detection. In: Proceedings of the 8th Annual International Conference (COCOON), pp. 279–289 (2002)
30. Metcalfe, R.M., Boggs, D.R.: Ethernet: Distributed packet switching for local computer networks. CACM **19**(7), 395–404 (1976)
31. Ogierman, A., Richa, A., Scheideler, C., Schmid, S., Zhang, J.: Competitive MAC under adversarial SINR. In: Proceedings of IEEE Conference on Computer Communications (INFOCOM), pp. 2751–2759 (2014)
32. Ramaiyan, V., Vaishakh, J.: An information theoretic point of view to contention resolution. In: 2014 Sixth International Conference on Communication Systems and Networks (COMSNETS), pp. 1–8. IEEE (2014)

33. Richa, A., Scheideler, C., Schmid, S., Zhang, J.: An efficient and fair MAC protocol robust to reactive interference. IEEE/ACM Trans. Netw. **21**(1), 760–771 (2013)
34. Richa, A., Scheideler, C., Schmid, S., Zhang, J.: Competitive throughput in multi-hop wireless networks despite adaptive jamming. Distrib. Comput. **26**(3), 159–171 (2013)
35. Willard, D.E.: Log-logarithmic selection resolution protocols in a multiple access channel. SIAM J. Comput. **15**(2), 468–477 (1986)
36. Zhang, Z.: Routing in intermittently connected mobile ad hoc networks and delay tolerant networks: overview and challenges. IEEE Commun. Surv. Tutor. **8**(1), 24–37 (2006)

Generating the Convergence Stairs
of the Collatz Program

Ali Ebnenasir[✉]

Department of Computer Science, Michigan Technological University,
Houghton, MI 49931, USA
aebnenas@mtu.edu

Abstract. For the first time in decades, this paper presents an algorithmic method that, given a positive integer j, generates the j-th convergence stair containing all natural numbers from where the Collatz conjecture holds by exactly j applications of the Collatz function. To this end, we present a novel formulation of the Collatz conjecture as a concurrent program, and provide the general case specification of the j-th convergence stair for any $j > 0$. The proposed specifications provide a layered and linearized orientation of Collatz numbers organized in an infinite set of infinite binary trees. Such a general specification can have significant applications in analyzing and testing the stability of complex non-linear systems. We have implemented this method as a software tool that generates the Collatz numbers of individual stairs. We also show that starting from any value in any convergence stair the conjecture holds. However, to prove the conjecture, one has to show that every natural number will appear in some stair; i.e., the union of all stairs is equal to the set of natural numbers, which remains an open problem.

Keywords: Collatz Conjecture · Self-Stabilization · Convergence stairs

1 Introduction

The Collatz conjecture is known as the simplest unsolved math problem because it is easy to state it, but has remained an open problem since 1930s. The objective of this paper is to present a systematic approach for the *specification* of Collatz numbers based on their distance to the set of powers of two. Consider the function f_c over a variable x whose domain and range include the set of positive integers, denoted $\mathcal{N}$: if x is an even value, then update x with $f_c(x) = x/2$; otherwise, assign $f_c(x) = 3x+1$ to x. Thus, starting from a positive integer n, one can define a sequence of values $n, f_c(n), f_c^2(n), f_c^3(n), \cdots$ obtained by repetitive application of the Collatz function f_c, called the *orbit* of n. Let $f_{min}(n)$ be the minimum value in $\{n, f_c(n), f_c^2(n), f_c^3(n), \cdots\}$. The Collatz conjecture simply asks whether $f_{min}(n) = 1$ for any positive integer n. Once reached one, the set $\mathcal{I}_{cltz} = \{1, 2, 4\}$ remains closed in f_c. More generally, the set $\mathcal{I}_u = \{2^k \mid k \geq 0\}$ is closed in f_c.

© The Author(s), under exclusive license to Springer Nature Switzerland AG 2025
T. Masuzawa et al. (Eds.): SSS 2024, LNCS 14931, pp. 417–431, 2025.
https://doi.org/10.1007/978-3-031-74498-3_30

The significance of the Collatz conjecture lies in its applications in several domains such as self-stabilization of infinite-state systems, image encryption [3], software watermarking [24], and in designing the stability of complex and non-linear systems [18] that have chaotic behaviors, yet ensuring eventual stability. The notion of eventual stability is common knowledge in Computer Science in general, and in the self-stabilization community in particular. However, the behaviors of Collatz function defy any known type of convergence-assurance approach (e.g., ranking functions [28], convergence stairs [15,16]) as we know it in self-stabilizing systems. Moreover, while there are other distributed programs with unbounded variables (e.g., Dijkstra's token passing [9]), the domain of such unbounded variables often increases as the network size grows; i.e., unbounded but finite. This is not the case in Collatz function, and the domain of x is infinite in a shared-memory multi-threaded program. Note that the Collatz program is not distributed, but it poses an interesting challenge for the self-stabilization community because convergence must be provided from positive integers to $\mathcal{I}_u$. To ensure convergence, divergence-freedom must be shown in addition to deadlock and livelock-freedom outside $\mathcal{I}_u$.

While there is a rich body of work in mathematics on the Collatz problem, the most recent result states that "Almost all Collatz orbits attain almost bounded values." [29], where the notion of "almost all" is defined in the context of logarithmic density. We take a different approach by first formulating the Collatz function as a shared memory program (Sect. 2). Then, we reformulate the Collatz problem as a problem of specifying and verifying the convergence of the *Collatz program* through the specification of an infinite number of convergence stairs (Sect. 3). Each *convergence stair* is in fact an infinite set of natural values from where the set $\mathcal{I}_u$ can be reached in $j > 0$ steps of applying the Collatz function f_c. Formally, the j-th convergence stair, denoted S_j, is equal to the set $\{n \mid f_c^j(n) \in \mathcal{I}_u\}$. (Notice that, $S_0 = \mathcal{I}_u$.)

Our objective is to devise a scheme where, given $j > 0$ one can compute all the values in the j-th convergence stair S_j without expanding and exploring the binary tree generated by backward reachability from $\{1, 2, 4\}$. This way, the Collatz conjecture would be reduced to proving that (1) $\cup_{j=0}^{\infty} S_j = \mathcal{N}$, and (2) from each stair $j > 0$ the Collatz program reaches a value in stair $j - 1$. This approach provides a different method of tackling the Collatz conjecture, and enables an algorithmic way for analyzing the behavior of every individual stair. Moreover, this approach presents a layered and linearized orientation of Collatz numbers organized in an infinite set of infinite binary trees. Such a linearization method can provide insight in addressing similar conjectures (e.g., Kakutani conjectures [7]). To the best of our knowledge, this is the first time that such a general specification is developed, which can have significant applications in understanding and testing of the convergence of infinite-state self-stabilizing systems. For example, designers can study how neighboring stairs interact. We study this method for convergence to I_{cltz} and $\mathcal{I}_u = \{2^k \mid k \geq 0\}$, and show that specifying convergence stairs for $\mathcal{I}_u$ is feasible and more useful for verification. Specifically, we present an algorithm that takes as input a value j and generates all the values belonging to the j-th stair of converging to $\mathcal{I}_u$. We then show that every value generated is in fact a correct Collatz number in the j-th stair, and prove that our algorithm does not miss any value. The proof of correctness is performed through attaching a Binary Verification Code (BVC) to each number during its specification. During the generation phase, we use the BVC to verify the correctness of the generated value. An implementation of the proposed algorithm is available at https://github.com/aebne/CollatzStairs.

Organization. Sect. 2 provides a characterization of the problem as a shared-memory program. Section 3 then investigates the specification and verification of convergence stairs. Section 4 discusses related work. Finally, Sect. 5 makes concluding remarks and discusses some open problems.

2   Preliminaries

Let P_{cltz} denote the Collatz program. We refer to x as the state variable of P_{cltz} and the value of x identifies the current state of P_{cltz}. Throughout this paper, we interchangeably use the terms 'state', 'value', and 'number'. The P_{cltz} program includes the following actions:

$$
\begin{aligned}
&P_1: \quad (x \mod 2) = 0 \quad &\to x := x/2 \\
&P_2: \quad (x \mod 2) \neq 0 \quad &\to x := 3x + 1
\end{aligned}
$$

An action is a guarded command, denoted $grd \to stmt$, where the grd is a Boolean condition in terms of x and the $stmt$ specifies how x can atomically be updated when the guard holds; i.e., the action is *enabled*. Each action belongs to a separate thread, namely P_1 and P_2. Notice that, there are no distribution constraints as the two threads can atomically read and write the program state (i.e., value of x). A *transition* of P_{cltz} is an ordered pair of integer values generated by either P_1 or P_2 starting from any value in $\mathcal{N}$. A *computation* of P_{cltz} is a sequence of transitions generated from any value in $\mathcal{N}$.

Invariant, closure and convergence. Starting in any state in the set $\mathcal{I}_{cltz} = \{1, 2, 4\}$, the computations of P_{cltz} remain in $\mathcal{I}_{cltz}$; i.e., *closure*. A weaker version of $\mathcal{I}_{cltz}$ is $\mathcal{I}_u = \{2^k \mid k \in \mathcal{N} \wedge k \geq 0\}$, which is also closed in P_{cltz} by successive execution of action P_1. We refer to $\mathcal{I}_u$ as the *unbounded* invariant of P_{cltz}. A program P *converges* to an invariant I *iff* (if and only if) any computation of P starting from any state in its state space eventually reaches I. The requirements of the *convergence* of P_{cltz} to $\mathcal{I}_u$ include: (1) *deadlock-freedom* outside $\mathcal{I}_u$: there is no value in $\mathcal{N} - \mathcal{I}_u$ where both actions of P_{cltz} are disabled; (2) *livelock-freedom* outside $\mathcal{I}_u$: there is no cycle $v_0, v_1, \cdots, v_k, v_0$ for $k \geq 0$ of values in $\mathcal{N} - \mathcal{I}_u$ such that $v_{i \oplus 1}$ can be obtained from v_i by actions of P_{cltz}, where $0 \leq i \leq k$ and $\oplus$ denotes addition modulo $k + 1$, and (3) *divergence-freedom*: there is no value from where the computations of P_{cltz} diverge to infinity. A program P is *self-stabilizing* to an invariant I *iff* (1) I is closed in P, and (2) P converges to I from any state in its state space.

Problem 1. Does program P_{cltz} self-stabilize to $\mathcal{I}_u$ from any value in $\mathcal{N}$?

3   Convergence Stairs of the Collatz Program

This section investigates the problem of specifying convergence stairs towards solving Problem 1. Notice that, the Collatz program is deadlock-free outside $\mathcal{I}_u$ because any value is either even or odd, which would respectively enable either action P_1 or P_2. To ensure convergence, one has to show livelock-freedom and divergence-freedom of P_{cltz} in $\mathcal{N} - \mathcal{I}_u$. In this section, we study convergence to $\mathcal{I}_{cltz}$ (Subsect. 3.1) and $\mathcal{I}_u$ (Subsect. 3.2) through the lenses of convergence stairs [15, 16]. The j-th stair includes the set of states from where an action of P_{cltz} can take its state to some state in the $(j-1)$-th stair, where $j > 0$. The stair zero includes invariant states. We note that our notion of a stair differs from that of [15, 16] in that stairs are disjoint sets of states; i.e., *state predicates*, in our work.

3.1   Convergence Stairs With Respect to $\mathcal{I}_{cltz}$

We consider the orientation of states with respect to the number of steps it takes for P_{cltz} to reach a state in $\mathcal{I}_{cltz}$; called the *stair* of a state. To this end, we perform a backward reachability analysis using the inverse of $f_c(x)$, specified as the relation $R(x)$.

$$
R(x) = \begin{cases} 2x \\ (x-1)/3 & \text{if } (x-1)/3 \text{ is an odd integer; otherwise, undefined.} \end{cases} \tag{1}
$$

Expansion of $R(x)$ from 1 results in an infinite binary tree (in Fig. 1), called the *computation tree* of P_{cltz}, whose root includes the cycle $4 \to 2 \to 1 \to 4$. We consider each level of this tree as a stair that converges to the next lower level. The first stair contains $\{8\}$, the second stair is $\{16\}$, the third stair is $\{5, 32\}$, the fourth stair has $\{10, 64\}$, and so on. The k-th *stair*, where $k \geq 1$, includes all states from where there is exactly k steps to $\mathcal{I}_{cltz}$. In this orientation of stairs, the largest value in the k-th stair is 2^{k+2}. The k-th stair can also be represented through the application of the relation $R(x)$ k times, denoted $R^k(x)$. For example, we have $R^0(8) = \{8\}, R^1(8) = \{16\}, R^2(8) = \{5, 32\}, R^3(8) = \{10, 64\}$, and $R^4(8) = \{3, 20, 21, 128\}$. Notice that, convergence from $R^k(8)$ to $R^{k-1}(8)$ through the actions of P_{cltz} is guaranteed, for any $k > 1$. By a misuse of notation, we apply $R(x)$ to a set of states/values too. That is, $R(S) = \{x' \mid \exists x_0 : x_0 \in S : x' = R(x_0)\}$.

Lemma 1. *For any value y in the $(k+1)$-th stair, there is some value x in the k-th stair such that $y \in R(x)$. (Proof straightforward; hence omitted.)*

Lemma 2. *The set of values in the k-th stair is complete, for $k \geq 1$. That is, there is no positive integer x_0 from where P_{cltz} can reach $\mathcal{I}_{cltz}$ by exactly k transitions such that x_0 is missing in the set of values in the k-th stair. (Proof by induction on k; details in [11].)*

Problem 2. The set of backward reachable states from $\mathcal{I}_{cltz}$ is complete. Formally, $\{1, 2, 4\} \cup (\cup_{k=0}^{\infty} R^k(8)) = \mathcal{N}$.

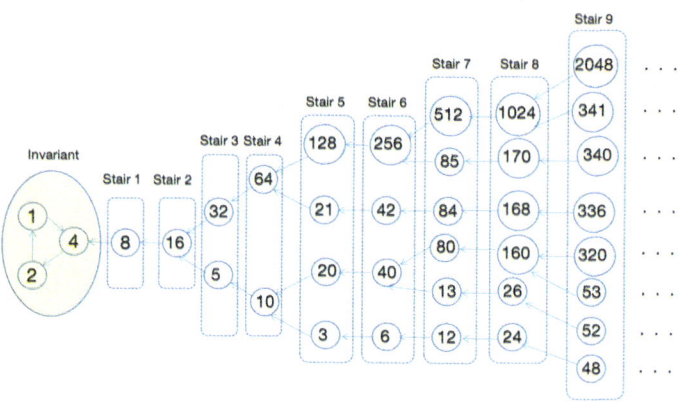

Fig. 1. Collatz computation tree and convergence stairs with respect to I_{cltz}

Solving Problem 2 amounts to solving the Collatz conjecture. To this end, one must prove that all positive integers can be generated by this backward reachability method; i.e., any value in $\neg I_{cltz}$ belongs to some stair.

Problem 3. Design a function *stair*: $\mathcal{N} \rightarrow 2^{\mathcal{N}}$ that takes the index of a stair, and returns the set of states in that stair.

This is an interesting problem, where $stair(k)$ is the specification of the k-th stair. Looking at Fig. 1, we observe that the states in $stair(k)$ have an upper bound of 2^{k+2}, but it is unclear how one can specify $stair(k)$ so it accurately identifies all states of the k-th stair without performing a backward reachability analysis from I_{cltz} using $R(x)$. Another interesting problem is as follows:

Problem 4. Given a positive integer $n \in \neg I_{cltz}$, in which stair would n be located? That is, design a function $stairIndex(n) : \mathcal{N} \rightarrow \mathcal{N}$ that takes an integer n and returns another integer $stairIndex(n)$ which determines the number of steps required to reach I_{cltz} by the actions of P_{cltz} from n.

3.2   Convergence Stairs With Respect to $\mathcal{I}_u$

Figure 2 illustrates a different way of thinking about the infinite computation tree of the Collatz program, where the invariant is the unbounded set $\mathcal{I}_u$ instead of the finite invariant I_{cltz}. The first stair in Figure 2 includes all the green states that can be specified formally as $(2^{2k} - 1)/3$, for $k > 1$ (shown by Lemma 3). Lemma 4 shows that the green children in Fig. 2 only exists for even powers of 2; i.e., $2^4, 2^6, 2^8, \cdots$.

Lemma 3. $(2^{2k} - 1)$ *is divisible by 3 and* $(2^{2k} - 1)/3$ *is an odd value, for* $k > 0$. *(Proof by induction on k; details in [11].)*

Lemma 4. $2^m - 1$ *is not divisible by 3 for odd values of* $m > 1$. *(Proof by induction on m; details in [11].)*

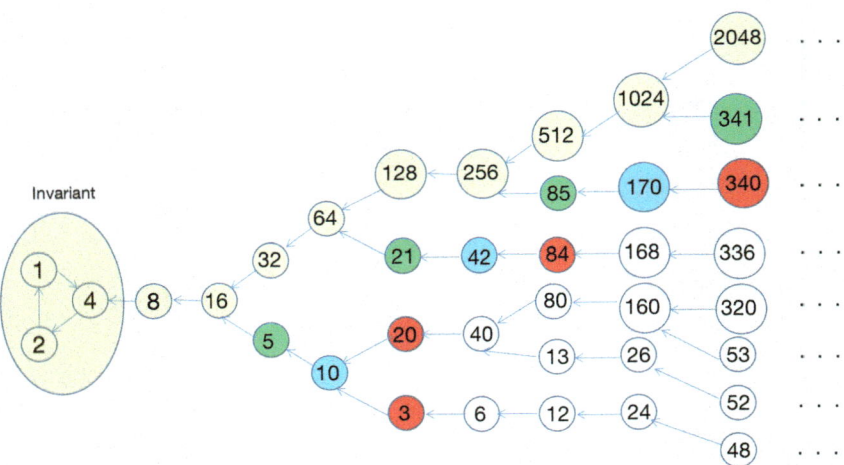

Fig. 2. Collatz computation tree and convergence stairs with respect to $\mathcal{I}_u = \{1, 2, 4\} \cup \{2^k \mid k \in \mathcal{N} \wedge (k > 2)\}$. Green states represent Stair 1, Blue states capture Stair 2, and Red states illustrate Stair 3 with respect to $\mathcal{I}_u$

To simplify the specification problem, let $Y_k = (2^{2k} - 1)$. Then, $Y_k/3$ is a nice specification of the first stair (the green nodes in Fig. 2) that gives us every single state in it, for any integer $k > 1$. For instance, for $k = 2$ we have $Y_2/3 = (2^4 - 1)/3 = 5$, and $Y_3/3 = (2^6 - 1)/3 = 21$, and so on. This is an achievement with respect to the first orientation of stairs (in Sect. 3.1) because it solves Problem 3 for the first stair. Since the elements of the first stair are odd values (by Lemma 3), applying the first rule of $R(x)$ would give us $2Y_k/3$ as a subset of elements in the second stair. The second rule of $R(x)$ would subtract 1 from $Y_k/3$ and divide it by three.

Lemma 5. $(Y_k/3 - 1)/3$ *is not a valid Collatz number.*

Proof.. Lemma 3 shows that Y_k is divisible by three and $Y_k/3$ is an odd value. Thus, $Y_k/3 - 1$ is even. If $Y_k/3 - 1$ is divisible by three, then dividing $Y_k/3 - 1$ by three must give us an even value; otherwise, $Y_k/3 - 1$ would have been odd. Thus, $(Y_k/3 - 1)/3 = (Y_k - 3)/3^2$ is an even value. This means that the Collatz function f_c would do a divide-by-two operation on $(Y_k - 3)/3^2$ instead of applying the $3x + 1$ rule; i.e., $f_c((Y_k - 3)/3^2) \neq Y_k/3$. Therefore, starting from $Y_k/3$, the relation $R(x)$ would give us only one child in the computation tree, and that is equal to $2Y_k/3$; i.e., all green nodes have only a single blue child.   □

Based on Lemma 5, the only members of the second stairs include $2Y_k/3$ for $k > 1$ (see the Blue nodes in Fig. 2). Thus, we just solved Problem 3 for the second stair too. The specification of the third stair (Red states in Fig. 2) can be obtained by applying $R(x)$ on the blue states. The first subset of the third stair includes states $2^2Y_k/3$ due to applying the first rule of $R(x)$, and the second subset includes $(2Y_k - 3)/3^2$. Continuing this way, in the fourth stair, we have the following values: $2^3 Y_k/3, (2^2 Y_k - 3)/3^2, (2^2 Y_k - 2 \times 3)/3^2, (2Y_k - 3 - 3^2)/3^3$. Figure 3 illustrates the structure of the complete subtree rooted at $Y_k/3$ (for $k > 1$) up to its fifth stair. Applying $R(x)$ further would give us an infinite binary tree. Such a binary tree is an over-approximation because some of its nodes may not be valid Collatz numbers for the same reason stated in the proof of Lemma 5. The main question is: *Given a specific $j > 1$, how do we compute the Collatz numbers at the j-th stair/level in the infinite tree?*

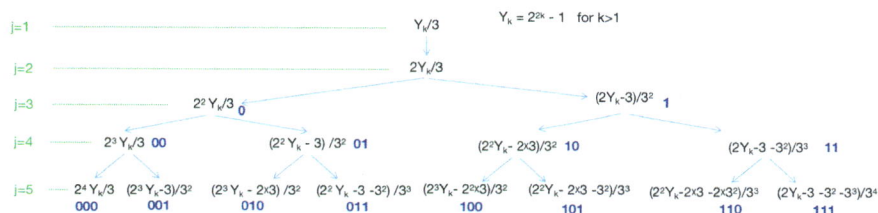

Fig. 3. Structure of a subtree rooted at $Y_k/3$, labelled with binary verification codes

Notice that, for each specific value of $k > 1$, we get a green value $Y_k/3$ in Figure 2, which is the root of a subtree by itself. Every stair $j > 1$ of a subtree rooted at $Y_k/3$ contains a single term $2^{j-1} Y_k/3$ corresponding to the leftmost branch of the tree rooted at $Y_k/3$. For example, for $j = 4$ in Fig. 3, we have a term $2^3 Y_k/3$. The remaining terms in the j-th stair are in the form of $(2^{f(j)} Y_k - (\sum_{k=1}^{q(j)-1} 2^{g(j)} \times 3^{h(j)}))/3^{q(j)}$, where $f(j) + q(j) = j$, $1 \le h(j) \le q(j) - 1$ and $0 \le g(j) \le q(j)$ depending on the value of $q(j)$. For instance, if $j = 1$, we have the value $Y_k/3$, where $f(j) = 0$ and $q(j) = 1$.

Lemma 6. *No two subtrees intersect. (Proof in [11].)*

Now, the question is *how do we identify and specify the functions $f(j), g(j), h(j), q(j)$ for values in each stair j and some $k > 1$?*

Lemma 7. *For every term in the j-th stair, $f(j) + q(j) = j$ holds. (Proof by induction on j; details in [11].)*

At each stair j, $f(j)$ start with $j - 1$ and decrements down to 1. In turn, $q(j)$ starts at 1 and increments to $j - 1$. That is, $1 \le q(j) \le j - 1$. We now analyze the specification of the members of the j-th stair for different values of $q(j)$. This analysis will be the basis for designing an algorithm that takes j and k, and generates the members of the j-th stair in the subtree rooted at $Y_k/3$. Simultaneously, we present a scheme for verifying whether each individual term in the j-th stair is actually a valid Collatz number. The reason behind this is that the subtree rooted at $Y_k/3$ is an over-approximation, and not all terms have acceptable values based on the Collatz functions. For instance, in the absence of such verification, the subtree rooted at 5 in Fig. 2 would include 4 as a child of 13 too, which is incorrect because $f_c(4) = 2$. To enable such verification, we attach a binary string, called the *Binary Verification Code* (BVC), to each node of the tree, where a '0' indicates multiplication by two, and '1' represents subtraction of one and division by three. For example, the term $(2^2 Y_k - 2 \times 3 - 3^2)/3^3$ is obtained by the following three operations performed on $2Y_k/3$: (1) subtract one from $2Y_k/3$ and divide it by three, which results in $(2Y_k - 3)/3^2$; (2) multiply by two, which gives $(2^2 Y_k - 2 \times 3)/3^2$, and (3) finally, subtract one from $(2^2 Y_k - 2 \times 3)/3^2$ and divide it by three, hence $(2^2 Y_k - 2 \times 3 - 3^2)/3^3$. Another way to think about this is to attach 0 or 1 to the right side of the BVC of some node v if you respectively go to the left or right child of v. For instance, to reach $(2^2 Y_k - 2 \times 3 - 3^2)/3^3$ from $2Y_k/3$, we go right-left-right; hence the binary code 101 for $(2^2 Y_k - 2 \times 3 - 3^2)/3^3$. Now, imagine we start with $(2^2 Y_k - 2 \times 3 - 3^2)/3^3$ and would like to verify whether it is an acceptable Collatz number. We scan its BVC from right to left. If we observe a '1', we multiply the current term by three and add one unit to derive its parent. Otherwise, we have a '0' and we divide the current term by two, and generate its parent. Subsequently, consider the derived parent as the current term and repeat scanning until all bits are processed. In each step, we can verify whether the child and the parent nodes meet the constraints of f_c. We note that, the notion of BVC is similar to the concept of parity vectors/sequences in [13] (also defined in [23]). In the following analysis, we discuss the BVCs of the generated terms in *italic*.

(1) For $q(j) = 1$, we have only one term $2^{j-1} Y_k/3$ in the j-th stair. Since $q(j) = 1$, we have $f(j) = j - 1$ due to $f(j) + q(j) = j$. (That is why 2 is raised to $j - 1$.) *The corresponding BVC is the string $\langle 00 \cdots 0 \rangle$ of length $j - 2$.* (See Lines 3–13 in Algorithm 1)

(2) For terms with $q(j) = j - 1$ where $j > 1$, we have one term $(2Y_k - (\Sigma_{i=1}^{j-2} 3^i))/3^{j-1}$ in the j-th stair. *We have one single BVC $\langle 11 \cdots 1 \rangle$ of length $j - 2$ bits.* (See Lines 14–24 in Algorithm 1)

(3) For $q(j) = 2$ where $j > 2$, we have $j - 2$ terms in the j-th stair as follows: $(2^{j-2} Y_k - 2^i \times 3)/3^2$ for $0 \le i < j - 2$. *For each one of these terms, we have the following BVCs: $\langle 0 \cdots 1 \rangle, \langle 0 \cdots 10 \rangle, \langle 0 \cdots 100 \rangle, \cdots, \langle 10 \cdots 0 \rangle$ each of length $j - 2$ bits.* (See Lines 25–42 in Algorithm 1)

(4) For terms with $q(j) = j - 2$ where $j > 2$, we have $j - 2$ terms in the j-th stair, each of the form $(2^2 Y_k - (\Sigma_{i=1}^{m} 2 \times 3^i + \Sigma_{i=m+1}^{j-3} 3^i))/3^{j-2}$ where $0 \le m \le j - 3$. Note that, in this case, $j - 3 = q(j) - 1$. *The corresponding $j - 2$ BVCs include $\langle 01 \cdots 1 \rangle, \langle 101 \cdots 1 \rangle, \langle 1101 \cdots 1 \rangle, \cdots, \langle 11 \cdots 01 \rangle, \langle 11 \cdots 10 \rangle$ of length $j - 2$ bits each. In all these strings, there is a single 0 that moves from msb to lsb.* (See Lines 43–68 in Algorithm 1.)

(5) For $2 < q(j) < j - 2$ where $j > 5$, the general form of the expressions includes $(2^{j-q(j)} Y_k - (\Sigma_{k=1}^{q(j)-1} 2^i 3^k))/3^{q(j)}$, where the sequence of exponents in powers of two is a non-increasing sequence out of the space of $(j - q(j) - 1)^{q(j)-1}$ possible sequences. The maximum value of any exponent in powers of two is $(j - q(j) - 1)$ and there are $q(j) - 1$ terms in $\Sigma_{k=1}^{q(j)-1} 2^i 3^k$. We use Algorithm 3 (invoked in Line 70 of Algorithm 1) to compute such sequences recursively. The recursive nature of Algorithm 3 is due to the fact that the number of terms in

---

Algorithm 1 An algorithm for specifying the members of the j-th stair for a given value $k > 1$, along with constructing their binary verification code.

Require: k and j are positive integers where $k > 1$ and $j > 0$

1: $Y_k \leftarrow 2^{2k} - 1$
2: **for** $1 \leq q_j \leq j - 1$ **do**   ▷ q_j *in this algorithm denotes* $q(j)$.
3:      **if** $q_j = 1$ **then**
4:          $x \leftarrow 2^{j-1} Y_k / 3$
5:          bvc $\leftarrow$ null   ▷ *Construct the binary verification code (bvc) as a string.*
6:          **for** $1 \leq t \leq j - 2$ **do** bvc $\leftarrow$ concat(bvc, '0')   ▷ *Attach 0 to bvc from right.*
7:          **end for**   ▷ *End of bvc construction.*
8:          **if** verify(x,bvc) = True **then**
9:              print(x,bvc);
10:         **else**
11:             print("Invalid Collatz number: Verification failed!");
12:         **end if**
13:     **end if**
14:     **if** $q_j = j - 1$ **then**
15:         $x \leftarrow 2Y_k - (\Sigma_{i=1}^{j-2} 3^i)/3^{j-1}$
16:         bvc $\leftarrow$ null   ▷ *Construct the bvc.*
17:         **for** $1 \leq t \leq j - 2$ **do** bvc $\leftarrow$ concat(bvc, '1')   ▷ *Attach 1 to bvc from right.*
18:         **end for**   ▷ *End of bvc construction.*
19:         **if** verify(x,bvc) = True **then**
20:             print(x,bvc);
21:         **else**
22:             print("Invalid Collatz number: Verification failed!");
23:         **end if**
24:     **end if**
25:     **if** $q_j = 2$ **then**
26:         **for** $0 \leq i < j - 2$ **do**
27:             $x \leftarrow (2^{j-2} Y_k - 3 \times 2^i)/3^2$
28:             bvc $\leftarrow$ null   ▷ *Construct the bvc.*
29:             **for** $0 \leq t < j - 2$ **do**
30:                 **if** $(t = i)$ **then**
31:                     bvc $\leftarrow$ concat('1',bvc)   ▷ *Attach 1 to bvc from left.*
32:                 **else**
33:                     bvc $\leftarrow$ concat('0',bvc)   ▷ *Attach 0 to bvc from left.*
34:                 **end if**
35:             **end for**   ▷ *End of bvc construction.*
36:             **if** verify(x,bvc) = True **then**
37:                 print(x,bvc);
38:             **else**
39:                 print("Invalid Collatz number: Verification failed!");
40:             **end if**
41:         **end for**
42:     **end if**

```
43:        if q_j = j - 2 then
44:            for 0 ≤ m < j - 2 do
45:                x ← (2²Y_k − (Σ_{i=1}^{m} 2 × 3^i + Σ_{i=m+1}^{j-3} 3^i))/3^{j-2}
46:                if (m = 0) then
47:                    bvc ← null                                    ▷ Construct the bvc.
48:                    for 1 ≤ t < j − 2 do
49:                        bvc ← concat(bvc, '1')
50:                    end for
51:                    print bvc;                                    ▷ End of bvc construction.
52:                else
53:                    bvc ← null                                    ▷ Construct the bvc.
54:                    for m + 1 ≤ t < j − 2 do
55:                        bvc ← concat('1',bvc)
56:                    end for
57:                    bvc ← concat('0',bvc)
58:                    for 1 ≤ t < m + 1 do
59:                        bvc ← concat('1',bvc)
60:                    end for                                       ▷ End of bvc construction.
61:                    if verify(x,bvc) = True then
62:                        print(x,bvc);
63:                    else
64:                        print("Invalid Collatz number: Verification failed!");
65:                    end if
66:                end if
67:            end for
68:        end if
69:        if 2 < q_j < j − 2 then
70:            recursiveFor(q_j − 1, j − q_j − 1, q_j − 1, j, q_j, Y_k)   ▷ Recursively compute the
       remaining numbers.
71:        end if
72: end for
```

$\Sigma_{k=1}^{q(j)-1} 2^i 3^k$ depends on $q(j)$ and their formation depends upon the exponents of powers of two. These two parameters both change from one stair to another. Thus, we need an algorithmic structure that dynamically changes; hence the recursion. Algorithm 3 has six parameters. The first one captures the number of terms in $\Sigma_{k=1}^{q(j)-1} 2^i 3^k$; i.e., $q(j)-1$, the second parameter holds the largest possible exponent of two, the third one is a copy of the first one, and the remaining three parameters j, q_j, Y_k are passed for the calculation of the expression $(2^{j-q(j)}Y_k - (\Sigma_{k=1}^{q(j)-1} 2^i 3^k))/3^{q(j)}$.

The core of Algorithm 3 includes Lines 2 to 5 where a nested for-loop is dynamically formed with the depth of n. During each recursion, we insert a value $0 \le x \le l$ into a vector array, denoted *list*, which holds a permutation (with repetition) of $n = q(j) - 1$ values in the domain $[0, ..., l]$. Algorithm 3 returns only those permutations that are non-increasing (stored in *list[]*). In Line 18, Algorithm 3 uses the contents of *list[]* to compute $Num = (2^{j-q(j)}Y_k -$

$(\Sigma_{r=1,t=list[r-1]}^{q(j)-1} 2^t \times 3^r))/3^{q(j)}$. The remaining lines then use Algorithm verify(x,s) to verify the acceptability of Num as a Collatz number. The algorithm verify(x,s) (pseudocode omitted due to space constraint [11]) verifies whether a value x is a legitimate Collatz number with the help of the binary string s as the BVC of x. Initially, verify(x,s) performs some sanity checks to ensure that x is an integer greater than 1 and not a power of 2. Then, it simply scans s from its least significant bit (lsb) and checks the scenarios where the rules of $f_c(x)$ are applied incorrectly. When the current bit of s is 1, the value of x cannot be even because the $3x + 1$ rule has been applied to derive its parent y in the tree. Moreover, if the current bit of s is 1, then x and y cannot both be odd. When the current bit of s is 0, x cannot be odd because the $x/2$ rule is applied (under f_c) to derive y. Then, verify(x,s) considers y as the current node in the tree, and repeats the aforementioned process.

Theorem 1. (Soundness and completeness) *Algorithm 1 is sound and complete. That is, Algorithm 1 correctly explores the members of each stair j, for $j > 0$, and does not miss any value. (Proof in [11].)*

Theorem 2. (Complexity) *The asymptotic time complexity of Algorithm 1 for the j-th stair is $O(j^{j+1})$. (Proof in [11].)*

Theorem 3. *Every state in the j-th stair with respect to $\mathcal{I}_u$ will reach a state in the $(j-1)$-th stair through the execution of the Collatz program. (Proof by induction on j; details in [11].)*

Theorem 4. *Starting from any state/value in any subtree rooted at $Y_k/3$ for some arbitrary $k > 1$, the Collatz program will eventually reach a state in $\mathcal{I}_u$, and will subsequently reach I_{cltz}. (Proof follows from Theorem 3.)*

Remark. Theorem 4 proves livelock-freedom and divergence-freedom starting from any state in any subtree rooted at $Y_k/3$ for $k > 1$. However, to prove the conjecture, we have to show that the union of $\mathcal{I}_u$ and all the values in all stairs of all subtrees is equal to the set of positive integers, which remains an open problem. One way to tackle this problem is to look for any natural value outside $\mathcal{I}_u$ that fails to be placed in any subtree; i.e., solve Problem 4.

Optimization. Since the time complexity of Algorithm 1 is exponential (based on Theorem 2), we discuss some potential optimizations one can make to enhance the efficiency of the algorithm. First, we observe that the computation of the values of the j-th stair has great potential for parallelization, where we instantiate several instances of Algorithm 1 for distinct values of j and k in an embarrassingly parallel way. Second, each case in the body of the main loop for different values of q_j can also be computed in parallel as they do not depend on each other. Third, Algorithm 3 can be unrolled and implemented iteratively for fixed values of j and k. Our recursive design of Algorithm 3 is mainly for presenting the dynamic nature of nested for-loop inside Algorithm 3. This is not the only way that the exponents of two in Algorithm 3 can be computed.

Applications. We would also like to emphasize the application of Algorithm 1 in the reachability analysis of infinite-state systems where program variables

---

Algorithm 2 Algorithm recursiveFor implements a nested for-loop with variable depth n for computing Collatz numbers of the j-th stair where $2 < q_j < j - 2$.

Require: n represents the length of the summation $\Sigma_{r,s} 2^r 3^s$ in $Num = (2^{j-q_j} Y_k - (\Sigma_{r,s} 2^r 3^s))/3^{q_j}$, which is the same as the depth of the nested for-loop. (q_j in this algorithm denotes $q(j)$.)

Require: l denotes the maximum exponent for the powers of 2 in $\Sigma_{r,s} 2^r 3^s$ used in Num.

Require: n_c is a copy of n. The values j, q_j, Y_k are passed to this function for the computation of the members of the j-th stair.

```
 1: if (n > 1) then
 2:     for x := l down to 0 do
 3:         list.push(x)                              ▷ Add x to the end of the list.
 4:         recursiveFor(n − 1, l, nc, j, qj, Yk)
 5:         list.pop(nc − n)                  ▷ Return the element at position nc − n.
 6:     end for
 7: else
 8:     for x := l down to 0 do
 9:         list.push(x)
10:         accept = True
11:         for i := 0 to nc − 1 do   ▷ Powers of 2 in Num must be non-increasing in
terms of their exponents.
12:             if (list[i] < list[i+1]) then
13:                 accept = False
14:                 break
15:             end if
16:         end for
17:         if (accept == True) then
18:             Num = (2^{j−q_j} Y_k − (Σ_{r=1,t=list[r−1]}^{q_j−1} 2^t × 3^r))/3^{q_j}   ▷ Use the contents of
list[] as powers of two in Num.
19:             Compute the corresponding bvc.
20:             if (verify(Num,bvc) = True) then
21:                 print(x,bvc);
22:             else
23:                 print("Invalid Collatz number: Verification failed!");
24:             end if
25:         end if
26:     end for
27: end if
```

---

have a domain equal to natural numbers. Moreover, our approach can be used for tackling other similar conjectures as well as for analyzing the convergence of complex non-linear systems to stability. Another important application is in blockchain technology [6] where Collatz orbits (and respectively stairs) provide a pseudo-randomness used for generating proof-of-work.

4   Related Work

There is a rich body of work on the Collatz conjecture, which can broadly be classified into theoretical, computational and representation in other domains (e.g., term rewriting or graph theory). On the theoretical front, some mathematicians prove [21, 29] weaker statements than the conjecture itself. For instance, Tao [29] shows that "Almost all Collatz orbits attain almost bounded values." Leventides and Poulios [23] formulate the conjecture through bounded linear operators, and study the properties of these operators and their relation with the Collatz orbits.

Computational methods investigate the limits of natural values that actually convergence to I_{cltz} through running the Collatz program from initial values to the extent their computational resources permit. For example, Lagarias [22] computationally verifies the conjecture for values up to 5.78×10^{18}. Barina [5] then improves this result by verifying the convergence of values up to 1.5×2^{70} using both a single-threaded and a parallel implementation. Barghout et al. [4] analyze the Collatz conjecture probabilistically and reason that chances of not converging is low. Another class of computational methods focuses on searching for a livelock outside I_{cltz}. For instance, Eliahou [12] computationally verifies that there are no livelocks with a length up to 1.7×10^7. Due to the limited computational resources, these methods can verify only a finite scope of integers.

Many existing methods reduce the Collatz conjecture to problems in other domains. For example, Stérin [27] improves algorithmic methods for the representation of ancestors of any value x in the Collatz tree as a regular expression $reg_k(x)$, which captures the set of binary representations of any ancestor y of x from where x can be reached in k application of the $3x + 1$ rule. Briscese and Calogero study conjectures similar to Collatz's [7]. Hernandez [19] uses modular arithmetic to show that each orbit can be captured by a word in a regular language accepted by a DFA. However, their proof of convergence is not rigorous. Yolcu et al. [30] develop a term/string rewriting system that terminates *if and only if* the conjecture is true. This method provides an alternative way of reasoning about the conjecture. Orús-Lacort and Jouis [25] present a manual proof by induction, whereas there are methods [14] that use theorem provers to mechanically verify the conjecture. Rahn et al. [26] present an approach for proving convergence of Collatz function by linearizing odd numbers in a bottom up fashion. By contrast, the proposed notion of convergence stairs and Algorithm 1 provide an algorithmic method for exact generation of Collatz values in each stair without actually generating the numbers in lower-level stairs.

While program verification methods [1,2,8,17,20] inspire us, they lack sufficient machinery to tackle the Collatz conjecture. Specifically, verifying the self-stabilization of Collatz program requires convergence from every single concrete state in an infinite state space, whereas existing verification methods prove correctness from a set of initial states. Abstraction techniques will be of little help because convergence must be guaranteed from every concrete state/value. Finally, existing techniques for the verification and synthesis of self-stabilizing programs with unbounded variables [10] would not apply either because it is unclear how we can capture the transitions of the Collatz program as semilinear sets, representing sets of periodic integer vectors.

5 Conclusions and Future Work

For the first time, this paper presented an algorithmic method for the specification and generation of all natural values from where Collatz conjecture holds through exactly $j > 0$ steps of applying the Collatz function, called the j-th convergence stair (Problem 3). The proposed specification is an over-approximation of the Collatz numbers in the j-th stair, which is then fine tuned through a verification step embedded in the generation algorithm. We showed that the reachability of the set of powers of two is guaranteed from any value in any stair $j > 0$. The proposed approach can shed light on the self-stabilization of infinite-state systems as well as providing insight on similar conjectures [7]. While one can generate all valid Collatz numbers belonging to any j-th stair (where $j > 0$) using the algorithmic method of this paper and its implementation (available at https://github.com/aebne/CollatzStairs), another equally important problem remains open where the stair of a given natural number should be determined (Problem 4). Solving this problem will help us prove the Collatz conjecture by showing that the union of stairs is equal to the set of natural numbers. Moreover, the mechanical verification of Algorithm 1 will increase confidence in its correctness, and will provide reusable proof strategies for similar problems.

References

1. Abdulla, P., Haziza, F., Holík, L.: Parameterized verification through view abstraction. Int. J. Softw. Tools Technol. Transfer **18**(5), 495–516 (2016)
2. Ball, T., Majumdar, R., Millstein, T., Rajamani, S.K.: Automatic predicate abstraction of c programs. In: Proceedings of the ACM SIGPLAN 2001 conference on Programming language design and implementation, pp. 203–213 (2001)
3. Ballesteros, D.M., Peña, J., Renza, D.: A novel image encryption scheme based on Collatz conjecture. Entropy **20**(12), 901 (2018)
4. Barghout, K., Hajji, W., Abu-Libdeh, N., Al-Jamal, M.: Statistical analysis of descending open cycles of Collatz function. Mathematics **11**(3), 675 (2023)
5. Barina, D.: Convergence verification of the Collatz problem. J. Supercomput. **77**(3), 2681–2688 (2021)
6. Bocart, F.: Inflation propensity of Collatz orbits: a new proof-of-work for blockchain applications. J. Risk Fin. Manage. **11**(4), 83 (2018)
7. Briscese, F., Calogero, F.: Conjectures analogous to the Collatz conjecture. Open Commun. Nonlin. Math. Phys. **4**(1) (2024)
8. Bultan, T., Gerber, R., Pugh, W.: Model-checking concurrent systems with unbounded integer variables: symbolic representations, approximations, and experimental results. ACM Trans. Programm. Lang. Syst. (TOPLAS) **21**(4), 747–789 (1999)
9. Dijkstra, E.W.: Self-stabilizing systems in spite of distributed control. Commun. ACM **17**(11), 643–644 (1974)

10. Ebnenasir, A.: Synthesizing self-stabilizing parameterized protocols with unbounded variables. In: Conference on Formal Methods in Computer-Aided Design (FMCAD), pp. 245–254 (2022)

11. Ebnenasir, A.: Specifying and verifying the convergence stairs of the Collatz program, Mar 2024. arXiv:2403.04777

12. Eliahou, S.: The 3x + 1 problem: new lower bounds on nontrivial cycle lengths. Discret. Math. **118**(1–3), 45–56 (1993)

13. Everett, C.J.: Iteration of the number-theoretic function f(2n) = n, f(2n + 1) = 3n + 2. Adv. Math. **25**, 42–45 (1977)

14. Furuta, M.: Proof of Collatz conjecture using division sequence. Adv. Pure Math. **12**(2), 96–108 (2022)

15. Gouda, M.: Multiphase stabilization. IEEE Trans. Software Eng. **28**(2), 201–208 (2002)

16. Gouda, M.G., Multari, N.J.: Stabilizing communication protocols. IEEE Trans. Comput. **40**(4), 448–458 (1991)

17. Graf, S., Saidi, H.: Construction of abstract state graphs with PVS. CAV **97**, 72–83 (1997)

18. Grauer, J.A.: Analogy between the Collatz conjecture and sliding mode control. NASA/TM-20210019810 (2021)

19. Hernandez, J.: The Collatz Regular Language. ScienceOpen Preprints (2023)

20. Kaiser, A., Kroening, D., Wahl, T.: Dynamic cutoff detection in parameterized concurrent programs. In: 22nd International Conference on Computer Aided Verification, pp. 645–659. Springer (2010)

21. Krasikov, I., Lagarias, J.C.: Bounds for the $3x+1$ problem using difference inequalities. Acta Arith. **109**, 237–258 (2003)

22. Lagarias, J.C.: The ultimate challenge: The 3x + 1 problem. Am. Math. Soc. **10**, 12 (2010)

23. Leventides, J., Poulios, C.: Koopman operators and the 3x + 1-dynamical system. SIAM J. Appl. Dyn. Syst. **20**(4), 1773–1813 (2021)

24. Ma, H., Jia, C., Li, S., Zheng, W., Wu, D.: Xmark: dynamic software watermarking using Collatz conjecture. IEEE Trans. Inf. Forensics Secur. **14**(11), 2859–2874 (2019)

25. Orús-Lacort, M., Jouis, C.: Analyzing the Collatz conjecture using the mathematical complete induction method. Mathematics **10**(12), 1972 (2022)

26. Rahn, A., Sultanow, E., Henkel, M., Ghosh, S., Aberkane, I.J.: An algorithm for linearizing the Collatz convergence. Mathematics **9**(16), 1898 (2021)

27. Stérin, T.: Binary expression of ancestors in the Collatz graph. In: International Conference on Reachability Problems, pp. 115–130. Springer (2020)

28. Stomp, F.: Structured design of self-stabilizing programs. In: Proceedings of the 2nd Israel Symposium on Theory and Computing Systems, pp. 167–176 (1993)

29. Tao, T.: Almost all orbits of the Collatz map attain almost bounded values. In: Forum of Mathematics, Pi, vol. 10. Cambridge University Press (2022)

30. Yolcu, E., Aaronson, S., Heule, M.J.: An automated approach to the Collatz conjecture. In: CADE, pp. 468–484 (2021)

Consensus Through Knot Discovery in Asynchronous Dynamic Networks

Rachel Bricker(✉), Mikhail Nesterenko, and Gokarna Sharma

Kent State University, Kent, OH 44242, USA
{rbricke2,gsharma2}@kent.edu, mikhail@cs.kent.edu

Abstract. We state the Problem of Knot Identification as a way to achieve consensus in dynamic networks. The network adversary is asynchronous and not oblivious. The network may be disconnected throughout the computation. We determine the necessary and sufficient conditions for the existence of a solution to the Knot Identification Problem: the knots must be observable by all processes and the first observed knot must be the same for all processes. We present an algorithm *KIA* that solves it. We conduct *KIA* performance evaluation.

1 Introduction

In a dynamic network, the topology changes arbitrarily from one state of the computation to the next. Thus, it is one of the most general models for mobile networks. Moreover, these intermittent changes in topology may represent message losses. Hence, dynamic networks are a good model for an environment with low connectivity or high fault rates.

In such a hostile setting, the fundamental question of consensus among network processes is of interest. One approach to consensus is to require that processes in the network remain connected and mutually reachable long enough for them to exchange information and come to an agreement. However, this may be too restrictive. This is especially problematic if the network is asynchronous and there is no bound on the communication delay between processes.

The question arises whether it is possible to achieve consensus under less stringent connectivity requirements. In the extreme case, the network is never connected at all. Then, the processes may not rely on mutual communication for agreement.

An interesting approach to consensus is for the processes to use the topological features of the dynamic network itself as a basis for agreement. For example, the process with the smallest identifier or the oldest edge. However, as processes collect information about the network topology, due to network delays, such features may not be stable. Indeed, some process may discover another process with a smaller identifier. Therefore, basing consensus decisions on such unstable information may not be possible.

A knot is a strongly connected component with no incoming edges. We consider a knot to include edges across some time interval. That is,

T. Masuzawa et al. (Eds.): SSS 2024, LNCS 14931, pp. 432–445, 2025.
https://doi.org/10.1007/978-3-031-74498-3_31

knot processes may never be connected in a single state. In general, the presence of a knot in a dynamic network is not invariant. For example, a dynamic network may have more than one knot, or a knot may grow throughout a network computation. In this paper, we study the conditions under which knots may be used for consensus in asynchronous dynamic networks.

Related work. The impossibility of consensus in asynchronous systems [12] in case of a single faulty process has precipitated extensive research on the subject of consensus. Santoro and Widmayer [19] show that consensus is impossible even in a synchronous system subject to link failures. The original paper [12] uses knot determination as a topological feature of a computation to achieve consensus.

There are a number of models related to dynamic networks where consensus is studied. Charron-Bost and Schiper [9] introduce a heard-of (HO) model and consider consensus solvability there. In this model, the set of "heard-of" for each process is analyzed. The consensus is proved to be solvable only if there is an agreement between processes on these heard-of sets.

There is a related research direction of consensus with unknown participants [8], where participant detectors perform a similar role to links in dynamic networks. Rather than place restrictions on the dynamic network to enable a solution, Altisen et al. [4] relax the problem and consider an eventually stabilizing version of it.

Kuhn et al. [13,14] study consensus under a related model of directed networks. In their case, it is assumed that there exists a spanning subgraph of the network within T rounds. In the case $T = 1$, the network is always connected.

Afek and Gafni [3] introduce the concept of an adversary as a collection of allowed network topologies. An oblivious adversary [10] composes an allowed computation by selecting the topology of each state from a fixed set of allowed network topologies. There are a number of papers that study consensus under this adversary [7,11,21].

In the case of a non-oblivious adversary, no restrictions on the potential state topologies are placed. In this case, the adversary completely controls the connectivity of the network in any state and the changes in connectivity from state to state. In the work known to us, to solve consensus under such a powerful adversary, extra connectivity assumptions are assumed. Biely et al. [5] consider consensus under an eventually stabilizing connected root component. Some papers study the case where the network stays connected long enough to achieve consensus [6,20].

In the present work, we assume a non-oblivious adversary and consider the case where the system may remain disconnected in any state of the computation.

Our contribution. We use the non-oblivious adversary defined and studied previously [5,6,20]. We focus on knot identification under such adversary. We define the Knot Identification Problem and study it for directed dynamic networks. This problem requires the network processes to agree on a single knot. The solution to this problem can immediately be used to solve consensus.

We assume an asynchronous adversary. The asynchrony of the adversary allows it to delay the communication between any pair of processes for arbitrarily long. If the adversary is asynchronous, each process may not hope to gain additional information by waiting and must make the output decision on the basis of what it has observed so far.

We consider a *knot observation final* adversary. In such adversary, it is possible that the collection of knots observed by some process may not increase throughout the rest of the computation. Thus, the processes may not ignore any of the observed knots hoping to get others later. Instead, each process must make the decision on the basis of the knots seen so far.

Since the processes must agree on a knot, a process may output the knot only if it is observed by other processes. Hence, the same knot needs to be observed by all processes. Once a process observes a knot, it must determine if this knot is observed by everyone else. We call an adversary *knot opaque* if it does not allow a process to determine whether the knot is observed by other processes or not. We prove that there is no solution to the Knot Identification Problem for such a knot opaque adversary. We then consider adversaries that are *knot transparent* rather than knot opaque.

For an adversary that is asynchronous, knot transparent, and knot observation final, we prove that it is necessary and sufficient for all processes to observe the same first knot.

For sufficiency, we present a simple knot identification algorithm *KIA* that solves the Knot Identification Problem. We conduct performance evaluation to study *KIA* behavior. This evaluation studies the dynamics of knot detection under various parameters. It demonstrates the practicality of *KIA* and our approach to consensus for dynamic networks with little connectivity.

2   Notation and Problem Definitions

We state the notation to be used throughout the paper in this section. To simplify the exposition, we add further definitions in later sections, closer to place of their usage.

Links, states, computations. The network consists of N *processes*. The processes have unique identifiers. No process a priori knows N or the identifiers of the other processes.

A pair of processes may be connected by a unidirectional *link*. The network *state* s is a collection of such links that, together with the processes, form a *state communication graph* or just *state graph*. Thus, processes are nodes in this graph. One specific state is a *non-communicating state* whose state graph has no links. That is, all processes are disconnected in this state.

A *computation* σ is an infinite sequence of network states. An *adversary* is a set of allowed computations. Given an adversary, an algorithm attempts to solve a particular problem. We use the term computation for both the states allowed by the adversary and the operation of the algorithm in these states.

To aid in the solution, processes exchange information across existing links. If two processes are connected by a link in a particular state, the sender may transmit an unlimited amount of information to the receiver. This communication is reliable. The sender does not learn the receiver's identifier.

The processes do not fail. Alternatively, a process failure may be considered as permanent disconnection of the failed process from the rest of the network.

Causality, asynchrony, computation graphs, and knots. A computation *event* is any computation action or topological occurrence that happens in a computation. Examples of computation events are a process carrying out its local calculations or an appearance of a link. Given a particular computation, a communication event e_1 *causally precedes* another event e_2 if (i) both events occur in the same process and e_1 occurs before e_2; (ii) there is a communication link between processes p_1 and p_2 and e_1 occurs at p_1 before the link and e_2 occurs at p_2 after the link; (iii) there is another event e_3 such that e_1 causally precedes e_3 and e_3 causally precedes e_2. We consider the presence of a link in a particular state to be a single event. If the same link is present in the subsequent state, it is considered a separate event. This way, causal precedence is defined for links. Note that the insertion of a non-communicating state into a computation preserves all causality relations of the computation.

Consider a computation σ_1 allowed by some adversary $\mathcal{A}$. Let σ_2 be obtained from σ_1 by inserting a non-communicating state after an arbitrary state of σ_1. If the adversary $\mathcal{A}$ also allows σ_2, then $\mathcal{A}$ is *asynchronous*. Intuitively, an asynchronous adversary may delay process communication for arbitrarily long.

Given a computation σ, a *computation graph* $G(\sigma, i)$ is the union of all the state graphs up to and including state s_i. To put another way, the computation graph is formed by the processes and the links present in any state s_j, for $j \leq i$.

A *knot* is a strongly connected subgraph with no incoming links. A process p_i is in a knot if for every process p_j reachable from p_i, p_i is reachable from p_j. Given a graph G, this definition suggests a simple knot computation algorithm. For each process in G, compute a reachability set S. For a process p_i with reachability set S_i, if there is a process $p_j \in S_i$ such that $p_i \in S_j$ then, p_i and p_j are in the same knot.

When it is clear from the context, we use the term *knot* for both the subgraph and for the set of processes that form this subgraph. Any process that has not communicated yet is trivially a singleton knot. Therefore, we only consider knots of size at least two.

Computation σ contains a knot K if there is a state s_i, $i < \infty$, such that $G(\sigma, i)$ contains K. Note that there is no requirement that the edges of the knot in a computation are causally related, just that the union of all state graphs up to some state s_i contains a knot. As the computation progresses, edges are added to the computation graph of this computation. In general, a knot in this graph is not stable. If an incoming edge is added, the knot may disappear. Similarly, added links may expand the knot by joining mutually reachable processes.

Observability. A *local observation graph* $LG(p, \sigma, i)$ is all the links and adjacent processes that causally precede the events in p in state s_i of computation σ. A local observation graph $LG(p, \sigma, i)$ is thus a subgraph of the computation graph $G(\sigma, i)$. In effect, the local observation graph of p is what p sees of the computation so far. In the beginning of the computation LG of process p is empty and LG grows as p receives topological information from incoming links.

Two computations σ_1 and σ_2 are *observation graph identical* for process p up to state s_i if $LG(p, \sigma_1, i) = LG(p, \sigma_2, i)$.

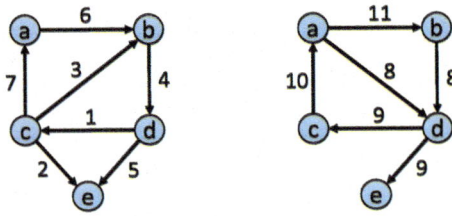

Fig. 1. Knot formation example. Edge labels denote states when the edges are present. Process e observes knot $K_1 = \{b, c, d\}$; process d is the first to observe knot $K_2 = \{a, b, c, d\}$.

Let us illustrate these concepts with an example shown in Fig. 1. In state 4, a knot $K_1 = \{b, c, d\}$ is formed due to the links $d \rightarrow c$, $c \rightarrow b$ and $b \rightarrow d$. In state 5, due to the link $d \rightarrow e$, process e observes K_1. In state 6, K_1 is destroyed because of an incoming link: $a \rightarrow b$. In state 7, link $c \rightarrow a$ creates knot $K_2 = \{a, b, c, d\}$. In state 8, process d observes K_2. In the remaining states, all processes observe K_2.

In general, a knot may exist in the computation graph but may not be visible to any of the processes that belong to this knot or even any of the processes in the network at all. Indeed, processes that belong to a knot may not see said knot because it does not belong to their local observation graphs. For example, in Fig. 1, if no more links appear after state 7 in the computation, none of the processes in knot K_2, or even in the entire network, observe K_2, yet it exists in the computation graph.

A knot K is *observable* in computation σ by process p if there is a state s_i such that $K \subset LG(p, \sigma, i)$. A knot is *globally observable* in a computation if it is observable by every process in the network. That is, a knot is globally observable if every process eventually sees it.

Consider the earliest state in the computation σ where p observes knot K. This state contains an incoming link (or links) to p that brings additional topological information to $LG(p, \sigma, i)$ to complete the knot K. This link is the *observation event* at process p for this knot. For example, in Fig. 1, link $d \rightarrow e$ is the observation event for K_1 at process e.

We consider algorithms that are deterministic in the following way. If two computations σ_1 and σ_2 are observation graph identical for process p up to state s_i, then all the outputs of p up to state s_i for algorithm $\mathcal{S}$ in the two computations are identical. Put another way, in such an algorithm, each process makes its decisions only on the basis of its local observation graph.

The Knot Identification and Consensus Problems.

Definition 1 (Consensus). Given that every process is input a binary value v, a consensus algorithm requires each process to output a decision value following the three properties.
    Consensus Validity: *if all processes are input the same value v, then output decision is v;*
    Consensus Agreement: *if one process outputs v, then every output decision is v;*
    Consensus Termination: *every process decides.*

Definition 2 (Knot Identification). A solution to the Knot Identification Problem requires that given a computation, each process outputs the set of processes K that form a knot in this computation. The output is subject to the following properties.
    KI Agreement: *if one process outputs a knot K, then every output knot is also K;*
    KI Termination: *every process outputs a knot.*

An adversary is *consensual* if there exists an algorithm that solves Consensus on every computation allowed by this adversary. Similarly, a *knot-identification* adversary admits an algorithm that solves this problem on every allowed computation.

Once the Knot Identification Problem is solved, consensus follows. Indeed, if all processes agree on a knot, they may use it to determine the consensus value to be output. For example, the consensus value may be the input to the knot process with the highest identifier, or the process incident to the oldest link, etc. We state this observation in the below proposition.

Proposition 1. *A knot-identification adversary is also a consensual adversary.*

In the remainder of the paper, we focus on the Knot Identification Problem.

3   Necessary and Sufficient Conditions for Knot Identification

Knot opacity. The KI Agreement property requires that every process outputs the same knot. A process may output only a knot that it observes. Hence, the following proposition.

Proposition 2. *In a solution to the Knot Identification Problem, every process outputs only a globally observable knot.*

However, even if an adversary has a globally observable knot in every computation, it does not guarantee that this adversary admits a solution to the Knot Identification Problem. A process observing a particular knot must also know whether or not this specific knot is globally observable. Let us discuss this in detail.

An adversary $\mathcal{A}$ is *knot opaque* if there is a process p and a computation $\sigma_1 \in \mathcal{A}$ such that for every state s_i of σ_1 and every knot K observed by p in states up to s_i, there is another computation σ_2 that is local observation graph identical to σ_1 for p up to s_i, yet K is not globally observable in σ_2. Intuitively, a knot opaque adversary does not allow a process p to distinguish whether or not any knot K that p observes is also observed by all other processes, i.e. this knot is globally observable. An adversary is *knot transparent* if it is not knot opaque.

Lemma 1. *There does not exist a solution to the Knot Identification Problem for a knot opaque adversary.*

Proof. Assume the opposite. Suppose there exists a knot opaque adversary $\mathcal{A}$. Also, let $\mathcal{S}$ be the algorithm that solves the Knot Identification Problem in $\mathcal{A}$. Since $\mathcal{A}$ is knot opaque, there exists a computation σ_1 and a process p_1 such that for every knot that p_1 observes, it is unclear to p_1 whether or not this knot is globally observable.

Algorithm $\mathcal{S}$ is assumed to be a solution to the Knot Identification Problem. According to the KI Termination property, p_1 in σ_1 must output one of its observed knots. Let p_1 output knot K in some state s_i of σ_1. Since $\mathcal{A}$ is knot opaque, it contains a computation σ_2 that is observation graph identical to σ_1 for p_1 up to state s_i, yet knot K is not globally observable in σ_2. If σ_2 is observation graph identical to σ_1 for p_1 up to state s_i, then process p_1 in algorithm $\mathcal{S}$ outputs K in σ_2 just like it does in σ_1.

If knot K is not globally observable in σ_2, then there is a process p_2 that does not observe K in σ_2. If so, p_2 in σ_2 either outputs a knot different from K or none at all. In the first case, $\mathcal{S}$ violates the KI Agreement property that requires that every process outputs the same knot. In the second case, if p_2 does not output a knot in σ_2, $\mathcal{S}$ violates KI Termination Property requiring every process to output a knot.

In either case $\mathcal{S}$ does not comply with the properties of the Knot Identification Problem. This means that, contrary to our initial assumption, $\mathcal{S}$ may not be a solution to this problem. Hence the lemma.   □

Knot finality. Lemma 1 restricts the adversary from hiding whether a particular knot a process observes is globally observable or not. However, even if each process knows if the knot is globally observable, it may still be insufficient to ensure the existence of a solution.

Consider an arbitrary computation σ_1 and an arbitrary process p of some adversary $\mathcal{A}$. An adversary $\mathcal{A}$ is *knot observation final* if it contains a computation σ_1 where there is a process p such that, for every state s_i

of σ_1, there is a computation σ_2 which is observation graph identical to σ_1 for p up to state s_i such that, after state s_i, it does not contain any more knot observations by p. Intuitively, in such an adversary, a process may not gain additional knot information by delaying its decision.

A knot is *primary* for some process p in computation σ if it is the first observed knot by p in σ.

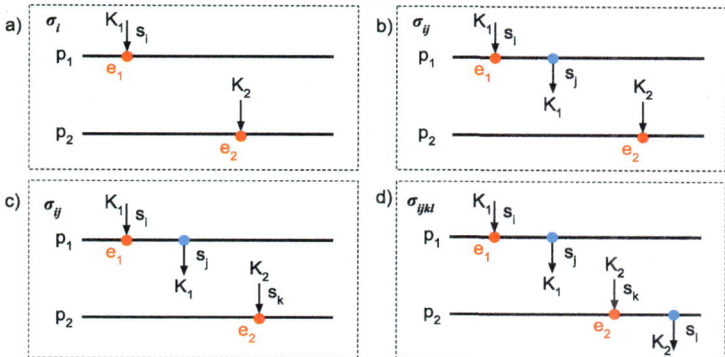

Fig. 2. Illustration for the proof of Lemma 2. In figure a), in computation σ_i, process p_1 observes knot K_1 with event e_1 in state s_i. In figure b), in computation σ_{ij}, process p_1 outputs knot K_1 in state s_{ij}. In figure c), in the same computation σ_{ij}, process p_2 observes knot K_2 in state s_k with event e_2. In figure d), in computation σ_{ijkl}, process p_2 outputs K_2 in state s_l.

Lemma 2. *Consider an observation final, asynchronous, knot transparent adversary $\mathcal{A}$. If $\mathcal{A}$ contains a computation σ such that a pair of processes observe two different primary knots, then this adversary does not have a knot identification solution even though this adversary is knot transparent.*

Proof. Consider the adversary $\mathcal{A}$ that conforms to the conditions of the lemma. Yet, there is an algorithm $\mathcal{S}$ that solves the Knot Identification Problem on $\mathcal{A}$. According to the lemma conditions, $\mathcal{A}$ allows some computation σ with a pair of processes p_1 and p_2 that observe different primary knots K_1 and K_2, respectively. Since $\mathcal{A}$ is knot transparent, K_1 and K_2 may be globally observable. Refer to Fig. 2 for illustration.

Let e_1 and e_2 be the corresponding knot observation events in σ. The two events may be either concurrent or causally dependent. In the latter case, we assume, without loss of generality, that e_1 causally precedes e_2. Let event e_1 occur in state s_i.

Since $\mathcal{A}$ is knot observation final, it allows a computation σ_i' that is observation graph identical for p_1 to σ up to state s_i, yet p_1 does not observe any knots after state s_i in σ_i'. That is, the only knot p_1 observes is K_1. (We denote computations where a process does not observe any more knots with the prime symbol.)

Since $\mathcal{S}$ is assumed to be a solution to the Knot Identification Problem, each process, including p_1, must output a knot in σ_i'. The only knot that p_1 observes in σ_i' is K_1. Hence, p_1 outputs K_1. It may output it in state s_i, or in some later state. We consider the case where p_1 outputs K_1 later.

Since K_1 is primary for p_1, the observation event e_1 for K_1 at p_1 in σ causally precedes observation events of other knots at p_1 if such observations ever happen. We construct a computation σ_{i1} from σ by adding a non-communication state after state s_i. Since $\mathcal{A}$ is an asynchronous adversary, $\mathcal{A}$ allows σ_{i1}. Note that $\mathcal{A}$ also allows a computation σ_{i1}' which is observation identical to σ_{i1} for p_1 for states up to s_{i+1} but where p_1 observes no other knots besides K_1. Similarly, $\mathcal{S}$ must have p_1 output K_1 in σ_{i1}'. This output occurs in state s_{i+1} or later.

Note that the purported solution to the Knot Identification Problem $\mathcal{S}$ has to comply with its Termination Property. This means that each process must eventually output a knot. Therefore, as we continue this process of adding non-communication states past s_i, we find computation $\sigma_{ij} \in \mathcal{A}$ where p_1 outputs K_1 in state s_j following state s_i.

Let us examine σ_{ij}. In this computation, p_2 observes its primary knot K_2 with observation event e_2. By construction, e_2 happens in some state s_k following s_j. Similar to the above procedure, we continue adding non-communication states past s_k until we obtain computation σ_{ijkl} where p_2 outputs knot K_2 in state s_l following state s_k.

Let us now examine σ_{ijkl}. In this computation, in algorithm $\mathcal{S}$, process p_1 outputs knot K_1 while process p_2 outputs knot K_2. However, these two knots are different. Therefore, $\mathcal{S}$ violates the Agreement Property of the Knot Identification Problem requiring every process to output the same knot. Yet, this means that $\mathcal{S}$ may not be a solution to this problem and our initial assumption is incorrect. This proves the lemma.   □

An adversary $\mathcal{A}$ is *primary uniform* if the following conditions hold for every computation $\sigma \in \mathcal{A}$: (i) each process observes at least one knot; (ii) if some process p_1 observes its primary knot K_1 and another process observes its primary knot K_2, then $K_1 = K_2$. To put another way, in a single computation of a primary uniform adversary, all processes observe the same primary knot.

Theorem 1. *For a knot observation final asynchronous knot transparent adversary $\mathcal{A}$ to allow a solution to the Knot Identification Problem, it is necessary and sufficient for $\mathcal{A}$ to be primary uniform.*

Proof. The necessity part of the theorem follows from Lemma 2. We prove the sufficiency by presenting the algorithm *KIA* below that solves the Knot Identification Problem under $\mathcal{A}$.   □

4   Knot Identification Algorithm *KIA*

Description. The knot identification algorithm *KIA* operates as follows. See Algorithm 1. Across every available outgoing link, each process p relays all the connectivity data that it has observed so far. That is, if process p communicates with process q at state s_i of computation σ, then p transmits its entire local observation graph, $LG(p, \sigma, i)$, to q.

Once a process p detects a knot in $LG(p, \sigma, i)$, it outputs it. Since the adversary is primary uniform, each process is guaranteed to eventually observe a primary knot and this knot is the same for every process. That is, *KIA* solves the Knot Identification Problem.

Algorithm 1: Knot Identification Algorithm *KIA*

---

1 **Constants:**
2   p ▷ process identifier

3 **Variables:**
4   $LG(p, \sigma, i)$ ▷ local observation graph of process p

5 **Actions:**
6 **if** *exist outgoing links* **then**
7     $\lfloor$ **send** $LG(p, \sigma, i)$ *to every outgoing link*
8 **if** **receive** $LG(q, \sigma, i)$ *from process* q **then**
9     $\lfloor$ $LG(p, \sigma, i) = LG(p, \sigma, i) \cup LG(q, \sigma, i)$ ▷ merge graphs
10 **if** $\exists$ *knot* K: $K \subset LG(p, \sigma, i)$ **then**
11     $\lfloor$ **output** K ▷ report knot

Complexity estimation. Let us estimate the number of states it takes for each process of *KIA* to output its decision. This estimate is tricky since the algorithm may not do anything productive if no edges appear. Hence, we only count states where information spreads. To put another way, we compute the worst case number of causally related links before every process outputs a knot.

Let n be the number of processes in the network. The algorithm operation can be divided into two parts: (i) knot formation and (ii) knot data propagation. In the worst case, these two parts run consecutively. Suppose the last process, p, that participates in the knot is the first to observe it. Then, p is the only process that informs the other processes of the knot. To put another way, the knot observations at the other processes are causally preceded by the knot observation at p.

The knot with the longest causally related links is a cycle of n edges. The knot data propagation part requires $n - 1$ edges if all processes are informed sequentially. Hence, the worst case *KIA* complexity is $2n - 1$, which is in $O(n)$.

5   *KIA* Performance Evaluation

We studied the performance of our Knot Identification Algorithm *KIA* using an abstract algorithm simulator QUANTAS [16]. The QUANTAS code for *KIA* as well as our performance evaluation data is available

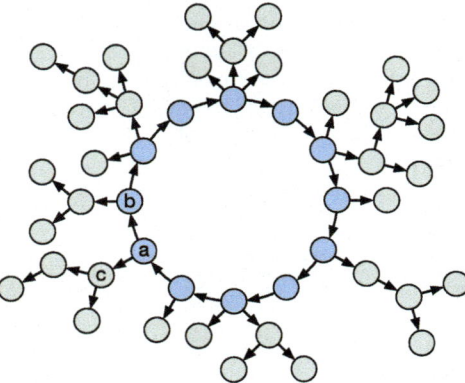

Fig. 3. Intermittent Connectivity Topology Example. The underlying topology contains a single knot: the cycle.

online [1,2]. The computations were selected as follows. First, we generated the underlying *backbone* topology. In the backbone, a certain number of nodes are jointed in a cycle. Each remaining node is randomly attached with a single edge to an already selected connected node. See Fig. 3 for an example of such a topology. In each round of a computation, a fixed number of backbone edges appear. The edges to appear are selected uniformly at random. Thus, each computation contains a single knot—the backbone cycle—while the whole network is unlikely to be connected in a single round. Moreover, the information about this cycle is eventually propagated to all nodes in the network. That is, all generated computations contain exactly one globally observable knot.

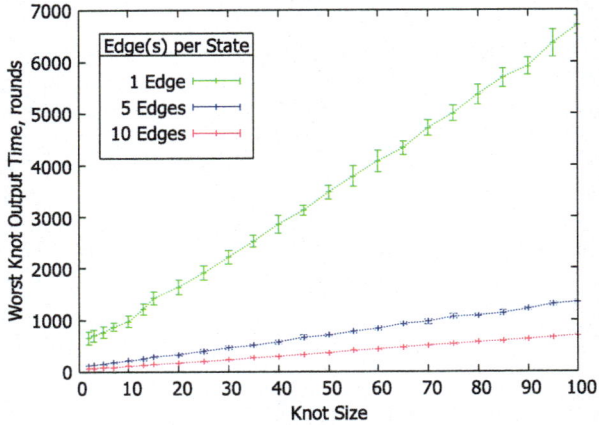

Fig. 4. Longest knot output time as a function of the knot size.

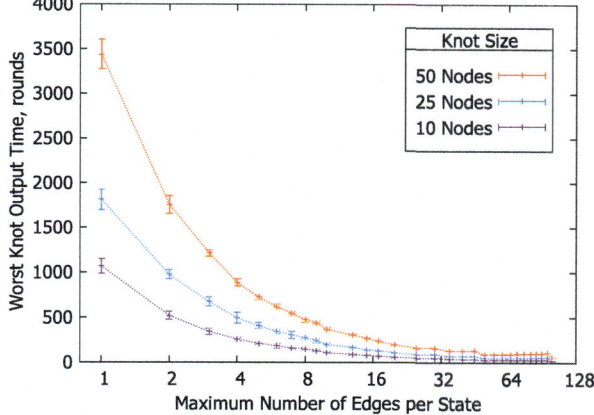

Fig. 5. Longest knot output time as a function of the maximum number of edges per state.

We implemented *KIA* and measured its performance. We measured the speed of knot detection expressed as the longest number of rounds it takes for any process in the network to output the knot.

In the first experiment, we fixed the number of random edges appearing per round and varied the knot (cycle) size. The results are shown in Fig. 4. We set the network size to 100 nodes. The cycle size varies from 2 to 100. That is, the largest cycle comprises the whole network. The computation length is set at 6000 rounds. The knot output time is averaged across 10 computations. We plot *KIA* performance for the case of 1, 5 and 10 backbone edges appearing per round. The data shows that smaller knots are detected quicker by all the nodes in the network.

In the second experiment, we fixed the knot, i.e. cycle, size and varied the number of random edges per round. The results are shown in Fig. 5. Intuitively, it shows that a greater number of edges appearing in one round, even if the network remains disconnected, provides greater overall connectivity and accelerates knot detection.

Our experiments demonstrate the practicality of our knot identification approach to agreement in dynamic networks.

6   Extensions of Knot-Based Consensus

Distinguished knots. In this paper, we treated the problem of knot-based consensus as generally as possible. However, it may be adapted to particular systems: certain topologies may be significant to the system and the processes could be programmed to distinguish such knots. For example, the processes would reject all cycles with fewer than 10 nodes or accept only knots which are completely connected subgraphs.

Expiring links. In the communication model, it is assumed that the sender process transfers its entire communication history to the receiver process across the communication link. This may require extensive communication and resources.

Our algorithm may be adapted to limit resource usage. For example, the algorithm may discard the links older than some pre-determined period, say P. To put another way, the links and topological information expire after P states. This model would nicely represent the network with moving topology or changing membership. In this case, the necessary conditions of Theorem 1 must apply for the links within this period P.

Future research. In this paper, we apply knot identification to the problem of agreement in dynamic networks. In the future, it would be interesting to study what other topological features can be effectively used for consensus and related tasks. Alternatively, it would be interesting to determine communication environments that naturally yield the dynamic graphs that comply with the adversary conditions allowing the solution the Knot Identification Problem.

Another promising research direction is implementing our knot identification algorithm in a complete system and testing its performance in practical environments such as Internet-of-Things networks.

The computation model we consider can address message loss and process failure as special topologies. However, these faults are benign. It is interesting to address solvability of Knot Identification and similar problems in the presence of Byzantine faults where faulty processes may behave arbitrarily [15, 17, 18].

References

1. Kia implementation in *QUANTAS* (2024). https://github.com/QuantasSupport/Quantas/tree/master/quantas/CycleOfTreesPeer
2. Knot Perfromance Evaluation Data (2024). http://www.cs.kent.edu/~mikhail/Research/knot.zip
3. Afek, Y., Gafni, E.: Asynchrony from synchrony. In: Frey, D., Raynal, M., Sarkar, S., Shyamasundar, R.K., Sinha, P. (eds.) ICDCN 2013. LNCS, vol. 7730, pp. 225–239. Springer, Heidelberg (2013). https://doi.org/10.1007/978-3-642-35668-1_16
4. Altisen, K., Devismes, S., Durand, A., Johnen, C., Petit, F.: On implementing stabilizing leader election with weak assumptions on network dynamics. In: Proceedings of the 2021 ACM Symposium on Principles of Distributed Computing, pp. 21–31 (2021)
5. Biely, M., Robinson, P., Schmid, U.: Agreement in directed dynamic networks. In: International Colloquium on Structural Information and Communication Complexity, pp. 73–84. Springer (2012)
6. Biely, M., Robinson, P., Schmid, U., Schwarz, M., Winkler, K.: Gracefully degrading consensus and k-set agreement in directed dynamic networks. Theor. Comput. Sci. **726**, 41–77 (2018)
7. Castañeda, A., Fraigniaud, P., Paz, A., Rajsbaum, S., Roy, M., Travers, C.: A topological perspective on distributed network algorithms. Theor. Comput. Sci. **849**, 121–137 (2021)

8. Cavin, D., Sasson, Y., Schiper, A.: Consensus with unknown participants or fundamental self-organization. In: International Conference on Ad-Hoc Networks and Wireless, pp. 135–148. Springer (2004)
9. Charron-Bost, B., Schiper, A.: The heard-of model: computing in distributed systems with benign faults. Distrib. Comput. **22**, 49–71 (2009)
10. Coulouma, É., Godard, E., Peters, J.: A characterization of oblivious message adversaries for which consensus is solvable. Theoret. Comput. Sci. **584**, 80–90 (2015)
11. Fevat, T., Godard, E.: Minimal obstructions for the coordinated attack problem and beyond. In: 2011 IEEE International Parallel & Distributed Processing Symposium, pp. 1001–1011. IEEE (2011)
12. Fischer, M.J., Lynch, N.A., Paterson, M.S.: Impossibility of distributed consensus with one faulty process. J. ACM, **32**(2), 374–382 (1985)
13. Kuhn, F., Lynch, N., Oshman, R.: Distributed computation in dynamic networks. In: Proceedings of Forty-Second ACM Symposium on Theory of Computing, pp. 513–522 (2010)
14. Kuhn, F., Moses, Y., Oshman, R.: Coordinated consensus in dynamic networks. In: Proceedings of the 30th Annual ACM SIGACT-SIGOPS Symposium on Principles of Distributed Computing, pp. 1–10 (2011)
15. Lamport, L., Shostak, R., Pease, M.: The byzantine generals problem. ACM Trans. Program. Lang. Syst. **4**(3), 382–401 (1982)
16. Oglio, J., Hood, K., Nesterenko, M., Tixeuil, S.: Quantas: quantitative user-friendly adaptable networked things abstract simulator. In: Proceedings of the 2022 Workshop on Advanced tools, Programming Languages, and Platforms for Implementing and Evaluating Algorithms for Distributed systems, pp. 40–46 (2022)
17. Oglio, J., Hood, K., Sharma, G., Nesterenko. M.: Consensus on an unknown torus with dense byzantine faults. In: International Conference on Networked Systems, pp. 105–121. Springer (2023)
18. Pease, M., Shostak, R., Lamport, L.: Reaching agreement in the presence of faults. J. ACM **27**(2), 228–234 (1980)
19. Santoro, N., Widmayer, P.: Time is not a healer. In: Monien, B., Cori, R. (eds.) STACS 1989. LNCS, vol. 349, pp. 304–313. Springer, Heidelberg (1989). https://doi.org/10.1007/BFb0028994
20. Schwarz, M., Winkler, K., Schmid, U.: Fast consensus under eventually stabilizing message adversaries. In: Proceedings of the 17th International Conference on Distributed Computing and Networking, pp. 1–10 (2016)
21. Winkler, K., Paz, A., Galeana, H.R., Schmid, S., Schmid, U.: The time complexity of consensus under oblivious message adversaries. In: 14th Innovations in Theoretical Computer Science Conference (ITCS 2023). Schloss Dagstuhl-Leibniz-Zentrum für Informatik (2023)

A Self-stabilizing Algorithm for the 1-Minimal Minus Domination Problem

Tota Yamada$^{(\boxtimes)}$ and Yonghwan Kim

Nagoya Institute of Technology, Aichi, Japan
t.yamada.620@stn.nitech.ac.jp, kim@nitech.ac.jp

Abstract. A *Minus Dominating (MD) Function* of a graph $G = (V, E)$
($|V| = n$) is a function that assigns a value from $\{-1, 0, 1\}$ to each node
$i \in V$ such that the sum of the values of node i and all its neighboring nodes is positive (i.e., equal to or greater than 1). An MD function is
minimal if decreasing the value of any node by 1 causes a violation of the
conditions of the MD function. As an extension of the MD function, we
introduce the k-*Minimal Minus Dominating (MMD) Function* ($k \geq 0$),
which is a minimal MD function such that no other MD function can
be obtained by increasing the values for some nodes by k in total and
decreasing the values for some nodes by at least $k + 1$ in total. Note
that any minimal MD function can be referred to as a 0-MMD function. In this paper, we propose a silent self-stabilizing algorithm to solve
the 1-*Minimal Minus Domination Problem* on an arbitrary graph, using
a composition technique that repeatedly applies several self-stabilizing
algorithms in order, known as loop composition. It converges within
$\mathcal{O}(n(\Delta^2 + D))$ rounds, where D is the diameter and Δ is the maximum
degree of a graph, and each node requires $\mathcal{O}(\Delta^4 \log n)$ bits of memory.

Keywords: Distributed System · Self-Stabilizing Algorithm · Loop
Composition · Minus Domination

1 Introduction

A distributed system is a network of multiple computational entities (called
nodes) that are connected to each other through communication links. Even there
are many difficulties in the design of distributed systems, they are widely used
due to their numerous advantages, such as *load balancing* or *high availability*. In
a distributed algorithm, which is specifically designed for distributed systems,
nodes cooperate by exchanging information (if necessary) to achieve a specific
goal. Many distributed algorithms guarantee correctness only when the network
(i.e., distributed system) is properly initialized. This implies that the entire
system must be reinitialized to execute the algorithm again if some nodes fail or
the topology changes.

This work was supported in part by JSPS KAKENHI Grant Numbers 20KK0232 and
23K24825.

Dijkstra introduces a new paradigm for designing distributed algorithms, called a self-stabilizing algorithm [1]. A self-stabilizing algorithm is a type of distributed algorithm that can eventually reach a desired configuration (referred to as a *legitimate configuration*) from any initial configuration. This property ensures that the system can recover to the legitimate configuration within a finite time, even if it encounters any inconsistent configurations due to transient failures (i.e., temporary changes in the variables, program counters of the nodes, or the network topology).

A Minus Dominating (MD) function of a graph $G = (V, E)$, where V is the set of nodes and E is the set of edges, is a function $f : V \rightarrow \{-1, 0, 1\}$ such that each node $i \in V$ satisfies $\sum_{j \in N(i) \cup \{i\}} f(j) \geq 1$. Here, $N(i) = \{j \in V \mid (i, j) \in E\}$ represents the set of neighbors of node i. An MD function f is said to be *minimum* if there is no other MD function f' for which $\sum_{i \in V} f'(i) < \sum_{i \in V} f(i)$. Additionally, an MD function f is *minimal* if there is no other MD function f' ($f \neq f'$) such that $f'(i) \leq f(i)$ for every $i \in V$.

The problem of finding an MD function is called the *Minus Domination Problem*. The problem of finding a minimum (resp. minimal) MD function is referred to as the *Minimum* (resp. *Minimal*) *Minus Domination Problem*. It is worthwhile to noting that the minimal minus domination problem generalizes the *minimal dominating set problem*, where each node is assigned a value of either 0 or 1, such that the sum of the values of each node and all its adjacent nodes (called a *local sum*) is positive. In the minus domination problem, a value -1 can also be assigned to any node. Since the minimum minus domination problem is known to be NP-hard [3], research on the minus domination problem primarily focuses on finding a minimal minus dominating (MMD) function. However, an MMD function does not always ensure that its $\sum_{i \in V} f(i)$ is relatively small among all possible MMD functions. This issue is particularly notable in any star graph. For example, in a star graph with n nodes, the $\sum_{i \in V} f(i)$ for the minimum MD function clearly equals to 1. However, the MMD function assigning each leaf to 1 (and the other node to 0) exists and its $\sum_{i \in V} f(i)$ equals to $n - 1$, which is almost the worst case for any MD function in this graph. Given this observation, we aim to find an MMD function with the higher minimality, meaning that its $\sum_{i \in V} f(i)$ is relatively small among all possible MMD functions (the precise definition of "relatively small" will be provided later).

We define a k-*Minimal Minus Dominating (k-MMD) Function* as follows: an MD function f is k-*minimal* ($k \geq 0$) if no other MD function f' can be obtained by increasing the values of some nodes by k in total and decreasing the values of some nodes by at least $k + 1$ in total. Here, the aforementioned MMD function can be considered as a 0-MMD function (i.e., k-MMD function when $k = 0$) because no other MD function can be obtained by decreasing any value of any node without increasing the value of any node. Clearly, a $(k + 1)$-MMD function also satisfies the conditions of the k-MMD function, which means that the larger values of k impose stronger constraints.

Lemma 1. *Let $S_k(G)$ ($k \geq 0$) be the maximum value of $\sum_{i \in V} f(i)$ among all k-MMD functions f of a graph $G = (V, E)$ ($|V| = n$) and let $S(G)$ be the value of*

$\sum_{i \in V} f(i)$ *for the minimum MD function of* G. *The following inequality holds:*

$$S(G) \leq S_{n-1}(G) \leq \cdots \leq S_k(G) \leq \cdots \leq S_1(G) \leq S_0(G).$$

From this lemma, a k-MMD function provides a better solution than a $(k-1)$-MMD function in many cases. For example, there is only one 1-MMD function in MMD functions of any star graph, whose $\sum_{i \in V} f(i)$ equals to 1 (i.e., minimum). Therefore, we focus on finding a 1-MMD function to obtain an MMD function with a relatively small sum. The problem of finding a 1-MMD function is called the 1-*Minimal Minus Domination Problem*.

Related work: In 1996, Dunbar et al. introduced the concept of a minus dominating (MD) function as a generalization of the dominating set in [2,3]. In [3], they also described the decision problem associated with MD functions, which involves determining whether a given graph $G = (V, E)$ has an MD function f such that $j = \sum_{i \in V} f(i)$ for a given integer j. The authors proved that this problem is NP-complete. Additionally, they proposed a sequential algorithm to solve the minus domination problem for graphs restricted to rooted trees in a linear time.

Several extensions of the minus domination problem have also been introduced. In 2007, Xu et al. presented the *Minus Edge Domination Problem* [4], which involves finding a function $f : E \rightarrow \{-1, 0, 1\}$ such that $\sum_{e \in E(i)} f(e) \geq 1$ for each node $i \in V$, where $E(i)$ denotes the set of incident edges of i. Xu et al. further introduced the *Mixed Minus Domination Problem* [5], which assigns values to both the nodes and the edges.

In more recent work, the *Minus* (L, K, Z)-*Domination* problem was introduced by Kakugawa et al. in 2022 [6]. This problem is a natural generalization of the minus domination problem and involves finding a function, named a *Minimal Minus (L,K,Z)-Dominating* function, of the form $f : V \rightarrow \{L, L + 1, \ldots, 0, \ldots, K - 1, K\}$ satisfying $\sum_{j \in N(i) \cup \{i\}} f(j) \geq Z$ for each $i \in V$. The authors also proposed a self-stabilizing algorithm to find an appropriate minimal minus (L,K,Z)-dominating function under an unfair central daemon based on the distance-2 shared memory model where each process can read the local variables of processes within distance 2.

The *loop composition* framework [7] is designed to repeatedly apply several self-stabilizing algorithms to refine the current legal configuration until a legitimate configuration is achieved. This technique allows for the iterative application of self-stabilizing algorithms starting from a legal configuration (i.e., one that satisfies predetermined conditions), continually seeking a better configuration (e.g., one with higher minimality) until no further improvements can be made. It is worthwhile to noting that the self-stabilizing algorithm used in the loop composition ensures correctness only when the current configuration is legal. Therefore, an error-detecting predicate and another self-stabilizing algorithm to initialize the system are necessary.

Contribution: We address the 1-minimal minus domination problem and propose a silent self-stabilizing algorithm to solve it. The proposed algorithm terminates within $\mathcal{O}(n(\Delta^2 + D))$ rounds, where D is the diameter and Δ is the maximum degree of the given graph. Each process requires $\mathcal{O}(\Delta^4 \log n)$ bits of memory.

To the best of our knowledge, the only existing self-stabilizing algorithm for minus domination was proposed in [6]. Our proposed algorithm nct only achieves a relatively better MD function (based on the definition of k-MMD), but also assumes a weaker model in terms of the scheduler and communication model compared to the existing work. Specifically, we assume a distributed daemon rather than a central daemon and the distance-1 locally shared memory model (also known as the state reading model) rather than the distance-2 model. Furthermore, we utilize the loop composition technique to construct a silent self-stabilizing algorithm, marking the first instance of its application to the minus domination problem.

2 Preliminary

2.1 System Model

We assume a distributed system consisting of n processes and communication links between them, represented as an undirected graph $G = (V, E)$ where V is the set of processes and E is the set of edges representing communication links between two processes. The set of processes within distance d from process $P_i \in V$ is denoted as $N^d[P_i] = \{P_j \mid dist(P_i, P_j) \le d\}$ (where $dist(P_i, P_j)$ is the distance between P_i and P_j), and is called as the d-closed neighborhood of process P_i. The diameter $D(G)$ of G is defined as the maximum distance between any two processes in G (i.e., $\max\{dist(P_i, P_j) \mid P_i, P_j \in V\}$), and we denote simply as D. Each process has a unique non-negative integer identifier (ID), denoted as i for process P_i. We assume the locally shared memory model (alsc known as the state reading model) for communication between processes as follows:

Definition 1. (Locally shared memory model) Let s_i be a state of process P_i and $N^1[P_i] = \{P_{i1}, P_{i2}, \ldots, P_{ik}\}(k = |N^1[P_i]|)$ be the 1-closed neighborhood of process P_i. Each process can read the state of any process in its 1-closed neighborhood. The state transition function δ_i takes as input the set of states of all processes in $N^1[P_i]$ and outputs the next state of process P_i, denoted as s_i'. The **locally shared memory model** is a communication mcdel where the next state s_i' is given by $s_i' = \delta_i(s_{i1}, s_{i2}, \ldots, s_{ik})$.

2.2 Notation of Algorithm

A self-stabilizing algorithm can be described as a set of rules in the form $\langle label \rangle$: $\langle guard \rangle \Rightarrow \langle action \rangle$. The $\langle label \rangle$ serves as a reference of the rules and indicates their priority. The $\langle guard \rangle$ is a predicate for process P_i that invclves variables from the 1-closed neighborhood of P_i. The $\langle action \rangle$ specifies the update to the variables of process P_i when $\langle guard \rangle$ is satisfied.

When the $\langle guard \rangle$ of any rule for process P_i holds true in configuration C_t, we call the rule is *executable* at process P_i in configuration C_t. If there is at least one executable rule for process P_i in configuration C_t, we say that process P_i is *enabled* in configuration C_t.

To simplicity, any rule of the form $x_i \neq f(P_i) \Rightarrow x_i \leftarrow f(P_i)$ can be denoted as $x_i \leftarrow f(P_i)$. This notation implies that the rule $x_i \leftarrow f(P_i)$ is executable only when $x_i \neq f(P_i)$. That is, when a rule is written as $x_i \leftarrow f(P_i)$, we omit the $\langle guard \rangle$ $x_i \neq f(P_i)$.

2.3 Configuration, Scheduler, and Execution

A *configuration* of system is represented as a tuple of states s_i for each process. Specifically, the configuration is denoted as $\gamma = (s_0, s_1, \ldots, s_{n-1})$ where s_i is the state of process P_i. Let x_i be the value of variable x for process P_i and $x_i(\gamma)$ be x_i in configuration γ. The set of all possible network states is denoted as Γ. We denote the set of possible states for process P_i as S_i, Thus $\Gamma = S_0 \times S_1 \times \cdots \times S_{n-1}$.

Definition 2. (Scheduler and Execution) Let $\mathcal{A}$ be an algorithm and V' be a non-empty subset of V. We denote $\gamma \mapsto_{(V',\mathcal{A})} \gamma'$ if a configuration $\gamma \in \Gamma$ is obtained when each node in V' simultaneously performs an atomic action of $\mathcal{A}$ in configuration γ. A **schedule** is presented as an infinite sequence $\varrho = (V_0, V_1, V_2, \ldots)$, where each V_i is a non-empty subset of V. An **execution** $\Xi_{\mathcal{A}}(\varrho, \gamma_0)$ of algorithm $\mathcal{A}$ along schedule $\varrho = (V_0, V_1, V_2, \ldots)$ starting from an initial configuration γ_0 is uniquely defined as the infinite sequence $(\gamma_0, \gamma_1, \gamma_2, \ldots)$ of configurations such that $\gamma_i \mapsto_{(V_i, \mathcal{A})} \gamma_{i+1}$ for any $i \geq 0$.

Definition 3. (Maximal execution and Fair execution) If Ξ is a finite sequence of configurations $(\gamma_0, \gamma_1, \gamma_2, \ldots, \gamma_f)$ and no process is enabled in configuration γ_f, We say that an execution Ξ terminates at configuration γ_f. An execution Ξ is **maximal** if Ξ terminates at some specific configuration or Ξ is an infinite sequence.

A maximal execution Ξ is (weakly) fair if there is no process P_i such that P_i is always enabled but never activated in any suffix of Ξ. This means that a fair execution guarantees that every continuously enabled process will be eventually activated.

In the locally shared-memory model, several assumptions are based on the number of processes selected by the scheduler. We say that a scheduler $\varrho = (V_0, V_1, V_2, \ldots)$ is a *central scheduler* (or a *central daemon*) if $|V_i| = 1$ holds for all $i \geq 0$. Otherwise, a scheduler ϱ is a *distributed scheduler* (or a *distributed daemon*), which means $0 < |V_i| \leq n$. In this paper, we assume a distributed daemon and fair execution.

2.4 Self-stabilizing Algorithm

Let $\mathcal{L}$ be a predicate for configurations and $\Gamma^{\mathcal{L}}$ be a set of configurations satisfying $\mathcal{L}$. Additonally, let $\Xi_{\mathcal{A}} = (\gamma_0, \gamma_1, \ldots)$ be a maximal execution of algorithm $\mathcal{A}$. An algorithm $\mathcal{A}$ is a self-stabilizing algorithm for predicate $\mathcal{L}$ if it satisfies the following conditions:

- **Convergence:** Every fair execution of $\mathcal{A}$ starting from any configuration eventually reaches a configuration in $\Gamma^{\mathcal{L}}$.
- **Closure:** A configuration in $\Gamma^{\mathcal{L}}$ never transitions to a configuration outside $\Gamma^{\mathcal{L}}$ under algorithm $\mathcal{A}$. That is, there is no $\gamma_i \in \Gamma^{\mathcal{L}}$, $\gamma_j \notin \Gamma^{\mathcal{L}}$, and a non-empty set $V' \subseteq V$ such that $\gamma_i \mapsto_{(V',\mathcal{A})} \gamma_j$.

A configuration in $\Gamma^{\mathcal{L}}$ is called *legitimate*, whereas a configuration not in $\Gamma^{\mathcal{L}}$ is called *illegitimate*. If every maximal execution of algorithm $\mathcal{A}$ terminates, we say that algorithm $\mathcal{A}$ is *silent*. For simplicity, here we denote $\Gamma^{\mathcal{L}}$ by just $\mathcal{L}$, implying that we treat $\mathcal{L}$ as a set of configurations satisfying predicate $\mathcal{L}$. Thus, $\gamma \in \mathcal{L}$ means $\mathcal{L}(\gamma) = \mathsf{TRUE}$.

2.5   Time Complexity

We measure the time complexity using the number of *rounds*. Let $\Xi_{\mathcal{A}}(\varrho, \gamma_0) = (\gamma_0, \gamma_1, \ldots, \gamma_f)$ be a maximal execution of algorithm $\mathcal{A}$ starting from an initial configuration γ_0 with a schedule $\varrho = (V_0, V_1, \ldots, V_{f-1})$. The first round of execution Ξ is defined as the smallest prefix $\gamma_0, \gamma_1, \ldots, \gamma_t$ of Ξ such that every enabled process in γ_0 executes at least one action or becomes disabled by state changes of its 1-closed neighborhood within the first t steps. The second round of Ξ is defined as the first round of Ξ', where Ξ' is the suffix of Ξ starting from γ_t, that is, $\Xi' = (\gamma_t, \gamma_{t+1}, \ldots)$, and so on.

2.6   Specification of Problem

First, we formally define an MD function and an MMD function as follows:

Definition 4. (Minus Dominating (MD) function) A **Minus Dominating function** f of a graph $G = (V, E)$ is a function of the form $f : V \to \{-1, 0, 1\}$ such that $\sum_{P_j \in N^1[P_i]} f(P_j) \geq 1$ for each process $P_i \in V$.

Definition 5. (Minimal Minus Dominating (MMD) function) An MD function f of a graph $G = (V, E)$ is **minimal** if there is no other MD function f' ($f \neq f'$) such that $f'(P_i) \leq f(P_i)$ for each process $P_i \in V$.

Now we define a *k-Minimal Minus Dominating function* and introduce the *1-Minimal Minus Domination problem* which is a main concern of this paper.

Definition 6. (k-Minimal Minus Dominating (k-MMD) function) An MD function f of a graph $G = (V, E)$ is **k-minimal** ($k \geq 0$) if there do not exists two functions $f^+ : V \to \{0, 1, 2\}$ and $f^- : V \to \{0, 1, 2\}$ that satisfy all the following conditions:

1. $\sum f^+(P_i) \leq k$
2. $\sum f^-(P_i) > k$
3. $f'(P_i) = f(P_i) + f^+(P_i) - f^-(P_i)$ is an MD function.

Intuitively, an MMD function f is called k-minimal if there are no other MD function f' can be obtained by increasing the values of some nodes by k in total and decreasing the values of some nodes by at least $k + 1$ in total. In this paper, we consider the *1-Minimal Minus Domination problem* defined as follows.

Definition 7. (1-Minimal Minus Domination problem) Let $G = (V, E)$ be an undirected graph representing a distributed system consisting of a set of processes V and a set of edges E between them. Each process $P_i \in V$ maintains its own local variable $\mathtt{num}_i \in \{-1, 0, 1\}$. Let f_O be a function of the form $f_O : V \to \{-1, 0, 1\}$ such that $f_O(P_i) = \mathtt{num}_i$. The **1-Minimal Minus Domination problem** is a problem of determining the value of $\mathtt{num}_i$ for each process P_i ensuring that the function f_O is a 1-MMD function.

Let $\mathcal{L}_{1MMD}$ be a predicate for configurations that is satisfied if and only if the function f_O is a 1-MMD function. In this paper, we propose a silent self-stabilizing algorithm for $\mathcal{L}_{1MMD}$.

3   Loop Composition: Framework for Self-stabilization

The *loop composition* [7] is a novel framework for constructing a silent self-stabilizing algorithm, denoted as $\mathsf{Loop}(\mathcal{A}, E, \mathcal{P})$ for a given predicate $\mathcal{L}$. This framework requires two algorithms, a base algorithm $\mathcal{A}$, an initializing algorithm $\mathcal{P}$, and an error detecting predicate E. The resulting algorithm terminates within $O(n + T_{\mathcal{P}} + R_{\mathcal{A}} + L_{\mathcal{A}}D)$ rounds, where $T_{\mathcal{P}}$ is the upper bound on the number of rounds of any maximal execution of $\mathcal{P}$, $L_{\mathcal{A}}$ is the upper bound on the number of iterations of $\mathcal{A}$'s executions, $R_{\mathcal{A}}$ is the upper bound on the number of rounds of those (iterated) executions in $\mathcal{A}$.

A predicate E serves as an *error detecting predicate* which can be locally evaluated to determine whether any error exists in the current configuration at each process, $E(P_i) \in \{\mathsf{TRUE}, \mathsf{FALSE}\}$, where $P_i \in V$. A configuration γ is called *erroneous* for E if $\bigvee_{P_i \in V} E(P_i)$ holds in γ. Thus, we define the predicate $\mathcal{E}$ for γ such that $\mathcal{E}(\gamma)$ holds if and only if $\bigvee_{P_i \in V} E(P_i)$ holds in γ.

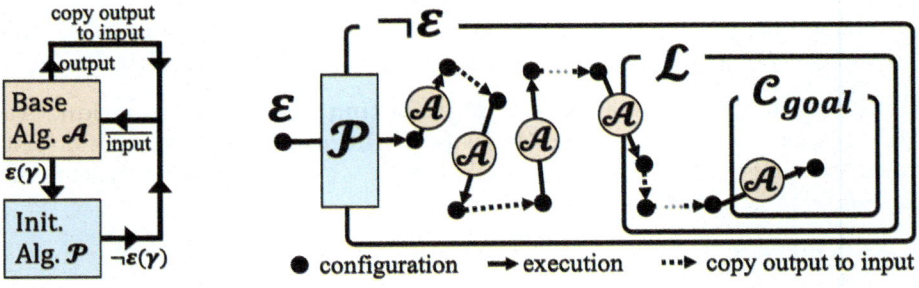

Fig. 1. An overview   **Fig. 2.** An execution of $\mathsf{Loop}(\mathcal{A}, E, \mathcal{P})$

Algorithm $\mathcal{P}$ is a silent self-stabilizing algorithm, named an *initializing algorithm*, designed to transition from any erroneous configuration to a non-erroneous configuration. When a process detects that the current configuration is erroneous, the initializing algorithm $\mathcal{P}$ is executed, transitioning the configuration to a non-erroneous one (refer to Fig. 1). Any maximal execution of $\mathcal{P}$ terminates at a configuration satisfying $\neg\mathcal{E}$ within $T_\mathcal{P}$ rounds. Figure 2 illustrates an execution example of loop composition: each black circle represents a configuration, and an erroneous configuration in $\mathcal{E}$ (i.e., there is a process P_i such that $E(P_i) = \mathsf{TRUE}$) converges to a non-erroneous configuration by the initializing algorithm $\mathcal{P}$.

A base algorithm $\mathcal{A}$ can be executed when the configuration is non-erroneous, generating another non-erroneous configuration through its execution. The base algorithm $\mathcal{A}$ repeatedly executes by taking the resulting (non-erroneous) configuration as input and producing another configuration (refer to Fig. 1). If, after repeated executions of the base algorithm $\mathcal{A}$, no new configuration is obtained (i.e., the input and output configurations are identical), the execution of the algorithm terminates. Figure 2 shows the transitions of the configurations by repeated executions of algorithm $\mathcal{A}$ to reach a goal configuration.

A base algorithm $\mathcal{A}$ must be designed to satisfy the following three conditions:

Shiftable Convergence Any maximal execution of $\mathcal{A}$ that starts from a configuration satisfying $\neg\mathcal{E}$ terminates at a configuration γ such that $\gamma^{copy} \in \neg\mathcal{E}$.
Loop Convergence There exists two integers $L_\mathcal{A}$ and $R_\mathcal{A}$ that satisfy the following proposition: if $\Xi_0, \Xi_1, \ldots$ is an infinite sequence of maximal executions of $\mathcal{A}$ where $\Xi_i = (\gamma_{i,0}, \gamma_{i,1}, \ldots, \gamma_{i,s_i})$, $\gamma_{0,0} \in \neg\mathcal{E}$, and $\gamma_{i+1,0} = \gamma_{i,s_i}^{copy}$ hold for each $i \geq 0$, then $\gamma_{j,s_j} \in \mathcal{C}_{goal}$ and $R(\Xi_0) + R(\Xi_1) + \cdots + R(\Xi_j) \leq R_\mathcal{A}$ hold for some $j < L_\mathcal{A}$ (where $R(\Xi_i)$ is the number of rounds of Ξ_i).
Correctness $\gamma \in \mathcal{C}_{goal} \implies \gamma \in \mathcal{L}$ holds for any configuration γ.

Let $O_\mathcal{A}$ be the set of variables used in $\mathcal{A}$ whose values can be updated by actions of $\mathcal{A}$, and $I_\mathcal{A}$ be the set of variables of $\mathcal{A}$ whose values are never updated (only read) by actions of $\mathcal{A}$. Here, we assume that $O_\mathcal{A} \cap O_\mathcal{P} = \emptyset$.

In the base algorithm $\mathcal{A}$, it is assumed that all processes maintain a *copy variable* $\bar{x} \in I_\mathcal{A}$ for any variable $x \in O_\mathcal{A}$. Thus, let γ^{copy} be a configuration where the value of each variable x_i is replaced by the value of $\bar{x}_i$ for any variable $x \in O_\mathcal{A}$ of each process $P_i \in V$ in configuration γ. Moreover, we define the predicate $\mathcal{C}_{goal}(\mathcal{A}, E)$ ($\mathcal{C}_{goal}$) for the configuration γ such that $\gamma \in \mathcal{C}_{goal}$ holds if and only if $\gamma \in \neg\mathcal{E}$ and $\gamma^{copy} = \gamma$ hold and there is no enabled process for $\mathcal{A}$.

By designing $\mathcal{A}$, E, and $\mathcal{P}$ to satisfy the above conditions, it has been proven in [7] that the self-stabilizing algorithm $\mathsf{Loop}(\mathcal{A}, E, \mathcal{P})$ can be constructed. The following Theorem 1 holds according to [7].

Theorem 1. *[7] Algorithm $\mathsf{Loop}(\mathcal{A}, E, \mathcal{P})$ is a silent and self-stabilizing algorithm for predicate $\mathcal{L}$. Every execution of $\mathsf{Loop}(\mathcal{A}, E, \mathcal{P})$ terminates within $\mathcal{O}(n + T_\mathcal{P} + R_\mathcal{A} + L_\mathcal{A}D)$ rounds. Its space complexity is $\mathcal{O}(S_\mathcal{A} + S_\mathcal{P} + \log n)$ bits per process, where $S_\mathcal{A}$ (resp. $S_\mathcal{P}$) is space complexity of $\mathcal{A}$ (resp. $\mathcal{P}$) in bits per process.*

Table 1. Input and output variables in the base algorithm DEC

variables (output)	copy variables (input)
$\mathsf{num}_i \in \{-1, 0, 1\}$	$\overline{\mathsf{num}_i} \in \{-1, 0, 1\}$
$\mathsf{sum}_i \in \{-1, 0, 1\}$	$\overline{\mathsf{sum}_i} \in \{-1, 0, 1\}$
$\mathsf{change}_i = (\mathsf{cNum}, \mathsf{cDec}, \mathsf{cInc}) \in \{-1, 0, 1\} \times 2^V \times V$	
$\mathsf{decCount}_i \in \mathbb{N}$	
$\mathsf{semiInc}_i \in V$	
$\mathsf{tellSemiInc}_i \in V$	
$\mathsf{minSemiInc}_i \in V$	
$\mathsf{shareSemiInc}_i \in V$	
$\mathsf{isSemiInc}_i \in \{\mathsf{TRUE}, \mathsf{FALSE}\}$	
$\mathsf{isInc}_i \in \{\mathsf{TRUE}, \mathsf{FALSE}\}$	
$\mathsf{isDec}_i \in \{\mathsf{TRUE}, \mathsf{FALSE}\}$	
$\mathsf{CANINC}_i \subseteq V$	
$\mathsf{INCCAND}_i \subseteq V$	
$\mathsf{CANDMAP}_i \subseteq V \times 2^V$	
$\mathsf{DECCNTMAP}_i \subseteq V \times \mathbb{N}$	

4   The Proposed Algorithm

We propose the self-stabilizing algorithm $\mathsf{Loop}(\mathsf{DEC}, E_{MMD}, \mathsf{INIT})$ for the predicate $\mathcal{L}_{1MMD}$. Each process P_i maintains several variables and the corresponding copy variables, as shown in Table 1. Note that, in algorithm DEC, the variables can be read and updated by the processes, whereas the copy variables can only be read (they serve as the input to the algorithm). We named each element used in the 3-tuple variable change_i as cNum, cDec, and cInc respectively to refer to each element in the variable.

Let f_I be a function of the form $f_I : V \to \{-1, 0, 1\}$ such that $f_I(P_i) = \overline{\mathsf{num}_i}$. Let $\mathcal{L}_{MMD}$ be a predicate for configurations that is satisfied if and only if f_I is an MMD function. Note that the predicate $\mathcal{L}_{MMD}$ corresponds to $\neg \mathcal{E}$ from the previous section. The algorithms DEC and INIT, and the predicate E_{MMD} have the following properties:

- $E_{MMD}(P_i)$ holds at least one process P_i if f_I is not an MMD function (i.e., the current configuration does not satisfy $\mathcal{L}_{MMD}$). This means that $\gamma \in \mathcal{L}_{MMD}$ holds if and only if $\neg \bigvee_{P_i \in V} E_{MMD}(P_i)$ holds in configuration γ.
- Any execution of INIT starting from any configuration terminates at a configuration $\gamma \in \mathcal{L}_{MMD}$ within $O(n)$ rounds.
- Any execution of DEC starting from a configuration satisfying $\mathcal{L}_{MMD}$ but where f_I is not a 1-MMD function terminates at a configuration γ where f_O is another MMD function and $\sum_{P_i \in V} \overline{\mathsf{num}_i}(\gamma) > \sum_{P_i \in V} \mathsf{num}_i(\gamma)$ holds within $O(\Delta^2)$ rounds. Here, Δ is the maximum degree of the given graph G.

Algorithm 1 Initializing algorithm INIT

$$\mathbf{R1} : \overline{\text{num}_i} \leftarrow \begin{cases} 1 & \text{if } \forall P_j \in N^1[P_i] : ((\overline{\text{num}_j} \neq 1) \vee (i < j)) \\ 0 & \textbf{otherwise} \end{cases}$$

$$\mathbf{R2} : \overline{\text{sum}_i} \leftarrow \sum_{P_j \in N^1[P_i]} \overline{\text{num}_j}$$

- Any execution of DEC starting from a configuration satisfying $\mathcal{L}_{1MMD}$ (which also satisfies $\mathcal{L}_{MMD}$) terminates at a configuration γ where f_O is a 1-MMD function and $\sum_{P_i \in V} \overline{\text{num}_i}(\gamma) = \sum_{P_i \in V} \text{num}_i(\gamma)$ holds within $C(1)$ rounds.

If the above conditions are satisfied, DEC, E_{MMD}, and INIT will satisfy all the requirements of the loop composition for $\mathcal{L}_{1MMD}$ with $T_{\mathsf{INIT}} = O(n)$, $R_{\mathsf{DEC}} = O(n\Delta^2)$, $L_{\mathsf{DEC}} = O(n)$, $S_{\mathsf{INIT}} = O(\log n)$, and $S_{\mathsf{DEC}} = O(\Delta^4 \log n)$. Hence, $\mathsf{Loop}(\mathsf{DEC}, E_{MMD}, \mathsf{INIT})$ is a silent self-stabilizing algorithm for $\mathcal{L}_{1MMD}$, and its time complexity is $O(n + T_{\mathsf{INIT}} + R_{\mathsf{DEC}} + L_{\mathsf{DEC}}D) = O(n(\Delta^2 + D))$ rounds with $O(S_{\mathsf{DEC}} + S_{\mathsf{INIT}} + \log n) = O(\Delta^4 \log n)$ bits of memory.

4.1 Error Detecting Predicate E_{MMD}

Each process $P_i \in V$ checks the following two conditions: (i) whether the conditions of the MMD function are satisfied (in process P_i's local view), and (ii) $\overline{\text{sum}_i} = \sum_{P_j \in N^1[P_i]} \overline{\text{num}_j}$. Thus, we can define the error detecting predicate E_{MMD} as follows:

$$E_{MMD}(P_i) \equiv ((\overline{\text{num}_i} > -1) \wedge (\forall P_j \in N^1[P_i] : \overline{\text{sum}_j} > 1))$$

$$\vee (\sum_{P_j \in N^1[P_i]} \overline{\text{num}_j} \neq \overline{\text{sum}_i}) \vee (\overline{\text{sum}_i} < 1)$$

The following lemma about the correctness of the properties of E_{MMD} holds.

Lemma 2. *For any configuration γ, $\gamma \in \mathcal{L}_{MMD}$ holds if and only if $\neg \bigvee_{P_i \in V} E_{MMD}(P_i)$ is satisfied in γ.*

4.2 Initializing Algorithm INIT

Algorithm INIT is an initializing algorithm to obtain a non-erroneous configuration from any configuration (i.e., INIT is an algorithm to find an MMD function). The details of Algorithm INIT are provided in Algorithm 1.

Algorithm INIT constructs a maximal independent set (MIS) S and assigns a value of 1 to the $\overline{\text{num}}$ of processes in S (and 0 to those not in S). Note that an MIS S is a subset of processes in a graph G such that processes in S are not adjacent to each other, and each process $P_i \in V \setminus S$ is adjacent to at least one process in S. The algorithm for constructing an MIS is also provided in [8].

The following lemma on the properties of the MIS holds.

Algorithm 2 Base algorithm DEC

R1: CANINC$_i$	$\leftarrow$	$\{P_j \in N^1[P_i] \mid \overline{\text{num}_j} < 1\}$
R2: INCCAND$_i$	$\leftarrow$	FindIncCand(P_i)
R3: CANDMAP$_i$	$\leftarrow$	$\{(P_j, \text{DecSet}(P_j)) \mid P_j \in \bigcup_{P_x \in N^1[P_i]} \text{INCCAND}_x\}$
R4: decCount$_i$	$\leftarrow$	$\|\text{FindDec}(P_i)\|$
R5: DECCNTMAP$_i$	$\leftarrow$	$\{(P_j, \text{decCount}_j) \mid P_j \in N^1[P_i]\}$
R6: semiInc$_i$	$\leftarrow$	$\min(\text{FindSemiInc}(P_i))$
R7: tellSemiInc$_i$	$\leftarrow$	$\min(\{\text{semiInc}_j \mid P_j \in N^1[P_i]\})$
R8: isSemiInc$_i$	$\leftarrow$	$\exists P_j \in N^1[P_i] : \text{tellSemiInc}_j = P_i$
R9: shareSemiInc$_i$	$\leftarrow$	$\min(\{P_j \in N^1[P_i] \mid \text{isSemiInc}_j\})$
R10: minSemiInc$_i$	$\leftarrow$	$\min(\{\text{shareSemiInc}_j \mid P_j \in N^1[P_i]\})$
R11: isDec$_i$	$\leftarrow$	$\forall P_j \in N^1[P_i] : \text{tellSemiInc}_j = \text{semiInc}_i$
R12: isInc$_i$	$\leftarrow$	JudgeInc(P_i)
R13: change$_i$	$\leftarrow$	$(\text{NumChange}(P_i), \text{FindDecProc}(P_i), \text{FindIncProc}(P_i))$
R14: num$_i$	$\leftarrow$	$\overline{\text{num}_i} + \text{change}_i.\text{cNum}$
R15: sum$_i$	$\leftarrow$	$\sum_{P_j \in N^1[P_i]} \text{num}_j$

Lemma 3. *Let S be an MIS of a graph $G = (V, E)$ $(S \subseteq V)$. The function f defined by $f(P_i) = \begin{cases} 1 & \textbf{if } P_i \in S \\ 0 & \textbf{otherwise} \end{cases}$ is an MMD function.*

The following corollary holds by Lemma 3 and the literature [8].

Corollary 1. *Any execution of* INIT *from any configuration terminates at a configuration $\gamma \in \mathcal{L}_{MMD}$ within $\mathcal{O}(n)$ rounds.*

4.3   Base Algorithm DEC

The base algorithm DEC is repeatedly executed to reach a configuration $\gamma \in \mathcal{L}_{1MMD}$. Specifically, DEC is executed from a configuration satisfying an MMD function and iteratively finds another MMD function by increasing the values of some processes by 1, and for each process increasing the value, decreasing the values of at least two other processes by 1. This process continues until the algorithm can no longer find such a configuration (i.e., a configuration satisfying $\mathcal{L}_{1MMD}$). Note that DEC operates under the assumption that f_I is an MMD function. Algorithm 2 shows the details of DEC. The functions and macros used in Algorithm 2 are presented in Table 2.

We say that a rule Rx *converges* if there is no process that can execute R1, R2, ..., Rx. It is important to note that any rule Rx in DEC only uses input variables or variables determined by converged rules R1, R2, ... R$(x-1)$. Consequently, the values of the variables determined in Rx may not be correct until R$(x-1)$ has converged, as some variables set by these rules may not be accurate. This property ensures that the rules converge in the order R1, R2, ..., R15, which guarantees that the algorithm DEC will eventually terminate.

We provide a detailed explanation of each rule in the algorithm DEC. In rules R1 and R2, each process P_i searches for all processes P_j $(\neq P_i)$ in the 2-closed

Table 2. Functions and macros for algorithm DEC

$$\mathsf{FindIncCand}(P_i) = \begin{cases} \bigcap_{P_j \in \{P_x \in N^1[P_i] | \overline{\mathsf{sum}_x} = 1\}} \mathsf{CANINC}_j \setminus \{P_i\} & \text{if } \overline{\mathsf{num}_i} > -1 \\ \emptyset & \text{otherwise} \end{cases}$$

$\mathsf{DecSet}(P_j) = \{P_k \in N^1[P_i] \mid P_j \in \mathsf{INCCAND}_k\}$

$\mathsf{FindDec}(P_k) = \bigcup_{P_j \in N^1[P_i]} \mathsf{GetCandMap}(P_j, P_k)$

$$\mathsf{GetCandMap}(P_j, P_k) = \begin{cases} S & \text{if } (P_k, S) \in \mathsf{CANDMAP}_j \text{ exists} \\ \emptyset & \text{otherwise} \end{cases}$$

$\mathsf{FindSemiInc}(P_i) = \{P_k \in \mathsf{INCCAND}_i \mid (|\mathsf{FindDec}(P_k)| < \mathsf{FindDecCount}(P_k)) \vee \mathsf{Sim}(P_k)\}$

$\mathsf{FindDecCount}(P_k) = \max(\{\mathsf{GetDecCount}(P_j, P_k) \mid P_j \in N^1[P_i]\})$

$$\mathsf{GetDecCount}(P_j, P_k) = \begin{cases} \mathsf{decCount}_k & \text{if } (P_k, \mathsf{decCount}_k) \in \mathsf{DECCNTMAP}_j \text{ exists} \\ 0 & \text{otherwise} \end{cases}$$

$\mathsf{Sim}(P_k) \equiv \exists P_l \in \mathsf{FindDec}(P_k) \; \forall P_m \in \mathsf{DupSet}(P_k) : (\overline{\mathsf{sum}_m} + \mathsf{IsAdjacent}(P_k, P_m)) \geq 3$

$\mathsf{DupSet}(P_k) = \{P_j \in N^1[P_i] \mid P_l \in \mathsf{GetCandMap}(P_j, P_k)\}$

$$\mathsf{IsAdjacent}(P_k, P_m) = \begin{cases} 1 & \text{if } P_k \in \mathsf{CANINC}_m \text{ exists} \\ 0 & \text{otherwise} \end{cases}$$

$\mathsf{JudgeInc}(P_i) \equiv \forall P_j \in N^1[P_i] : (((\mathsf{semiInc}_j \neq \mathsf{minSemiInc}_i) \vee \mathsf{isDec}_j) \wedge$
$(\mathsf{minSemiInc}_j = \mathsf{minSemiInc}_i))$

$$\mathsf{NumChange}(P_i) = \begin{cases} 1 & \text{if } \forall P_j \in N^1[P_i] : (\mathsf{isInc}_i \wedge (\mathsf{minSemiInc}_j = P_i)) \\ -1 & \text{if } \forall P_j \in N^1[P_i] : \mathsf{OkMD}(P_j) \\ 0 & \text{otherwise} \end{cases}$$

$\mathsf{OkMD}(P_j) \equiv (\overline{\mathsf{sum}_j} - |\{P_k \in \mathsf{change}_j.\mathsf{cDec}) \mid k < i\}| + \mathsf{ConfInc}(P_j)) \geq 2$

$$\mathsf{ConfInc}(P_j) = \begin{cases} 1 & \text{if } \mathsf{change}_j.\mathsf{cInc} = \mathsf{semiInc}_i \\ 0 & \text{otherwise} \end{cases}$$

$\mathsf{FindDecProc}(P_i) = \{P_j \in N^1[P_i] \mid \mathsf{change}_j.\mathsf{cNum} = -1\}$

$\mathsf{FindIncProc}(P_i) = \max(\{P_j \in N^1[P_i] \mid \mathsf{change}_j.\mathsf{cNum} = 1\})$

neighborhood of P_i, $N^2[P_i]$, that satisfy the following conditions: (1) $\overline{\mathsf{num}_j}$ is less than 1, and (2) any process in 1-closed neighborhood of P_i will have the local sum of 2 (or more) if the value of $\overline{\mathsf{num}_j}$ is increased by 1.

These conditions ensure that process P_i can safely decrease $\overline{\mathsf{num}_i}$ by 1 without violating the conditions of the MD function if process P_j increases $\overline{\mathsf{num}_j}$ by 1. The process P_i then records all processes that meet these conditions in the set $\mathsf{INCCAND}_i$.

A process P_j is called an *increase candidate* if it is included in the $\mathsf{INCCAND}$ of two or more processes. The processes that include P_j in their $\mathsf{INCCAND}$ are referred to as *decrease candidate* for P_j. This implies that at least two decrease candidates can decrease their $\overline{\mathsf{num}}$ if P_j (i.e., increase candidate) increases $\overline{\mathsf{num}_j}$ by 1.

Figure 3 illustrates an example of the execution of DEC. After the convergence of rule R2, processes 1, 3, and 4 become increase candidates, as shown in Fig. 3(2). In this scenario, the decrease candidates for process 1 (resp. 4) are

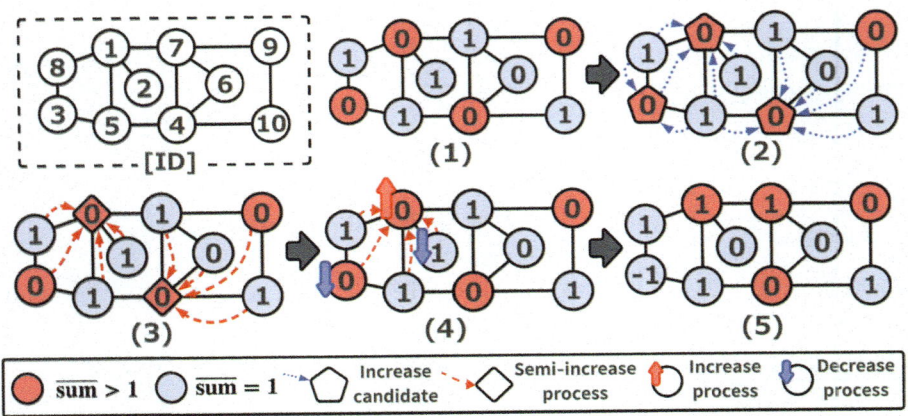

Fig. 3. An example of execution of DEC

processes 2, 3, 5, and 8 (resp. 5, 6, 7, 9, and 10). The decrease candidates for process 3 are only processes 5 and 8. Note that even if there are two or more decrease candidates, not all of them can decrease their $\overline{num}$ (while the increase candidate's $\overline{num}$ increases) as this may violate the conditions of the MD function. For example, in Fig. 3(2), increasing $\overline{num}$ of process 1 and decreasing $\overline{num}$ of processes 5 and 8 violate the conditions of the MD function at process 3. Consequently, processes 5 and 8 cannot decrease their $\overline{num}$.

The following lemma on the properties of the 1-MMD function holds.

Lemma 4. *There is at least one increase candidate if f_I is not a 1-MMD function.*

In rules R3 to R6, each decrease candidate P_i identifies an increase candidate P_j such that at least two decrease candidates for P_j can decrease their $\overline{num}$ by 1 without violating the conditions of the MD function if P_j increases $\overline{num}_j$ by 1. If there are two or more such increase candidates, the one with the smallest ID will be selected. We call such a process P_j a *semi-increase process*. Specifically, each process P_i identifies all increase candidates P_j that satisfy the above condition by checking whether they satisfy one of the following conditions:

(1) There are some decrease candidates for P_j that are not in $N^2[P_i]$.
(2) There is a decrease candidate $P_k \in N^2[P_i]$ for P_j, and processes P_i and P_k can decrease $\overline{num}_i$ and $\overline{num}_k$ by 1 respectively without violating the conditions of the MD function when P_j increases $\overline{num}_j$ by 1.

For example, in Fig. 3(2), the decrease candidates for process 4 are processes 5, 6, 7, 9, and 10. Among them, processes 6 and 9 satisfy condition (1), and processes 7 and 10 satisfy condition (2). Thus, increase candidate 4 becomes a semi-increase process. Moreover, process 5 satisfies condition (1) as a decrease candidate for process 4, and also condition (2) for another increase candidate 1. In this case, process 5 selects process 1 as a semi-increase process because $1 < 5$. Note that a decrease candidate satisfying condition (2) is evidently a semi-increase process. The following lemma shows that a decrease candidate satisfying condition (1) is also a semi-increase process.

Lemma 5. *Let S_i be the set of decrease candidates for P_i. If $\exists P_j, P_k \in S_i$: $dist(P_j, P_k) > 2$, another MD function can be obtained by increasing $\overline{num_i}$ by 1 and decreasing $\overline{num_j}$ and $\overline{num_k}$ by 1 respectively.*

The following two lemmas hold based on the properties of the 1-MMD function.

Lemma 6. *There is at least one semi-increase process if f_I is an MMD function but not a 1-MMD function.*

Lemma 7. *There is no semi-increase process if f_I is a 1-MMD function.*

To prevent interference among the semi-increase processes (or decrease candidates) that could violate the conditions of the MMD function, some are eliminated through rules R7 to R12. Specifically, if the distance between any two semi-increase processes is within 4, the one with the smaller ID will remain. Additionally, if a decrease candidate finds another decrease candidate for a different semi-increase process with the smaller ID (i.e., having a smaller semiInc) within distance 2, it will not remain. The remaining semi-increase process (resp. decrease candidate) is called an *increase process* (resp. a *semi-decrease process*). In Fig. 3(3), process 4 cannot become an increase process because another process (process 1) within distance 4 becomes an increase process.

Rules R13 and R14 determine whether each process can increase (or decrease) num. Each increase process P_i can increase its num_i by 1 unless even one process among decrease candidates for P_i is not a semi-decrease process (i.e., cannot remain). Subsequently, the rules identify all processes P_j that can decrease num_j among the semi-decrease processes for P_i, called *decrease processes*, based on their IDs. Finally, each process updates the variable sum according to rule R15.

In Fig. 3(4), the semi-decrease processes for process 1 are processes $2, 3, 5$, and 8. Among them, processes 2 and 3 eventually become decrease processes because they have the smallest IDs.

When algorithm DEC terminates, a new configuration, depicted in Fig. 3(5) is obtained by copying variables. Subsequently, algorithm DEC is executed again from this new configuration. Here, any execution of DEC satisfies the following lemmas.

Lemma 8. *Any execution of DEC starting from $\gamma \in \mathcal{L}_{MMD}$ terminates at a configuration γ' where f_O is an MMD function.*

Lemma 9. *Any execution of DEC starting from a configuration where f_I is an MMD function but not a 1-MMD function terminates at a configuration such that $\sum_{P_i \in V} \overline{num_i} > \sum_{P_i \in V} num_i$ within $\mathcal{O}(\Delta^2)$ rounds.*

Lemma 10. *Any execution of DEC starting from a configuration where f_I is a 1-MMD function terminates at a configuration such that $\sum_{P_i \in V} \overline{num_i} = \sum_{P_i \in V} num_i$ within $\mathcal{O}(1)$ rounds.*

The following requirements of the loop composition for $\mathcal{L}_{1MMD}$ hold by the above lemmas.

Lemma 11. *(Shiftable Convergence) Any maximal execution of* DEC *that starts from a configuration satisfying* $\mathcal{L}_{MMD}$ *terminates at a configuration* γ *such that* $\gamma^{copy} \in \mathcal{L}_{MMD}$.

Lemma 12. *(Loop Convergence) If* $\Xi_0, \Xi_1, \ldots$ *is an infinite sequence of maximal executions of* $\mathcal{A}$ *where* $\Xi_i = (\gamma_{i,0}, \gamma_{i,1}, \ldots, \gamma_{i,s_i})$, $\gamma_{0,0} \in \mathcal{L}_{MMD}$, *and* $\gamma_{i+1,0} = \gamma_{i,s_i}^{copy}$ *hold for each* $i \geq 0$, *then* $\gamma_{j,s_j} \in \mathcal{C}_{goal}$ *and* $R(\Xi_0) + R(\Xi_1) + \cdots + R(\Xi_j) \leq \mathcal{O}(n\Delta^2)$ *hold for some* $j \leq 2n$.

Lemma 13. *(Correctness) For any configuration* γ, $\gamma \in \mathcal{C}_{goal} \implies \gamma \in \mathcal{L}_{1MMD}$.

Next, we discuss the space complexity of the proposed algorithm. As shown in Table 1, each process maintains a constant number of variables, and CANDMAP$_i$ requires the most space. The domain of the first element of CANDMAP is within $N^3[P_i]$ and the second is within $2^{N^1[P_i]}$, thus this requires at most $\Delta^4 \log n$ bits of memory. Therefore, the following corollary holds.

Corollary 2. *Algorithm* Loop(DEC, E_{MMD}, INIT) *requires* $\mathcal{O}(\Delta^4 \log n)$ *bits of memory (per process).*

The following theorem holds from Lemmas 11, 12, and 13, and Corollary 2.

Theorem 2. *Algorithm* Loop(DEC, E_{MMD}, INIT) *is a silent self-stabilizing algorithm for predicate* $\mathcal{L}_{1MMD}$, *which converges within* $\mathcal{O}(n(\Delta^2 + D))$ *rounds with* $\mathcal{O}(\Delta^4 \log n)$ *bits of memory (per process).*

References

1. Dijkstra, E.W.: Self-stabilizing systems in spite of distributed control. Commun. ACM **17**(11), 643–644 (1974)
2. Dunbar, J., Hedetniemi, S., Henning, M.A., McRae, A.: Minus domination in graphs. Disc. Math. **199**(1), 35–47 (1999)
3. Dunbar, J., Goddard, W., Hedetniemi, S., McRae, A., Henning, M.A.: The algorithmic complexity of minus domination in graphs. Disc. Appl. Math. **68**(1), 73–84 (1996)
4. Xu, B., Zhou, S.: On minus edge domination in graphs. J. Jiangxi Normal Univ. **1**, 21–24 (2007). (In Chinese)
5. Xu, B., Kong, X.: On the mixed minus domination in graphs. J. Oper. Res. Soc. China **1**, 385–391 (2013)
6. Kakugawa, H., Kamei, S.: A linear-time self-stabilizing distributed algorithm for the minimal minus (L,K,Z)-domination problem under the distance-2 model. In: 14th International Workshop on Parallel and Distributed Algorithms and Applications (PDAA), pp. 168–173 (2022)
7. Datta, A.K., Larmore, L.L., Masuzawa, T., Sudo, Y.: A self-stabilizing minimal k-grouping algorithm. In: 18th International Conference on Distributed Computing and Networking (ICDCN) (2019)
8. Tanaka, H., Sudo, Y., Kakugawa, H., Masuzawa, T., Datta, A.K.: A self-stabilizing 1-maximal independent set algorithm. J. Inf. Process. **29**, 247–255 (2021)
9. Maruyama, S., Sudo, Y., Kamei, S., Kakugawa, H.: Self-stabilizing 2-minimal dominating set algorithms based on loop composition. Theor. Comput. Sci. **983** (2024)

Brief Announcement: A Self-* and Persistent Hub Sampling Service

Mohamed Amine Legheraba[1]([✉]), Maria Potop-Butucaru[1],
and Sébastien Tixeuil[1,2]

[1] Sorbonne Université, CNRS, LIP6, 75005 Paris, France
{mohamed.legheraba,maria.potop-butucaru,sebastien.tixeuil}@lip6.fr
[2] Institut Universitaire de France, Paris, France

Abstract. We present Elevator, a novel algorithm for hub sampling in peer-to-peer networks. Elevator constructs overlays whose topology lies between a random graph and a star network. Our approach makes use of preferential attachment, forming hubs spontaneously, and offering a decentralized solution for use cases that require networks with both low diameter and resilience to failures.

Keywords: Peer-to-peer networks · Peer sampling service · Hub sampling · Resilient networks · System design · Algorithms · Simulations

1 Introduction

In recent years, the growing prevalence of decentralized systems has generated significant interest in peer-to-peer (P2P) protocols. P2P overlay networks are generally classified as either structured [8] or unstructured [3]. While structured overlays can be efficient in certain scenarios, they are more vulnerable to Byzantine attacks [10] and churn [7]. On the other hand, unstructured networks offer greater resilience [5], though evaluating the quality of services built upon them remains a challenge. In unstructured overlays, peers rely on a mechanism known as peer sampling [5], which allows nodes to collect and exchange information about other nodes within the network, thereby influencing the network's topology. The peer sampling algorithms discussed in the literature typically produce two types of topologies-random and power-law-that exhibit desirable networking characteristics.

However, when applying these concepts to federated learning [9], certain limitations become apparent: *(i)* Gossip learning, which relies on gossip-based peer sampling, has a slower convergence rate compared to centralized federated learning approaches [4], and *(ii)* although power-law topologies theoretically enhance

The work presented in this document has received funding from the EU Horizon Europe research and innovation Programme under Grant Agreement No. 101070118.

T. Masuzawa et al. (Eds.): SSS 2024, LNCS 14931, pp. 461–465, 2025.
https://doi.org/10.1007/978-3-031-74498-3_33

convergence efficiency, most previous research has concentrated on constructing networks that strictly conform to power-law distributions [2,12], thereby
restricting the number of hubs. In the context of federated learning, however,
the presence of hubs is beneficial, as these hubs enable the swift relay of machine
learning models throughout the network, thus accelerating convergence rates.

Therefore, there is a clear need for a protocol that encourages the natural
formation of hubs within networks. The service proposed in this article is specifically designed to address this need, allowing certain nodes to organically rise to
hub status through a process we refer to as "hub sampling."

Our contribution. Our main objective is to create a protocol, named Elevator,
that autonomously elevates certain nodes to function as hubs within unstructured peer-to-peer networks. To accomplish this, we combine two key principles: *preferential attachment* and *random attachment*. By integrating these concepts, our protocol fosters a balanced network structure in which hubs naturally
emerge according to connectivity patterns, while also maintaining adaptability
to dynamic changes within the network.

2   Elevator Protocol

In our study, we consider an overlay network of interconnected nodes modeled
as a directed graph. Communication within this network is bidirectional, corresponding to an underlying undirected graph that represents the physical network.
Each node in this network possesses a unique address, akin to an IP address in
the context of the Internet, serving as an abstract identifier of its identity. Nodes
maintain a local list called *cache*, which contains addresses of other nodes, and
represents their partial knowledge of the network's node set. The maximum size
of this cache, denoted by parameter c, is uniform across all nodes. The cache
is pivotal for peer sampling, as it serves as the basis for neighbor selection and
information exchange. At the network's inception, nodes are initially connected
to a random subset of nodes, forming what is known as a random k-out graph.
Subsequently, new nodes joining the network also establish connections with a
random subset of existing nodes, a process that populates their cache and integrates them into the network. Given the decentralized nature of the network,
peer sampling algorithms are designed to operate asynchronously, as it is the
case for Elevator, but to help the evaluation of protocols during simulations, we
can refer to the idea of *cycles* of the protocol. During each cycle, every node
initiates one execution of the peer sampling protocol, potentially updating its
cache based on interactions with neighboring nodes. By leveraging cycles, we
can analyze the convergence, performance, and robustness of peer sampling protocols under varying conditions and scenarios within a decentralized network
environment.

2.1   Elevator Core Concepts

To achieve both robustness and a low network diameter, we incorporate two
fundamental concepts: preferential attachment [1] and random attachment [5,

11]. Each of these concepts plays a distinct yet complementary role in shaping the network topology (see the extended version of this paper [6]).

Our goal is to design a network topology that satisfies the following properties: *(i)* There are h designated hubs, where h is a predefined parameter known to all nodes before the network begins, *(ii)* excluding the hubs, the distribution of the remaining connections is random, and *(iii)* each node has c connections, comprising h connections to hubs and $c-h$ connections to randomly selected nodes.

2.2 Elevator Detailed Description

The algorithm uses the following parameters and data structures:

– *Parameter c*: The maximum number of outgoing connections (its default value for all nodes is 20).
– *Parameter h*: The number of preferential attachment connections (its default value for all nodes is $c/2$).
– *Parameter maxsize_buffer_backward*: The maximum number of backward connections to send (its default value for all nodes is 100).
– *Structure cache*: The list of outgoing connections. The list is implemented as an array of size c. The list is initialized with random existing addresses (random connections to other nodes of the network).
– *Structure backward_peers*: The list of other nodes that have tried to connect to the node. The list is implemented as a linked list (initially empty).

Additionally, we have three temporary structures: *(i) frequency_map* holds the frequency of occurrences for all neighbors of neighbors, implemented as a map (*node → integer*), *(ii) preferred* holds the list of preferred nodes, implemented as a linked list, and *(iii) preferred_backward* holds the list of backward connections of the preferred nodes, implemented as a linked list.

The proposed protocol executes the following actions at each run: Each node retrieves the neighbor's list of their neighbors (*i.e.*, the neighbors at distance two). The node then builds an ordered list of the most frequent peers (the frequency map) and contacts the c most frequent nodes (called *preferred*). Each contacted node sends back to the contacting node a maximum of *maxsize_buffer_backward* addresses from its backward list, maintained in the structure *backward_peers*, and adds the contacting node to its backward list. The cache of the contacting node is then reset as an empty array. Then the node selects the h most frequent peers and $c-h$ random peers from the list of backward peers of all preferred peers to fill its cache. If the cache is not full, the node adds random peers from the frequency map to the cache until the size of the cache is c (see Algorithms 1 and 2 for detailed pseudocode of the algorithm).

In the companion technical report [6], we propose a detailed theoretical analysis of Elevator altogether with an extensive simulation campaign. Through simulation evaluation, we demonstrate the effectiveness and advantages of our protocol with respect to state-of-the-art algorithms.

Algorithm 1: Elevator Algorithm (active thread)

Data: initial peer list: *cache*
Data: cache size: *c*
Data: number of hubs desired: *h*
Data: initial backward list: *backward_peers* (empty)

1 **Loop**
2   **for** *peer* $\in$ *backward_peers* **do**
3   **if** *peer not responding* **then**
4   *backward_peers.remove(peer)*

5   **for** *peer* $\in$ *cache* **do**
6   **if** *peer not responding* **then**
7   *cache.remove(peer)*

8   *frequency_map* $\leftarrow$ {}
9   **for** *peer* $\in$ *cache* **do**
10   *peer_cache* $\leftarrow$ *send(CACHE_REQUEST, peer)*
11   *frequency_map* $\leftarrow$ *frequency_map* $\cup$ *peer_cache*

12   *preferred* $\leftarrow$ *frequency_map.sortByFrequency().select(number = c)*
13   *frequency_map.remove(preferred)*
14   *preferred_backward* $\leftarrow$ {}
15   **for** *peer* $\in$ *preferred* **do**
16   *peer_backward_peers* $\leftarrow$ *send(BACKWARD_REQUEST, peer)*
17   *preferred_backward* $\leftarrow$ *preferred_backward* $\cup$ *peer_backward_peers*

18   *preferred.shuffle()*
19   *preferred_backward.shuffle()*
20   *cache* $\leftarrow$ {}
21   *cache* $\leftarrow$ *preferred*[1..h] + *preferred_backward*[1..c − h]
22   **while** *cache.size()* < c **do**
23   *peer* $\leftarrow$ *frequency_map.selectRandom()*
24   *cache.append(peer)*

Algorithm 2: Elevator Algorithm (background thread)

Data: max number of backward connections to send:
       maxsize_buffer_backward

1 **Loop**
2   *request, peer* $\leftarrow$ *receive()*
3   **if** *request* = *CACHE_REQUEST* **then**
4   *send(cache, peer)*
5   *backward_peers.add(peer)*

6   **if** *request* = *BACKWARD_REQUEST* **then**
7   *backward_peers.shuffle()*
8   *send(backward_peers[: maxsize_buffer_backward], peer)*

---

3 Discussions and Conclusions

We proposed a novel peer sampling algorithm, Elevator, which facilitates the organic promotion of specific nodes to serve as hubs. Our extensive simulations presented in the companion technical report [6] confirm that the Elevator algorithm successfully maintains network connectivity, constructs networks with low diameters, reaches stability with a user-defined number of hubs (denoted as h), and demonstrates resilience against crashes, churn, and targeted attacks on hubs.

References

1. Barabâsi, A.L., Jeong, H., Néda, Z., Ravasz, E., Schubert, A., Vicsek, T.: Evolution of the social network of scientific collaborations. Phys. A **311**(3–4), 590–614 (2002)
2. Bulut, E., Szymanski, B.K.: Constructing limited scale-free topologies over peer-to-peer networks. IEEE Trans. Parallel Distrib. Syst. **25**(4), 919–928 (2013)
3. Frankel, J.: The gnutella protocol specification, v0. 4 (2003). http://www.clip2.com/gnutellaprotocol04.pdf
4. Hegedűs, I., Danner, G., Jelasity, M.: Decentralized learning works: an empirical comparison of gossip learning and federated learning. J. Parallel Distrib. Comput. **148**, 109–124 (2021)
5. Jelasity, M., Voulgaris, S., Guerraoui, R., Kermarrec, A.M., Van Steen, M.: Gossip-based peer sampling. ACM Trans. Comput. Syst. **25**(3), 8–es (2007)
6. Legheraba, M.A., Potop-Butucaru, M., Tixeuil, S.: Elevator: Self-* and persistent hub sampling service in unstructured peer-to-peer networks (2024). arXiv preprint arXiv:2406.07946
7. Malatras, A.: State-of-the-art survey on p2p overlay networks in pervasive computing environments. J. Netw. Comput. Appl. **55**, 1–23 (2015)
8. Maymounkov, P., Mazieres, D.: Kademlia: A peer-to-peer information system based on the xor metric. In: International Workshop on Peer-to-Peer Systems, pp. 53–65. Springer (2002)
9. McMahan, B., Moore, E., Ramage, D., Hampson, S., y Arcas, B.A.: Communication-efficient learning of deep networks from decentralized data. In: Artificial Intelligence and Statistics, pp. 1273–1282. PMLR (2017)
10. Naik, A.R., Keshavamurthy, B.N.: Next level peer-to-peer overlay networks under high churns: a survey. Peer-to-Peer Netw. Appl. **13**(3), 905–931 (2020)
11. Stavrou, A., Rubenstein, D., Sahu, S.: A lightweight, robust p2p system to handle flash crowds. IEEE J. Sel. Areas Commun. **22**(1), 6–17 (2004)
12. Xie, Y.B., Zhou, T., Wang, B.H.: Scale-free networks without growth. Phys. A **387**(7), 1683–1688 (2008)

Author Index

T. Masuzawa et al. (Eds.): SSS 2024, LNCS 14931, pp. 467–468, 2024.
https://doi.org/10.1007/978-3-031-74498-3

The manufacturer's authorised representative in the EU is Springer
Nature Customer Service Centre GmbH, Europaplatz 3, 69115 Heidelberg,
Germany. If you have any concerns regarding our products, please
contact ProductSafety@springernature.com

Printed and bound by CPI Group (UK) Ltd, Croydon, CR0 4YY
24/04/2026
02096372-0001